Y0-BXH-254

Methods in Enzymology

Volume 303

cDNA PREPARATION AND CHARACTERIZATION

METHODS IN ENZYMOLOGY

EDITORS-IN-CHIEF

John N. Abelson Melvin I. Simon

DIVISION OF BIOLOGY
CALIFORNIA INSTITUTE OF TECHNOLOGY
PASADENA, CALIFORNIA

FOUNDING EDITORS

Sidney P. Colowick and Nathan O. Kaplan

Methods in Enzymology

Volume 303

cDNA Preparation and Characterization

EDITED BY

Sherman M. Weissman

BOYER CENTER FOR MOLECULAR MEDICINE
YALE UNIVERSITY SCHOOL OF MEDICINE
NEW HAVEN, CONNECTICUT

QP601
C71
v.303
1999

ACADEMIC PRESS
San Diego London Boston New York Sydney Tokyo Toronto

This book is printed on acid-free paper. 

Copyright © 1999 by ACADEMIC PRESS

All Rights Reserved.
No part of this publication may be reproduced or transmitted in any form or by any means, electronic or mechanical, including photocopy, recording, or any information storage and retrieval system, without permission in writing from the Publisher.
The appearance of the code at the bottom of the first page of a chapter in this book indicates the Publisher's consent that copies of the chapter may be made for personal or internal use, or for the personal or internal use of specific clients. This consent is given on the condition, however, that the copier pay the stated per copy fee through the Copyright Clearance Center, Inc. (222 Rosewood Drive, Danvers, Massachusetts 01923) for copying beyond that permitted by Sections 107 or 108 of the U.S. Copyright Law. This consent does not extend to other kinds of copying, such as copying for general distribution, for advertising or promotional purposes, for creating new collective works, or for resale. Copy fees for pre-1999 chapters are as shown on the chapter title pages. If no fee code appears on the chapter title page, the copy fee is the same as for current chapters.
0076-6879/99 $30.00

Academic Press
a division of Harcourt Brace & Company
525 B Street, Suite 1900, San Diego, California 92101-4495, USA
http://www.academicpress.com

Academic Press Limited
24-28 Oval Road, London NW1 7DX, UK
http://www.hbuk.co.uk/ap/

International Standard Book Number: 0-12-182204-4

PRINTED IN THE UNITED STATES OF AMERICA
99 00 01 02 03 04 MM 9 8 7 6 5 4 3 2 1

Table of Contents

Section I. cDNA Preparation

Section II. Gene Identification

Section III. Patterns of mRNA Expression

Section IV. Functional Relationship among cDNA Translation Products

Contributors to Volume 303

Article numbers are in parentheses following the names of contributors.
Affiliations listed are current.

CHRISTOPHER ASTON (4), *Department of Chemistry, W. M. Keck Laboratory for Biomolecular Imaging, New York University, New York, New York 10003*

NAMADEV BASKARAN (3, 16), *Genome Therapeutics Corporation, Waltham, Massachusetts 02453*

ALASTAIR J. H. BROWN (22), *Trafford Centre for Medical Research, University of Sussex, Brighton, BN1 9RY, United Kingdom*

PATRICK O. BROWN (12), *Department of Biochemistry, Stanford University School of Medicine and Howard Hughes Medical Institute, Stanford, California 94305-5428*

ALAN J. BUCKLER (6), *Axys Pharmaceuticals, La Jolla, California 92037*

JULIAN F. BURKE (22), *School of Biological Sciences, University of Sussex, Brighton, BN1 9RY, United Kingdom*

KONRAD BÜSSOW (13), *Abt. Lehrach, Max Planck Institut für Molekulare Genetik, Berlin (Dahlem) D-14195, Germany*

DOLORES J. CAHILL (13), *Abt. Lehrach, Max Planck Institut für Molekulare Genetik, Berlin (Dahlem) D-14195, Germany*

PIERO CARNINCI (2), *Laboratory for Genome Exploration Research Project, Genomic Sciences Center (GSC), and Genome Science Laboratory, Riken Tsukuba Life Science Center, The Institute of Physical and Chemical Research (RIKEN), CREST, Japan Science and Technology Corporation (JST), Tsukuba, Ibaraki 305-0074, Japan*

DEANNA M. CHURCH (6), *Samuel Lunenfeld Research Institute, Mount Sinai Hospital, Toronto, Ontario M5G 1X5 Canada*

MATTHEW D. CLARK (13), *Abt. Lehrach, Max Planck Institut für Molekulare Genetik, Berlin (Dahlem) D-14195, Germany*

MAUREEN COLBERT (26), *Genetics Institute, Cambridge, Massachusetts 02140*

LISA A. COLLINS-RACIE (26), *Genetics Institute, Cambridge, Massachusetts 02140*

PETER B. CRINO (1), *Department of Neurology, University of Pennsylvania Medical Center, Philadelphia, Pennsylvania 19104*

JOHAN T. DEN DUNNEN (7), *Department of Human Genetics, Leiden University Medical Center, 2333 Al Leiden, The Netherlands*

LUDA DIATCHENKO (20), *CLONTECH Laboratories, Inc., Palo Alto, California 94303-4230*

RADOJE DRMANAC (11), *Hyseq, Inc., Sunnyvale, California 94086*

SNEZANA DRMANAC (11), *Hyseq, Inc., Sunnyvale, California 94086*

MCKEOUGH DUCKETT (26), *Genetics Institute, Cambridge, Massachusetts 02140*

JAMES EBERWINE (1), *Departments of Pharmacology and Psychiatry, University of Pennsylvania Medical Center, Philadelphia, Pennsylvania 19104*

MICHAEL B. EISEN (12), *Departments of Genetics and Biochemistry, Stanford University School of Medicine, Stanford, California 94305-5428*

JANET ESTEE KACHARMINA (1), *Departments of Pharmacology and Psychiatry, University of Pennsylvania Medical Center, Philadelphia, Pennsylvania 19104*

CHERYL EVANS (26), *Genetics Institute, Cambridge, Massachusetts 02140*

CARL FRIDDLE (29), *Department of Genetics, Yale University, New Haven, Connecticut 06520*

KATHELEEN GARDINER (10), *Eleanor Roosevelt Institute, Denver, Colorado 80206*

F. JOSEPH GERMINO (24), *University of Medical Dentistry of New Jersey, Clinical Institute of New Jersey, New Brunswick, New Jersey 08901*

MARGARET GOLDEN-FLEET (26), *Genetics Institute, Cambridge, Massachusetts 02140*

YOSHIHIDE HAYASHIZAKI (2), *Laboratory for Genome Exploration Research Project, Genomic Sciences Center (GSC), and Genome Science Laboratory, Riken Tsukuba Life Science Center, The Institute of Physical and Chemical Research (RIKEN), CREST, Japan Science and Technology Corporation (JST), Tsukuba, Ibaraki 305-0074, Japan*

CATHARINA HIORT (4), *Department of Chemistry, W. M. Keck Laboratory for Biomolecular Imaging, New York University, New York, New York 10003*

TASUKU HONJO (27), *Department of Medical Chemistry, Faculty of Medicine, Kyoto University, Sakyo-ku, Kyoto 606-8501, Japan*

MICHAEL HUBANK (19), *Trafford Centre for Medical Research, University of Sussex, Brighton, Sussex BN1 9RY, England*

CATHERINE HUTCHINGS (22), *Trafford Centre for Medical Research, University of Sussex, Brighton, Sussex BN1 9RY, United Kingdom*

TAKASHI ITO (17), *Human Genome Center, Institute of Medical Science, University of Tokyo, Minato-ku, Tokyo 108-8639, Japan*

KENNETH A. JACOBS (26), *Genetics Institute, Cambridge, Massachusetts 02140*

BARBARA JUNG (18), *Sidney Kimmel Cancer Center, San Diego, California 92121*

KERRY KELLEHER (26), *Genetics Institute, Cambridge, Massachusetts 02140*

RONALD KRIZ (26), *Genetics Institute, Cambridge, Massachusetts 02140*

MAGNUS LARSSON (28), *Department of Biochemistry and Biotechnology, KTH, Royal Institute of Technology, S-100 44 Stockholm, Sweden*

YUN-FAI CHRIS LAU (20), *Division of Cell and Developmental Genetics, Department of Medicine, VA Medical Center, University of California, San Francisco, San Francisco, California 94121*

EDWARD R. LAVALLIE (26), *Genetics Institute, Cambridge, Massachusetts 02140*

HANS LEHRACH (13), *Abt. Lehrach, Max Planck Institut für Molekulare Genetik, Berlin (Dahlem) D-14195, Germany*

MIN LI (25), *Departments of Physiology and Neuroscience, School of Medicine, Johns Hopkins University, Baltimore, Maryland 21205*

MENG LIU (3), *Department of Genetics, Boyer Center for Molecular Medicine, Yale University School of Medicine, New Haven, Connecticut 06510*

MICHAEL LOVETT (8), *The McDermott Center for Human Growth and Development, University of Texas Southwestern Medical Center at Dallas, Dallas, Texas 75235*

SERGEY LUKYANOV (20), *Shemyakin and Ovchinnikov Institute of Bioorganic Chemistry, Russian Academy of Science, V-437 Moscow 117871, Russia*

JOAKIM LUNDEBERG (28), *Department of Biochemistry and Biotechnology, KTH, Royal Institute of Technology, S-100 44 Stockholm, Sweden*

MARIE-CLAUDE MARSOLIER (23), *Service de Biochimie et Génétique Moléculaire, CEA/SACLAY, 91191 Gif Sur Yvette Cedex, France*

KATHERINE J. MARTIN (14), *Dana-Farber Cancer Institute, Boston, Massachusetts 02113*

FRANÇOISE MATHIEU-DAUDÉ (18, 21), *Sidney Kimmel Cancer Center, San Diego, California 92121*

LYNNE V. MAYNE (22), *Trafford Centre for Medical Research, University of Sussex, Brighton, BN1 9RY, United Kingdom*

MICHAEL MCCLELLAND (18, 21), *Sidney Kimmel Cancer Center, San Diego, California 92121*

JOHN M. MCCOY (26), *Genetics Institute, Cambridge, Massachusetts 02140*

DAVID MERBERG (26), *Astra Research Center, Boston, Cambridge, Massachusetts 02139-4239*

NEAL K. MOSKOWITZ (24), *Department of Molecular Genetics and Microbiology, University of Medical Dentistry of New Jersey, New Brunswick, New Jersey 08901*

RICHARD J. MURAL (5), *Computational Biology Section, Life Sciences Division, Oak Ridge National Laboratory, Oak Ridge, Tennessee 37831*

TOMOYUKI NAKAMURA (27), *Department of Medical Chemistry, Faculty of Medicine, Kyoto University, Sakyo-ku, Kyoto 606-8501, Japan*

PETER E. NEWBURGER (16), *University of Massachusetts Medical Center, Worcester, Massachusetts 01605*

JACOB ODEBERG (28), *Department of Biochemistry and Biotechnology, KTH, Royal Institute of Technology, S-100 44 Stockholm, Sweden*

GEORGIA D. PANOPOULOU (13), *Abt. Lehrach, Max Planck Institut für Molekulare Genetik, Berlin (Dahlem) D-14195, Germany*

ARTHUR B. PARDEE (14), *Dana-Farber Cancer Institute, Boston, Massachusetts 02113*

SATISH PARIMOO (9), *Skin Biology Research Center, Johnson & Johnson, Skillman, New Jersey 08558*

YATINDRA PRASHAR (15), *Gene Logic, Inc., Gaithersburg, Maryland 20878*

ANNE HANSEN REE (28), *Department of Tumor Biology, Institute of Cancer Research, The Norwegian Radium Hospital, N-0310 Oslo, Norway*

G. SHIRLEEN ROEDER (29), *Department of Biology, Yale University, New Haven, Connecticut 06520*

ØYSTEIN RØSOK (28), *Department of Immunology, Institute of Cancer Research, The Norwegian Radium Hospital, N-0310 Oslo, Norway*

PETRA ROSS-MACDONALD (29), *Department of Biology, Yale University, New Haven, Connecticut 06520*

YOSHIYUKI SAKAKI (17), *Human Genome Center, Institute of Medical Science, University of Tokyo, Minato-ku, Tokyo 108-8639, Japan*

DAVID G. SCHATZ (19), *Howard Hughes Medical Institute, Section of Immunobiology, Yale University School of Medicine, New Haven, Connecticut 06510*

PETER SCHATZ (25), *Affymax Research Institute, 4001 Miranda Avenue, Palo Alto, CA 94304*

DAVID C. SCHWARTZ (4), *Department of Chemistry, W. M. Keck Laboratory for Biomolecular Imaging, New York University, New York, New York 10003*

ANDRÉ SENTENAC (23), *Service de Biochimie et Génétique Moléculaire, CEA/SACLAY, 91191 Gif Sur Yvette Cedex, France*

AMY SHEEHAN (29), *Department of Biology, Yale University, New Haven, Connecticut 06520*

PAUL D. SIEBERT (20), *CLONTECH Laboratories, Inc., Palo Alto, California 94303-4230*

ANDREW D. SIMMONS (8), *The McDermott Center for Human Growth and Development, University of Texas Southwestern Medical Center at Dallas, Dallas, Texas 75235*

MICHAEL SNYDER (29), *Department of Biology, Yale University, New Haven, Connecticut 06520*

VIKKI SPAULDING (26), *Genetics Institute, Cambridge, Massachusetts 02140*

STEFAN STÅHL (28), *Department of Biochemistry and Biotechnology, KTH, Royal Institute of Technology, S-100 44 Stockholm, Sweden*

JEN STOVER (26), *Genetics Institute, Cambridge, Massachusetts 02140*

NICOLE L. STRICKER (25), *Department of Neuroscience, School of Medicine, Johns Hopkins University, Baltimore, Maryland 21205*

Y. V. B. K. SUBRAMANYAM (3, 16), *Department of Genetics, Boyer Center for Molecular Medicine, Yale University School of Medicine, New Haven, Connecticut 06536 and Gene Logic, Inc., Gaithersburg, Maryland 20878*

KEI TASHIRO (27), *Center for Molecular Biology and Genetics, Kyoto University, Sakyo-ku, Kyoto 606-8507, Japan*

THOMAS TRENKLE (18, 21), *Sidney Kimmel Cancer Center, San Diego, California 92121*

THOMAS VOGT (18), *University of Regensburg, 93042 Regensburg, Germany*

SHERMAN M. WEISSMAN (9, 15, 16), *Department of Genetics, Boyer Center for Molecular Medicine, Yale University School of Medicine, New Haven, Connecticut 06536*

JOHN WELSH (18, 21), *Sidney Kimmel Cancer Center, San Diego, California 92121*

MARK J. WILLIAMSON (26), *Millenium Pharmaceuticals, Inc., Cambridge, Massachusetts 02139*

Preface

Genomic sequences, now emerging at a rapid rate, are greatly expediting certain aspects of molecular biology. However, in more complex organisms, predicting mRNA structure from genomic sequences can often be difficult. Alternative splicing, the use of alternative promoters, and orphan genes without known analogues can all offer difficulties in the predictions of the structure of mRNAs or even in gene detection. Both computational and experimental methods remain useful for recognizing genes and transcript templates, even in sequenced DNA. Methods for producing full-length cDNAs are important for determining the structure of the proteins the mRNA encodes, the position of promoters, and the considerable regulatory information for translation that may be encoded in the 5′ untranslated regions of the mRNA.

Methods for studying levels of mRNA and their changes in different physiological circumstances are rapidly evolving, and the information from this area will rival the superabundance of information derived from genomic sequences. In particular, cDNAs can be prepared even from single cells, and this approach has already yielded valuable information in several areas. To the extent that reliable and reproducible information, both quantitative and qualitative, can be generated from very small numbers of cells, there are rather remarkable possibilities for complementing functional and genetic analysis of developmental patterns with descriptions of changes in mRNAs. Dense array analysis promises to be particularly valuable for the rapid expression pattern of known genes, while other methods such as gel display approaches offer the opportunity of discovering unidentified genes or for investigating species whose cDNAs or genomes have not been studied intensively.

Knowledge of mRNA structure, genomic location, and patterns of expression must be converted into information of the function of the encoded proteins. Each gene can become the subject of years of intensive study. Nevertheless, a number of methods are being developed that use cDNA to predict properties or permit the selective isolation of cDNAs encoding proteins with certain general properties such as subcellular location. This volume presents an update of a number of approaches relevant to the areas referred to above. The technology in this field is rapidly evolving and these contributions represent a "snapshot in time" of the number of currently available and useful approaches to the problems referred to above.

Sherman M. Weissman

METHODS IN ENZYMOLOGY

VOLUME XXXVI. Hormone Action (Part A: Steroid Hormones)
Edited by BERT W. O'MALLEY AND JOEL G. HARDMAN

VOLUME XXXVII. Hormone Action (Part B: Peptide Hormones)
Edited by BERT W. O'MALLEY AND JOEL G. HARDMAN

VOLUME XXXVIII. Hormone Action (Part C: Cyclic Nucleotides)
Edited by JOEL G. HARDMAN AND BERT W. O'MALLEY

VOLUME XXXIX. Hormone Action (Part D: Isolated Cells, Tissues, and Organ Systems)
Edited by JOEL G. HARDMAN AND BERT W. O'MALLEY

VOLUME XL. Hormone Action (Part E: Nuclear Structure and Function)
Edited by BERT W. O'MALLEY AND JOEL G. HARDMAN

VOLUME XLI. Carbohydrate Metabolism (Part B)
Edited by W. A. WOOD

VOLUME XLII. Carbohydrate Metabolism (Part C)
Edited by W. A. WOOD

VOLUME XLIII. Antibiotics
Edited by JOHN H. HASH

VOLUME XLIV. Immobilized Enzymes
Edited by KLAUS MOSBACH

VOLUME XLV. Proteolytic Enzymes (Part B)
Edited by LASZLO LORAND

VOLUME XLVI. Affinity Labeling
Edited by WILLIAM B. JAKOBY AND MEIR WILCHEK

VOLUME XLVII. Enzyme Structure (Part E)
Edited by C. H. W. HIRS AND SERGE N. TIMASHEFF

VOLUME XLVIII. Enzyme Structure (Part F)
Edited by C. H. W. HIRS AND SERGE N. TIMASHEFF

VOLUME XLIX. Enzyme Structure (Part G)
Edited by C. H. W. HIRS AND SERGE N. TIMASHEFF

VOLUME L. Complex Carbohydrates (Part C)
Edited by VICTOR GINSBURG

VOLUME LI. Purine and Pyrimidine Nucleotide Metabolism
Edited by PATRICIA A. HOFFEE AND MARY ELLEN JONES

VOLUME LII. Biomembranes (Part C: Biological Oxidations)
Edited by SIDNEY FLEISCHER AND LESTER PACKER

VOLUME LIII. Biomembranes (Part D: Biological Oxidations)
Edited by SIDNEY FLEISCHER AND LESTER PACKER

VOLUME LIV. Biomembranes (Part E: Biological Oxidations)
Edited by SIDNEY FLEISCHER AND LESTER PACKER

Volume 91. Enzyme Structure (Part I)
Edited by C. H. W. Hirs and Serge N. Timasheff

Volume 92. Immunochemical Techniques (Part E: Monoclonal Antibodies and General Immunoassay Methods)
Edited by John J. Langone and Helen Van Vunakis

Volume 93. Immunochemical Techniques (Part F: Conventional Antibodies, Fc Receptors, and Cytotoxicity)
Edited by John J. Langone and Helen Van Vunakis

Volume 94. Polyamines
Edited by Herbert Tabor and Celia White Tabor

Volume 95. Cumulative Subject Index Volumes 61–74, 76–80
Edited by Edward A. Dennis and Martha G. Dennis

Volume 96. Biomembranes [Part J: Membrane Biogenesis: Assembly and Targeting (General Methods; Eukaryotes)]
Edited by Sidney Fleischer and Becca Fleischer

Volume 97. Biomembranes [Part K: Membrane Biogenesis: Assembly and Targeting (Prokaryotes, Mitochondria, and Chloroplasts)]
Edited by Sidney Fleischer and Becca Fleischer

Volume 98. Biomembranes (Part L: Membrane Biogenesis: Processing and Recycling)
Edited by Sidney Fleischer and Becca Fleischer

Volume 99. Hormone Action (Part F: Protein Kinases)
Edited by Jackie D. Corbin and Joel G. Hardman

Volume 100. Recombinant DNA (Part B)
Edited by Ray Wu, Lawrence Grossman, and Kivie Moldave

Volume 101. Recombinant DNA (Part C)
Edited by Ray Wu, Lawrence Grossman, and Kivie Moldave

Volume 102. Hormone Action (Part G: Calmodulin and Calcium-Binding Proteins)
Edited by Anthony R. Means and Bert W. O'Malley

Volume 103. Hormone Action (Part H: Neuroendocrine Peptides)
Edited by P. Michael Conn

Volume 104. Enzyme Purification and Related Techniques (Part C)
Edited by William B. Jakoby

Volume 105. Oxygen Radicals in Biological Systems
Edited by Lester Packer

Volume 106. Posttranslational Modifications (Part A)
Edited by Finn Wold and Kivie Moldave

Volume 107. Posttranslational Modifications (Part B)
Edited by Finn Wold and Kivie Moldave

VOLUME 193. Mass Spectrometry
Edited by JAMES A. MCCLOSKEY

VOLUME 194. Guide to Yeast Genetics and Molecular Biology
Edited by CHRISTINE GUTHRIE AND GERALD R. FINK

VOLUME 195. Adenylyl Cyclase, G Proteins, and Guanylyl Cyclase
Edited by ROGER A. JOHNSON AND JACKIE D. CORBIN

VOLUME 196. Molecular Motors and the Cytoskeleton
Edited by RICHARD B. VALLEE

VOLUME 197. Phospholipases
Edited by EDWARD A. DENNIS

VOLUME 198. Peptide Growth Factors (Part C)
Edited by DAVID BARNES, J. P. MATHER, AND GORDON H. SATO

VOLUME 199. Cumulative Subject Index Volumes 168–174, 176–194

VOLUME 200. Protein Phosphorylation (Part A: Protein Kinases: Assays, Purification, Antibodies, Functional Analysis, Cloning, and Expression)
Edited by TONY HUNTER AND BARTHOLOMEW M. SEFTON

VOLUME 201. Protein Phosphorylation (Part B: Analysis of Protein Phosphorylation, Protein Kinase Inhibitors, and Protein Phosphatases)
Edited by TONY HUNTER AND BARTHOLOMEW M. SEFTON

VOLUME 202. Molecular Design and Modeling: Concepts and Applications (Part A: Proteins, Peptides, and Enzymes)
Edited by JOHN J. LANGONE

VOLUME 203. Molecular Design and Modeling: Concepts and Applications (Part B: Antibodies and Antigens, Nucleic Acids, Polysaccharides, and Drugs)
Edited by JOHN J. LANGONE

VOLUME 204. Bacterial Genetic Systems
Edited by JEFFREY H. MILLER

VOLUME 205. Metallobiochemistry (Part B: Metallothionein and Related Molecules)
Edited by JAMES F. RIORDAN AND BERT L. VALLEE

VOLUME 206. Cytochrome P450
Edited by MICHAEL R. WATERMAN AND ERIC F. JOHNSON

VOLUME 207. Ion Channels
Edited by BERNARDO RUDY AND LINDA E. IVERSON

VOLUME 208. Protein–DNA Interactions
Edited by ROBERT T. SAUER

VOLUME 209. Phospholipid Biosynthesis
Edited by EDWARD A. DENNIS AND DENNIS E. VANCE

VOLUME 210. Numerical Computer Methods
Edited by LUDWIG BRAND AND MICHAEL L. JOHNSON

VOLUME 263. Plasma Lipoproteins (Part C: Quantitation)
Edited by WILLIAM A. BRADLEY, SANDRA H. GIANTURCO, AND JERE P. SEGREST

VOLUME 264. Mitochondrial Biogenesis and Genetics (Part B)
Edited by GIUSEPPE M. ATTARDI AND ANNE CHOMYN

VOLUME 265. Cumulative Subject Index Volumes 228, 230–262

VOLUME 266. Computer Methods for Macromolecular Sequence Analysis
Edited by RUSSELL F. DOOLITTLE

VOLUME 267. Combinatorial Chemistry
Edited by JOHN N. ABELSON

VOLUME 268. Nitric Oxide (Part A: Sources and Detection of NO; NO Synthase)
Edited by LESTER PACKER

VOLUME 269. Nitric Oxide (Part B: Physiological and Pathological Processes)
Edited by LESTER PACKER

VOLUME 270. High Resolution Separation and Analysis of Biological Macromolecules (Part A: Fundamentals)
Edited by BARRY L. KARGER AND WILLIAM S. HANCOCK

VOLUME 271. High Resolution Separation and Analysis of Biological Macromolecules (Part B: Applications)
Edited by BARRY L. KARGER AND WILLIAM S. HANCOCK

VOLUME 272. Cytochrome P450 (Part B)
Edited by ERIC F. JOHNSON AND MICHAEL R. WATERMAN

VOLUME 273. RNA Polymerase and Associated Factors (Part A)
Edited by SANKAR ADHYA

VOLUME 274. RNA Polymerase and Associated Factors (Part B)
Edited by SANKAR ADHYA

VOLUME 275. Viral Polymerases and Related Proteins
Edited by LAWRENCE C. KUO, DAVID B. OLSEN, AND STEVEN S. CARROLL

VOLUME 276. Macromolecular Crystallography (Part A)
Edited by CHARLES W. CARTER, JR., AND ROBERT M. SWEET

VOLUME 277. Macromolecular Crystallography (Part B)
Edited by CHARLES W. CARTER, JR., AND ROBERT M. SWEET

VOLUME 278. Fluorescence Spectroscopy
Edited by LUDWIG BRAND AND MICHAEL L. JOHNSON

VOLUME 279. Vitamins and Coenzymes (Part I)
Edited by DONALD B. MCCORMICK, JOHN W. SUTTIE, AND CONRAD WAGNER

VOLUME 280. Vitamins and Coenzymes (Part J)
Edited by DONALD B. MCCORMICK, JOHN W. SUTTIE, AND CONRAD WAGNER

VOLUME 281. Vitamins and Coenzymes (Part K)
Edited by DONALD B. MCCORMICK, JOHN W. SUTTIE, AND CONRAD WAGNER

Section I

cDNA Preparation

[1] Preparation of cDNA from Single Cells and Subcellular Regions

By JANET ESTEE KACHARMINA, PETER B. CRINO, and JAMES EBERWINE

Introduction

The heterogeneity of cellular phenotypes in the brain has complicated attempts to characterize alterations in gene expression under normal conditions such as development and senescence, as well as in disease states. Historically, the classification of cellular identity has been ascertained on the basis of regional anatomic location (e.g., hippocampal CA1 neuron or cerebellar Purkinje cell), individual morphology (e.g., pyramidal or stellate), and electrophysiological properties as determined by whole-cell patch-clamp techniques (e.g., excitatory or inhibitory). Neurons may also be categorized on the basis of the proteins, neurotransmitters, or receptors they express, as determined by immunohistochemistry or receptor autoradiography. For example, neuronal progenitor cells express the embryonic intermediate filament protein nestin, mature cells express neurofilament proteins, and astroglial cells express glial fibrillary acidic protein and vimentin.

In addition, the analysis of mRNA abundances in individual neurons has been difficult because of the high level of cellular diversity in the brain. That is, two cells that seem to be morphologically identical may express different mRNAs at distinct abundances. Another complication in studying neuronal gene expression in single cells is that many important messages are expressed in low abundance. Acquiring sufficient quantities of mRNA to analyze gene expression has been difficult, as it is estimated that the amount of mRNA in a single cell is between 0.1 and 1.0 pg. Among the techniques used to study gene expression, *in situ* hybridization (ISH) can render a view of mRNA localization in many cells simultaneously, both in culture and in tissue sections.[1] However, one shortcoming of ISH is that only one or two cRNA probes can be applied to a selected tissue section at once, and thus, only one or two mRNAs can be assayed. Also, ISH may not be sensitive enough to detect low-abundance mRNAs. *In situ* transcription (IST), a technique that shares many features with ISH, involves the synthesis of cDNA via the reverse transcription of polyade-

[1] J. Eberwine, K. Valentino, and J. Barchas, "*In Situ* Hybridization in Neurobiology." Oxford Press, New York, 1994.

Copyright © 1999 by Academic Press
All rights of reproduction in any form reserved.

0076-6879/99 $30.00

nylated [poly(A)$^+$] mRNA in fixed tissues or live cells in culture.[2–5] There are two technologies that begin by an initial cDNA synthesis step and can be used to amplify mRNA from single cells, namely, reverse transcriptase-polymerase chain reaction (RT-PCR)[6–8] and antisense RNA (aRNA) amplification.[4,5] RT-PCR uses *Taq* DNA polymerase and is therefore an exponential amplification procedure, making accurate quantification of mRNA abundances problematic. In contrast, the aRNA amplification technique amplifies mRNA in a linear fashion, using T7 RNA polymerase, and permits direct quantification of the relative abundances of individual mRNA species. The ability to quantitate mRNA levels is critical, because expression analysis of multiple genes simultaneously in single cells provides an important avenue toward understanding neuronal functioning and the molecular pathophysiology of specific diseases. Indeed, because many mRNAs are expressed in only a few copies in single cells, and there is a high level of cellular diversity, it is imperative that studies of neuronal gene expression be accomplished at the single-cell level.

Overview of cDNA Synthesis and Antisense RNA Amplification

The analysis of mRNA populations within single cells became possible with the advent of the amplified, antisense (aRNA) amplification technique. The mRNA population in a single cell can be amplified such that its size and complexity are proportionally represented in the resultant amplified aRNA population.[4] A schematic diagram of the aRNA amplification method is shown in Fig. 1. The first step in the synthesis of cDNA is hybridization of an oligo(dT) primer, [oligo(dT)$_{24}$-T7] to the endogenous poly(A)$^+$ mRNA. The oligo(dT)$_{24}$-T7 contains the bacteriophage T7 RNA polymerase promoter sequence 5′ to the polythymidine segment. In addition, individual mRNAs can be amplified by using primers complementary to specific mRNAs of interest, and that have an attached T7 RNA polymerase promoter sequence, to prime cDNA synthesis. Avian myeloblastosis reverse transcriptase (AMV-RT; Seikagaku America, Ijamsville, MD) binds

[2] L. Tecott, J. Barchas, and J. Eberwine, *Science* **240,** 1661 (1988).

[3] I. Zangger, L. Tecott, and J. Eberwine, *Technique* **1,** 108 (1989).

[4] R. VanGelder, M. vonZastrow, A. Yool, W. Dement, J. Barchas, and J. Eberwine, *Proc. Natl. Acad. Sci. U.S.A.* **87,** 1663 (1990).

[5] J. Eberwine, H. Yeh, K. Miyashiro, Y. Cao, S. Nair, R. Finnell, M. Zettel, and P. Coleman, *Proc. Natl. Acad. Sci. U.S.A.* **89,** 3010 (1992).

[6] J. Sam-Singer, M. Robinson, A. Bellvue, M. Simon, and A. Riggs, *Nucleic Acids Res.* **18,** 1255 (1990).

[7] J. Robinson and M. Simon, *Nucleic Acids Res.* **19,** 1557 (1991).

[8] C. Owczarek, P. Enriquez-Harris, and N. Proudfoot, *Nucleic Acids Res.* **20,** 851 (1992).

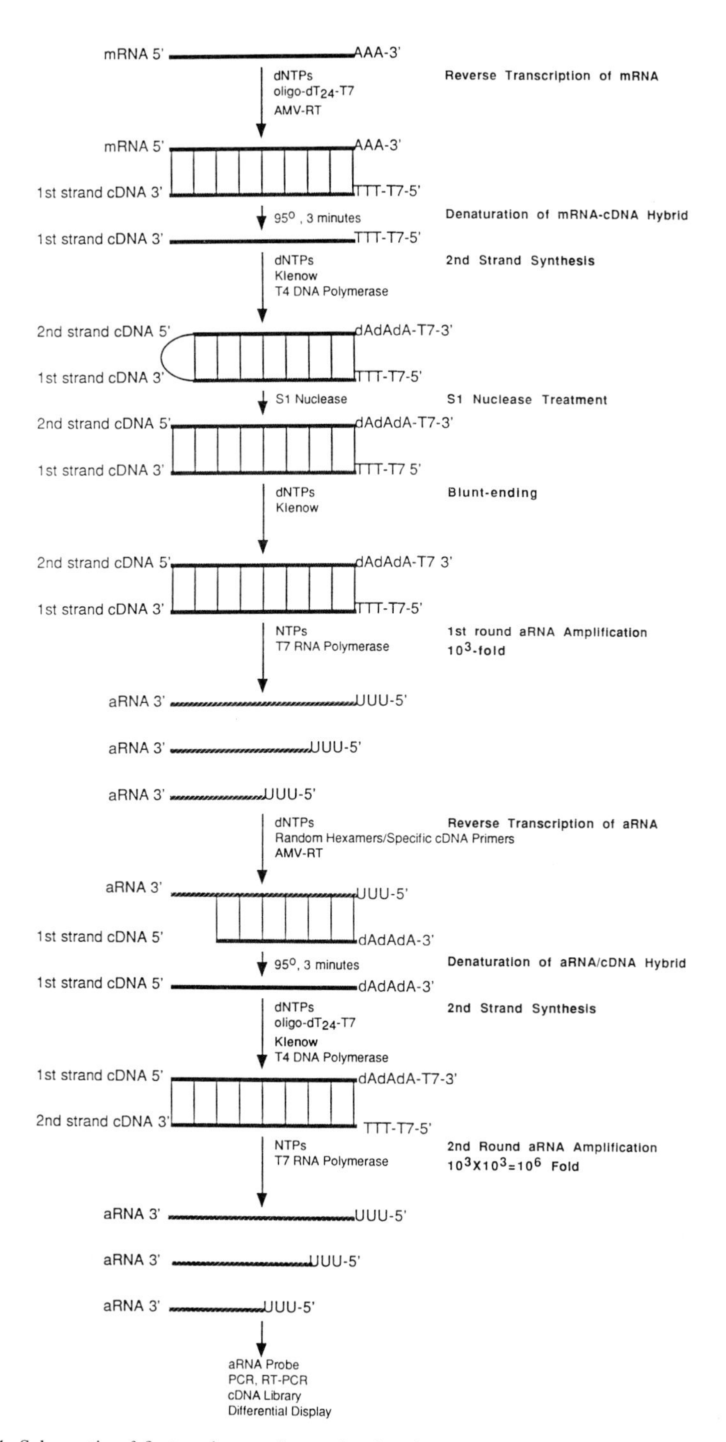

FIG. 1. Schematic of first and second rounds of antisense RNA amplification procedure.

to the primer/template [oligo(dT)$_{24}$-T7/mRNA] complex and copies single-stranded mRNA into cDNA, such that each cDNA contains a T7 RNA polymerase promoter site. This results in an mRNA/cDNA duplex that is then denatured. The cDNA is made double stranded by hairpin loop second-strand synthesis using Klenow fragment of *Escherichia coli* DNA polymerase I and T4 DNA polymerase. The hairpin loop of the double-stranded cDNA is cleaved with S1 nuclease and blunt ended with Klenow fragment. The T7 RNA polymerase promoter region, once double stranded, is functional, at which point the cDNA can serve as a template for aRNA synthesis with the addition of T7 RNA polymerase. To yield higher levels of aRNA, a second round of amplification is performed in which the aRNA serves as a template for cDNA synthesis, and is reverse transcribed with AMV-RT, resulting in an aRNA–cDNA hybrid. After heat denaturation, the cDNA is made double stranded, drop dialyzed, and reamplified using T7 RNA polymerase, as described above.

By incorporating a radiolabel, the second-round aRNA product can be used as a probe for reverse Northern blots to create mRNA expression profiles of single cells. The aRNA can also be used as a template for RT-PCR or cloned to construct a single-cell cDNA library. In addition, single- or double-stranded cDNA processed through one round of the aRNA procedure can serve as a template for differential display of single cells to identify novel, differentially expressed genes.[9] The aRNA amplification technique has been used to synthesize aRNA from mRNA in single dissociated live neurons in culture, subcellular regions such as dendrites[10] and growth cones,[11] single cells from fixed tissue,[12] and single cells from live slice preparations.[13,14] When using primary cultures of neurons the complex synaptic interactions required for proper CNS functioning are disrupted. The use of live slice preparations facilitates the study of neuronal gene expression in a system that preserves many synaptic and glial connections. Furthermore, whole-cell electrophysiological recordings can be performed in conjunction with the aRNA amplification procedure in single cells from dissociated cultures[4,5,15,16] or live slice preparations.[13,14] Electrophysiologi-

[9] P. Liang and A. Pardee, *Science* **257,** 967 (1992).

[10] K. Miyashiro, M. Dichter, and J. Eberwine, *Proc. Natl. Acad. Sci. U.S.A.* **91,** 10800 (1994).

[11] P. Crino and J. Eberwine, *Neuron* **17,** 1173 (1996).

[12] P. Crino, M. Dichter, J. Trojanowski, and J. Eberwine, *Proc. Natl. Acad. Sci. U.S.A.* **93,** 14152 (1996).

[13] S. Mackler, B. Brooks, and J. Eberwine, *Neuron* **9,** 539 (1992).

[14] S. Mackler and J. Eberwine, *Mol. Pharmacol.* **44,** 308 (1993).

[15] J. Surmeier, J. Eberwine, C. Wilson, Y. Cao, A. Stefani, and S. Kitai, *Proc. Natl. Acad. Sci. U.S.A.* **89,** 10178 (1992).

[16] J. Bargas, A. Howe, J. Eberwine, Y. Cao, and D. Surmeier, *J. Neurosci.* **14,** 6667 (1994).

cal data can then be correlated with coordinate changes in gene expression in single, morphologically identified cells.

Preparation of cDNA and Antisense RNA Amplification of mRNA from Single Dissociated Cells in Culture

In these studies a whole-cell patch electrode is utilized to harvest the mRNA from a single dissociated hippocampal neuron (embryonic day 20–21).[17] It is imperative that all reagents and the working environment be kept RNase free to prevent RNA degradation throughout the aRNA amplification procedure. Water should be treated with diethylpyrocarbonate (DEPC).[18] The microelectrode is backfilled with a solution containing the reagents necessary for first-strand cDNA synthesis in the following final concentrations: a 250 μM concentration each of deoxyadenosine triphosphate (dATP), deoxyguanosine triphosphate (dGTP), thymidine triphosphate (TTP), and deoxycytosine triphosphate (dCTP); 5 ng/μl oligo(dT)$_{24}$-T7 (100 ng/μl) primer containing a T7 RNA polymerase promoter region [AAACGACGGCCAGTGAATTGTAATACGACTCACTA-TAGGCGC(T)$_{24}$-T7]; and 0.5 U/μl AMV-RT (20 U/μl) in 1.0× electrode buffer (120 mM KCl, 1 mM $MgCl_2$, and 10 mM HEPES, pH 7.3). Whole-cell recordings may be done at this point if desired; however, modification of the electrode buffer may be necessary. The contents of a single, cultured, hippocampal cell are then gently aspirated into the microelectrode (Fig. 2). The entire contents of the patch pipette are then expelled into a microcentrifuge tube containing a 250 μM concentration each of dATP, dGTP, TTP, and dCTP; 5 mM dithiothreitol (DTT); 1 U/μl RNasin; and 0.5 U/μl AMV-RT in the above-mentioned buffer, adjusted to pH 8.3 to optimize for AMV-RT activity. First-strand cDNA synthesis is accomplished by incubating the preceding reagents with cellular contents for 60–90 min at temperatures between 37 and 50°.

Alternatively, oligo(dT)$_{24}$-T7 primers coupled to magnetic porous glass beads (CPG, Lincoln Park, NJ) that are attached via a primary amine linkage can be utilized.[19] The use of the magnetic beads obviates the need for phenol–chloroform extraction and ethanol precipitation. If the magnetic beads attached to the oligo(dT)$_{24}$-T7 are not used, the sample is phenol–chloroform extracted by adding 0.5 vol of TE [10 mM Tris-HCl (pH 8.0) and 1 mM EDTA]-saturated phenol, 0.5 vol of chloroform, and 0.1 vol of

[17] J. Buchalter and M. Dichter, *Brain Res. Bull.* **26,** 333 (1991).

[18] J. Sambrook, E. Fritsch, and T. Maniatis, "Molecular Cloning: A Laboratory Manual." Cold Spring Harbor Press, Cold Spring Harbor, New York, 1989.

[19] J. Eberwine, *Biotechniques* **20,** 584 (1996).

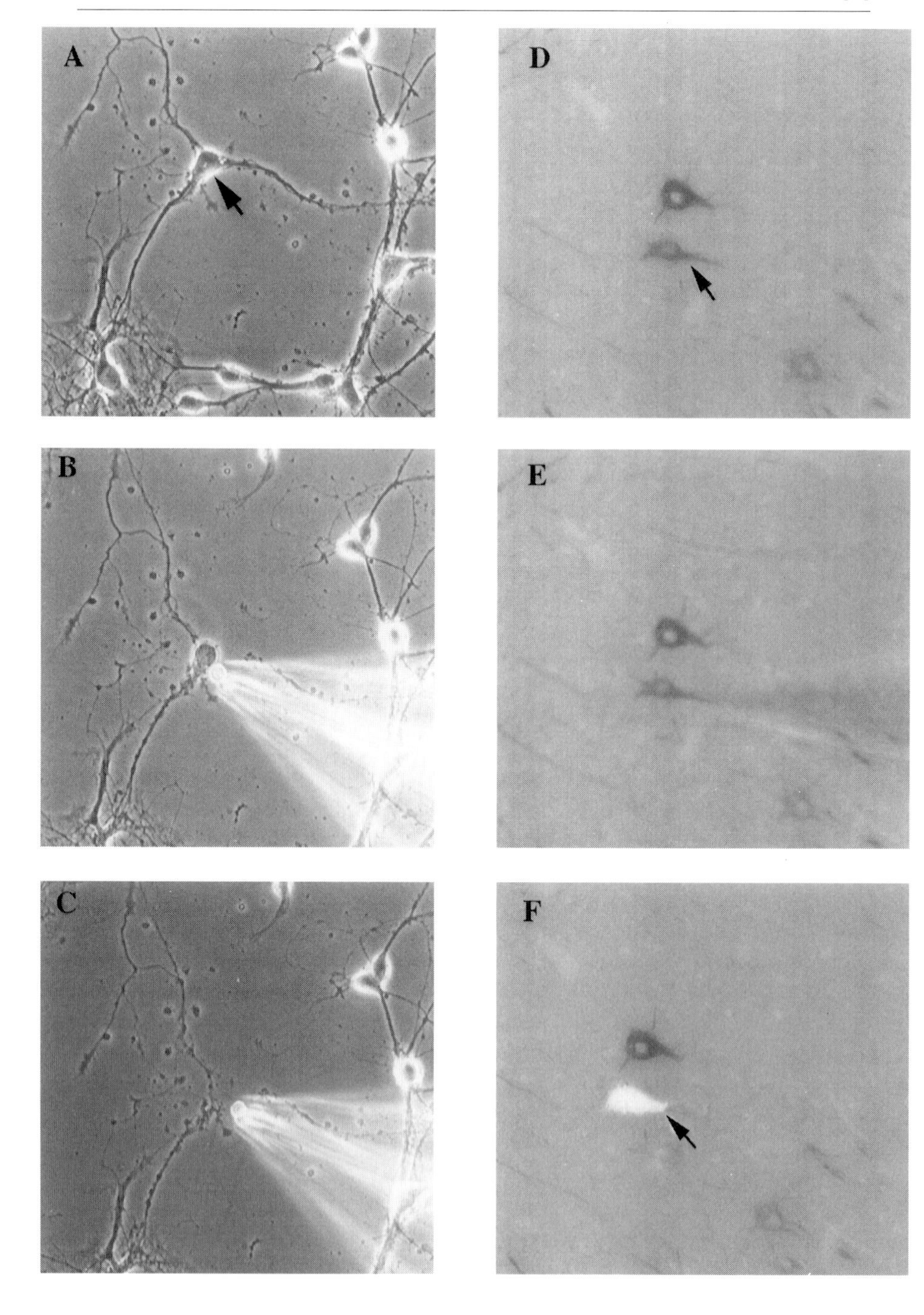
A
B
C
D
E
F

salt (5 *M* NaCl or 3*M* sodium acetate). The sample is vortexed, centrifuged, and the aqueous phase transferred to a new microcentrifuge tube. Ethanol precipitation is performed by the addition of 0.5 μg of *E. coli* transfer RNA (tRNA), which acts as a carrier, and 2–2.5 vol of ethanol, and placed on dry ice for at least 30 min. Glycogen or linear polyacrylamide in a final concentration of 1 μg/100 μl may be used in place of tRNA to aid in precipitation. The sample is pelleted, air dried, and resuspended in 20 μl of DEPC-treated water. The first-strand mRNA–cDNA hybrid is heat denatured at 95° for 3 min and placed on ice.

In this protocol, second-strand synthesis is accomplished by allowing the cDNA to serve as its own primer and to hairpin at its 3′ end.[20] However, alternative procedures such as that of Gubler and Hoffman,[21] or 3′-end tailing and specific priming,[22] can be used to synthesize double-stranded cDNA. For self-primed second-strand synthesis, the following reagents are added to the denatured cDNA in a final reaction volume of 40 μl: a 250 μM concentration each of dATP, dGTP, TTP, and dCTP; 1 U of T4 DNA polymerase; and 2 U of Klenow fragment in 1.0× second-strand buffer [100 m*M* Tris-HCl (pH 7.4), 20 m*M* KCl, 10 m*M* $MgCl_2$, 40 m*M* $(NH_4)_2SO_4$, and 5 m*M* DTT]. The reaction mixture is then incubated for at least 2 hr to overnight at 14°.

This results in double-stranded cDNA with a hairpin loop at the 5′-end. To cleave the hairpin loop that formed during the second-strand synthesis reaction and eliminate any remaining single-stranded cDNA, 1 U of S1 nuclease in 1.0× S1 buffer [200 m*M* NaCl, 1m*M* $ZnSO_4$, and 50 m*M* sodium acetate (pH 4.5)] is added to the sample in a final volume of 400 μl. The reaction mixture is incubated for 5 min at 37°. The sample is then phenol–chloroform extracted, ethanol precipitated, pelleted, air dried, and resuspended in 20 μl of DEPC-treated water.

[20] A. Efstriadis, F. Karatos, A. Maxam, and T. Maniatis, *Cell* **7,** 279 (1976).
[21] U. Gubler and B. Hoffman, *Gene* **25,** 263 (1983).
[22] S. Berger and A. Kimmel (eds.). *Methods Enzymol.* **152,** (1987).

FIG. 2. Aspiration of single neurons in culture and in fixed human brain tissue section. (A) Single hippocampal neuron at 1 week in culture. (B) Recording electrode apposed in close proximity to neuron. (C) After aspiration of neuron, dendrites and axons remain attached to the culture dish but the cell body is within the electrode. (D) Identification of fixed, normal temporal neocortical pyramidal neurons immunohistochemically labeled with antibodies that recognize the medium molecular weight form of neurofilament (NFM). The image is not fully in focus because the section is viewed under water. (E) Recording electrode apposed in close proximity to neuron. (F) After microdissection and aspiration of the selected neurons there is little disturbance or contamination by the surrounding neuropil. Note that the adjacent neurons have not been affected.

The double-stranded cDNA ends are then blunt ended in order to fill in gaps. This is accomplished by adding 1 U of Klenow, and a 250 μM concentration each of dATP, dGTP, TTP, and dCTP, in 1.0× Klenow filling-in (KFI) buffer [20 mM Tris-HCl (pH 7.5), 10 mM $MgCl_2$, 5 mM NaCl, and 5 mM DTT], in a final volume of 25 μl, and incubating the sample for 15 min at 37°. The sample is then phenol–chloroform extracted, ethanol precipitated, pelleted, air dried, and resuspended in 20 μl of DEPC-treated water.

To remove any excess salts and free deoxynucleotide triphosphates from first- and second-strand synthesis reactions, which can inhibit the efficiency of the T7 RNA polymerase, 10 μl of the double-stranded cDNA is drop dialyzed on a Millipore (Bedford, MA) membrane (0.025 μm) against DEPC-treated water for 30 min to overnight. The dialyzed sample is recovered and transferred to a new microcentrifuge tube. Antisense RNA can be synthesized using the double-stranded cDNA as a template by the addition of T7 RNA polymerase.

First-round aRNA amplification is initiated by adding the following to one-fifth of the double-stranded cDNA sample recovered from the drop dialysis, in a final volume of 20 μl: 2000 U of T7 RNA polymerase (2000 U/μl; Epicenter Technology, Madison, WI; 1 U/μl RNasin; 5 mM DTT; 250 μM ATP, GTP, and UTP; and varying concentrations of CTP, in 1.0× RNA amplification buffer [40 mM Tris-HCl (pH 7.5), 7 mM $MgCl_2$, 10 mM NaCl, and 2 mM spermidine]. In an unlabeled reaction, 250 μM CTP is added. In a radiolabeled reaction, 12.5 μM CTP, in addition to 3 μl of [α-^{32}P]CTP (1 mCi/100 μl; 3000 Ci/mmol), are added. This reaction is then incubated at 37° for 3.5 hr to overnight.

To survey the molecular weights of the radiolabeled aRNA products, they are heat denatured and electrophoresed on a 1% denaturing formaldehyde gel. The gel is run in 1.0× MOPS buffer {20 mM MOPS [3-(N-morpholino)propanesulfonic acid] (pH 7.0), 5 mM sodium acetate, and 1 mM EDTA}, precipitated with trichloroacetic acid (TCA), dried, and apposed to X-ray film. This Northern analysis indicates the aRNA size distribution. First-round aRNA amplification of the contents of a single cell results in a 10^3-fold amplification of the starting material. After first-round amplification, the aRNA product can be used as a template for RT-PCR or mRNA differential display, after reverse transcription of the aRNA into single- or double-stranded cDNA. However, if the desired use of the aRNA product is to create a cDNA library, or a gene expression profile using reverse Northern blotting, it is often necessary to do a second round of aRNA amplification, which would result in a $10^3 \times 10^3 = 10^6$-fold amplification of the starting material.

To process the aRNA through a second round of aRNA amplification, the sample must be phenol–chloroform extracted and ethanol precipitated. The sample is then heat denatured for 3 min at 85°, and first-strand synthesis is performed with the addition of the following reagents: 10–100 ng of random hexanucleotide primers, or specific cDNA primers; a 250 μM concentration each of dATP, dGTP, TTP, and dCTP; 10 mM DTT; 20 U of RNasin; and 40 U of AMV-RT in 1.0× reverse transcription buffer [50 mM Tris-HCl (pH 8.3); 120 mM KCl, 10 mM $MgCl_2$, and 0.05 mM sodium pyrophosphate]. This 30-μl reaction mixture is incubated at 37° for 60 min. The sample is then phenol–chloroform extracted, ethanol precipitated, and resuspended in 10 μl of DEPC-treated water. To synthesize second-strand cDNA, the aRNA–cDNA hybrid is heat denatured at 95° for 3 min, and placed on ice. The following reagents are added: a 250 μM concentration each of dATP, dGTP, TTP, and dCTP; 5 ng/μl oligo$(dT)_{24}$-T7; 1 U of T4 DNA polymerase; 2 U of Klenow fragment; and 5 mM DTT in 1.0× second-strand buffer, in a final volume of 20 μl. The reaction mixture is incubated for at least 2 hr at 14°. The sample is then phenol–chloroform extracted, ethanol precipitated, and resuspended in 20 μl of DEPC-treated water. The sample is then drop dialyzed and amplified as previously described.

The aRNA produced by a second round of aRNA amplification can be used to probe reverse Northern slot blots.[11,23] Linearized plasmid cDNAs are adhered to a solid support such as a nylon membrane (Hybond N; Amersham, Arlington Heights, IL) after heat denaturation for 10–15 min at 70°–95°. Individual cDNAs are selected by using a candidate gene approach, in which mRNAs of interest can be investigated. For example, several analyses have been performed in single live cultured neurons to study candidate cDNAs relevant to neuronal function, such as cytoskeletal elements, cell cycle markers, synthetic enzymes, neurotransmitter receptor subunits, neurotrophic factor receptors, or ion channel subunits. One microgram of each linearized plasmid cDNA is bound to the nylon membrane in 6× sodium citrate–sodium chloride (SSC) buffer and then immobilized with heat and vacuum or UV cross-linking.

aRNA can be labeled with [α-^{32}P]NTPs, [α-^{33}P]NTPs, digoxigenin, or other markers and used to probe reverse Northern blots. The blot is prehybridized in a shaking water bath for 8–12 hr at 42° in buffer [6× standard saline–phosphate–EDTA (SSPE) buffer, 5× Denhardt's solution, 50% formamide, 0.1% sodium dodecyl sulfate (SDS), 200 μg/ml salmon sperm DNA]. The riboprobe is heat denatured for 3 min at 95°, added to the

[23] Y. Cao, K. Wilcox, C. Martin, J. Eberwine, and M. Dichter, *Proc. Natl. Acad. Sci. U.S.A.* **93,** 9844 (1996).

prehybridization solution, and hybridized to the blots at 42° for 24–72 hr. Blots are washed in 2× SSC and 0.1% SDS twice for 30 min at room temperature and in 0.1× SSC, 0.5% SDS for 45 min at 37°. The radiolabeled blots are then dried and apposed to X-ray film for 24–48 hr to generate an autoradiograph or, alternatively, placed in a PhosphorImager cassette for 24 hr. Quantitation of hybridization intensity is performed by scanning densitometry. The intensity of the hybridization signal can be correlated with the relative abundances of specific mRNAs and used to determine differential gene expression profiles.

cDNA Synthesis and Antisense RNA Amplification of mRNA from Subcellular Regions

The subcellular localization of mRNA within the dendritic domain has raised the possibility that local protein synthesis can occur in dendrites. Translation of mRNAs into proteins within dendrites would provide a novel mechanism by which proteins necessary for local process outgrowth, growth cone navigation, synaptogenesis, and synaptic plasticity could be modulated without protein trafficking into the somatic cytoplasm. Initial evidence suggested that polyribosomes were located directly beneath synaptic spines in dendrites and in synaptodendrosomal preparations.[24] In addition, the identification of several mRNAs including microtubule-associated protein 2 (MAP2),[25–27] calcium/calmodulin-dependent protein kinase II (CaMKII),[10,28] brain-derived neurotrophic factor (BDNF),[29] several glutamate receptor subtypes,[10] and more than 20 distinct mRNAs in dendritic growth cones[11] provided evidence that specific mRNAs are localized within dendrites and raised the notion that dendritic mRNA-targeting mechanisms in the dendrite existed. Translation of a reporter-tagged mRNA construct within isolated dendrites in culture provided definitive evidence that local synthesis of proteins within dendrites occurred.[11]

Initial determination of the subcellular distribution of mRNAs in dendrites was made via ISH. To analyze mRNA expression levels in subcellular regions, such as dendrites[10] or growth cones,[11] the aRNA amplification method was applied to poly(A)$^+$ mRNA harvested from microdissected

[24] O. Steward and G. Banker, *Trends Neurosci.* **15,** 180 (1992).

[25] A. Caceres, G. Banker, O. Steward, L. Binder, and M. Payne, *Dev. Brain Res.* **13,** 314 (1984).

[26] L. Davis, G. Banker, and O. Steward, *Nature (London)* **330,** 477 (1987).

[27] C. Garner, R. Tucker, and A. Matus, *Nature (London)* **336,** 674 (1988).

[28] K. Burgin, M. Waxham, S. Rickling, S. Westgate, W. Mobley, and P. Kelly, *Neuroscience* **10,** 1788 (1990).

[29] M. Dugich-Djordjevic, G. Tocco, D. Willoghby, I. Najm, G. Pasinetti, R. Thompson, P. Lapchak, and F. Hefti, *Neuron* **8,** 1127 (1992).

cultured hippocampal neurons. Glass recording microelectrodes were filled with the reagents necessary for cDNA synthesis (as described above) including dNTPs, oligo(dT)$_{24}$-T7, and AMV-RT. In these experiments, single dendrites or growth cones were transected from their cognate cell bodies and aspirated into the microelectrode. cDNA synthesis was performed for 90 min at 37°, and the procedure for aRNA amplification was followed as described for single dissociated cells in culture.

The complexity of the mRNA population in such restricted subcellular domains was examined by converting the resultant aRNA from single dendrites or growth cones into single- or double-stranded cDNA and generating differential displays.[9] A differential display of the poly(A)$^+$ mRNA amplified from a single growth cone, as shown in Fig. 3, provided a broad view of the potential mRNA species within this region.[11] Further analysis, such as sequencing or subcloning, can be performed to distinguish between real and artifactual bands corresponding to mRNAs amplified by differential display. To generate gene expression profiles of mRNA populations within restricted subcellular regions, labeled aRNA can be used to probe reverse Northern blots containing a variety of candidate cDNAs, or cDNA arrays containing from just a few to several thousand candidate cDNAs.[30,31]

cDNA Synthesis and Antisense RNA Amplification of mRNA from Single Cells in Fixed Tissue

One primary restriction in studying the molecular basis of many neurologic and psychiatric diseases is the difficulty in identifying abnormal cells within a given brain region. In some disorders, such as Alzheimer's disease, there are characteristic pathologic features, such as neurofibrillary tangles, which can be definitively identified.[32] However, in other disorders, such as schizophrenia or epilepsy, clear pathologic changes can be difficult to discern. As a result, the precise anatomical or subcellular location of such alterations cannot be readily determined. Furthermore, in some neurological disorders, alterations in the expression of low-abundance genes occurs only in certain cell types. The lack of change in adjacent cell types can obscure the study of these modifications in gene expression, unless single-cell mRNA analysis is utilized. Finally, obtaining fresh-frozen brain tissue is often logistically difficult. To circumvent these difficulties, IST coupled

[30] D. Lockhart, H. Dong, M. Byrne, M. Follettie, M. Gallo, M. Chee, M. Mittmann, C. Wang, M. Kobayashi, H. Horton, and E. Brown, *Nature Biotech.* **14,** 1675 (1996).

[31] M. Chee, R. Yang, E. Hubbell, A. Berno, X. Huang, D. Stern, J. Winkler, D. Lockhart, M. Morris, and S. Fodor, *Science* **274,** 610 (1996).

[32] W. Wasco and R. Tanzi, "Molecular Mechanisms of Dementia." Humana Press, Clifton Heights, New Jersey, 1996.

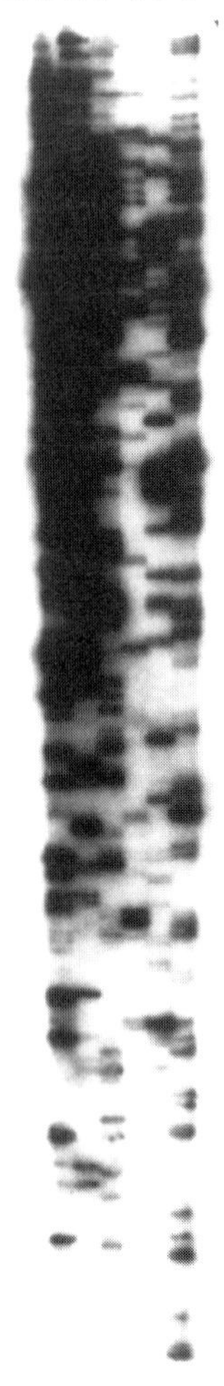

FIG. 3. Differential display of aRNA generated from single hippocampal dendritic growth cone mRNA. Lanes 1–6 reflect a single polymorphic 5′-primer (OPA-5, AGGGGTCTTG) and distinct 3′ primers [$(dT)_{11}$ primers with a 3′-dinucleotide extension] that were used for the differential display.

to the aRNA amplification technique (IST/aRNA), which has been used to study multiple gene products in fixed-tissue sections,[2,3] was used to assay gene expression within single, fixed, immunohistochemically labeled neurons.[12] With this approach, archival brain specimens can be studied to identify changes in gene expression, or to study the molecular pathogenesis of specific neurological disorders. Moreover, by immunohistochemically labeling tissue, individual neurons can be phenotypically identified on the basis of protein expression indicative of a specific disease. This approach provides a rapid and powerful means to quantify the relative abundances of multiple mRNAs in a uniform population of cells, all of which are expressing the disease phenotype. The use of a single-cell aRNA amplification has been applied to several brain disorders including tuberous sclerosis, Alzheimer's disease, schizophrenia, brain trauma, and epilepsy.

The first step in tissue preparation for IST/aRNA is fixation. Ideally, brain specimens with a postmortem interval of less than 4 hr should be selected. In preliminary experiments, fixation by immersion overnight in 70% ethanol–150 m*M* NaCl was optimal, although immersion in 4% paraformaldehyde–phosphate-buffered saline (PBS) was also effective.[33] Bouin's fixative, which contains picric acid, is not compatible with this approach. A detailed study of fixative compatibility with IST/aRNA has not been done.

Pathologic specimens are embedded in paraffin, sectioned with a microtome at less than 15 μm, and mounted onto glass slides coated with poly-L-lysine. The slides are then deparaffinized through xylenes and graded ethanols, prior to immunohistochemical labeling. This consists of three 5-min washes with xylenes, followed by two 5-min washes in 100, 90, and 70% ethanol and sterile water. Staining with the fluorochrome acridine orange can be done to confirm that mRNA is still present in the sections after extensive handling or storage.[34]

The immunohistochemical labeling is performed using conjugation methods such as those employing peroxidase–anti-peroxidase or avidin–biotin, and visualized with 3,3'-diaminobenzidine tetrahydrochloride (DAB).[33] After immunohistochemical labeling the fixed tissue sections are rinsed in DEPC-treated water. Tissue sections fixed in 4% paraformaldehyde are placed in 0.2*N* HCl for 20 min, washed briefly in PBS, pH 7.4, and treated with proteinase K (50 μg/ml) at 37° for 30 min. Pre-treatment of fixed tissue with proteinase K prior to cDNA synthesis enhances the ultimate yield of mRNA amplified from single cells. Oligo$(dT)_{24}$-T7 primer in hybridization buffer (50% formamide and 5× SSC buffer) is applied directly onto the fixed tissue section and placed in a humidified chamber. The sections are prehybridized overnight at 37° or room temperature in order to anneal the oligo$(dT)_{24}$-T7 to the cellular poly$(A)^+$ mRNA.[2] After

[33] S. Hockfield, S. Carlson, C. Evans, P. Levitt, J. Pintar, and L. Silberstein, "Selected Methods for Antibody and Nucleic Acid Probes." Cold Spring Harbor Press, Cold Spring Harbor, New York, 1993.

[34] H. Topaloglu and H. Sarnat, *Anat. Rec.* **224,** 88 (1989).

prehybridization, the sections are washed twice for 15 min each in 2× SSC buffer at room temperature, in 0.5× SSC at 40° for 1 hr, and briefly in 1.0× IST buffer [50 m*M* Tris-HCl (pH 8.3), 6 m*M* $MgCl_2$, and 120 m*M* KCl]. cDNA synthesis is then carried out directly on the tissue section by adding the following reagents and incubating for 60–90 min at 37°: a 250 μM concentration each of dATP, dGTP, TTP, and dCTP; 0.5 U/μl AMV-RT; 1 U/μl RNasin; and 7 m*M* DTT, in 1.0× IST buffer. [α-^{32}P]dCTP or [α-^{35}S]dCTP can be incorporated to determine the efficacy of the IST reaction. After IST, if radiolabels are utilized, sections are washed in 2× SSC for 1 hr and in 0.5× SSC for 6–18 hr at room temperature, in order to remove unincorporated dNTPs. Radiolabeled sections are then dried and apposed to film to generate an autoradiograph.

After cDNA synthesis has been carried out on the fixed tissue section, one of two possible directions can be taken. mRNA–cDNA hybrids from the entire section can be extracted by alkaline denaturation and amplified for further analysis. After incubating the sections in 0.2 *N* NaOH–1% SDS, the entire section can be transferred to a new microcentrifuge tube by repeated pipetting. The solution containing the mRNA–cDNA hybrids are then phenol–chloroform extracted and ethanol precipitated. Second-strand synthesis is performed and the sample is processed through the remainder of the aRNA amplification technique as described above for dissociated cells in culture. Using this method, mRNA levels from a selected brain region such as the hippocampus can be determined. This approach is especially useful if the objective is to compare regional changes in gene expression. In contrast, if a higher degree of experimental resolution is needed, the section is not alkaline denatured and dissection of single cells is performed in order to assay mRNA levels in individual neurons. Single neurons can be viewed under water following IHC and IST. Glass recording electrodes are used to gently aspirate individual immunohistochemically labeled cells from the fixed tissue (Fig. 2). The remainder of the aRNA procedure is followed as described above. After a desired number of individual cells is dissected from a tissue section, the cDNA can be extracted from the whole section using the alkaline denaturation procedure described above. This is a useful practice, since it permits comparison of amplified mRNA from whole tissue sections with mRNA from single cells in the same section. For example, if an mRNA cannot be detected within a single cell but can be detected within the mRNA population of the entire section, it provides evidence that absence or low abundance in the single cell may have biological significance.

In analyzing the aRNA amplification data generated from single neurons in fixed tissue, it is important to address a technical point. On the basis of a cDNA library screen,[12] it was estimated that approximately 20% of mRNAs within a dissected cell are amplified by the oligo(dT) approach. Thus, while many mRNAs can be assayed, some may not be detected, because they were not amplified. By comparison, in whole live cells, nearly 40–50% of genes may be assayed. It would be ideal to amplify the entire mRNA population, because genes that may be relevant to specific diseases may not be reflected in the limited aRNA pool described above.

Radiolabeled aRNA can be used to probe reverse Northern blots or cDNA arrays, which contain candidate genes of interest. Figure 4 shows a reverse Northern blot that was probed with ^{32}P-labeled aRNA from a single giant cell, immunohistochemically labeled with nestin, from a patient with tuberous sclerosis. The intensity of each hybridization signal reflects the quantity of aRNA and thus the abundance of the original mRNA species in the tissue sample.

Summary

Phenotypic characterization of cells in conjunction with single-cell mRNA analysis, which yields information regarding expression of multiple genes in individual neurons, facilitates a detailed and comprehensive view of neuronal cell biology. More specifically, the aRNA amplification method has provided an approach to analyze mRNA levels in single cells that have been phenotypically characterized on the basis of electrophysiology, morphology, and/or protein expression. In this way, relative mRNA abundances can be directly assayed from a well-defined population of neurons.

The concept of expression profiling led to the development of robotics methods for arraying thousands of cDNAs on microarrays. These cDNA arrays can be screened with labeled aRNA or cDNA to generate a molecular

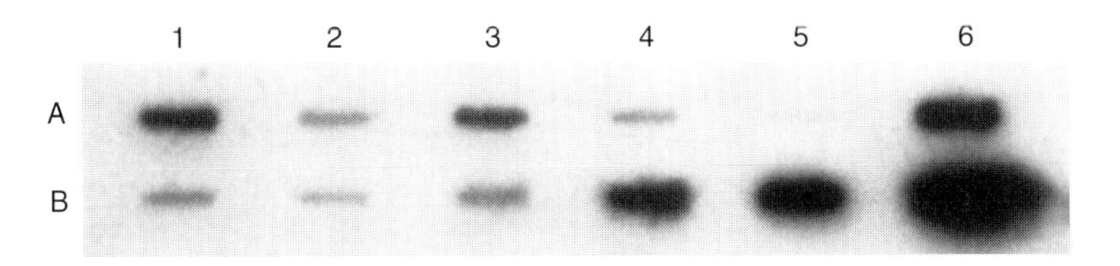

Fig. 4. Reverse Northern blot of a single abnormal neuron from a patient with the autosomal epilepsy syndrome, tuberous sclerosis. The pattern of hybridization intensities can be used to generate an mRNA expression profile. pBluescript plasmid cDNA was used to determine nonspecific hybridization. The following cDNAs were blotted onto the nylon membrane, and probed with ^{32}P-labeled aRNA from a single nestin-labeled cortical giant cell: A1, N-type calcium channel (Ca-N); A2, low molecular weight neurofilament (NFL); A3, medium molecular weight neurofilament (NFM); A4, α-internexin; A5, pBluescript; A6, glial fibrillary acidic protein (GFAP); B1, TrkA; B2, connexin 32 (CX32); B3, synthetic enzyme (GAD65); B4, human nestin; B5, microtubule-associated protein 2 (MAP2); B6, calcium/calmodulin-dependent protein kinase (CaMKII).

fingerprint of a specific cell type, disease state, or therapeutic efficacy. A broad view of how gene expression is altered in single neurons affected by a particular disease process may provide clues to pathogenetic disease mechanisms or avenues for therapeutic interventions. The use of mRNA profiles to produce diagnostics and therapeutics is called transcript-aided drug design (TADD).[15,35,36] When coupled with single-cell resolution, TADD promises to be an important tool in diagnosis of disease states, as well as provide a blueprint on which to develop therapeutic strategies. For example, mRNA abundances in an individual diseased cell may increase, decrease, or remain constant, and thus it is possible that a pharmaceutical alone or in combination with other drugs may be specifically designed to restore mRNA abundances to a normal state. Alternatively, if functional protein levels parallel the mRNA level changes, then drugs targeting the function of the proteins translated from these altered mRNAs may prove to be therapeutic. One promise of such an approach is that information about mRNA abundances that are altered in a diseased cell may provide new therapeutic indications for existing drugs. For example, if the abundance of mRNA for the β-adrenergic receptor is altered as shown by the microarrays for a particular disease, already available adrenergic receptor agonists or antagonists that had not previously been used in this particular disease paradigm may prove to be therapeutically efficacious.

The expression profile of a given cell is a measure of the potential for protein expression. Proteins are generally the functional entities within cells and differences in protein function often result in disease. The ability to monitor the coordinate changes in gene expression, in single phenotypically identified cells, that correlate with disease will provide unique insight into the expressed genetic variability of cells and will likely furnish unforeseen insight into the underlying cellular mechanisms that produce disease etiology.

[35] J. Eberwine, P. Crino, and M. Dichter, *Neuroscientist* **1,** 200 (1995).
[36] R. Finnell, M. Van Waes, G. Bennett, and J. Eberwine, *Dev. Genet.* **14,** 137 (1993).

[2] High-Efficiency Full-Length cDNA Cloning

By PIERO CARNINCI and YOSHIHIDE HAYASHIZAKI

Introduction

A full-length cDNA library is advantageous in that it allows cloning of a complete sequence in a single step. However, the representation of full-length cDNA clones has been low in cDNA libraries prepared using standard techniques. Full-length cDNA libraries have become available with a high proportion of the clones containing all of a particular coding sequence and its 3′ and 5′ untranslated regions (UTRs). Such libraries are particularly useful for large-scale sequencing projects (EST), in which the recovery of full-length clones from among truncated clones can be a formidable task.

In the preparation of full-length cDNA libraries, two problems have arisen. The first is the difficulty in reaching the cap site (5′ end of mRNA) in the first-strand synthesis with the reverse transcriptase (RT), the first enzyme involved in the preparation of a cDNA library. This is due to the presence of strong secondary structures of mRNAs.[1] These structures cause early termination of the reaction and detachment of the RT in a large number of clones, thus resulting in clones lacking the 5′ end, especially when priming the mRNA on the 3′ poly(A) tail. We have been able to overcome this drawback by introducing a disaccharide, trehalose, into the RT reaction. Trehalose can stabilize several enzymes, including RT, at temperatures as high as 60°[2] instead of the more usual 42° reaction temperature.[3] At 60°, the strength of the secondary structures of RNAs is thought to be greatly decreased, thus allowing much more efficient reverse transcription even of palindromic sequences, which are often encountered in the 5′ UTR of mRNAs. This results in longer cDNAs, higher representation of long, full-length cDNAs in the library, and an overall higher yield of the recovered full-length cDNA.

The other problem in library preparation is that there have not been effective methods for selection of full-length cDNAs from incompletely extended cDNAs. To solve this, we introduced a modified "biotinylated cap trapper" to select full-length cDNAs after biotinylation of the cap

[1] F. Payvar and R. T. Schimke, *J. Biol. Chem.* **254,** 7636 (1979).

[2] P. Carninci, Y. Nishiyama, A. Westover, M. Itoh, S. Nagaoka, N. Sasaki, Y. Okazaki, M. Muramatsu, and Y. Hayashizaki, *Proc. Natl. Acad. Sci. U.S.A.* **95,** 520 (1998).

[3] M. S. Krug and S. L. Berger, *Methods Enzymol.* **152,** 316 (1987).

Copyright © 1999 by Academic Press
All rights of reproduction in any form reserved.

0076-6879/99 $30.00

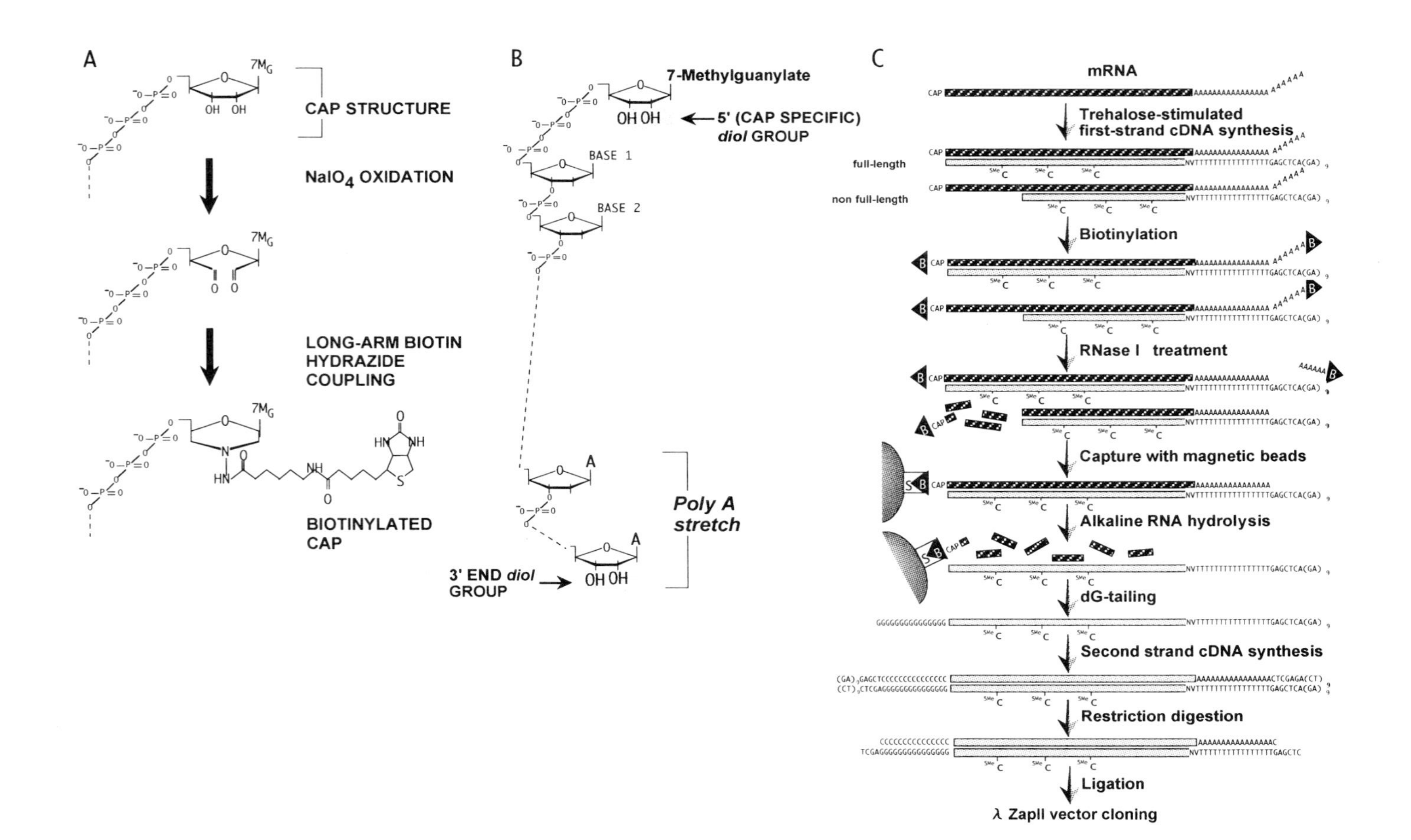

A
CAP STRUCTURE
NaIO4 OXIDATION
LONG-ARM BIOTIN HYDRAZIDE COUPLING
BIOTINYLATED CAP
B
7-Methylguanylate
5' (CAP SPECIFIC) diol GROUP
BASE 1
BASE 2
Poly A stretch
3' END diol GROUP
C
mRNA
Trehalose-stimulated first-strand cDNA synthesis
full-length
non full-length
Biotinylation
RNase I treatment
Capture with magnetic beads
Alkaline RNA hydrolysis
dG-tailing
Second strand cDNA synthesis
Restriction digestion
Ligation
λ ZapII vector cloning

structure (Fig. 1).[4,5] The method is based on a two-step chemical biotinylation of the two diol groups (Fig. 1A), one on the cap structure, specific to eukaryotic mRNAs, and the other on the terminal 3′ end of any RNA (Fig. 1B). The mRNA is biotinylated after the first-strand cDNA synthesis.[5] In the case of incomplete first-strand synthesis, there is a single-stranded RNA (ssRNA) that connects the biotinylated cap and the truncated cDNA–mRNA hybrid (Fig. 1C). This ssRNA can be digested by RNase I, an ssRNA-specific ribonuclease, in the case of truncated cDNAs, resulting in physical separation of the biotinylated cap from the cDNA. When the cDNA extends to the cap structure, RNase I cannot remove the cap from the hybrid, allowing its capture by steptavidin-coated magnetic beads (Fig. 1C). The second biotin group at the 3′ end is always removed by RNase I. In fact, the poly(A) tail is usually much longer than the part protected by the 16 Ts of the [5′$(GA)_8$ACTCGAG$(T)_{16}$VN-3′] oligonucleotide used to prime the first-strand cDNA. The degenerate V and N bases, where V is G, A, or C, and N is any base, are used for exact priming at the beginning of the poly(A), just after the cleavage and polyadenylation site. This priming thus allows for effective degradation of the unprotected single-stranded 3′ end of the poly(A) by RNase I (Fig. 1C), which is usually much longer than the fraction covered by the 16 Ts of the primer, and consequent removal of the biotin group at the 3′ end. The long sequence of repeated GAs at the 5′ end of the oligonucleotides (both first- and second-strand primers) is designed to provide a reasonably long DNA segment upstream of the restriction site to overcome the limitations encountered in the restriction cleavage close to the DNA ends.

The full-length cDNA is subsequently released from the beads by alkali treatment, followed by oligo(dG) tailing and second-strand synthesis mediated by a long-range thermostable DNA polymerase mixture,[6] which makes

[4] P. Carninci, C. Kvam, A. Kitamura, T. Ohsumi, Y. Okazaki, M. Itoh, M. Kamiya, K. Shibata, N. Sasaki, M. Izawa, M. Muramatsu, Y. Hayashizaki, and C. Schneider, *Genomics* **37,** 327 (1996).

[5] P. Carninci, A. Westover, Y. Nishiyama, T. Ohsumi, M. Itoh, S. Nagaoka, N. Sasaki, Y. Okazaki, M. Muramatsu, C. Schneider, and Y. Hayashizaki, *DNA Res.* **4,** 61 (1997).

[6] S. Cheng, C. Fockler, W. M. Barnes, and R. Higuchi, *Proc. Natl. Acad. Sci. U.S.A.* **91,** 5695 (1994).

FIG. 1. Overall strategy for preparation of a full-length cDNA library by modified biotinylated cap trapper. (A) Two-step coupling of the biotin hydrazide with diol groups. (B) Structure of mRNA and position of the two diol groups at the 5′ end(cap) and the 3′ end of the mRNA. (C) Strategy for the preparation of the cDNA library. FL, Full-length first-strand cDNA; NFL, non-full-length cDNA.

it possible to synthesize the second strand with high efficiency even in the case of long cDNAs. Because this protocol employs 5-methyl-dCTP instead of dCTP for the first-strand cDNA synthesis, the resulting cDNA is hemimethylated and is resistant to restriction digestion by the enzymes *Xho*I[7] and *Sst*I.[8] These enzymes can cleave the unmethylated sites only on the first- and second-strand primer adapters. The cDNAs is then cloned in λ vector at high efficiency. λ insertional vectors are used here because of their high efficiency and high capacity (0–10 kb).[9] In addition, λ ZAPII is useful for automatic *in vivo* excision into pBluescript plasmid, which does not require tedious subcloning from λ DNA.[10,11]

In this chapter we describe protocols using the powerful "thermostabilized" RT, which make it possible to prepare efficiently full-length cDNAs longer than 10 kb, combined with the selection of cDNA by biotinylated cap trapper to remove residual non-full-length cDNAs. Using these protocols, we can prepare full-length cDNA libraries at high yield (several to tens of millions of independent clones can be routinely produced) without using polymerase chain reaction (PCR), which introduces sequence bias and causes overrepresentation of short clones in a library. For a more extensive discussion about alternative procotols for full-length libraries, see Carninci *et al.*,[4] and the references therein.[12–15]

Materials. We routinely use the following materials to construct full-length cDNA libraries. Some alternatives are possible, especially relative to the cloning vectors and bacterial strains. Commonly used chemicals are not listed here. To avoid contamination due to "foreign" DNA and nucleases, all reagents for the preparation of cDNA libraries are stored separately from commonly used reagents.

Biological materials: Bacterial strain XL1 Blue *mrf′* (Stratagene, La Jolla, CA); cloning vector; λ ZapII (Stratagene).[10,11]

Enzymes and buffers: Restriction enzymes and reaction buffers, *Sst*I (GIBCO-BRL, Gaithersburg, MD) and *Xho*I [Takara (Ohtsu, Japan) or New England BioLabs (Beverly, MA)]; β-agarase and reac-

[7] P. S. Nelson, T. S. Papas, and C. W. Schweinfest, *Nucleic Acids Res.* **3,** 681 (1993).

[8] A. Kitamura and P. Carninci, unpublished observation (1997).

[9] J. Sambrook, E. F. Fritsch, and T. Maniatis, "Molecular Cloning." Cold Spring Harbor Laboratory Press, New York, 1989.

[10] J. M. Short, J. M. Fernandez, J. A. Sorge, and W. D. Huse, *Nucleic Acids Res.* **16,** 7583 (1992).

[11] J. M. Short and J. A. Sorge, *Methods Enzymol.* **216,** 495 (1992).

[12] K. Maruyama and S. Sugano, *Gene* **138,** 171 (1994).

[13] S. Kato, S. Sekine, S.-W. Oh, N.-S. Kim, Y. Umezawa, N. Abe, M. Yokoyama-Kobayashi, and T. Aoki, *Gene* **150,** 243 (1994).

[1] I. Ederly, L. L. Chu, N. Sonenberg, and J. Pelletier, *Mol. Cell. Biol.* **15,** 3363 (1995).

[1] CLONTECHniques technical bulletin, January (1996).

tion buffer (New England BioLabs); T4 DNA ligation and reaction buffer (New England BioLabs); mouse mammary leukemia virus (MMLV) reverse transcriptase and reaction buffer (Superscript II; GIBCO-BRL); placental RNase inhibitor (Wako, Osaka, Japan); proteinase K (Sigma, St. Louis, MO) at 10 $\mu g/\mu l$ in water, stored at $-20°$ in aliquots; RNase I and reaction buffer (Promega, Madison, WI); terminal deoxynucleotidyltransferase (TdT; Takara); EX-*Taq* polymerase (Takara)

Chemicals and biochemicals: Low melting point agarose (Sea Plaque; FMC, Philadelphia, PA); RNase-free cetyltrimethylammonium bromide (CTAB; Sigma); deoxynucleotide triphosphates and 5-methyl-dCTP, sodium salts, stored at $-80°$ as a 100 m*M* stock (Pharmacia, Piscataway, NJ); bovine serum albumin, DNase and RNase free (Takara); saturated trehalose [Fluka (Ronkonkoma, NY) or Sigma], approximately 80% (w/v), autoclaved; [α-^{32}P]dGTP, 3000 Ci/mmol (Amersham, Arlington Heights, IL); DE-81 chromatographic paper (Whatman, Clifton, NJ); $NaIO_4$ [Sigma or ICN (Irvine, CA)]; biotin hydrazide, long arm, also called biocytin hydrazide [Vector Laboratories (Burlingame, CA) or Sigma]; DNase- and RNase-free glycogen (Boehringer Mannheim, Indianapolis, IN); transfer RNA (tRNA, *E. coli*; Sigma). To ensure the absence of DNA, RNase-free DNase I (Promega) digestion is done, followed by SDS–proteinase K treatment, phenol–chloroform and chloroform extraction, and ethanol precipitation following standard procedures; first- and second-strand cDNA primers (see text for sequence and preparation)

Kits: λ DNA packaging extract (Max Plax; Epicentre, Madison, WI; mRNA extraction kit (Poly-A-Quick; Stratagene); magnetic porous glass (MPG) beads coated with streptavidin (CPG, Lincoln Park, NJ); CL-4B Spun column kit (Pharmacia)

Methods

λ Cloning Vector Preparation

Before starting the preparation of a cDNA library, prepare and test a high-efficiency, low-background cloning vector. As the cDNA fraction is radiolabeled, it becomes unstable and undergoes degradation if cloning is delayed. This will more dramatically affect the cloning of the longer than the shorter clones, resulting in a size bias and overrepresentation of shorter clones.

The oriented cDNA cloning protocol uses the λ ZAPII vector[10,11]; however, other vectors can be used provided that suitable restriction sites are present (see below). Here, alkaline phosphatase is not used, as it may dramatically decrease the cloning efficiency of the *Sst*I 5′ protruding ends. This drawback may be due to some exonuclease contamination, which is sometimes present in commercial sources of alkaline phosphatase. As we did not use alkaline phosphatase, there was no need for the formation of λ DNA concatamers by ligating the arms before the restriction, thus reducing the risk of mechanical shearing of the resulting very high molecular weight DNA.[9]

The highest quality restriction enzymes should be selected by checking the efficiency of religation after severalfold excess restriction digestion, as stated by the manufacturers.

Restrict the λ DNA in two steps, because the buffer requirements for *Sst*I and *Xho*I are different. First, cleave 15 μg of λ ZAPII in a large volume, such as 200 μl, by using 45 units of *Sst*I in the buffer provided by the manufacturer. Incubate for 2 hr at 37°. Notice that *Sst*I is an isoschizomer of *Sac*I, which cannot cleave hemimethylated cDNA. As it is difficult to separate the λ ZAPII arms by electrophoresis, a cut-check reaction should be done, by transferring 10 μl of the reaction to a separate tube containing 100 ng of a supercoiled plasmid in 0.5 μl of water. The plasmid should contain an *Sst*I restriction site (e.g., pBluescript; Stratagene). Incubate the cut-check reaction sample together with the main reaction. After incubating the sample for 2 hr, place the main reaction sample on ice and load the cut-check reaction sample on a 0.8% agarose minigel, together with a size marker and 100 ng of the starting uncut supercoiled plasmid. After electrophoresis, follow the extent of cleavage by checking the complete conversion from the supercoiled form of the plasmid to the linear form, which shows a slower migration.[9] If the reaction is incomplete, incubate the main reaction for an additional 1 or 2 hr and, if necessary, add more of the enzyme. When the reaction is complete, inactivate the *Sst*I at 65° for 10 min, change the buffer condition by adding 1.9 μl of 5 *M* NaCl and, after cooling the sample, add 45 units of *Xho*I. Repeat this procedure for the cut-check sample and incubate it for 2 hr at 37°. To reduce the background of the vector, a third cleavage can be performed at a third site on the λ DNA between *Xho*I and *Sst*I. For instance, we can add 30 units of *Not*I, the cut site of which lies between those of *Xho*I and *Sst*I in λ ZAPII, to the *Xho*I reaction. In this case, even if either *Xho*I or *Sst*I does not cleave completely, most of the termini that originate are incompatible for the ligation, thus reducing the background of the vector to a low level, usually less than 1–2%.

Finally, load the digested λ DNA in a 0.6% low melting point agarose minigel and run the electrophoresis for 60 min at 50 V. To follow the migration, instead of the usual high-energy 312-nm ultraviolet (UV) light transilluminator, use a mild 365-nm transilluminator, which is best if it is a hand-held type. Strong, short-wavelength UV light can cause DNA damage that will decrease the subsequent cloning efficiency, but a short exposure to 365-nm UV light will not. To minimize UV exposure, we usually do not take a picture of the gel at this stage and, instead, keep the previous cut-check data as a record of the experiment. The short stuffer is easily separated and the λ arms (which migrate together in the case of λ ZAPII) can be subsequently recovered from the gel by cutting with a sterile blade and transferral to a 2-ml Eppendorf tube. The tube is then briefly centrifuged to estimate the gel volume roughly, in order to calculate the number of units of β-agarase necessary to digest the agarose. To the gel, add the agarase buffer to the final 1× concentration and melt the gel at 65° for 10 min. Next, cool the gel to 40° for 5 min to equilibrate the temperature and add 3 units of β-agarase for each 100 μl of starting agarose gel. This concentration is in excess, but will completely remove all agarose, which would otherwise inhibit the ligation and the *in vitro* λ packaging reaction. Incubate the reaction for 2 to 4 hr at 40°. Finally, add 5 *M* NaCl to a final concentration of 1.2 *M*. The concentrated NaCl during the subsequent ethanol precipitation contributes to removal of the residual neutral polysaccharides.[16] Carefully perform phenol–chloroform and chloroform extraction (use a wide-bore pipette to avoid shearing of the DNA), then add 2 vol of ethanol. Incubate the reaction for 15 min on ice, then centrifuge at 12,000 rpm for 10 min to precipitate the DNA. Excessive centrifugation of λ DNA, which can lead to troublesome resuspension and the related risk of damage, should be avoided. After removing the supernatant, wash the pellet with 70% ethanol, centrifuge it for 2–3 min at 14,000 rpm, briefly air dry the pellet, and resuspend the DNA in 10–20 μl of 0.1× TE [1 m*M* Tris (pH 7.5 or 8.0) and 0.1 m*M* EDTA]. Measure the concentration by UV spectrophotometer and the polysaccharide contamination. A DNA preparation free of polysaccharides should have a 230/260 optical density (OD) ratio lower than 0.5, and a 260/280 OD ratio of approximately 1.8. Single-use aliquots (1–2 μg) of the vector can be conveniently stored for an indefinite period at −80°. Use one fraction of the vector to test the efficiency and background of the vector.

Testing Vector with Test Insert

Prepare a suitable test insert, for instance, any DNA of size 500–2000 bp cloned in pBluescript, by cleaving with *Sst*I and *Xho*I under standard

[16] G. Fang, S. Hammar, and R. Grumet, *Biotechniques* **13,** 52 (1992).

conditions. Purify the test insert from the agarose gel by standard silica-based methods,[17] or by using a kit such as GeneClean (Bio 101, La Jolla, CA) or Prep-A-Gene (Bio-Rad, Hercules, CA) as suggested by the manufacturers, or by adapting the preceding protocol using β-agarase.

Set up the following ligation reactions in a final volume of 5 μl:

λ DNA and test insert:

Reaction a: Only purified λ vector (100 ng)

Reaction b: λ vector (100 ng) and test insert, ratio 1:3

Reaction c: λ vector (100 ng) and test insert, ratio 1:1

Reaction d: λ vector (100 ng) and test insert, ratio 3:1

DNA ligase buffer: 0.5 μl

DNA ligase (200 Weiss units/μl): 0.4 μl

Incubate the reactions overnight at 16°. On the same day, inoculate a single colony of XL1-Blue *mrf'* in LB broth–0.2% maltose–10 m*M* $MgSO_4$, and incubate at 37° for 6–8 hr to measure the efficiency on the next day. Prepare the bacterial cells using standard procedures.[9] The next day, package half (2.5 μl) of the ligation reactions. At the same time, package 50 ng of the cut, but not religated, λ DNA (tube e), and 50 ng of the starting noncleaved λ DNA (tube f) in a 2.5-μl volume.

Thaw one Max Plax packaging extract[18] quickly by holding it between the fingers and promptly add 7.5 μl of the extract to each aliquot (tubes a–f). Conduct the reaction at 22° for 1.5 to 2 hr, then add 500 μl of SM buffer[9] and 20 μl of chloroform ($CHCl_3$) and titer as described by Sambrook *et al.*[9] After 6 to 8 hr, count the plaques and calculate the relative titer expected for 1 μg of DNA.

A satisfactory vector may give about 10^7 PFU/μg of λ DNA for the positive reactions b and c. The cloning background is given by reaction a. The titer of this reaction should be less than 5% for an acceptably low background when using the cDNA. As a reference value, a typical titer for the positive control is around 10^8 PFU/μg of λ DNA (reaction f), and less than 10^4 PFU/μg of λ DNA for the cleaved unligated sample (reaction e).

Working with RNA

For details on setting up an RNase-free environment, refer to more detailed laboratory guides.[9,19] As a rule, when working with RNA samples, gloves should be worn at all times; it is best to work off a bench dedicated to RNA work. Plasticware and solutions should be autoclaved (except

[17] R. Boom, C. J. Sol, M. M. M. Salimans, C. L. Jansen, P. M. E. Wertheim-vanDillen, and J. van der Noordaa, *J. Clin. Micribiol.* **28,** 495 (1990).

[18] E. J. Gunther, N. E. Murray, and P. Glazer, *Nucleic Acids Res.* **21,** 3903 (1993).

[19] D. D. Blumberg, *Methods Enzymol.* **152,** 20 (1987).

for solution D, phenol–chloroform, CTAB–urea, and 7 *M* guanidinium chloride solutions; see below), and glassware should be baked at 250° before use. Even if not specified, all reagents should be RNase free, at least until the end of the biotinylation of the mRNA–cDNA hybrid (see below). From our experience, using diethylpyrocarbonate (DEPC) is not recommended because autoclaving is sufficient to remove any RNase contamination if dedicated materials are used. We did find that the RNA was sometimes degraded after incubation at 65° in DEPC-treated water, which has a moderately acidic pH. To prepare RNase-free water, autoclaved double-distilled water or water of higher quality (for instance, MilliQ; Millipore, Bedford, MA), without DEPC treatment, is usually acceptable.

RNA Preparation

Total RNA is commonly prepared from mouse tissues by modification of a standard procedure[20] with adaption of the CTAB precipitation method[21,22] for selective removal of polysaccharides. This protocol also has been used for cultured cells and tissues of other mammals. We have found that the presence of polysaccharides, which may be subsequently biotinylated, inhibits the binding of full-length cDNA to streptavidin-coated magnetic beads (see below), perhaps by competing for the available streptavidin sites. Under the conditions described, the cationic micelles formed by CTAB undergo selective formation of complexes by charging the anionic nucleic acids, thus leaving in solution all neutral polysaccharides and residual proteins, which are easily eliminated in a single step.

This protocol is suitable for approximately 0.5–1 g of tissue. Volumes should be scaled up or down if the amount of tissue differs considerably. Quickly homogenize the fresh tissue in 10 ml of solution D [4 *M* guanidinium thiocyanate, 25 m*M* sodium citrate (pH 7.0), 100 m*M* 2-mercaptoethanol, 0.5% *N*-laurylsarcosine],[20] followed by the addition of 1 ml of 2 *M* sodium acetate, pH 4.0, 8 ml of water-equilibrated phenol,[9,20] and 2 ml of chloroform. High-quality RNA is obtained at consistently high yield with freshly equilibrated phenol (no earlier than a few weeks before use) or from frozen aliquots. After 15 min on ice, samples are centrifuged at 7500*g* for 15 min. The upper, aqueous phase containing the RNA is then gently separated and transferred into a new tube by pipetting, taking care to avoid the precipitated material at the layer between the two phases, which contains genomic DNA and proteins. Next, precipitate the RNA from the aqueous

[20] P. Chomczynski and N. Sacchi, *Anal. Biochem.* **162,** 156 (1987).

[21] G. Del Sal, G. Manfioletti, and C. Schneider, *BioTechniques* **7,** 514 (1989).

[22] S. Gustincich, P. Carninci, G. Del Sal, G. Manfioletti, and C. Schneider, *Biotechniques* **11,** 298 (1991).

phase by adding 1 vol of isopropanol. The sample is then incubated for 1 hr on ice, after which the RNA is pelleted by centrifugation at 7500*g* for 15 min. The pellet is washed twice with 70% ethanol, each time followed by centrifugation at 7500 rpm for 2 min, in order to remove the SCN salts, which would interfere with the following CTAB precipitation by forming a chemical precipitate. Selective CTAB precipitation of mRNA is performed after complete RNA resuspension in 4 ml of water. Subsequently, 1.3 ml of 5 *M* NaCl is added and the RNA is selectively precipitated by adding 16 ml of a CTAB–urea solution [1% CTAB, 4 *M* Urea, 50 m*M* Tris (pH 7.0), 1 m*M* EDTA (pH 8.0)]. The CTAB solution should not be autoclaved but prepared with RNase-free reagents (RNase-free CTAB is available from Sigma), water, and glassware. After 15 min of centrifugation at 7500 rpm at room temperature, the RNA is resuspended in 4 ml of 7 *M* guanidine chloride. The resuspension of RNA using a high salt concentration is necessary to remove the CTAB by ionic exchange.[21,22] Resuspended RNA is finally precipitated by adding 8 ml of ethanol, and after 1 hr on ice, the RNA is pelleted by centrifuging at 7500 rpm for 15 min. The pellet is washed with 70% ethanol and resuspended in water. RNA purity is monitored by reading the OD ratio at 230, 260, and 280 nm. Removal of polysaccharides is successful when the 230/260 ratio is lower than 0.5, and effective removal of proteins is obtained when the 260/280 ratio is higher than 1.8 and possibly ~2.0.

The messenger RNA is subsequently prepared by using commercial kits based on oligo(dT)-cellulose and starting from the guanidinium isothiocyanate–CTAB-purified RNA.[9] We have found Poly-A-Quick (Stratagene) to provide a satisfactory yield of mRNA under the conditions recommended by the manufacturer. For the latter use, we redissolve the poly$(A)^+$ RNA at a high concentration of 1 to 2 μg/μl (see [3] in this volume).[22a] From 500 μg of total RNA, 10 to 20 μg of mRNA can be routinely purified. Only one oligo(dT) selection of the mRNA is done; extensive pretreatment of the samples may cause underrepresentation of long mRNAs. Moreover, traces of cDNA derived from ribosomal RNAs (that lack the cap structure) are lost during selection of the full-length cDNA.

Preparation of Primers

First-strand cDNA primers should be thoroughly purified and rendered RNase free. Primer adapters containing the *Xho*I site [5′$(GA)_8$ ACTCGAG$(T)_{16}$VN-3′] are synthesized by adding V as a degenerate base in synthesis as G, A, or C, and N as G, A, T, or C, by synthesizing four

[22a] M. Liu, Y. V. B. K. Subramanyam, and N. Baskaran, *Methods Enzymol.* **303,** [3], 1999 (this volume).

different oligonucleotides from the four different CPG matrices. After ammonia deblocking, oligonucleotides are precipitated twice with 10 vol of butanol,[23] and further purified by acrylamide gel electrophoresis following standard techniques.[9] We use 12% acrylamide, 8 *M* urea, and 1× TBE (Tris–borate–EDTA). Alternatively, primers can be purified by high-performance liquid chromatography (HPLC). To ensure that purified primers are RNase free, the primers should be extracted with phenol–chloroform and chloroform and precipitated by addition of 0.2 *M* NaCl and 2.5 vol of ethanol, incubated at −20° for 1 hr, and centrifuged for 20 min at 15,000 rpm. After washing the pellet with 80% ethanol, primers are resuspended in water at a high concentration (>2 μg/μl). After checking the OD and mixing together the four primers in equal parts, the primer mixture is ready to prime the first-strand cDNA synthesis.

cDNA Preparation: First-Strand Synthesis

The average size of the first-strand cDNA when the RT is thermostabilized by addition of trehalose is greater than that of cDNA synthesized under standard conditions. The highest performance of the thermostabilized RT is obtained with temperature cycling between 55 and 60°.[2] Before starting the reaction, a thermal cycler with a hot lid (e.g., MJ Research, Watertown, MA) is set with the following first-strand cDNA synthesis program:

Step 1: 45° for 2 min (hot start)
Step 2: Negative ramp: go to 35° in 1 min (gradient annealing)
Step 3: 35° for 2 min (complete annealing)
Step 4: 45° for 5 min
Step 5: Positive ramp: +15° (until 60°) at +0.1°/sec
Step 6: 55° for 2 min
Step 7: 60° for 2 min
Step 8: Go to step 6 for 10 additional times
Step 9: +4° forever

The total time required is about 60 min. The ramp from 45 to 35° is designed specifically to anneal the primer at the beginning of the poly(A) of the mRNA.

To prepare the first-strand cDNA, put together the following reagents in three different 0.5-ml PCR tubes (A, B, and C):

Tube A: In a final volume of 24 μl, add the following:

mRNA:	5 to 10 μg
First-strand primer mixture:	5 μg
Glycerol (80%):	11.2 μl

[23] M. Sawadogo and M. W. Van Dyke, *Nucleic Acids Res.* **19,** 674 (1991).

The concentration of primers should be scaled up or down depending on the size, if different primer adapters are used. See below for a discussion of the possible primer adapter sites that can be used in the oligo primers. Heat the mixture (mRNA, primer, and glycerol) at 65° for 10 min to dissolve the secondary structures of mRNA. During the incubation, quickly prepare the reagent mix (tube B) and tube C:

Tube B: In a final volume of 76 μl, add the following:

First-strand buffer (5×):	18.2 μl
Dithiothreitol (DTT, 0.1 *M*):	9.1 μl
dATP, dTTP, dGTP, and 5-methyl-dCTP (instead of dCTP), 10 m*M* each:	6.0 μl
Bovine serum albumin (BSA, 2.5 γ/λ):	2.3 μl
Saturated trehalose (approximately 80%):	29.6 μl
Placental RNase inhibitor:	1.0 μl
Superscript II reverse transcriptase (200 U/μl):	10.0 μl

In tube C, place 1.0 μl of [α-^{32}P]dGTP or, alternatively, [α-^{32}P]dTTP or [α-^{32}P]dATP. Do not use [α-^{32}P]dCTP as a tracer; hemimethylated cDNA containing a fraction of unmethylated C becomes partially sensitive to the restriction digestion, thus leading to internal cleavage of the cDNA before cloning. If the mixture in tube B is not ready, after incubation at 65°, tube A can be left on ice for only a short time, to minimize the reformation of secondary structures of RNA.

Quickly put tubes A, B, and C on the thermal cycler to begin step 1, the hot-start incubation at 45°. After 15 sec, to equilibrate the temperature, transfer the contents of tube B to A, mix quickly but thoroughly, and transfer 25 μl of the resulting A + B mixture into tube C and mix. Complete the manipulation before the beginning of step 2 for an exact estimation of the incorporation of [α-^{32}P]dGTP that reflects the yield of the synthesized cDNA. Alternatively, the contents of tubes A, B, and C can be quickly mixed on ice instead of on the thermal cycler, and transferred immediately to 45° to begin step 1 of the cycling program. We usually label only 25% of the cDNA; the remaining 75% is unlabeled and thus does not undergo radiodegradation even if cloning is delayed.

At the end of the reaction (step 9), take 0.5 μl of the reaction in tube C and spot it onto a small square of DE-81 paper; keep a 0.5-μl aliquot in a separate tube for subsequent alkaline gel analysis. Measure the radioactivity of the spot before and after three 10-min washings with 50 ml of 0.5 *M* sodium phosphate, pH 7.0, followed by a brief washing with water, and 70% ethanol and quick air drying. Typical incorporation rates range between 3 to 8%, depending on the actual concentration of mRNA, which might have been overestimated by OD reading owing to contamination of ribosomal RNA. Alternatively, some trouble may have occurred in the first-strand synthesis, mainly owing to RNase contamination or decreased enzymatic activity due to the mishandling of the RT. In the first case, the cDNA, if analyzed by alkaline gel, will have an average size of 2 kb or higher, and the longest cDNA will be longer than 10 kb. In the second case, the cDNA will be notably shorter than 2 kb. In this case, the procedure should be stopped and the reagents tested.

Calculate the total yield (μg) of the synthesized cDNA, starting from the quantity of the dNTPs used in the reaction. The four dNTPs in the stock solution are present at 10 mM each, for a total concentration of 40 mM. The concentration of the tracer, [α-^{32}P]dGTP, is not relevant to this calculation. Because 6 μl is used in the reaction, it gives 6×10^{-6} (liter) $\times$ 40×10^{-3} (moles/liter) $= 2.4 \times 10^{-7}$ mol of dNTPs in the reaction. Because the average molecular weight of a residue is 340, in the reaction there are $340 \times 2.4 \times 10^{-7} = 8.17 \times 10^{-5}$ g of dNTPs, that is 81.7 μg. Consequently, the final yield is given by

$$81.7\ \mu\text{g} \times (\%\ \text{of incorporation})/100 = \text{micrograms of first-strand cDNA synthesized}$$

Organic Phase Extraction and cDNA Precipitation. Transfer both the "hot" and "cold" first-strand synthesis (tubes B and C) to a centrifuge tube (PCR tubes cannot be used for the centrifugation at 15,000 rpm with organic solvents, because they may be crushed) and add

EDTA (0.5 M) to a final concentration of 10 mM (2 μl)
Sodium dodecyl sulfate (SDS, 10%) to a final concentration of 0.2% (2 μl)
Proteinase K (10 μg/μl) to a final concentration of 100 ng/μl (1 μl)

Incubate at 45° for at least 15 min. Although the proteinase K incubation is time consuming, it contributes greatly to increased sample purity. The proteins (especially BSA) must be completely removed to facilitate subsequent redissolution of the pellet. If this is not done, the cDNA becomes "sticky." Moreover, in the absence of proteinase K treatment, cDNA is partially entrapped in the precipitated material that is sometimes observed in the layer between the organic and aqueous phase (see below), which appears "dirty" and contains some cDNA (radioactive) that cannot be back extracted.

Finally, add, in a final volume of 200 μl:

Ammonium acetate (10 M):	60 μl
Water	37 μl

Perform phenol–chloroform and chloroform extraction and back extraction:

1. Add 0.5 vol (100 μl) of phenol.
2. Add 0.5 vol (100 μl) of chloroform.
3. Vortex moderately until the two phases mix.
4. Leave on ice for 1–2 min.
5. Centrifuge for 2 min at 15,000 rpm.
6. Carefully remove the aqueous phase (upper layer), using a P-200 pipette, and transfer to a fresh 1.5-ml tube. Transfer as much as possible of the upper phase but avoid touching or transferring any contaminant material from the lower, organic phase. Keep the tube with phenol for the back extraction (see below).
7. Mix the phenol–chloroform-extracted cDNA and 200 μl of chloroform.
8. Vortex gently.
9. Centrifuge for 2 min at 15,000 rpm.
10. Transfer the aqueous phase (with a P-200 pipette) to a fresh, clean tube. Keep the tube with chloroform for the back extraction.

Perform back extraction to recover the residual cDNA from the residual aqueous phase that could not be recovered after the first extraction. This procedure helps increase the yield by 10–15% at each step.

1. Add 50 μl of water to the phenol–chloroform tube.
2. Vortex gently.
3. Centrifuge again for 2 min at 15,000 rpm.
4. Transfer the upper phase to the chloroform tube; discard the phenol–chloroform organic phase.
5. Vortex and centrifuge for 2 min at 15,000 rpm.
6. Transfer the upper phase to the chloroform tube.
7. Vortex and centrifuge for 2 min at 15,000 rpm.
8. Transfer the upper, aqueous phase to the previously extracted fraction of cDNA.

Checking the efficiency of cDNA recovery by hand-held monitor is not particularly effective at this stage owing to the large excess of free nucleotides. This should be done instead for all subsequent extraction/back extraction procedures, in the absence of free radioactive nucleotides. Refer to this protocol (except for the addition of ammonium acetate) for proteinase K incubation and extraction/back extraction at subsequent stages.

Finally, precipitate the cDNA with 2.5 vol of absolute ethanol.

1. Add 625 μl of absolute ethanol.
2. Incubate at $-80°$ for 20 min, or at $-20°$ or on ice for 30–60 min.

The ethanol precipitation step is always a convenient stopping point; the procedure can be interrupted for hours or overnight, if necessary.

After the incubation, obtain the cDNA by centrifugation at 15,000 rpm for 15 min. Remove the supernatant (*caution:* very hot) and wash the pellet twice with 800 μl of 80% ethanol. Each time, add 80% ethanol to the tube on the opposite side from where the cDNA is being pelleted and centrifuge for 2–3 min at 15,000 rpm. Finally, resuspend the cDNA in 47 μl of water. The extent of resuspension can be determined by checking the counts in the tube and in the solution by hand-held monitor. When the resuspension is complete, less than 5% of the total count will remain on the tube wall. Check that there are no clusters of nonredissolved cDNA in the solution. Do not use TE (Tris–EDTA) buffer at this stage to resuspend the cDNA, because the polyhydroxyl groups of TE are oxidized in the subsequent step and may quench the biotinylation reaction.

Biotinylation Reaction

OXIDATION OF DIOL GROUPS OF mRNA. In a final volume of 50 μl, add the following:

Resuspended cDNA sample
Sodium acetate buffer, pH 4.5 (1 *M*): 3.3 μl
Freshly prepared solution of $NaIO_4$ to a final concentration of 5 m*M*

Incubate on ice in the dark for 45 min. A note of caution: as exposure to light and high temperature makes the oxidation steps less specific, wrap the tube with aluminum foil. Finally, precipitate the cDNA by adding

SDS (10%): 0.5 μl
NaCl: 11 μl
Isopropanol: 61 μl

Incubate on ice for 45 min, or at -20 to $-80°$ for 30 min, in the dark. Centrifuge for 10 min at 15,000 rpm. Rinse the pellet twice with 70% ethanol, and each time recentrifuge at 15,000 rpm for 2 min. Resuspend the cDNA-pellet in 50 μl of water. The resuspension can be partially inhibited by the acidic pH of the acetate buffer used before the precipitation.

DERIVATIZATION OF OXIDIZED DIOL GROUPS. Prepare 10 m*M* biotin hydrazide long arm (MW 371.51) in water. Note that this reagent requires a long time and extensive mixing for complete solubilization in water. Always use fresh solutions of biotin hydrazide; frozen aliquots are sometimes not reactive.

To the cDNA add, in a final volume of 210 μl:

Sodium acetate buffer, pH 6.1 (1 *M*):	5 μl
SDS (10%):	5 μl
Biotin hydrazide long arm (10 m*M*):	150 μl

Incubate overnight (10–16 hr) at room temperature (22–26°). On the next day, to precipitate the biotinylated cDNA, add

Sodium acetate, pH 6.1 (1 *M*):	75 μl
NaCl (5 *M*):	5 μl
Absolute ethanol:	750 μl

Incubate on ice for 1 hr or at −80 to −20° for 30 min. Centrifuge the sample at 15,000 rpm for 10 min. Wash the precipitate once with 70% ethanol and once with 80% ethanol. Accurate washing of the pellet will help to remove the free biotin hydrazide. If not removed, it will compete with the biotinylated cDNA to bind the streptavidin beads. Redissolve the sample in 70 μl of 0.1× TE [1 m*M* Tris (pH 7.5), 0.1 m*M* EDTA]. Monitor for complete resuspension using a hand-held monitor.

Magnetic Bead Preparation. Streptavidin-coated porous glass beads (MPG) were selected because of the low nonspecific binding of the nucleic acids to the glass matrix compared with latex magnetic beads, and because of their excellent binding capacity.[24] To minimize further the nonspecific binding of nucleic acids, the beads are further preincubated with DNA-free tRNA. Start the blocking of the magnetic beads with DNA-free tRNA, and subsequent washings, just prior to the RNase I treatment as described below. Careful washing at this stage will remove all free streptavidin, which may have been released from the beads. Any trace of free streptavidin will decrease the efficiency of cDNA capture. Once washed, the beads should be used within 1 hr. Streptavidin does not remain stable longer than this in the BSA-containing storage buffer.

1. To 500 μl of streptavidin-coated MPG beads add 100 μg of DNA-free tRNA (at a concentration of 10–50 μg/μl).
2. Incubate on ice for 30 min with occasional mixing.
3. Separate the beads with a magnetic stand (for 3 min) and remove the supernatant.
4. Wash three times with 500 μl of washing/binding solution [2 *M* NaCl, 50 m*M* EDTA (pH 8.0)].

Washings, here and later, are performed by gentle redissolving of the beads after their capture by the appropriate wash solution, followed by their recapture for 3 min on a magnetic stand and removal of the solution

[24] P. Carninci, unpublished observation (1995).

by pipetting. Vortexing should be gentle, especially during the capture and washing of the cDNA.

RNase I Treatment. RNase I was selected because it is the only enzyme in this category that cleaves RNA at any base in a sequence-independent manner.[25] Consequently, no RNase cocktail is needed for the cleavage of any sequence. Also, RNase I can be inactivated simply by discarding it in solutions containing SDS at a concentration of at least 0.1% or higher. Thus, RNase I can be quite safely used in a RNA-dedicated laboratory.

To the cDNA sample (70 μl) add, in a final volume of 200 μl:

RNase I buffer (Promega): 20 μl

RNase I (5 U/μl; Promega): 200 units; glycerol concentration up to 10% does not cause a problem

Incubate at 37° for 15 min. Avoid prolonged incubation that may decrease the yield. To stop the reaction, put the sample on ice and add 100 μg of tRNA and 100μl of 5 *M* NaCl. The addition of tRNA at this stage prevents the nonspecific binding of cDNA to the beads (see the next section).

Capture of Full-Length cDNA. cDNA capture is performed in a high sodium chloride concentration. High-salt buffers help the binding of long cDNAs[26] and minimize the nonspecific interaction of nucleic acids with streptavidin. To capture the full-length cDNA, mix the RNase I-treated cDNA and the washed beads as follows:

1. Resuspend the beads in 400 μl of the washing/binding solution.
2. Transfer the beads into the tube containing the biotinylated first-strand cDNA.
3. After mixing, gently rotate the tube for 30 min at room temperature.

Full-length cDNA remains on the beads, but the shortened cDNAs do not. Separate the beads from the supernatant on a magnetic stirrer. The uncaptured fraction can be precipitated with 0.6 vol of isopropanol for further analysis by alkaline gel (optional step).

Washing Beads. Gently wash the beads in the indicated buffer to remove the nonspecifically absorbed cDNAs.

1. Twice with washing/binding solution
2. Once with 0.4% SDS, 50 μg/ml tRNA
3. Once with 10 m*M* Tris-HCl (pH 7.5), 0.2 m*M* EDTA, 40 μg/ml tRNA, 10 m*M* NaCl, 20% glycerol
4. Once with 50 μg/ml tRNA in water

[25] J. Meador III, B. Cannon, V. J. Cannistraro, and D. Kennel, *Eur. J. Biochem.* **187,** 549 (1990).
[26] S. T. Kostopulos and A. P. Shuber, *Biotechniques* **20,** 199 (1996).

cDNA Release from Beads. After the washings, release the cDNA from the beads by alkali treatment, which denatures the cDNA–mRNA hybrid and hydrolyzes the mRNA. To the tube containing the beads, add 50 μl of 50 m*M* NaOH and 5 m*M* EDTA, then briefly stir and incubate 10 min at room temperature with occasional mixing. Separate the magnetic beads and transfer the eluted cDNA on ice into a separate tube containing 50 μl of 1 *M* Tris-HCl, pH 7.5. Repeat the elution cycle with 50 μl of 50 m*M* NaOH, 5 m*M* EDTA once or twice until most of the cDNA, 80–90% as measured by monitoring the counts per minute with a hand-held monitor, can be recovered from the beads. Pool the eluted fractions.

As done for the first-strand reaction, treat the cDNA with proteinase K and phenol–chloroform extraction, including back extraction. At this stage, chloroform extraction is not necessary. All the precipitations are carried out from this step onward by using 0.2 *M* sodium chloride instead of 2 *M* ammonium acetate. To increase the efficiency of the precipitation, add 3 μg of glycogen and precipitate the sample with 1 vol of isopropanol, following standard procedures, then wash with 70% ethanol. Resuspend the pellet in 50 μl of 0.1× TE. Check the resuspension of cDNA by hand-held monitor and visual inspection for the presence of clusters of nonresuspended cDNA.

Siliconized tubes may be used to minimize the adsorption of the ssDNA to the plastic wall. The presence of carrier and the high purity of cDNA after proteinase K digestion and organic extraction can solve most of the trouble encountered due to the stickiness of single-stranded cDNA, which can usually be redissolved completely. To prevent the cDNA from sticking to the tube, do not dry the pellet under vacuum.

CL-4B Spun Column Fractionation of cDNA. Prior to the tailing reaction, traces of the first-strand primer, which become nonspecifically absorbed to the beads and later released, should be removed by gel filtration. The protocol described here uses a Sepharose CL-4B spun column (Pharmacia), essentially as described by the manufacturer. Depending on the experience and choice of the operator, spinning the column may be substituted with gravity column separation. We also successfully tested matrices other than CL-4B, such as S400.

For the spun column protocol:

1. Shake the column several times (until the matrix is completely redissolved), then let it stand upright.
2. Remove the upper cap, then remove the bottom one.
3. Drain the buffer from the column. If air bubbles enter the column, add column storage buffer again and go back to step 1.

4. Apply 2 ml of the buffer [10 m*M* Tris (pH 8.0), 1 m*M* EDTA, 0.1% SDS, and 100 m*M* NaCl] and drain twice by gravity.
5. Put the column into a 15-ml centrifuge tube, then centrifuge at 400*g* for 2 min in a swing-out rotor at room temperature. If cracks appear in the matrix, suspend it in the buffer defined in step 4; shake and repeat the procedure from step 4.
6. Apply 100 μl of buffer to the column, then centrifuge at 400*g* for 2 min. Check the eluted volume. If it is different from the input volume (100 μl), repeat this step until the eluted volume is the same as the added volume.
7. Set a 1.5-ml tube (after cutting off the cap) into the 15-ml centrifuge tube, then apply the sample to the column. Centrifuge at 400*g* for 2 min.
8. Collect the eluted fraction in a separate tube. Apply 50 μl of buffer to the column, repeat the centrifugation, and collect the fraction in a separate tube.
9. Repeat step 8 three to five more times; keep the eluted fractions separate.

Collected fractions should be counted by scintillation. The bulk of the cDNA (70% or more of the counts) is usually in the first three or four fractions, which contain the longer cDNAs and are free of contaminating first-strand primer. The subsequent fractions contain only short cDNAs and first-strand primer and should be discarded to prevent introducing short cDNAs and first strand–second primer dimers into the library. Pool the selected fractions (usually three or four), add 1 μg of carrier tRNA and NaCl to a final concentration of 0.2 *M*, and precipitate the cDNA by adding 1 vol of isopropanol. After precipitation and washing, dissolve the pellet in 31 μl of water.

Oligo(dG) Tailing of First-Strand cDNA. Oligo(dG) tailing is adopted here owing to its proven capacity to clone the 5′ ends of cDNAs from the first transcribed base.[27] Deoxyguanosine triphosphate is the selected nucleotide, because the reaction stops after the addition of approximately 15 G residues. Other methods to provide a sequence suitable for priming the second-strand synthesis, such as the RNA ligase-mediated protocols, yield too low a level of efficiency and sequence bias to be used in this strategy.[12,13]

To check the tailing reaction, prepare in a separate tube 0.5 ng of an oligonucleotide 20–40 bases long in a volume of 0.5 μl. This oligonucleotide

[27] P. L. Deininger, *Methods Enzymol.* **152,** 371 (1987).

should be 5′ end labeled with ^{32}P by polynucleotide kinase (PNK), following standard procedures, extracted with phenol–chloroform and chloroform, and precipitated with ethanol by using 2 μg of glycogen as a carrier.[9] Dilute the oligonucleotide at a concentration of approximately 1 ng/μl. The labeled oligonucleotide should be stored at −20° and can be used for no more than 4 weeks after labeling. The function of this oligonucleotide, the sequence of which is not really important unless there are no strong internal secondary structures, is to check the efficiency and length of the G tail in a parallel reaction.

Just prior to tailing, the cDNA is heated at 65° for 2 min to melt any secondary structure of the single-strand cDNA, which would decrease the tailing efficiency in some cases. After heating, transfer the sample to ice to prepare it for the reaction.

To the cDNA sample, in a final volume of 50 μl, add

- 10× TdT buffer [2 *M* potassium cacodylate (pH 7.2), 10 m*M* $MgCl_2$, 10 m*M* 2-mercaptoethanol]: 5 μl
- dGTP (50 μM): 5 μl, to a final concentration of 5 μM
- $CoCl_2$ (10 m*M*): 5 μl
- Terminal deoxynucleotidyltransferase: 40 units

Mix and transfer 3 μl to the control, tail-check tube, containing 0.5 μl of the labeled oligonucleotide. Incubate both cDNA and tail-check reaction for 30 min at 37°.

Stopping Tailing Reaction and cDNA Extraction–Precipitation. At the end of the tailing reaction, stop the main reaction (47 μl) by adding 1 μl of 0.5 *M* EDTA to the sample. If the reaction is not controlled, the resulting tail may be excessively long, causing difficulties in sequencing. Perform the proteinase K digestion and phenol–chloroform and chloroform extraction and back extraction as previously described for the first-strand cDNA reaction, and precipitate with 2.5 vol of ethanol after adding 0.2 *M* NaCl, as previously described. Further addition of carrier is not necessary because the glycogen and tRNA are still present from the previous steps. Finally, resuspend the pellet in 39 μl of 0.1× TE. In a parallel procedure, load the check-gel as explained in the next section.

Tailing Reaction Check. Use a small aliquot of the reaction to check the tailing efficiency. We usually prepare gels with 10% acrylamide, 8 *M* urea, and 1× TBE, and of size 400 × 150–200 × 0.25 mm (thickness). One side of the glass is treated with Sigmacote (Sigma) for easy removal of the gel after the electrophoresis. Thirty-five milliliters of acrylamide is polymerized by adding 35 μl of *N,N,N′,N′*-tetramethylethylenediamine (TEMED) and 70 μl of 25% ammonium persulfate (APS). For more details about the acrylamide gel electrophoresis, refer to Sambrook *et al.*[9]

Load the sample after adding 1 vol (3.5 μl) of sequence loading buffer

(95% deionized formamide, 0.05% bromophenol blue, 0.05% xylene cyanol, 10 m*M* EDTA) and denature at 65° for 3 min. Usually half of the control tailing reaction (3.5-μl aliquot) mixture is loaded with an equivalent amount of nontailed, kinased starting oligonucleotide (0.5 μl plus 3 μl of water plus 3.5 μl of loading buffer). A marker such as P-Bluescript, *Hpa*II cleaved and radiolabeled, can also be loaded using the same loading buffer. Convenient labeling can be done by filling in the *Hpa*II site with Klenow fragment in the restriction buffer.[9] The gel can be run at constant power (35 W) until the bromphenol dye approximates the bottom of the gel; usually 60 min of running is sufficient. Finally, dry the gel for 20–30 min at 80° with a gel dryer. The exposure time is as short as 30 min for an image analyzer [for instance, using a Fuji (Tokyo, Japan) BAS 2000 image analyzer] or a few hours if conventional films are used.[9]

By measuring the difference in length of the tail-check reaction and the starting oligonucleotide, the efficiency and length of tailing can be estimated. A satisfactory tailing reaction shows a shift in mobility of the oligonucleotide of approximately 15 bases (± two or three), with a narrow distribution of the G tail. G-stretches of such a length do not usually interfere with the sequencing operations.

cDNA Preparation: Second-Strand Synthesis

As a primer for the second-strand cDNA, an oligonucleotide of sequence 5′-$(GA)_9$(GAGCTCACTAGTC_{11}-3′ and containing the *Sst*I site is prepared and purified by commonly used techniques, for instance, as for the first-strand cDNA primer. To optimize the priming of the second-strand cDNA just at the end of the oligo(dG) stretch and to reconstitute the restriction sites, a temperature gradient is employed for annealing. In fact, by using a temperature ramp, mismatches in the oligo(dG) or oligo(dC) can be destabilized to obtain perfectly annealed primer/dG tails.[4] Hot-start priming is thus performed after setting the second-strand program on the thermal cycler as follows:

Step 1: 5 min at 55°
Step 2: Negative ramp of −20° (until 35°), with slope of −0.3°/1 min
Step 3: 35° for 10 min
Step 4: 68° for 20 min; repeat from step 3 twice (total, three times)
Step 5: +4°

Prepare tubes A and B containing:

Tube A:

Oligo(dG)-tailed cDNA: 39 μl
Second-strand *Sst*I primer adapter (100 ng/μl): 6 μl

Second-strand buffer [200 mM Tris (pH 8.93), 350 mM KCl, 50 mM $(NH_4)_2SO_4$, 10 mM $MgSO_4$, 10 mM $MgCl_2$, 0.5% Triton X-100, BSA (0.5 mg/ml]: 6 μl

dNTPs (2.5 mM each): 6 μl

Tube B:

[α-^{32}P]dGTP: 0.5 μl

After starting the second-strand program, put tubes A and B on the thermal cycler. Add to tube A 3 μl of Ex-*Taq* polymerase (5 U/μl) when the samples are at 55°, during the first step. Mix quickly but thoroughly, and immediately transfer a 5-μl aliquot to a new tube containing 0.5 μl of [α-^{32}P]dGTP. The transfer should be immediate to avoid the start of polymerization before mixing with the radioisotope-labeled nucleotide, which may lead to underestimation of the final cDNA yield. At this stage, any [α-^{32}P]dNTP can be used instead of [α-^{32}P]dGTP.

Spot a 0.5-μl aliquot from the cold 55-μl reaction sample and the hot 5-μl reaction sample onto DEAE paper (DE-81). Count the radioactivity before and after the washing with 0.5 M sodium phosphate and calculate the yield of the second strand after subtracting the contribution of the first stand (as obtained from the 0.5-μl spot from the 55-μl cold reaction sample). Because 20.4 μg of dNTP, calculated as for the first-strand reaction, is used in the above reaction, the yield of the second-strand cDNA is obtained by

$$\text{Second strand } (\mu g) = 20.4 \times (\% \text{ of incorporation})/100$$

Typical yields may vary between 100 and 500 ng, but even if the yield is as low as 50 ng there is still enough cDNA to make a representative library. The total double-strand cDNA quantity is determined simply by multiplying by 2 the quantity of second-strand cDNA. Keep a record of the counts per minute obtained at this point to use in a later stage to calculate the units of enzyme to restrict the cDNA and the ligation conditions.

A 0.5-μl aliquot of the second strand may be run on an alkaline gel to check the size after the second strand. It should approximately reflect the size of the first-strand cDNA and may help to estimate the average size to calculate the vector-to-insert ratio for the ligation (see below). The remaining aliquot of the radioactive reaction and the 55-μl main reaction can be combined if the CPM contribution of the first strand (counts per minute) seems to be too low to be followed in the later purification stages (less than 5000–10,000 cpm), or if there is some reason to suspect that the yield will be low.

The cDNA is treated as described above with proteinase K, organic extraction and back extraction, precipitation with ethanol using NaCl as a salt, and washing with 70% ethanol. Finally, resuspend the cDNA in 45 μl of 0.1× TE.

Cleaving cDNA. Calculate the units of restriction enzymes to be used to cleave the cDNA, based on the yield of the second-strand synthesis and assuming that no consistent fraction is lost during the precipitation. cDNA is restricted by using 25 U/μg of both *Sst*I and *Xho*I. These units have been tested and do not cleave the hemimethylated sites of the cDNA.[8] The restriction enzymes must be of high quality to prevent problems with overdigestion.

To the redissolved cDNA add, in a final volume of 50 μl:

Restriction buffer M (10×; BRL): 5 μl

25 U of *Sst*I per μg of double-stranded cDNA; in most cases, this quantity will be scaled down (for instance, for 200 ng of double-stranded cDNA use 5 U of enzyme)

Perform the following steps:

1. Incubate for 1 hr at 37°
2. Inactivate the restriction enzyme *Sst*I by incubating at 65° for 10 min.
3. Cool the sample on ice for 2 min and add additional NaCl to a final colume of 100 m*M* (1 μl of 2.5 *M* NaCl) to prepare high buffer conditions for *Xho*I.
4. Add the same amount (units) of *Xho*I as previously used for *Sst*I.
5. Incubate for 1 hr at 37°.

During the reaction, equilibrate a CL-4B column as previously for the purification of cDNA before tailing.

Stop the reaction by treating as described above with proteinase K, then conduct phenol–chloroform extraction once and back extraction. Chloroform extraction is not necessary at this stage. Keep the back-extracted aliquot separated until the CL-4B column is equilibrated. Apply the main fraction of cDNA, followed by the back-extracted aliquot, to the CL-4B column and centrifuge for 2 min at 400*g*. Collect the eluted cDNA in a new tube and apply 50 μl of the column buffer; centrifuge again at 400*g* for 2 min. Collect the fraction and transfer to a separate tube. Repeat this step two or three more times, keeping the fractions separate at all times.

Count by scintillation the tubes containing the cDNA. In most cases, the bulk of the cDNA is contained in the first two or three fractions. Subsequent fractions, which usually contribute only a small part of the cDNA (but that may contain first-strand/second-strand cDNA primer dimers), can be discarded. When precautions were not taken, we observed contamination by the dimer of the oligo(dG)-tailed first-strand cDNA primer and second primer. By selecting only the initial fractions, this risk can be greatly minimized.

Finally, precipitate with ethanol the selected, pooled fractions, as previously described, and wash with 70% ethanol. Before resuspending the cDNA, count the pellet by scintillation. This will allow calculation of the cDNA recovery after the CL-4B column and ethanol precipitation, for the next ligation step.

Ligation of cDNA into λ DNA Purified Arms. Roughly estimate the average size of the cDNA from the second-strand alkaline gel. If the size of the second-strand cDNA has not been checked by electrophoresis, assume that it has an average size of 2 kb when calculating the optimal ratio between cDNA and vector. A high vector-to-cDNA ratio will increase the background but decrease the formation of chimeric clones. Consequently, a high vector-to-cDNA (2:1) ratio should be used only in the case of a very low background vector. Notice that an increase in the vector concentration does not increase the yield of cloning, because for the packaging reaction the concatameric form of λ DNA has a higher efficiency than do single arms ligated with an insert. The highest efficiency of packaging is thus obtained with a 1:1 ratio between the vector and insert.

If the background of the vector is not very low (higher than 2.5%), the ratio of vector to cDNA should be between 1:0.8 and 1:1. When the vector-to-cDNA ratio is higher than 1:2 (to overcome the background of the vector), the contamination of chimeric clones is usually not acceptable. Alternatively, small aliquots of cDNA can be ligated at different cDNA-to-vector ratios (for instance, 1:3, 1:1, and 3:1) to search for the best compromise between the efficiency, background, and possibility of formation of chimeric clones.

Set up the ligations in a final volume of 5 μl (use up to 5 μg of vector), with the amount of cDNA selected. Ligate overnight at 16°, and subsequently conduct the *in vitro* packaging. An additional control experiment on the vector background can be done by ligating an aliquot of the λ vector without the cDNA. Usually the *in vitro* packaging reaction is done with at least half (25 μl) of the packaging extract Max Plax when using up to 5 μg of λ vector. Standard protocols are followed, as suggested by the manufacturer, for all of the subsequent manipulations.[9] Usually, the entire procedure can be expected to yield at least 10^6–10^7 primary plaque-forming units (PFU).

The library is then ready for the next analysis or amplification.[9]

To test the quality of the library, check the size of 20–30 clones by one of several methods: (1) direct PCR of the lysate by long PCR[6]; (2) λ miniprep followed by restriction digestion[21]; (3) *in vivo* excision, followed by checking the size of the excised plasmid by restriction digestion.[10,11]

The average size of a satisfactory library is 1.5 kb, and 2 kb in the case of an excellent library.

Additional controls to ensure the quality of the library may include random sequencing of 50 clones from the 5′ end (*Sst*I side). The sequences of several clones will be present in GenBank and can be checked if they are full length; more than 90–95% of identified clones should be full-length.

Alternatively, to quickly check the full-length content, two replica filters containing 50,000 PFU can be hybridized with probes 3′ and 5′ to a given cDNA, as full-length clones will hybridize to both. For instance, EF-1α (1.7 kb) and GAPDH (1.2 kb) are ubiquitously and highly expressed and

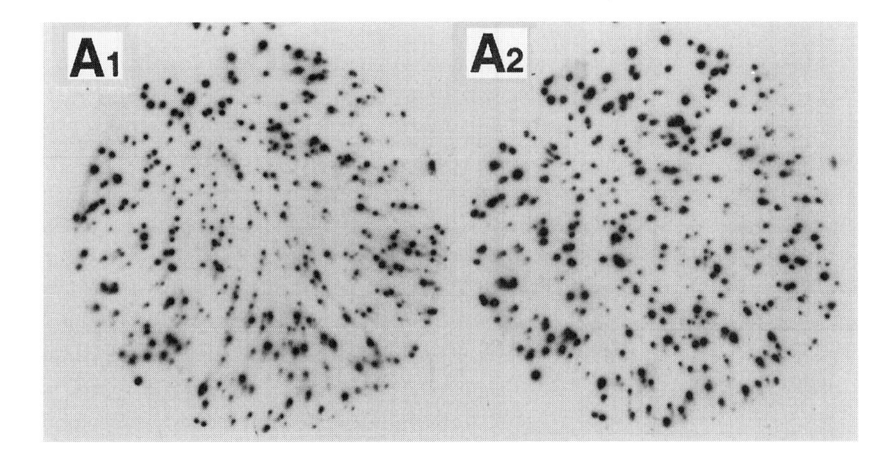

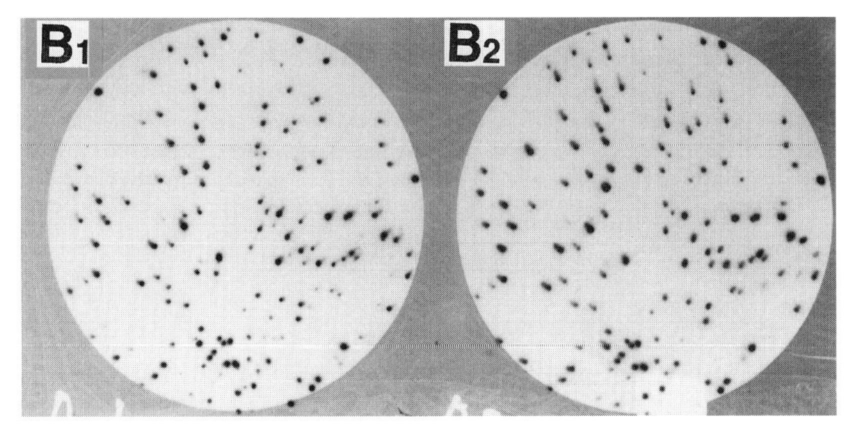

FIG. 2. Assessment of full-length cDNA content of a library by hybridization. A1 and A2, Hybridization with probes, respectively, at the 5′ end (from nucleotide 27 to 455) and 3′ end (from nucleotide 880 to 1175). About 98% of clones positive for the 3′-end probe are also positive for the 5′-end probe. B1 and B2, Hybridization with probes, respectively, at the 5′ end (from nucleotide 34 to 365) and 3′ end (from nucleotide 1264 to 1643) of EF-1α. More than 95% of the clones positive for the 3′ probe are also positive for the 5′ probe.

give, respectively, more than 95 and 98% of plaques positive for both 3′ and 5′ probes, in the case of a satisfactory full-length library (Fig. 2).[4]

Alternative Cloning Strategies

Our methodology can in principle be applied to a wide range of cloning vectors, including different λ vectors and plasmids, provided that they contain restriction sites that are sensitive to hemimethylation. When *Sst*I and/or *Xho*I need to substituted, different restriction sites can be designed on the primer adapters that are suitable for other vectors. We tested the resistance of hemimethylated cDNA to restriction digestion at a high enzyme-to-substrate ratio and identified several enzymes that are suitable for such a strategy and are present in many multiple cloning sites, such as *Bam*HI, *Not*I, *Apa*I, *Acc*I, *Bst*XI, *Eag*I, and *Sal*I. However, we do not recommend using *Not*I or *Eag*I next to the dC stretch of the second-strand primer because of subsequent difficulties in sequencing.[8] See also Nelson *et al.*[7] for an extensive list of enzymes, although we found, in contrast to what was described, that *Sac*I can partially cleave hemimethylated cDNA.[8]

Our method, using the powerful thermostabilized RT and the biotinylated cap trapper, makes possible efficient preparation of full-length cDNAs longer than 10 kb. The availability of full-length cDNA libraries can greatly facilitate cloning work, particularly in large-scale sequencing projects.

Acknowledgments

This study was supported by Special Coordination Funds and a Research Grant for the Genome Exploration Research Project from the Science and Technology Agency of the Japanese Government, and by a Grant-in-Aid for Scientific Research on Priority Areas and the Human Genome Program, from the Ministry of Education and Culture, Japan, to Y.H. This work was also supported by funding from the Core Research of the Evolutional Science and Technology Program from JST. Support also came from a Grant for Research on Aging and Health, and a Grant-in-Aid for a Second Term Comprehensive 10-Year Strategy for Cancer Control from the Ministry of Health and Welfare of the Japanese Government to Y.H.

[3] Preparation and Analysis of cDNA from a Small Number of Hematopoietic Cells

By MENG LIU, Y. V. B. K. SUBRAMANYAM, and NAMADEV BASKARAN

Introduction

Amplification of cDNA from a small number of cells or even a single cell has been accomplished in several laboratories and has led to the disclosure of interesting biological results. These methods used either polymerase chain reaction (PCR) amplification of cDNAs by DNA polymerase or amplification of cDNA templates by transcription with T7 or T3 RNA polymerase followed by conversion of the amplified RNA to cDNA.[1,2] Many potential applications of these methods exist, particularly when the source materials are scant. In that case, a method that can uniformly amplify cDNA from a small number of cells or a single cell would be needed. Elegant studies by Eberwine and colleagues have demonstrated the feasibility of obtaining display patterns of cDNA derived from mRNA of a single cell.[3] However, when working with small amounts of material, distortions of ratios of products and random dropout of representations of some mRNAs may become a major problem.[4]

We have been engaged in characterizing the messages from early hematopoietic precursors including stem cells with long-term repopulating potential. These are relatively small cells in the G_0 phase of the cell cycle, and contain less RNA than many larger cells. To obtain display patterns of cDNA from these cells, it was necessary first to explore the conditions for relatively uniform amplification of a mixture of cDNA fragments that differ in base composition and size from a small number of cells. Because our previous observations have suggested that T7 and T3 RNA polymerases had difficulty in transcribing some of the GC-rich segments of DNA in the HTF islands of mammalian genes, we looked for other methods for amplification. Quaternary ammonium salts have been used to decrease the differences in melting points among DNA molecules of different base

[1] G. Brady and N. N. Iscove, *Methods Enzymol.* **225,** 611 (1993).

[2] J. Eberwine, H. Yeh, K. Miyashiro, Y. Cao, S. Nair, R. Finnell, M. Zettel, and P. Coleman, *Proc. Natl. Acad. Sci. U.S.A.* **89,** 3010 (1992).

[3] K. Miyashiro, M. Dichter, and J. Eberwine, *Proc. Natl. Acad. Sci. U.S.A.* **91,** 10800 (1994).

[4] E. E. Karrer, J. E. Lincoln, S. Hogenhout, A. B. Bennett, R. M. Bostock, B. Martineau, W. J. Lucas, D. G. Gilchrist, and D. Alexander, *Proc. Natl. Acad. Sci. U.S.A.* **92,** 3814 (1995)

Copyright © 1999 by Academic Press
All rights of reproduction in any form reserved.

0076-6879/99 $30.00

composition.[5] However, simple quaternary amines at concentrations (>2.5 *M*) necessary to exert this effect inhibit thermostable DNA polymerases as well. To avoid this inhibition, we explored the use of zwitterionic forms of tetraalkyl ammonium compounds. Although other zwitterions may yield similar results, we chose betaine (trimethylglycine). This is because betaine is a naturally occurring substance that may serve as a major osmoprotectant in bacteria.[6-8] Dimethyl sulfoxide (DMSO) has previously been used to obtain PCR amplification of GC-rich fragments. At the concentrations necessary to amplify these fragments, a combination of DMSO and betaine proved effective in permitting simultaneous, almost equal amplification of a mixture of test fragments of different GC contents.[9]

In this chapter we have used the betaine–DMSO PCR method[9] to amplify cDNA derived from a small number of mouse hematopoietic cells. The faithfulness of the representation of the amplified cDNA was examined by a modified cDNA differential display method as described.[10-12]

Materials and Reagents

A MicroPoly(A)Pure mRNA isolation kit purchased from Ambion (Austin, TX) is used for mRNA isolation. All of the reagents for cDNA synthesis are obtained from Life Technologies (Gaithersburg, MD). Klentaq1 DNA polymerase (25 U/μl) is from Ab Peptides (St. Louis, MO). Native *Pfu* DNA polymerase (2.5 U/μl) is purchased from Stratagene (La Jolla, CA). Betaine monohydrate is from Fluka BioChemica (Buchs, Switzerland) and dimethyl sulfoxide (DMSO) is from Sigma Chemical Company (St. Louis, MO). Deoxynucleoside triphosphates (dNTPs, 100 m*M*) and bovine serum albumin (BSA, 10 mg/ml) are purchased from New England BioLabs (Beverly, MA). Qiaquick PCR purification kit (Qiagen, Chatsworth, CA) is used to purify the amplified PCR products. The oligonucleotides used in our experiments (listed in Table I) were synthesized and gel purified in the DNA synthesis laboratory (Department of Pathology, Yale University School of Medicine, New Haven, CT).

[5] W. B. Melchior, Jr. and P. H. von Hippel, *Proc. Natl. Acad. Sci. U.S.A.* **70,** 298 (1973).
[6] L. N. Csonka, *Microbiol. Rev.* **53,** 121 (1989).
[7] W. A. Rees, T. D. Yager, J. Korte, and P. H. von Hippel. *Biochemistry* **32,** 137 (1993).
[8] M. M. Santoro, Y. Liu, S. M. Khan, L. X. Hou, and D. W. Bolen, *Biochemistry* **31,** 5278 (1992).
[9] N. Baskaran, R. P. Kandpal, A. K. Bhargava, M. W. Glynn, A. Bale, and S. M. Weissman, *Genome Res.* **6,** 633 (1996).
[10] P. Liang and A. B. Pardee, *Science* **257,** 967 (1992).
[11] Y. Prashar and S. M. Weissman, *Proc. Natl. Acad. Sci. U.S.A.* **93,** 659 (1996).
[12] Y. V. B. K. Subrahmanyam, N. Baskaran, P. Newburger, and S. M. Weissman, *Methods Enzymol.* **303,** [16], 1999 (this volume).

TABLE I
SEQUENCES OF OLIGONUCLEOTIDES USED IN THIS CHAPTER

Oligonucleotide	Sequence
T7-*Sal*I-oligo-d(T)V	5′-ACG TAA TAC GAC TCA CTA TAG GGC GAA TTG GGT CGA C-$d(T)_{18}$V-3′, where V = A, C, G
Anti-*Not*I-long	5′-CTT ACA GCG GCC GCT TGG ACG-3′
*Not*I-short	5′-AGC GGC CGC TGT AAG-3′
*Not*I/RI primer	5′-GCG GAA TTC CGT CCA AGC GGC CGC TGT AAG-3′

Methods

Preparation of mRNA

It is important to make sure that all of the reagents and tubes are free of RNase while working with RNA. A MicroPoly(A)Pure mRNA isolation kit is used for the isolation of $poly(A)^+$ RNA, following the kit instructions. This method involves minimal exposure of mRNAs to nucleases, resulting in a good quality of RNA. mRNA from a small number of mouse hematopoietic cells (5000–10,000 cells) is extracted, eluted from the column, and precipitated by adding 0.1 vol of 5 *M* ammonium acetate and 2.5 vol of chilled ethanol with 2 μg of glycogen as carrier. The tubes are left at −20° overnight. The pellets are collected by centrifugation at top speed for 30 min, washed with 70% ethanol, and air dried at room temperature. The pellets are resuspended in 10 μl of H_2O–0.1 m*M* EDTA solution. We observed that the dissolved mRNA solution is cloudy, owing to the leaching of column materials; therefore the samples are centrifuged at 4° for 5 min. The supernatant is collected for further use.

cDNA Synthesis

First-Strand cDNA Synthesis. The cDNA synthesis reaction (final reaction volume, 20 μl) is carried out as described in the instruction manual (Superscript Choice system) provided by Life Technologies. For the first-strand cDNA synthesis, mRNA (10 μl) isolated from a small number of cells is annealed with 200 ng (1 μl) of T7-SalI-oligo-d(T)V primer (see Table I) in a 0.5-ml microcentrifuge tube (no-stick; USA Scientific Plastics, Ocala, FL) by heating the tubes at 65° for 5 min, followed by quick chilling on ice for 5 min. This step is repeated once and the contents are collected at the bottom of the tube by a brief centrifugation. The following components are added to the primer-annealed mRNA on ice prior to initiating the reaction: 1 μl of 10 m*M* dNTPs, 4 μl of 5× first-strand buffer [250 m*M* Tris-HCl (pH 8.3), 375 m*M* KCl, 15 m*M* $MgCl_2$], 2 μl of 100 m*M* dithiothreitol (DTT), and 1 μl of RNase inhibitor (40 U/μl). All of the contents are mixed gently and the tubes are prewarmed at 45° for 2 min. The cDNA synthesis is initiated by adding 200 units (1 μl) of Superscript II reverse transcriptase and the incubation is continued at 45° for 1 hr.

Second-Strand cDNA Synthesis. At the end of first-strand cDNA synthesis, the tubes are kept on ice. Second-strand cDNA synthesis reaction (final volume, 150 μl) is set up in the same tube on ice by adding 91 μl of nuclease-free water, 30 μl of 5× second-strand buffer [100 m*M* Tris-HCl (pH 6.9), 23 m*M* $MgCl_2$, 450 m*M* KCl, 0.75 m*M* β-NAD^+, and 50 m*M* ammonium sulfate], 3 μl of 10 m*M* dNTPs, 1 μl of *Escherichia coli* DNA ligase (10 U/μl), 4 μl of *E. coli* DNA polymerase I (10 U/μl), and 1 μl of *E. coli* RNase H (2 U/μl). The contents are mixed gently and the tubes are incubated at 16° for 2 hr. Following the incubation, the tubes are kept on ice, 2 μl of T4 DNA polymerase (3 U/μl) is added, and the incubation is continued for another 5 min at 16°. The reaction is stopped by the addition of 10 μl of 0.5 *M* EDTA (pH 8.0) and extracted once with an equal volume of phenol–chloroform (1:1, v/v) and once with chloroform. The aqueous phase is then transferred to a new tube and precipitated by adding 0.5 vol of 7.5 *M* ammonium acetate (pH 7.6), 2 μg of glycogen (as carrier), and 2.5 vol of chilled ethanol. The samples are left at −20° for overnight and the cDNA pellets are collected by centrifugation at top speed for 20 min. The pellets are washed once with 70% ethanol, air dried, and dissolved in 14 μl of nuclease-free water.

Amplification of cDNA

As the amount of cDNA derived from a small number of cells is low, it is necessary to amplify the cDNA for further analysis. To uniformly amplify the cDNA, an adaptor (*Not*I adaptor) is first ligated to both ends of the cDNA. Following adaptor ligation, the cDNAs are amplified with *Not*I/RI primer (see Table I), by a modified PCR method using betaine and DMSO.[9] Figure 1 shows a schematic representation of the procedure.

Ligation of cDNA with NotI Adaptor

PREPARATION OF *Not*I ADAPTOR. The *Not*I adaptor is prepared by annealing *Not*I-short and anti-*Not*I-long oligonucleotides (see Table I). The anti-*Not*I-long oligonucleotide is phosphorylated to ensure that both of the adaptor oligonucleotides are ligated to the cDNA. One microgram of anti-*Not*I-long is mixed with 1 μl of 10× T4 polynucleotide kinase buffer [700 m*M* Tris-HCl (pH 7.6), 100 m*M* $MgCl_2$, and 50 m*M* DTT] and 1 μl of 10 m*M* adenosine triphosphate (ATP), the volume is adjusted to 9 μl with

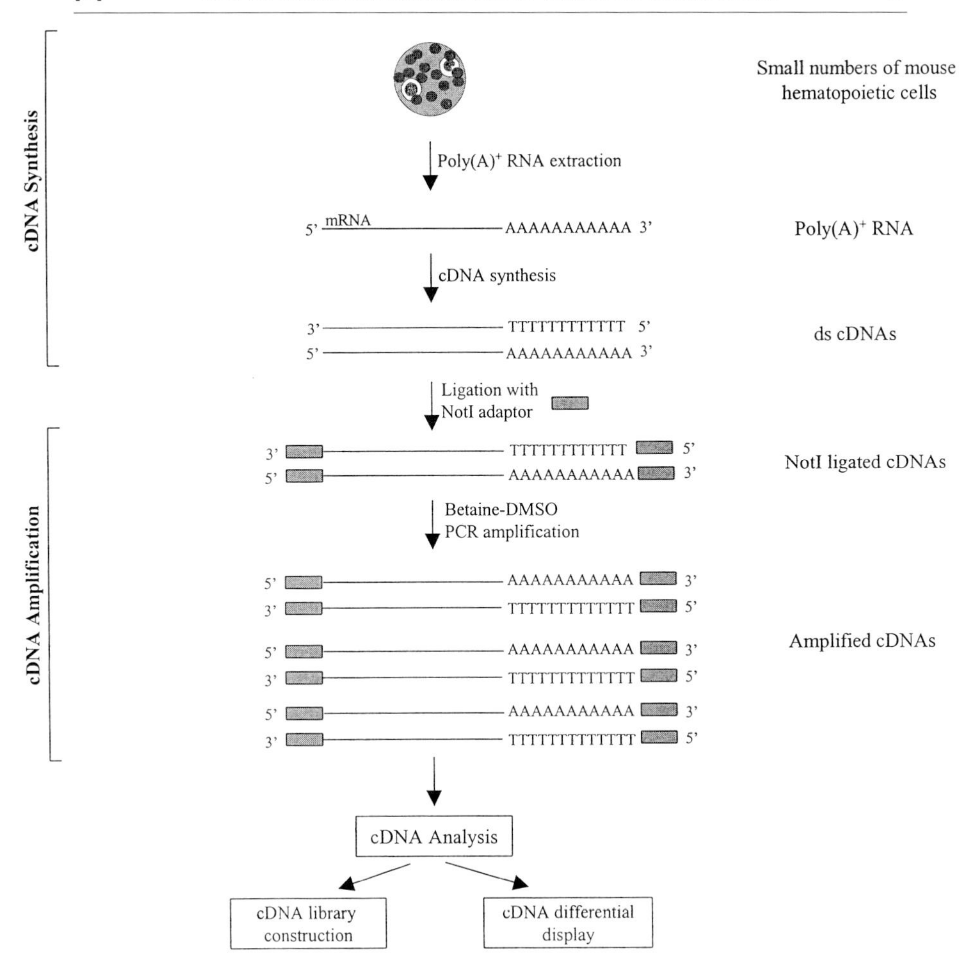

FIG. 1. Schematic representation of cDNA amplification by modified PCR method using betaine–DMSO.

water, and the reaction is initiated by adding 1 μl of T4 polynucleotide kinase (10 U/μl). The tubes are incubated at 37° for 30 min and then the enzyme is inactivated at 65° for 20 min. The annealing is carried out by adding the following components to the above-phosphorylated anti-*Not*I-long: 1 μg of *Not*I-short, 2 μl of 10× oligo annealing buffer [100 m*M* Tris-HCl (pH 8.0), 10 m*M* EDTA (pH 8.0), and 1 *M* NaCl] and water to adjust the final volume to 20 μl. The sample is heated at 65° for 10 min and allowed to cool to room temperature. The annealed adaptor is stored at −20°.

LIGATION OF cDNA WITH ANNEALED *Not*I ADAPTOR. To set up this reaction, 14 μl of cDNA is mixed with 100 ng of annealed *Not*I adaptor in a 0.5-ml microcentrifuge tube. To this mixture 2 μl of 10× T4 DNA ligase buffer [500 m*M* Tris-HCl (pH 7.8), 100 m*M* $MgCl_2$, 100 mM DTT, 10 m*M* ATP, and BSA (250 μg/ml)] is added and the volume is adjusted with water to 18 μl and mixed gently. The reaction is initiated by adding 2 μl of T4 DNA ligase (400 U/μl) and incubated at 16° overnight.

cDNA Amplification. A modified betaine–DMSO PCR method[9] is used to uniformly amplify the cDNA with different GC content. This method uses the LA system, which combines a highly thermostable form of *Taq* DNA polymerase (KlenTaq1, which is devoid of 5′-exonuclease activity) and a proofreading enzyme (*Pfu* DNA polymerase, which has 3′-exonuclease activity). The LA16 enzyme consists of 1 part of *Pfu* DNA polymerase and 15 parts of KlenTaq1 DNA polymerase (v/v). The *Not*I adaptor-ligated cDNA is diluted 10-fold with water. Two microliters of this diluted cDNA is used as the template for PCR. The PCR reaction (50-μl final volume) is set up with the following components: 5 μl of 10× PCR buffer [200 m*M* Tris-HCl (pH 9.0), 160 m*M* ammonium sulfate, and 25 m*M* $MgCl_2$], 16 μl of water, 0.8 μl of BSA (10 mg/ml), 1 μl of *Not*I/RI PCR primer (100 ng/μl), 5 μl of 50% DMSO (v/v), 15 μl of 5 *M* betaine, and 0.2 μl of LA16 enzyme. These components are mixed gently on ice and then heated to 95° for 15 sec on a PCR machine, and held at 80° while 5 μl of 2 m*M* dNTPs is added to start the reaction. The PCR conditions are as follows: stage 1: 95° for 15 sec, 55 ° for 1 min, 68° for 5 min, 5 cycles; stage 2: 95° for 15 sec, 60° for 1 min, 68° for 5 min, 15 cycles.

After amplification, cDNA is purified with the Qiaquick PCR purification kit (following the instructions provided by the supplier). The purified cDNA is eluted in the desired volume of water.

Analysis of Polymerase Chain Reaction-Amplified cDNA

Amplified cDNA from a small number of cells can have many potential applications, such as cDNA library construction, differential display, and representational differential analysis of cDNAs from two different cell types. To examine the faithfulness of representation of the amplified cDNAs, we used a modified differential display approach.[11,12]

The amplified cDNA preparations are analyzed on an agarose gel, to determine the size range of the amplified products (Fig. 2A). They are also tested for the presence of several representative messages such as β-actin, GATA-1, GATA-2, GATA-3, interleukin 5 receptor α subunit (IL-5Rα),

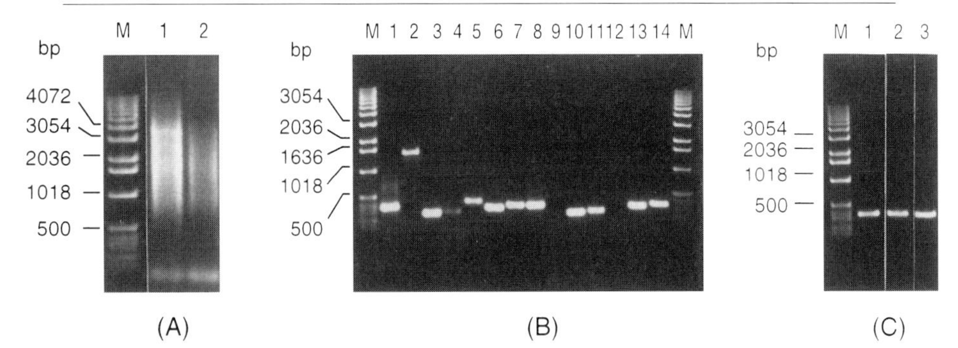

FIG. 2. Analysis of PCR-amplified cDNA from hematopoietic cells. (*A*) Agarose gel analysis. The PCR-amplified cDNAs from Lin^+ cells (lane 1) and LRH (lineage depleted and $\text{rhodamine}^{\text{low}}$ Hoechst 33342^{low} stained) cells (lane 2) were analyzed on a 1% agarose gel. (*B*) PCR analysis of representative messages using gene-specific primers. PCR-amplified cDNAs (10 ng) from Lin^+ and LRH cells were analyzed for the presence of β-actin (lanes 1, 8), GATA-1 (lanes 2, 9), GATA-2 (lanes 3, 10), GATA-3 (lanes 4, 11), IL-5R-α (lanes 5, 12), IL-6R-α (lanes 6, 13), and c-Kit receptor 3′ end (lanes 7, 14). (*C*) PCR analysis of different regions of the c-Kit receptor. PCR-amplified cDNA (10 ng) from Lin^+ cells was analyzed, using specific primers designed to the regions (described in text) spanning the 5′ end (lane 1), the middle region (lane 2), and the 3′ end (lane 3) of the message. A 1-kb ladder (GIBCO-BRL) was used as the molecular weight marker (lane M) for all of the panels.

interleukin 6 receptor α subunit (IL-6Rα), and c-Kit receptor by PCR using gene-specific primers[13] (Fig. 2B and C).

We also tested whether PCR amplification of cDNA causes any distortions in its representation. Unamplified cDNA (*Not*I adaptor ligated) from mouse bone marrow lineage committed cells (Lin^+) is used as a template to generate the first-round PCR product (I round), and the first-round PCR product (5 ng per reaction) is reamplified under identical conditions to generate second-round PCR product (II round). The PCR products are purified with a Qiaquick PCR purification kit. Equal amounts (600 ng) of these first-round, second-round amplified (Lin^+) cDNA samples and unamplified *Not*I adapter-ligated Lin^+ cDNA (2 μl) are digested with *Bgl*II restriction enzyme followed by ligation with fly adaptor, which has a GATC overhang. These cDNAs are PCR amplified with the ^{32}P-labeled fly adaptor-specific primer as 5′ primer and an oligo-d(T)AC as 3′ primer. The PCR products are separated on a 6% polyacrylamide sequencing gel (Fig. 3).

To achieve an accurate representation of various messages in a single

[13] D. Orlic, S. Anderson, L. G. Biesecker, B. P. Sorrentino, and D. M. Bodine, *Proc. Natl. Acad. Sci. U.S.A.* **92,** 4601 (1995).

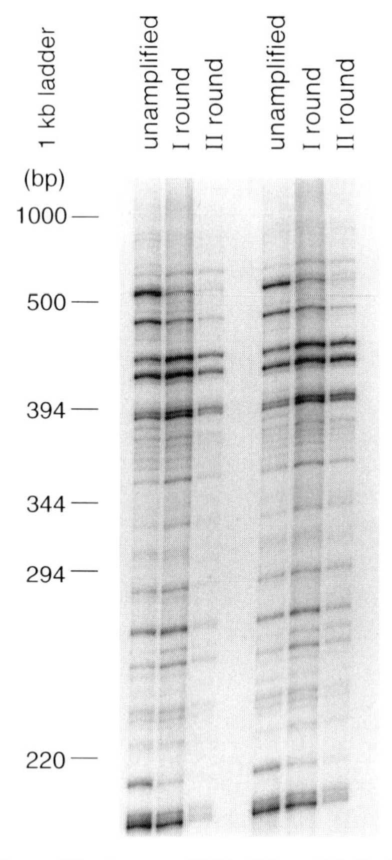

FIG. 3. Effect of PCR amplification on cDNA differential display pattern. Unamplified cDNA (*Not*I adaptor ligated, lane labeled *unamplified*) from mouse bone marrow lineage committed cells (Lin^+) was used as a template to generate the first-round PCR products (lane labeled *I round*), and the first-round PCR product (5 ng/reaction) was reamplified to generate second-round PCR products (lane labeled *II round*). The cDNAs were digested with *Bgl*II and ligated with a fly adaptor that has a GATC overhang. These cDNAs were amplified by a ^{32}P-labeled fly adaptor-specific primer as 5′ primer and an oligo-d(T)AC primer (AC as anchor bases) as 3′ primer and separated in a 6% polyacrylamide gel. At the end of the electrophoresis, the gel was dried and subjected to autoradiography.

cell, a dilution experiment is conducted. This is done to determine the lowest amount of cDNA template that can be amplified with good representation of various mRNAs without any distortion. The *Not*I adaptor-ligated cDNAs made from two different batches of Lin^+ cell preparations [Lin^+ (I) and Lin^+ (II)] are amplified separately and these are used as the starting material in this experiment. The starting amount of cDNA is 120 ng;

this is then serially diluted to obtain 12 ng, 1.2 ng, 120 pg, and 12 pg. These diluted cDNAs are reamplified by the betaine–DMSO PCR method. For the 12-pg sample, it is necessary to amplify a second time to obtain enough product to do the differential display. The differential display of the amplification products is performed similarly as described above (Fig. 4).

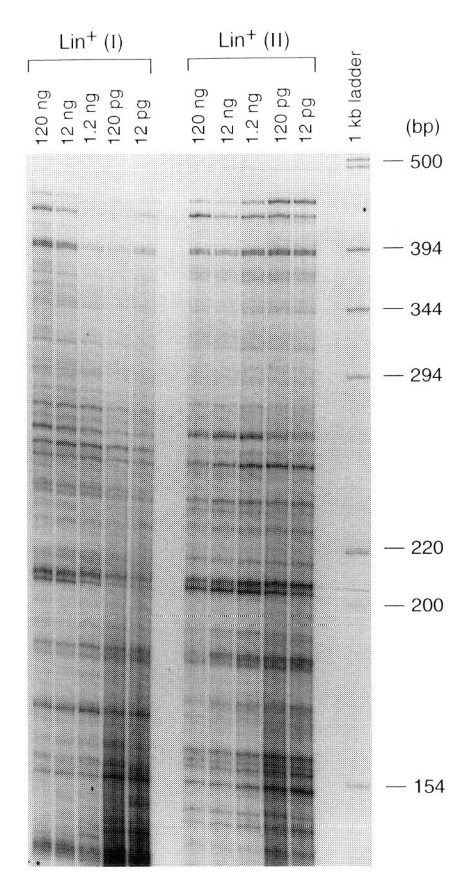

FIG. 4. cDNA differential display analysis of serially diluted and amplified cDNAs. The first-round amplified cDNAs from two different preparations of Lin^+ cells [lanes labeled *Lin*$^+$(*I*) and *Lin*$^+$(*II*)] were the starting material in this experiment. The starting amount of cDNA, 120 ng, was then serially diluted to obtain 12 ng, 1.2 ng, 120 pg, and 12 pg. These diluted cDNAs were reamplified by the betaine–DMSO PCR method. For the 12-pg sample, it was necessary to amplify a second time to obtain enough product. The differential display conditions are as described in the caption to Fig. 3.

Results and Discussion

The agarose gel analysis of the amplified cDNA preparations made from Lin^+ and LRH (lineage depleted and $rhodamine^{low}$ Hoechst 33342^{low} stained) cells showed that the size ranged from 0.5 to 5 kilobases (kb) (Fig. 2A). PCR analysis using gene-specific primers[13] for several representative cDNAs, such as β-actin, GATA-1, GATA-2, GATA-3, IL-5Rα, IL-6Rα, and c-Kit receptor 3′ end (Fig. 2B), showed that these messages were indeed represented in the amplified cDNA made from Lin^+ cells. As the LRH cells do not express GATA-1 and IL-5Rα message,[13] the amplified cDNA made from these cells did not result in any PCR products for GATA-1 and IL-5Rα. This suggested that the modified betaine–DMSO PCR method faithfully amplified the original cDNA. We also analyzed the presence of complete message for c-Kit receptor (Y00864, 5098 nucleotides) by PCR using primers corresponding to its 5′ end (nucleotides 1068–1442), middle region (nucleotides 2348–2698), and 3′ end (nucleotides 3675–4025) (Fig. 2C). These results indicated that the cDNA amplification maintained good representation and nearly full-length cDNA.

Analysis of the unamplified, first-round, and second-round amplified Lin^+ cDNAs by differential display is shown in Fig. 3. Most bands were of similar intensity before and after PCR amplification, although some variations were visible. This indicates that the betaine–DMSO PCR amplification procedure can uniformly amplify various cDNAs without any major distortion in the representation.

The display patterns of the cDNAs obtained from two batches of Lin^+ cell preparations [Lin^+(I) and Lin^+(II)] were similar at different dilutions and amplifications. Remarkably, 12 pg of cDNAs can still yield consistent patterns after a second round of PCR amplification (Fig. 4). Twelve picograms of mRNAs may represent the average mRNA levels in about 10–100 cells. Our data suggest that the betaine–DMSO PCR method can amplify as little as 1 pg of cDNA.

Several steps are necessary to provide a reasonably accurate representation of the relative levels of most or all messages in a single cell in various physiological states. The present results indicate that, once double-stranded cDNA is prepared with appropriate PCR adaptors at both ends, amplification of cDNA that can be obtained from the RNA of a single cell may be performed by betaine–DMSO PCR without major distortions in the representation of the various mRNAs. Amplified cDNAs prepared in this fashion can also be used to prepare cDNA libraries with average size inserts of well over 1 kb in length.

Although the amplification of cDNA did not produce major distortions in expression patterns, some variations in the abundance of single bands were seen when mRNA was prepared from mouse hematopoietic stem cells, converted to cDNA, and amplified. At least a part of this effect could be due to cellular stress during the cell purification procedure (e.g., heat shock proteins). This raises the general precaution that, regardless of the number of cells involved, isolation procedures for display of physiologic mRNA expression patterns should be rapid, isothermal, and otherwise nontraumatic to the cells.

In addition to erratic effects due to stress on cells during isolation, undoubtedly variations in the representation of mRNA from a small number of cells arise because of stochastic effects at the level of cDNA synthesis and adapter ligation. In principle, this could be overcome by adding a carrier that could be removed before cDNA display or other modes of expression analysis are applied. With such precautions, it would seem feasible to qualitatively study the levels of mRNA expression in a single cell, and such approaches could be quite instructive for developmental biology.

Acknowledgment

We acknowledge Dr. Sherman M. Weissman for direct supervision and critical reading of this manuscript. We are grateful to Dr. Le Wang for thoughtful discussion.

[4] Optical Mapping: An Approach for Fine Mapping

By CHRISTOPHER ASTON, CATHARINA HIORT, and DAVID C. SCHWARTZ

Introduction

The initial goal of the human genome project is to create an integrated genetic, physical, and transcriptional map of the genome. Identifying coding sequences within complex genomes has proved both technically challenging and time consuming, primarily because only small regions of the human genome, on the order of 0.06%, are transcribed. Indeed, if purely exonic regions are considered, the figure drops to 0.007%.[1] Clones from cDNA libraries are being used to generate expressed sequence tags (ESTs), which are partial cDNA sequences derived from a single exon and are represented only once in the genome. About 800,000 human ESTs have been reported, making it probable that almost every 1 of the 100,000 human genes is

[1] S. B. Primrose, "Principles of Genome Analysis: A Guide to Mapping and Sequencing DNA from Different Organisms." Blackwell Science, Oxford, UK, 1995.

Copyright © 1999 by Academic Press
All rights of reproduction in any form reserved.

0076-6879/99 $30.00

hallmarked by an EST. However, most of the ESTs from the human genome have not yet been mapped to a chromosome or even to a contiguous sequence of DNA (contig). Only about 10,000 ESTs have been given a chromosomal assignment based on mapping to radiation hybrids, yeast artificial chromosomes (YACs), or bacterial artificial chromosomes (BACs). Although EST mapping is proceeding at a rapid pace, the identified sequences are not generally being mapped with precise distances noted between markers. The establishment of a high-resolution map with a well-defined metric based on ordered restriction sites would serve as a unifying means for the alignment of disparate types of maps, including those positioning ESTs. Essentially, large-scale maps developed from clones or genomic DNA, annotated with hybridization data, could serve as a scaffold for the construction of detailed composite maps from sequencing efforts, clone contigs, cytogenetic data, and, with effort, genetic maps. To meet this need, our laboratory has developed an approach for the construction of ordered restriction maps from a small population of individual molecules, termed *optical mapping.*

A number of single-molecule approaches to characterizing expressed genes have been developed. These generally involve hybridization of probes followed by probe detection using either optical (fluorescence) or tactile (scanning probe) means. Ordered regions of metaphase chromosomes have been identified by fluorescent *in situ* hybridization (FISH). Probes were observed by near-field scanning optical microscopy and the order of the probes was confirmed by scanning probe techniques.[2] Positional FISH on naked DNA fibers (fiber FISH) has been used to map exonic fragments and cDNAs with sizes greater than 200 bp to their cognate cosmid clone.[3,4] Surface-mounted, λ phage DNA has been mapped by a FISH approach with a 1-kb resolution.[5] Deletions of genes in genomic DNA have been detected by determining the distance between two hybridized probes with a resolution of 7 kb.[6] Even though genomic DNA is amenable to FISH analysis, prior knowledge of substrate sequence is required in order to make probes and its resolution, information content, and throughput are limited.

[2] M. H. Moers, W. H. Kalle, A. G. Ruiter, J. C. Wiegant, A. K. Raap, J. Greve, B. G. de Grooth, and N. F. van Hulst, *J. Microsc.* **182,** 40 (1996).

[3] R. J. Florijn, F. M. van de Rijke, H. Vrolijk, L. A. Blonden, M. H. Hofker, J. T. den Dunnen, H. J. Tanke, G. J. van Ommen, and A. K. Raap, *Genomics* **38,** 277 (1996).

[4] R. J. Florijn, L. A. Bonden, H. Vrolijk, J. Wiegant, J. W. Vaandrager, F. Baas, J. T. den Dunnen, H. J. Tanken, G. J. van Ommen GJ, and A. K. Raap, *Hum. Mol. Genet.* **4,** 831 (1995).

[5] H. U. Weier, M. Wang, J. C. Mullikin, Y. Zhu, J. F. Cheng, K. M. Greulich, A. Bensimon, and J. W. Gray, *Hum. Mol. Genet.* **4,** 1903 (1995).

[6] X. Michalet, R. Ekong, F. Fougerousse, S. Rousseaux, C. Schurra, N. Hornigold, M. van Slegtenhorst, J. Wolfe, S. Povey, J. S. Beckmann, and A. Bensimon, *Science* **277,** 1518 (1977).

Optical mapping is a single-molecule approach for the rapid production of ordered restriction maps from DNA molecules. Optical mapping was developed in 1991 to map centromeric regions by analyzing patterns of DNA restriction digestion on individual molecules, using fluorescence microscopy. These regions contain tandem repeats and are therefore difficult to map by conventional means. Originally, ensembles of fluorescently labeled DNA molecules were elongated in a flow of molten agarose generated between a coverslip and microscope slide.[7,8] The DNA was fixed in an elongated state when the agarose gelled. Restriction endonucleases were present in the molten agarose and the DNA was cut when magnesium ions diffused into the gel. As the DNA relaxed, the cut sites in individual molecules were resolved by imaging the receding cut ends by time-lapse video microscopy. Ordered restriction maps were constructed by measuring the relative fluorescence intensity and apparent length of a number of DNA fragments, followed by averaging. Fragment sizes determined by optical mapping correlated with sizes obtained by pulsed-field gel electrophoresis (PFGE) and were more informative because the fragment order was determined concurrently. The approach had other advantages over PFGE in that problems associated with disambiguation of comigrating bands on pulsed-field gels were obviated.

We have improved this approach in a number of areas and can now map a wide variety of clone types including λ phage, BACs, YACs, as well as genomic DNA.[9-11] We have developed simple and reliable approaches to prepare large DNA molecules for optical mapping with usable distribution of molecular extension and minimal breakage, by mounting the DNA on open derivatized surfaces. We have expanded the range of effective restriction enzymes and increased the resolution of small fragments. Molecules as small as 800 bp[12] can be imaged and sized, representing a major improvement from our prior agarose gel-based optical mapping approach, which had a limit of resolution of 60 kb.[7] Furthermore, we have accelerated image acquisition and refined data analysis. These developments have increased throughput and provide the foundation for automated approaches to mapping DNA. It is our hope that approaches to genome-wide physical

[7] D. C. Schwartz, X. Li, L. Hernandez, S. Ramnarain, E. Huff, and Y. Wang, *Science* **262,** 110 (1993).

[8] X. Guo, E. J. Huff, and D. C. Schwartz, *Nature* (*London*) **359,** 783 (1992).

[9] A. H. Samad, W. W. Cai, X. Hu, B. Irvin, J. Jing, J. Reed, X. Meng, J. Huang, E. Huff, B. Porter, A. Shenkar, T. Anantharaman, B. Mishra, V. Clarke, E. Dimalanta, J. Edington, C. Hiort, R. Rabbah, J. Skiadas, and D. C. Schwartz, *Nature* (*London*) **378,** 516 (1995).

[10] Y. Wang, E. Huff, and D. C. Schwartz, *Proc. Natl. Acad. Sci. U.S.A.* **92,** 165 (1995).

[11] A. Samad, E. J. Huff, W. Cai, and D. C. Schwartz, *Genome Res.* **5,** 1 (1995).

[12] X. Meng, K. Benson, K. Chada, E. Huff, and D. C. Schwartz, *Nature Genet.* **9,** 432 (1995).

mapping of DNA, such as optical mapping, will find widespread future use in genomics, diagnostics, population genetics, and forensics, complementing information derived from single-gene analyses by traditional methods. This chapter describes some of our most recent protocols for mapping of DNA clones and genomic DNA and for localizing regions of gene expression. Finally, consideration is also given to the general limitations of these approaches.

Principle

Molecular fixation underlies a number of techniques for genomic analysis.[13–15] For our analysis of DNA molecules, samples are fixed onto derivatized glass surfaces either by sandwiching a DNA solution between a treated coverslip and glass slide or by using fluid flows generated in drying droplets to elongate the DNA. Restriction enzyme is added to the surface and cleaves the DNA at specific sites, forming discernible fragments. Fragmented DNA molecules are stained and imaged using a fluorescence microscope (Fig. 1). Optical Map Maker (OMM) software, developed in this laboratory, integrates all of the microscope workstation functions, such as movement of the stage, focus, and image collection, using a cooled charge-coupled device (CCD) camera, enabling fully automatic optical mapping from image acquisition to image analysis and map making. Figure 2 shows a schematic representation of our optical mapping system. Figure 3 shows an image of surface-mounted human genomic DNA digested with *Pac*I. Data are acquired using a series of image-processing algorithms that we have developed for optical data, and restriction maps are constructed using a rigorous statistical analysis.[16]

Maps made using specific restriction enzymes are overlaid and aligned with each other to create multiple enzyme maps. Overlaying of maps increases the number of cleavage sites, and yields informationally rich maps, as compared with single digests. The resulting high-resolution maps have restriction enzyme recognition sequence landmarks at 1-kb intervals and offer relatively unambiguous clone characterization. At this resolution, sequence lesions such as moderately sized deletions, inversions, duplications, and other rearrangements can be noted. Depending on the choice

[13] A. Bensimon, A. Simon, A. Chiffaudel, V. Croquette, F. Heslot, and D. Bensimon, *Science* **265,** 2096 (1994).

[14] J. Jing, J. Reed, J. Wang, X. Hu, V. Clarke, J. Edington, D. Housman, T. S. Anathraman, E. J. Huff, B. Mishra, B. Porter, A. Shenker, E. Wolfson, C. Hiort, R. Kantor, C. Aston, and D. C. Schwartz, *Proc. Natl. Acad. Sci. U.S.A.* **95,** 8046 (1998).

[15] D. Erie, G. Yang, H. Schultz, and C. Bustamante, *Science* **266,** 1562 (1994).

[16] T. Anantharaman, B. Mishra, and D. C. Schwartz, *J. Comput. Biol.* **4,** 91 (1997).

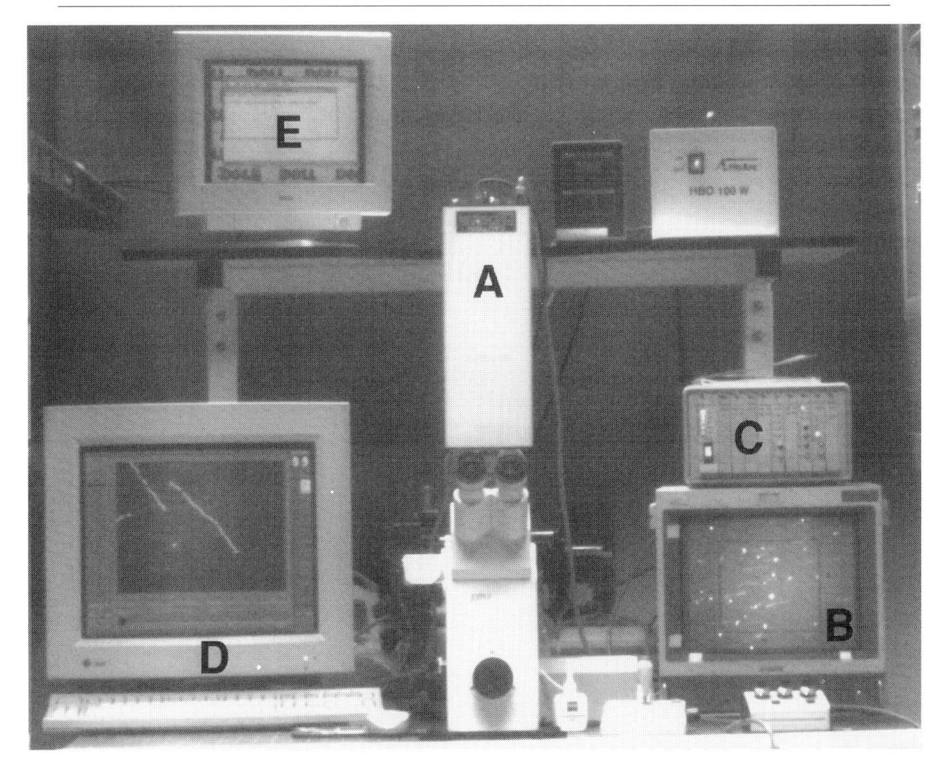

FIG. 1. An optical mapping workstation. The station is built around a Zeiss 135 Axiovert inverted microscope. The microscope is equipped with a SIT video camera (A) and accompanying monitor (B) for automatic and manual focusing, and a cooled CCD camera (attached to the bottom port of the microscope and therefore not visible in the picture) for acquiring high-resolution digital images. Automatic features such as movement of the microscope stage, image collection, light shutter operation, and video autofocusing are controlled by a Ludl Electronics box (C), which is interfaced to a Sun workstation (D). The CCD camera is also interfaced, via a microcomputer controller, to the Sun workstation. DNA molecules are imaged using the Optical Map Maker (OMM) software, operated through a collection of user-friendly interfaces, and the image collection process can be followed on the microcomputer monitor (E) as well as on the Sun workstation monitor (D). The Sun workstation is part of a distributed network on which interactive and automatic image processing is performed. The optical mapping workstation is mounted on a floating optical table.

of restriction enzyme, repeated regions can also be identified. The high-resolution map is useful for establishment of minimal tiling paths for sequencing, as a scaffold on which to assemble sequenced contigs and as a means of sequence verification, but at this resolution individual sequence reads or EST sequence information cannot be placed on this map with any degree of certainty. However, a high-resolution map can be used to rapidly

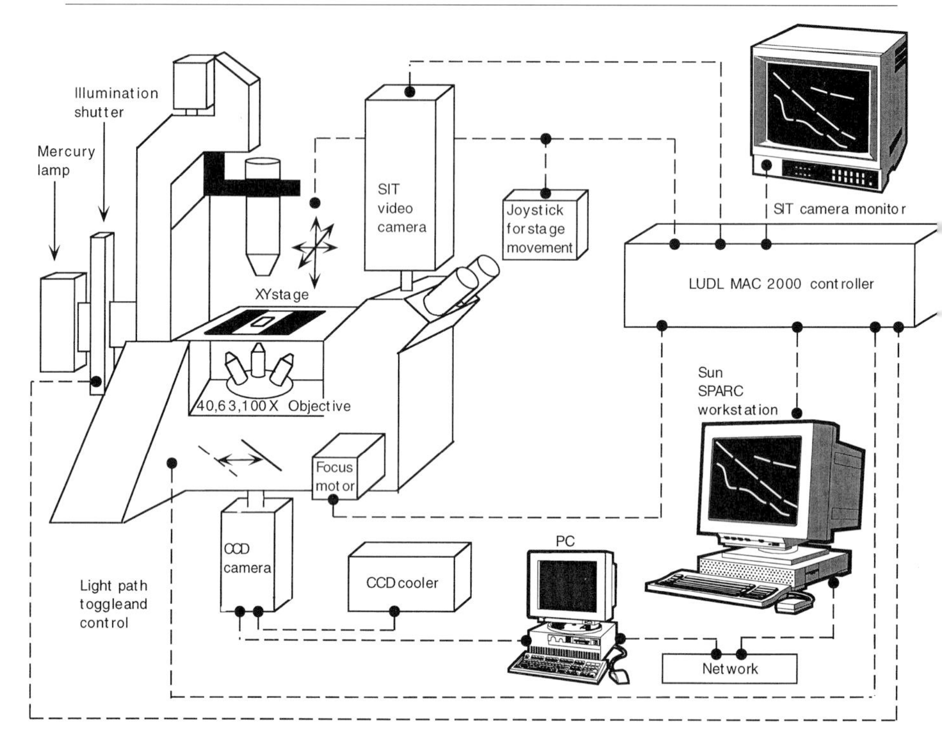

FIG. 2. Schematic of optical mapping workstation instrumentation. The station is built around a Zeiss 135 microscope equipped for epifluorescence, with several different objectives (e.g., ×40, ×63, and ×100) Zeiss plan-neofluor oil immersion) allowing for imaging of biomolecules of a wide range of sizes. The ×100 objective is most commonly used, whereas the ×63 objective is used for clones larger than 200 kb. Filter packs allow for imaging of biomolecules labeled with fluorescent dyes of different spectral properties. The microscope is also equipped with a Dage SIT68GL low light-level video camera for acquiring focus, and a cooled CCD digital camera (1316 × 1032 pixels, KAF 1400 chip, 12-bit digitization) (Princeton Instruments) for high-resolution imaging and photometry. A Ludl Electronics x–y microscope stage with 0.1-μm resolution is used for sample translation. Control of automation accessories, such as x–y stage, light shutters, and motor for movement of beam splitting prism, is accomplished by a Ludl Electronics MAC 2000 interface bus, housing all interface cards (PSSYST 200, MCMSE 500, MDMSP 503, AFCMS 801, FWSC 800, and RS232INT 400). The Ludl MAC 2000 is interfaced via an RS232 serial connection to a Sun Microsystems SPARC 20 dual-processor computer workstation. The Princeton Instruments CCD camera is also interfaced, via a Pentium-based microcomputer controller and a distributed network, to a Sun workstation. Software for control of the above peripherals is written in the C programming language.

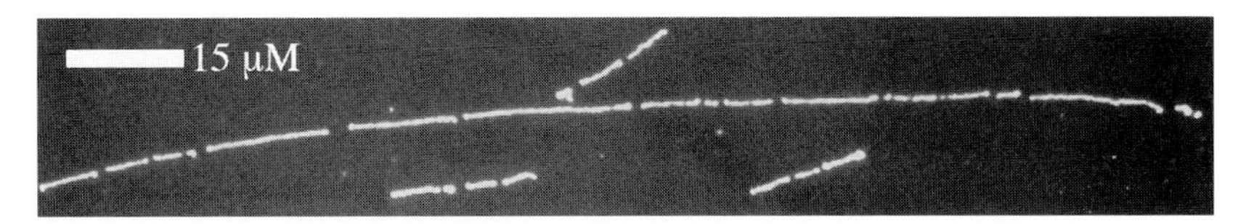

FIG. 3. Image of human genomic DNA from peripheral blood leukocytes digested with *Pac*I. Molecules shown are approximately 350 μm (1 Mb) long. Comounted cosmid DNA was used as an internal size standard (45 kb). This cosmid is of known sequence and is cut into four fragments by *Pac*I. The number of cuts therefore also provides an estimate of cutting efficiency.

scan the genomic landscape for genic regions based on indirect means, such as identification of CpG islands.

Application of Optical Mapping to Identifying Genic Regions

Optical mapping can be used to screen for genes in mammalian genomes. Expressed mammalian genes are characterized by a CG motif (CpG islands) that normally occurs at a low frequency in noncoding regions. CpG islands constitute a distinctive fraction of the genome because, unlike bulk DNA, they are nonmethylated and contain the dinucleotide CpG at its expected frequency. The average length of CpG islands is 1 kb and the haploid human genome is estimated to contain about 45,000 islands.[17] All human housekeeping genes and 40% of tissue-restricted genes are associated with CpG islands.[18] In total, only 57% of human genes have CpG islands, so unfortunately not all genes can be identified by CpG islands. Putative genes can be identified by optical mapping with enzymes such as *Not*I, *Mlu*I, *Eag*I, *Nar*I, *Bss*HII, *Sma*I, *Sal*I, *Sac*II, and *Cla*I, which all have CpG in their recognition sequences. However, CpG islands may not be of such value in locating genes in microbial genomes because these often possess strikingly different A + T and G + C content compared with human DNA.

Repeated regions hamper large-scale sequence assembly endeavors. When physical maps are prepared by gel electrophoresis, repeated regions comigrate and their numbers cannot be estimated. Optical mapping is consummate for identifying exonic repeats because the number of repeats can be determined if the restriction enzyme used cuts within the repeat. We have mapped human Y chromosome BAC clones in collaboration with D. Page at the Massachusetts Institute of Technology (Cambridge, MA).

[17] F. Antequera and A. Bird, *Proc. Natl. Acad. Sci. U.S.A.* **90,** 11995 (1993).

[18] F. Larsen, G. Gundersen, R. Lopez, and H. Prydz, *Genomics* **13,** 1095 (1992).

The AZF region of the Y chromosome at Yq11.23 exhibits *de novo* deletions in males with nonobstructive azoospermia and severe oligospermia. Presumably, there exists a gene or genes that are critical in sperm formation or development. To date, the only known transcriptional unit in the deleted region is the *DAZ* (deleted in azoospermia) gene.[19] We have identified repeated sequences that characterize the *DAZ* gene family by cutting of BAC clones with *Eco*RV and *Spe*I, which gave repetitive 2-kb fragments. Having identified such *DAZ* repeats, optical mapping can be used to screen for their absence in infertile male individuals.

Repeated regions have been used to identify genic regions in genomic DNA. The ribosomal RNA genes comprise large, numerous, and distinctive loci with about 400 copies of the rDNA repeat per diploid genome. Human rDNA accounts for about 0.3% of the total human genome. Human genomic DNA has been optically mapped with *Bss*HII and *Ssp*I, which identify ribosomal RNA genes. The large size of the human genome required genomic DNA to be predigested with *Apa*LI or *Eco*RV, which cleaves outside of the large tandem repeats. After searching through a few thousand molecules, some molecules were identified as matching the expected cleavage pattern for the repeat, based on a map generated from sequence information.[20] Maps were constructed from these molecules and the number of repeats could be determined, which would not have been possible by sequencing alone.

Optical restriction maps of clones provide markers that facilitate contig formation. BAC clones can be aligned to form a minimal tiling path. This enables ordering of ESTs on the contig of BAC clones. Optical restriction maps also provide markers that facilitate sequence read alignment, following a shotgun sequencing approach. A low-resolution map of a BAC clone made using one or two restriction endonucleases enables islands of sequence to be anchored to the optical map, whereas a high-resolution map made using 8–12 different restriction endonucleases provides sufficient information to anchor smaller islands of sequence as short as 3 or 4 reads or potentially even single sequence reads derived from each end of a plasmid subclone. Optical restriction maps provide accurate gap distances between the terminal sequences of adjacent islands. These gaps can then be bridged using direction gap closure, employing primer-based sequencing techniques such as end walking.

[19] R. Saxena, L. G. Brown, T. Hawkins, R.K. Alagappan, H. Skaletsky, M. P. Reeve, R. Reijo, S. Rozen, M. B. Dinulos, C. M. Disteche, and D. C. Page, *Nature Genet.* **14,** 292 (1996).

[20] I. L. Gonzalez and J. E. Sylvester, *Genomics* **27,** 320 (1995).

Methods

Cleaning of Surfaces

Optical mapping surfaces must be cleaned thoroughly before use to ensure uniform derivatization and optical transparency. We use a custom-made acid boiling system that is built from Pyrex glass and Teflon tubing, and assembled in a fume hood. No vacuum grease is used to seal the joints. A sodium hydroxide scrubbing system neutralizes acid vapors.

Coverslips (18×18 mm^2; Fisher, Pittsburgh, PA) are racked in custom-made Teflon racks, which hold the coverslips snugly on three edges, and cleaned by boiling in concentrated nitric acid (HNO_3) for 24 hr. The system is rinsed extensively with deionized, dust-free water until the effluent attains pH 6. The cleaning procedure is repeated with concentrated hydrochloric acid (HCl), which hydrolyzes the glass surface, preparing it for subsequent derivatization. The racked coverslips are rinsed extensively and then sonicated in deionized water to remove residual acid that collects between the Teflon parts in the racks. Residual water is then removed by sonicating the racks in distilled ethanol. The cleaned coverslips are stored in the racks under distilled ethanol in polypropylene containers (Qorpak, Fisher) at room temperature.

Distillation of 3-Aminopropyltriethoxysilane

3-Aminopropyltriethoxysilane (APTES; Aldrich, Milwaukee, WI) is distilled under argon at lowered pressure of approximately 5 mmHg. Teflon sleeves are used for the joints instead of vacuum grease in the distillation apparatus.

Derivatization of Glass Coverslips

A 2% APTES solution is made by weighing out 20 mg of distilled APTES and 980 mg of distilled water in an Eppendorf tube. The APTES is hydrolyzed in the tube for 6 hr at room temperature while shaking. Cleaned coverslips are derivatized in slotted Teflon racks that hold the coverslips by one edge. Twenty to 80 mg of the hydrolyzed 2% APTES solution is weighed in a polypropylene bottle (Qorpak) and 250 ml of distilled ethanol is added. The mixture is shaken vigorously and poured over a rack of 40 coverslips in a polypropylene or Teflon container. The APTES concentration in the treatment solution ranges from 7 to 28 μM. The coverslips are derivatized for 48 hr at room temperature on a rotary shaker. Macroscopic derivatization parameters such as silane concentration and duration of silane treatment alter the surface charge density, which plays a critical role in surface-based optical mapping. We are currently determining the rate of APTES hydrolysis and identifying factors that may play a role in determining this rate, but until these studies are completed the treatment time is only approximate. The surfaces must be checked for optimal performance by estimating the hydrophobicity by measuring the contact angle and assaying the degree of DNA stretching and fragment retention. Derivatized surfaces can be stored in treatment solution for a few weeks. We are currently investigating the use of 3-aminopropyldiethoxymethylsilane (APDEMS; Gelest, Tullytown, PA) as a derivatizing agent. This appears to give more reproducible surface properties in preliminary experiments.

Preparation of DNA

We have mapped DNA from a variety of clone types including YACs, BACs, cosmids, λ phage and polymerase chain reaction (PCR) products. We have also mapped whole genomic DNAs.

Cloned DNA is prepared by standard protocols. BAC and cosmid molecules are prepared by alkaline lysis[21] or plasmid preparation columns (Qiagen, Chatsworth, CA). Circular molecules such as BACs and cosmids are linearized with λ terminase prior to surface mounting (8 units/μg DNA; Amersham Pharmacia Biotech, Piscataway, NJ). λ terminase is an endonuclease that cleaves at the λ *cos* site. Occasionally, circular DNA is linearized by restriction digest utilizing a *Not*I site in the polylinker of the cloning vector when no internal *Not*I sites are present in the insert.

Increasing the size of measurable molecules decreases the number required for mapping a complete genome.[22] To minimize shearing we have borrowed approaches developed for preparing large DNA molecules for PFGE.[23,24] Genomic DNA is isolated from cells embedded in low melting point agarose gel inserts, which are treated with proteinase K and detergents. Using this approach we have extracted essentially intact DNA from blood leukocytes. Physical manipulations such as pipetting can shear large DNA molecules, so these are kept to a minimum; for example, by using wide-bore pipetting devices with a slow rate of delivery. We have mounted molecules of more than 2 Mb on optical mapping surfaces, spanning numer-

[21] Y. T. J. Silhavy, M. L. Berman, and L. W. Enquest, "Experiments with Gene Fusion." Cold Spring Harbor Laboratory Press, Cold Spring Harbor, New York, 1984.

[22] E. S. Lander and M. S. Waterman, *Genomics* **2,** 231 (1988).

[23] B. Birren and E. Lai, "Pulsed Field Gel Electrophoresis: A Practical Guide." Academic Press, San Diego, California, 1993.

[24] D. C. Schwartz and C. R. Cantor, *Cell* **37,** 67 (1984).

ous microscope fields. To further minimize shearing of chromosomal DNA, we have developed a protocol for YAC purification based on spermine condensation.[25] Gel inserts made from low melting point agarose are incubated in 0.01 m*M* spermine tetrachloride for 2 hr. This collapses DNA coils embedded within agarose so that subsequently molten agarose can be subjected to manipulations as severe as vortexing, which would ordinarily shear DNA. The DNA particles are decondensed by washing of gel inserts in spermine-free Tris–EDTA (TE) buffer supplemented with 100 m*M* NaCl. This protocol should also be applicable to extracting minimally sheared genomic DNA from a variety of organisms.

Mounting of DNA onto Surfaces

Our efforts to develop optical mapping approaches for analysis of large insert clones, such as BACs and YACs, have taught us how to mount large DNA molecules without significant breakage.[25,26] For example, 80% of BAC molecules can routinely be mounted without breakage.

The exact mechanism of how DNA molecules interact with derivatized surfaces is not known, but we postulate that electrostatic interactions predominate between the anionic DNA and cationic surfaces. Insufficient charge density leads to DNA fragment detachment; fragments smaller than about 1 kb sometimes desorb from the charged optical mapping surface. Too high a surface charge density generally leads to poor stretching or stretched but intact DNA molecules even after prolonged digestion.[12,25] The latter may be due to severe adsorption of the molecule to the surface, thereby occluding the enzyme-binding sites. Alternatively, cleavage may occur, but insufficient relaxation occurs to produce an observable gap.[27] Indeed, both mechanisms may apply. Larger molecules (BACs) require lower surface charge density than smaller molecules (cosmids) and molecules mounted by the peeling method require lower surface charge density than those mounted by the spotting method (see below). Efficient binding and analysis of molecules rely on a balance between flow and electrostatic forces. High fluid flow combined with a high surface charge density causes overstretching of the molecules, sometimes leading to breakage of molecules, during and after mounting. Low fluid flow combined with a high surface charge density causes understretching of the molecules, leading to coils of DNA deposited on the surface. Because optical mapping uses

[25] W. Cai, H. Aburatani, D. Housman, Y. Wang, and D. C. Schwartz, *Proc. Natl. Acad. Sci. U.S.A.* **92,** 5164 (1995).

[26] W. Cai, J. Jing, B. Irvin, L. Ohler, E. Rose, U. Kim, M. Simon, T. Anantharaman, B. Mishra, and D. C. Schwartz, *Proc. Natl. Acad. Sci. U.S.A.* **95,** 3390 (1998).

[27] J. Reed, E. Singer, G. Kresbach, and D. C. Schwartz, *Anal. Biochem.* **259,** 80 (1998).

fluorescence intensity measurements for mass determination, complete elongation of the DNA molecules is not required for mapping, but it does facilitate the data analysis and provides more mapping information because more cutting sites are resolved. DNA is mounted by either "peeling" or "spotting" protocols.

Mounting of DNA by Peeling. The amount of DNA mounted on a surface is titrated on a surface-by-surface basis to give optimum numbers of DNA molecules per microscope field of view. A high density of mounted molecules gives more information but also leads to crossed molecules and an inability to resolve groups of fragments from different molecules. Six to 8 μl of DNA solution is pipetted onto a microscope slide at room temperature. A derivatized coverslip is carefully placed on top of the sample to spread out the drop. The coverslip is left to sit for 20 sec and then it is pressed down lightly with a pair of forceps. This increases the degree of stretching of large DNA molecules such as BACs. The coverslip is removed carefully with forceps and rinsed in TE buffer. Other investigators have fixed DNA molecules by motor-driven movement of a meniscus held at a 45° angle between a coverslip and a derivatized slide.[28]

Mounting of DNA by Spotting. Densely arrayed samples on a single surface enable parallel processing. Current examples of this approach include gridding of DNA samples onto charged membrane surfaces[1] and modification of technologies from the photolithography and semiconductor industries generating DNA microarrays on "chips."[29,30] Such microarrays are used for hybridization-based approaches for mutation detection and nucleic acid quantitation.[31] We have developed a technique to apply many DNA spots in arrays in order to be digested and analyzed in parallel.[14] The fluid flows within evaporating droplets bring the DNA molecules closer to the surface, where they bind in an elongated conformation. Robotic application of samples to surfaces at high deposition rates is a critical requirement of such a high-throughput, fixation-based approach. Grids of spots are generated using a modified commercially available Biomek laboratory automation robot (Beckman, Fullerton, CA) equipped with a specialized workspace deck capable of holding multiple 96-well microtiter plates and up to 12 optical mapping surfaces in a vacuum chuck. In this configuration, the robot is able to deposit one sample approximately every

[28] H. Yokoda, F. Johnson, H. Lu, R. M. Robinson, A. M. Belu, M. D. Garrison, B. D. Ratner, B. J. Trask, and D. L. Miller, *Nucleic Acids Res.* **25,** 1064 (1997).

[29] M. Schena, D. Shalon, R. Davis, and P. O. Brown, *Science* **270,** 467 (1995).

[30] R. Lipshutz, D. Morris, M. Chee, E. Hubbell, M. Kozal, N. Shah, R. Yang, and S. P. Fodor, *Biotechniques* **19,** 442 (1995).

[31] M. Chee, R. Yang, E. Hubbell, A. Berno, X. C. Huang, D. Stern, J. Winkler, D. J. Lockhart, M. S. Morris, and S. P. Fodor, *Science* **274,** 610 (1996).

10 sec. Spotting tools have included a glass capillary, a pipette tip, and a solid tip. Fluid droplets (5–50 pg/μl DNA in TE buffer) of 10–20 nl are spotted onto open glass surfaces at room temperature, using several programs that alter spotting patterns and volumes. Such automated high-density deposition techniques make optical mapping ideal for large-scale genome analysis as well as a potential platform for new types of biochemical investigations.

Digestion of DNA Molecules on Surfaces

The optimal restriction enzyme digestion conditions, in terms of digestion time and enzyme concentration, depend on the enzyme being used and the surface characteristics. Typically, 50–200 μl of a 1 : 25–1 : 50 dilution of the enzyme (5–20 units) in the appropriate buffer containing 0.02% Triton X-100 is pipetted onto the surface. Digestion is carried out at room temperature for 10 min to 2 hr in a humidified closed chamber, which can be anything from an elaborate Plexiglas construction to an empty pipette tip box. The digestion reaction is terminated by aspirating off the enzyme solution and washing the coverslip twice with TE buffer, pH 8.0, and then with distilled water. DNA molecules are stained with 0.2 μM YOYO-1 in TE buffer containing 20% 2-mercaptoethanol. YOYO-1 is a fluorescent dye that bis-intercalates between the DNA bases at one molecule (two fluorophores) per four base pairs.[32] The YOYO-1 solution is filtered through sterile Acrodisc filters (0.2-μm pore size, HT Tuffryn membrane, low protein binding; Gelman Sciences, Ann Arbor, MI) prior to staining. Six to 8 μl of YOYO-1 solution is pipetted onto a microscope slide and the sample coverslip is placed on top of it. The edges of the coverslip are sealed onto the slide with transparent nail polish. Alternatively, the coverslip is covered with 100–200 μl of YOYO-1 solution and rinsed with distilled water after 1 min to decrease background staining.

Optical Mapping Workstation

Figure 2 shows a schematic of the Optical Mapping automatic imaging station used for imaging DNA molecules.

Microscope Automation Software

The OMM software includes full computer control of all the workstation functions such as focusing, sample translation (x–y position), switching

[32] H. S. Rye, S. Yue, D. E. Wemmer, M. A. Quesada, R. P. Haugland, R. A. Mathies, and A. N. Glazer, *Nucleic Acids Res.* **20,** 2803 (1992).

between silicon intensified target (SIT) and CCD cameras, and image collection. Manual and fully automatic modes are incorporated. The OMM software has user-friendly interfaces, permitting convenient variation of the image acquisition parameters; imaging time, sample grid position and size, and spot diameters. A sample image is collected by the CCD camera and the final imaging time is then set by the operator so that the pixel value on any part of the molecules is below saturation (4095 gray levels). Digital images can be acquired by the workstation at the rate of four per minute (using 5-sec imaging time). The development and implementation of automated microscope operations have enabled operator-free imaging of optical mapping surfaces. An operator can enter the coordinates of a series of gridded spots and images are subsequently collected and tiled together automatically, only from the areas on the surface where DNA has been spotted.[14] As part of a developmental path for large clone mapping with automatic image collection, we have developed the program Mark and Collect. This software saves the operators time spent waiting for images to be collected and transferred to the file system, which normally takes about 10 sec per image. Microscope fields of interest are selected by the operator and the coordinates stored. When enough areas of the optical mapping surface have been identified the program subsequently moves the stage and collects the images. The image acquisition output is linked to the automatic map-making routines, so that after the images have been acquired and stored on hard disk arrays, they can be either directly submitted for automatic image processing and extraction of restriction map data or saved for later processing. The OMM system runs on a network of 16 identical dual-processor Sun SPARC 20 workstations with a networked file system (Fig. 4). Access to all aspects of the OMM data and processing is made through one shared directory hierarchy. This file system structure and the accompanying software libraries provide uniform controlled access to all collection and processing activities and data. Several specially designed "viewers" permit the user to visually follow the different steps of the image analysis, so that convenient assessment and troubleshooting can be performed. A distributed processing system developed in this laboratory allows all the available computational resources on the network to be shared.

Image Processing

A restriction map is made from a group of images containing one or several molecules. The processing of the group comprises three major steps: image correction, image analysis, and map making. In the two first steps, each image is processed separately, and in the third step, data from all the

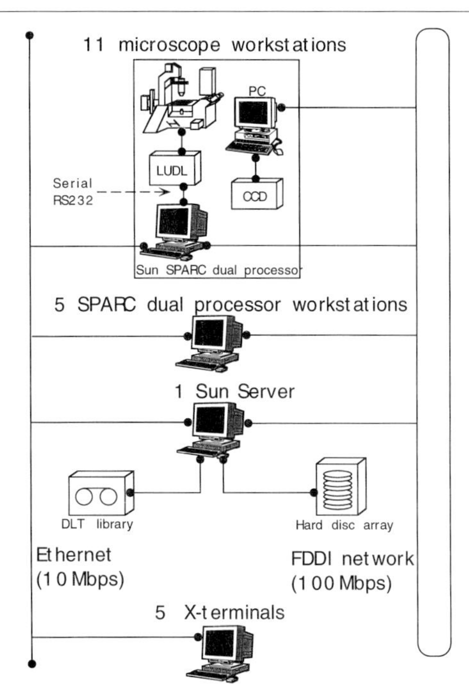

FIG. 4. Optical mapping infrastructure. Image files, typically 1.5 Mb each, acquired at the optical mapping stations are distributed among selected workstations, which have local storage volumes, including a Sun SPARC Storage Array 200. Files from Optical Map Maker (OMM) software are shared through a network file system, which runs on top of a 100-Mbps FDDI network. The FDDI network encompases all of the Sun workstations and the PCs, which are directly connected to CCD cameras through a high-speed serial protocol. The Sun workstations are dual homed, with a second network interface to a 10-Mbps ethernet. For archiving purposes, a 3-terabyte digital linear tape (DLT) library is used.

images (the outputs of the second step) are gathered for the calculation of the map.

Image Correction

The CCD camera produces images composed of 1317 × 1035 pixels of 12 bits each, which is read out and stored as an array of unsigned 16 bit values. The intensity value for a pixel at no light is the dark value, and an image taken with the microscope shutter closed is a dark image. This dark image, which represents the electronic noise in the CCD chip, varies from camera to camera and with the parameters used for taking the image (exposure time, etc.). The first correction that is made to an optical mapping image is to subtract the corresponding dark image. The second correction, flatfield correction, is carried out to adjust for differences in the amount of light registered by a particular pixel that are not due to the difference in distribution of dye in the sample. Such variations can be due to nonuniformity of the light source, differences in the light path (smudges on optical components, etc.), and possibly other factors in the CCD and readout electronics. The correction involves dividing each pixel value in the image with the corresponding pixel value in a bright image, an image of a thin, in-focus and uniformly fluorescent sample. Because we are interested in the fluorescence emanating from dye bound to DNA molecules, other sources of fluorescence are considered background and the subtraction of this background contribution is the last correction made. On the basis of the assumption that the major source of background is unbound dye on the surface, and that this does not vary rapidly, the background value for each pixel can be estimated. First, we find the median pixel values for pixels on concentric circles around the pixel in question. Then, a median of medians is taken, a median image calculated, and a gaussian smoothing operation applied to give the final background image. This image is subtracted from the flatfield corrected image.

Image Analysis

The purpose of the analysis of an optical mapping image is to find the molecules in the image and for each molecule to calculate a signal representing the intensity in the image along the length of the molecule. DNA molecules are found by looking for long, thin, bright objects which have a dominant direction. In the first phase, an algorithm identifies these isolated regions in the image, using both the fluorescence intensity and local directionality properties at each pixel, where each region contains the fragments of a parental molecule not overlapping with other molecules. This is done by first applying a ridge filter to the input corrected image, which produces 16 new images. Each image corresponds to 1 of 16 different directions, and the value of a pixel in one of these images represents a calculation of the degree to which the pixel appears to lie on a ridge in the particular direction. An image is then constructed that contains, at each pixel, the highest of the 16 values for that pixel. A threshold value is picked so that the peak on the histogram for an area containing the segment to be measured is the midpoint between the threshold value and the lowest background value. A threshold is calculated for the new image and objects are identified as contiguous collections of pixels whose values exceed the threshold. At this point the objects (collections of pixels with high numerical values) consist of pixels that are on ridges in some of the 16 directions, but because all 16 directions are included, and because the ridge filter tends to swell up in regions that have no predominant direction (e.g., at the end of molecules), this image has a tendency to merge molecules that happen to be close to each other. To separate these molecules, the ridge filtering process is repeated for nine, then five, and finally three directions so that in the final image, the objects consist of pixels lying along a ridge in one of three directions. This process has been shown to find and separate DNA molecules. In the second phase, a path down the center of the molecule (the "backbone") is computed using Dijkstra's "All paths least cost" algorithm. A pixel in the middle of the molecule is chosen, and the algorithm is applied to find the lowest cost path (traversing the molecule one step at a time from a pixel to one of its eight neighboring pixels) from the first pixel to each other pixel in the object. The cost of a path is the sum of the costs of the individual steps from pixel to pixel, and the cost function applied to each step favors moving to a pixel that is bright enough to be in the center of the molecule, but does not favor bright pixels so strongly that the path wanders all over the middle of the molecule looking for the very brightest pixels despite the extra length. The paths from the first pixel to all boundary pixels are evaluated to identify an end-point pixel, and the process is repeated using the end-point pixel instead of the first pixel. This finds the other end-point pixel and the path to it is the backbone. A smooth intensity signal along the backbone is then computed; for each position along the backbone, the intensity is calculated by summing the intensities for a set of pixels that are close to the backbone and lying along a line orthogonal to the backbone at that position. This intensity signal is then used to determine potential cut sites and estimated masses of all the potential fragments in that molecule. The estimated fragment mass data are normalized, and then, together with the information about the order of the fragments, collected for statistical analysis by the map-making program.

The program Visionade is an interactive visual editing tool that allows a user to quickly process a group of optical mapping images, identifying fragments and molecules in the images with a few point-and-click actions. In the optical mapping system this step is done after a group of images has been collected and automatically flattened. The outputs of Visionade are mask images locating the fragments and molecules. These outputs are combined with the flattened images to measure the sizes of the fragments in the molecules, and these data are the input to the statistical analysis program Lcluster, which is a novel computational approach to map construction from optical data and finds the consensus map.

Map Making

The map-making program lcluster gathers all of the per-molecule computations from all images and operates on them together. Here, the complete restriction map is computed by testing a set of plausible hypotheses about the restriction cleavage sites and choosing the most likely hypothesis supporting the previously accumulated data. The program starts by filtering out unsuitable molecules based on several criteria: (1) A molecule must have a total length in pixels between a specified minimum and maximum; (2) a molecule must not be within 10 pixels of the image boundary; (3) the maximum angle of curvature over the entire length must be below a specified angle (usually 30°); and (4) the total intensity of the molecule due to pixels outside a specified distance from the backbone must be less than a specified threshold. These criteria tend to exclude most partial and overlapping molecules. Bayesian estimation is then used to determine the optimum way of accounting for each of the fragments in the filtered data set. A hypothetical model of the map is formulated, which specifies the following parameters: cut locations, the variance of cut site measurements, the cutting efficiency, and the sample purity. Suitable probability distributions are applied to each of the parameters in the model. The cut locations follow

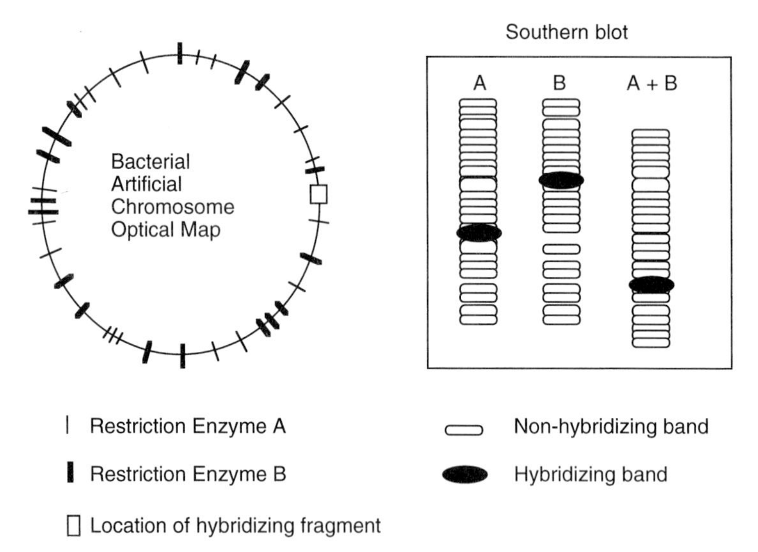

FIG. 5. Schematic of a Southern blotting approach to fine-map cDNAs and ESTs to bacterial artificial chromosomes (BACs). cDNAs or ESTs are used as probes on Southern blots of BAC DNA that has been tube digested with the same restriction enzyme used to prepare the optical map, which cuts the BAC into many fragments. Fragment sizes seen on Southern blots are correlated to fragment sizes on the optical map. Fine mapping can be made more accurate by the use of double digests. Multiple lanes can be electrophoresed and blotted. The resulting filters can be cut into their respective lanes. Both rare and frequent cutter blots can be probed together, yielding large amounts of information from relatively modest amounts of work.

a gaussian distribution, false cuts are assumed to be uniformly distributed over the length of the molecule, the number of false cuts follows a Poisson distribution, and the cutting frequency is a simple function of the number of cuts. The object of the analysis is to find the maximum-probability model of the map, given the data. The program considers about 8000 models for each model size, i.e., maps of 2, 3, 4, ... fragments, the picks the 32 best models of each size, whose parameters are then optimized locally to converge to the nearest maximum in probability space. The best model of any size is accepted as the solution map if its probability density is sigificantly better than that of any other model.

Sizing and Orientation

An internal standard of known length is used to accurately size fragments. In the case of cloned DNA this can be the vector that has *Not*I restriction sites in the polylinker at either end.[26] The pBEloBACII vector, for example, gives a 7.5-kb sizing standard. Circular molecules can therefore be mounted on surfaces and sized in a circular form. In thie case of genomic DNA, λ phage DNA (48.5 kb) or cosmid DNA (about 45 kb) can be comounted with the sample. Sizing of intact molecules eliminates potential sizing errors caused by fragmentation of a linearized molecule. Cutting efficiency can also be estimated if the comounted standard possesses a restriction site for the enzyme used for mapping.

High-resolution maps are made by overlaying and aligning single enzyme maps to create multiple enzyme maps. The orientation and alignment are determined by partial double digestion or by noting the position of the cloning vector fragment that is present on each clone.

Further Studies

Although optical mapping has not been used thus far to directly map cDNAs or ESTs to clones or genomic DNA, simple approaches to high-resolution mapping can be envisioned using a collection of EST markers binned using a pooling approach from radiation hybrids or BAC clones. For example, high-resolution restriction maps, provided by optical mapping, constructed from pools of BAC clones create a minimum tiling path and the necessary substrate for Southern blot analysis, using EST markers as probes (Fig. 5). The sizes of the hybridized bands from Southern blots are easily correlated with fragment sizes on optical maps when the same enzyme is used. Multiple enzyme maps and digests examined by Southern blotting will provide greater discrimination of closely sized fragments in the map, and allow for more precise mapping of any probe.

Section II

Gene Identification

[5] Current Status of Computational Gene Finding: A Perspective

By RICHARD J. MURAL

The current (circa early 1998) worldwide rate of sequencing human genomic DNA is on the order of 10 megabases per month. To complete the sequencing of the human genome by 2005, the stated goal of the U.S. human genome effort, this rate will have to increase about fivefold. Even at the current rate of DNA sequencing, the data flow is well beyond our current capabilities to experimentally identify and characterize the genes embedded in these sequences. Our first look at the biological content of the genomic sequence being generated in the Human Genome Project will, almost certainly, come from computationally based gene-finding programs that will produce, in a highly automated fashion, a "first pass" at the annotation of the human genome. Because computational gene finding will generate this first look at the human genome it is reasonable to look at the current state of these programs, their relative strengths and weaknesses, and what future developments we might expect.

There are several points that must be kept clear in our discussion of computational approaches to gene finding and its utility. First, it must be stressed that a gene predicted by computational methods must be viewed as a hypothesis subject to experimental verification. In some cases there may be strong evidence to support the model, the transcript of the model gene is an exact match to the sequence of a cDNA in GenBank, for example, but in many cases confirmation of the prediction will come only from the generation of further data. Another distinction that needs to be made is between finding a gene and providing a complete and accurate description (model) of it. Often the correct prediction of a single exon is sufficient to identify, and properly locate, a gene in a genomic sequence even if its complete intron–exon structure is not properly represented. On the other hand, even a "correct" gene model may not reflect the full repertoire of a gene because no program can currently predict alternatively spliced transcripts.

In the time that computational gene finding[1–3] has been evolving, considerable progress has been made; however, none of the current systems

[1] E. C. Uberbacher and R. J. Mural, *Proc. Natl. Acad. Sci. U.S.A.* **88,** 11261 (1991).

[2] C. A. Fields and C. A. Soderlund, *Comput. Appl. Biosci.* **6,** 263 (1990).

[3] M. S. Gelfand, *Nucleic Acids Res.* **18,** 5865 (1990).

Copyright © 1999 by Academic Press
All rights of reproduction in any form reserved.

0076-6879/99 $30.00

produces the "correct" result in all cases. Some of the limitations of current approaches, and their causes, are also discussed below.

Impact of Current Sequence Databases on Computational Gene Finding

All gene-finding programs approach the problem in one of two ways: by identifying statistical patterns that distinguish coding from noncoding DNA sequences, or by finding coding regions by their homology to expressed sequence tags (ESTs) or cDNA sequences present in available databases. Some programs combine both approaches.[4,5] The limitation for both approaches comes from different aspects of the same source, namely the amount and quality of data in current sequence databases.

The fact that current databases are incomplete is the primary limitation on methods that attempt to find genes on the basis of homology methods. As the genomic sequence of an increasing number of organisms is completed and we have complete gene inventories of these organisms, it becomes increasingly clear that we have identified and/or characterized less than half of the existent gene families. This fact, that many newly sequenced genes do not have close relatives of known sequence, places a severe limitation on gene finding based strictly on methods involving homology. When a newly sequenced gene does have a relative in a current database, homology-based methods not only help to identify the gene of interest, but (particularly when the relationship between the query and target sequence is close) can help to provide an accurate gene model and, ideally, information about the function of the identified gene.

Identifying genes on the basis of similarity to sequences present in EST databases comes with other caveats. Using EST data will identify and assist in modeling more genes than will be found by methods based solely on protein/cDNA homologies. Current estimates are that the public EST databases of human and mouse ESTs, which contain in total more than 1 million sequences, represent >80% of the genes expressed in these organisms. Therefore the coverage of ESTs is higher than the coverage based on well-characterized proteins. Finding good matches to ESTs is a strong suggestion that the region of interest is expressed and, because the data in EST databases represent cDNA made from a large number of tissues, can give some indication as to the expression profiles of the located genes. The problem of inferring function still remains, as an EST that represents a protein without known relatives will not be informative about gene function.

[4] Y. Xu and E. C. Uberbacher, *J. Comput. Biol.* **4**(3), 325 (1997).

[5] Y. Xu, R. J. Mural, and E. C. Uberbacher, *in* "Fifth International Conference on Intelligent Systems for Molecular Biology," pp. 344–353. AAAI Press, 1997.

Approaches that attempt to find patterns in DNA sequence and infer the presence of genes or other features of biological interest from these patterns are impacted by both the quantity and quality of data in current databases. Curation of data is a prerequisite to developing pattern recognition algorithms for identifying features of biological interest. For example, redundant data can cause a number of problems. When a sequence, or members of a closely related family of sequences, are represented multiple times in a sequence database, sequence motifs found in these sequences will be overrepresented and will therefore bias statistical parameters, such as n-mer frequencies, that may be derived from such data. This kind of bias presents real problems even in data sets that are used as "standards" in the sequence analysis community. For example, the data set of Burset and Guigo,[6] widely used for both training and testing gene identification programs, contains multiple hemoglobin genes that, although they come from a number of different organisms, are sufficiently similar to bias various pattern statistics that are used in building gene-finding programs. There are currently no well-cleansed, nonredundant databases available for investigators building gene-finding software.

Inadequacy of Current Annotation

The other class of data problems that confront investigators building computational tools for gene finding has to do with inaccurate or incomplete annotation associated with genomic sequence records. For example, the majority of non-consensus splice sites are the result of misannotation in the database entry.[7] Filtering out this kind of noise from the data that contribute to gene prediction algorithms is critical if such programs are to be successful. Equally troublesome is the case of incomplete annotation. It is often the case that the database entry for a region of genomic sequence may not include annotation of one or more genes present in the region. The reason for this is usually quite simple. Before the current large-scale sequencing phase of the Human Genome Project was undertaken, most regions of genomic sequence were generated to describe the genome organization of one gene of interest to the laboratory doing the sequencing. If the region happened to include other genes, they were likely to be overlooked because there was no particular reason to look for them, nor were there any gene-predicting programs available to suggest their presence. Why do unannotated genes present a problem? In general, statistical differences used to discriminate regions with protein-encoding potential from other

[6] M. Burset and R. Guigo, *Genomics* **34**(3), 353 (1996).

[7] I. J. Jackson, *Nucleic Acids Res.* **19**(14), 3795 (1991).

DNA sequences are derived by comparing the frequencies of short nucleotide "words" found in coding and noncoding regions of DNA sequences. The noncoding regions are, operationally, those parts of the sequence that are not annotated as having protein-coding exons, so the protein-coding exons of a nonannotated gene are counted as noncoding regions of the genome. This can bias the statistics used in developing the gene-finding programs. Unannotated genes also cause problems when evaluating gene-finding programs because, if the program labels them as genes, or as protein-coding exons, they will be scored as false positives when in fact they are correct predictions.

Computational Gene Predictions as Hypotheses

It is critically important to view computational annotation as a hypothesis. For programs that have combined the results of pattern recognition with search-based methods or from methods based solely on homology, the hypothesis is quite easily tested because the gene model makes strong predictions about the primary structure of the mRNA. For other predictions, testing may be more complicated. The final verification of a gene model should involve comparison with the full-length cDNA isolated from the organism of interest. In addition to verifying gene models, comparison with full-length cDNAs is necessary for the correct annotation of pseudogenes. Fortunately, a number of projects are underway to clone and sequence full-length cDNAs.

Danger of Transitive Catastrophes

There are a number of sources of potential "transitive catastrophes."[8] In general, relationships among proteins are transitive, that is: if protein *a* is a member of protein family *x* and protein *c* is closely related to protein *a*, then it usually follows that protein *c* is in fact a member of protein family *x*. Transitive catastrophes occur when the relationship of protein *a* to protein family *x* is tenuous but is entered into a database as an assertion (which may well be false). Then when protein *c* is discovered by a database search routine to be related to protein *a* the assertion as to membership in protein family *x* is picked up as an annotation for protein *c*. Or consider a protein with two domains A and B. This protein is attributed function *z*, but that function is carried out by the A domain. A newly described protein, C, may be only homologous to domain B of the first protein and yet receive annotation stating that it has function *z* even though it completely lacks domain A.

[8] O. White, R. Clayton, A. Kerlavage, and J. C. Venter, *Microsc. Comp. Genomics* **3,** 59 (1998).

The danger of circular annotation by gene-finding programs comes from having database annotation based on, but not clearly identified as, the prediction of a gene-finding program. In such a case when the sequence is again analyzed by the same gene-finding program, the predicted gene model matches the annotation exactly, which of course it should. This potential problem in automated first-pass annotation can be easily avoided if all genes predicted by gene-finding programs are clearly labeled as such. (In a GenBank feature table this means that the "evidence" field must contain "nonexperimental" and a miscellaneous comment should be made identifying the gene-finding program used.)

Current Gene-Finding Tools and How to Access Them

GRAIL[1,9] is, perhaps, the most widely used gene-finding program. It has been instrumental in identifying a number of genes in genomic DNA sequence. It is accessible by e-mail (*grail@ornl.gov*), through a client-server system for UNIX machines (*ftp://compbio.ornl.gov/pub/xgrail/*) and through the World Wide Web (*http://compbio.ornl.gov/tools/index.stml*).

A new version of GRAIL, GRAIL-EXP,[4,5] combines the pattern-recognition power of GRAIL with homology-based methods using the EST database to automatically produce high-quality gene models. Because this program uses homology data it can properly parse genes from long genomic DNA sequences. GRAIL-EXP is not available as a stand-alone system but is one of the gene-finding programs in the Genome Channel (see below, *http://compbio.ornl.gov/tools/channel/index.html*).

Genscan[10] is a program based on hidden Markov models that produces accurate gene models from genomic DNA sequence. Genscan also attempts to find the beginning and end of genes and to parse multiple genes from long genomic sequences. The accuracy of this parsing is variable because it is rule-based rather than homology based. Sequences can be analyzed using Genscan via e-mail (*genscan@gnomic.stanford.edu*), or through the Web (*http://gnomic.stanford.edu/GENSCANW.html*).

Genie[11] is another program using hidden Markov models to predict gene models. Genie can be accessed at *http://www-hgc.lbl.gov/inf/genie.html.*

FgeneH[12] predicts all potential exons using linear discriminant func-

[9] R. J. Mural, Y. Xu, M. Shah, X. Guan, and E. C. Uberbacher, *Curr. Protocols Hum. Genet.* **2,** 6.5.1 (1995).

[10] C. Burge and S. Karlin, *J. Mol. Biol.* **268,** 78 (1997).

[11] D. Kulp, D. Haussler, M. G. Reese, and F. H. Eeckman, *in* "Proc. Conf. on Intelligent Systems in Molecular Biology, '96" (D. J. States, P. Agarwal, T. Gaasterland, L. Hunter, and R. Smith, eds.), pp. 134–142. AAAI/MIT Press, St. Louis, Missouri, 1996.

[12] V. V. Solovyev, A. A. Salamov, and C. B. Lawrence, *Nucleic Acids Res.* **22**(24), 5156 (1994).

tions, then uses dynamic programming to construct gene models. This program can be accessed by e-mail (*service@bchs.uh.edu*), or through the World Wide Web at *http://genomic.sanger.ac.uk/gf/gf.html.*

MZEF[13] uses quadratic discriminant analysis for gene finding; it is available from *http://clio.cshl.org/genefinder/.*

Other programs currently in use include Geneparser[14] (*http: beagle. colorado.edu/~eesynder/GeneParser.html*), Xpound[15] (*alun@myriad. com, http://bioweb.pasteur.fr/seqanal/interfaces/xpound-simple.html*), and GeneID[16] (*http://www.imim.es/GeneIdentification/Geneid/geneid_input. html*).

Need for Automated "First-Pass" Annotation of Human Genomic DNA

Several million bases per day of human genomic sequence will soon have to be generated by the Human Genome Project in order to meet the goal of completing the human sequence by the year 2005. This will present a deluge of data that no one will be able to track or annotate in a timely fashion using current models of DNA sequence analysis. Only by applying several of the current gene-finding programs to these data in an automated way, and by presenting the results to the biomedical community in a simple and direct fashion, will we begin to appreciate the biological riches embedded in our DNA sequence. Computational gene-finding must evolve beyond an intellectual exercise into a robust analytical tool that presents gene models and their underlying evidence for testing in the laboratory.

Genome Channel: The Future

A prototype of a system that combines the outputs of several of the gene-finding programs described above into a user-friendly system that permits browsing of long stretches of human genomic sequence has been developed by the Genome Annotation Consortium (*http://compbio.ornl. gov/gac/index.stml*). This browser, called the Genome Channel (*http:// compbio.ornl.gov/tools/channel/index.html*), provides precomputed views of large regions of human sequence. In addition, conceptual translations of gene models generated by GRAIL-EXP and Genscan have been searched against protein databases and these results have been stored and are retrievable by the user. In this way it is possible to quickly inspect millions of bases of genomic DNA sequence in a timely fashion and truly begin to

[13] M. Q. Zhang, *Proc. Natl. Acad. Sci. U.S.A.* **94,** 565 (1997).
[14] E. E. Synder and G. D. Stormo, *J. Mol. Biol.* **248,** 1 (1995).
[15] T. Alun and M. H. Skolnick, *IMA J. Math. Appl. Med. Biol.* **11,** 149 (1994).
[16] R. Guigo, S. Knudsen, N. Drake, and T. Smith, *J. Mol. Biol.* **226,** 141 (1992).

appreciate the richness of the data being generated by high-throughput sequencing efforts. Integration of tools of this sort will represent the next wave of computational genomics.

Note

An excellent bibliography of computational gene-finding methods can be found at *http://linkage.rockefeller.edu/wli/gene/.*

Acknowledgment

This research was supported by the Office of Biological and Environmental Research, U.S. Department of Energy, under Contract DE-Ac05-96OR22464 with Lockheed Martin Energy Research Corporation.

[6] Gene Identification by Exon Amplification

By DEANNA M. CHURCH and ALAN J. BUCKLER

Introduction

The development of comprehensive gene maps will greatly facilitate study of the structure, function, and organization of the genome of an organism. Position information afforded by these maps will often be the first criterion used in assessing the potential role of a gene in the etiology of phenotypic variations that are genetically linked to specific genomic regions. Highly representative gene maps, developed for several prokaryotes and simple eukaryotes, are proving to be of great utility for genetic analysis and other areas of biological study.

For humans as well as other organisms with more complex genomes, development of such maps would provide a powerful tool in positional cloning efforts, but this is currently a daunting task. Whole-genome, chromosome-specific, and region-specific endeavors are ongoing, and they employ a variety of approaches that focus on characteristics that are common to most genes. These strategies fall into three general categories: (1) study of the evolutionary conservation of DNA sequences, (2) analysis of transcription, or (3) identification of functional elements associated with most transcription units. For each of the strategies, overlapping sets of experimental methods are utilized, and these are mainly centered around DNA sequencing, hybridization, or functional selection. Evidence from the analysis of various regions of the genome suggests that a combination of these

Copyright © 1999 by Academic Press
All rights of reproduction in any form reserved.

0076-6879/99 $30.00

approaches is the most effective way to identify all of the potential coding elements within a given region of the genome.[1–3]

Initially, searches for evolutionary sequence conservation involved the isolation of DNA fragments that cross-hybridize with DNA from diverse species, a technique commonly known as "zoo blotting."[4–6] This method, although quite effective, is rather cumbersome even for relatively small genomic stretches, and is no longer a primary method of choice. Analysis of sequence conservation is now moving from hybridization to comparative DNA sequencing, and is likely to again receive greater attention as genomic sequence information from different species grows in availability.

Expression-based identification of genes has historically entailed hybridization screening of cDNA libraries constructed from various tissues or cells. In some cases, complex probes such as yeast artificial chromsomes (YACs) or plasmid-based genomic clones [cosmids, bacterial artificial chromosomes (BACs)] have been used to identify cDNA clones according to the methods of Snell *et al.*[7] and Benton and Davis.[8] Screening for cDNAs has increased dramatically in efficiency with the development of solution hybridization methods such as cDNA selection or direct selection.[9,10] Selection methods use biotinylated DNA probes, typically genomic clones, to enrich for cDNAs from a given region. The techniques are described in detail elsewhere and are effective means to identify transcribed sequences

[1] M. L. Yaspo, L. Gellen, R. Mott, B. Korn, D. Nizetic, A. M. Poustka, and H. Lehrach, *Hum. Mol. Genet.* **4,** 1291 (1995).

[2] D. A. Ruddy, G. S. Kronmal, V. K. Lee, G. A. Mintier, L. Quintant, R. Domingo, Jr., N. C. Meyer, A. Irrinki, E. E. McClelland, A. Fullan, F. A. Mapa, T. Moore, W. Thomas, D. B. Loeb, C. Harmon, Z. Tsuchihashi, R. K. Wolff, R. C. Schatzman, and J. N. Feder, *Genome Res.* **7,** 441 (1997).

[3] I. Pribill, G. T. Barnes, J. Chen, D. Church, A. Buckler, G. P. Bates, H. Lehrach, M. J. Gusella, M. P. Duyao, C. M. Ambrose, M. E. MacDonald, and J. F. Gusella, *Somat. Cell Mol. Genet.* **23,** 413 (1997).

[4] A. P. Monaco, R. L. Neve, C. Colletti-Feener, C. J. Bertelson, D. M. Kurnit, and L. M. Kunkel, *Nature* (*London*) **323,** 646 (1986).

[5] J. M. Rommens, M. C. Iannuzzi, B. Kerem, M. L. Drumm, G. Melmer, M. Dean, R. Rozmahel, J. L. Cole, D. Kennedy, N. Hidaka, M. Buchwald, J. R. Riordan, L.-C. Tsui, and F. S. Collins, *Science* **245,** 1059 (1989).

[6] K. M. Call, T. Glaser, C. Y. Ito, A. J. Buckler, J. Pelletier, D. A. Haber, E. A. Rose, A. Kral, H. Yeger, W. H. Lewis, C. Jones, and D. E. Housman, *Cell* **60,** 509 (1990).

[7] R. G. Snell, L. A. Doucette-Stamm, K. M. Gillespie, S. A. Taylor, L. Riba, G. P. Bates, M. R. Altherr, M. E. MacDonald, J. F. Gusella, J. J. Wasmuth, H. Lehrach, D. E. Housman, P. S. Harper, and D. J. Shaw, *Hum. Mol. Genet.* **2,** 305 (1993).

[8] W. D. Benton and R. W. Davis, *Science* **196,** 180 (1977).

[9] S. Parimoo, S. R. Patanjali, H. Shukla, D. D. Chaplin, and S. M. Weissman, *Proc. Natl. Acad. Sci. U.S.A.* **88,** 9623 (1991).

[10] M. Lovett, J. Kere, and L. M. Hinton, *Proc. Natl. Acad. Sci. U.S.A.* **88,** 9628 (1991).

from genomic regions as large as entire chromosomes.[11,12] Expression-based gene identification also includes large-scale sequence analysis of randomly selected clones from cDNA libraries, which has resulted in the establishment of vast databases of expressed sequence tags (ESTs).[13–15] EST maps of the human genome are rapidly evolving[16] and will be extremely valuable for positional cloning, but whether these maps will be comprehensive remains unclear.

Searches for functional motifs in genomic DNA, the third approach to gene identification, is initially independent of gene expression. One methodology in this category involves the identification of transcription units based on the presence of hypomethylated CpG islands, which are associated with the 5′ ends of 50–70% of mammalian transcription units.[17,18] This procedure has been used to construct libraries of DNA fragments containing CpG islands.[19] Computer modeling methods for analysis of large stretches of mammalian genomic sequence have also been utilized to identify exons on the basis of the presence of open reading frames and

[11] R. G. Del Mastro, L. Wang, A. D. Simmons, T. D. Gallardo, G. A. Clines, J. A. Ashley, C. J. Hilliard, J. J. Wasmuth, J. D. McPherson, and M. Lovett, *Genome Res.* **5,** 185 (1995).

[12] J. W. Touchman, G. G. Bouffard, L. A. Weintraub, J. R. Idol, L. Wang, C. M. Robbins, J. C. Nussbaum, M. Lovett, and E. D. Green, *Genome Res.* **7,** 281 (1997).

[13] M. D. Adams, J. M. Kelley, J. D. Gocayne, M. Dubnick, M. H. Polymeropoulos, H. Xiao, C. R. Merril, A. Wu, B. Olde, R. F. Moreno, A. R. Kerlavage, W. R. McCombie, and J. C. Venter, *Science* **252,** 1651 (1991).

[14] A. S. Khan, A. S. Wilcox, M. H. Polymeropoulos, J. A. Hopkins, T. J. Stevens, M. Robinson, A. K. Orpana, and J. Sikela, *Nature Genet.* **2,** 180 (1992).

[15] K. Okubo, N. Hori, R. Matoba, T. Niiyama, A. Fukushima, Y. Kojima, and K. Matsubara, *Nature Genet.* **2,** 173 (1992).

[16] G. D. Schuler, M. S. Boguski, E. A. Stewart, L. D. Stein, G. Gyapay, K. Rice, R. E. White, P. Rodriguez-Tomé, A. Aggarwal, E. Bajorek, S. Bentolila, B. B. Birren, A. Butler, A. B. Castle, N. Chiannilkulchai, A. Chu, C. Clee, S. Cowles, P. J. R. Day, T. Dibling, N. Drouot, I. Dunham, S. Duprat, C. East, C. Edwards, J.-B. Fan, N. Fang, C. Fizames, C. Garrett, L. Green, D. Hadley, M. Harris, P. Harrison, S. Brady, A. Hicks, E. Holloway, L. Hui, S. Hussain, C. Louis-Dit-Sully, J. Ma, A. MacGilvery, C. Mader, A. Maratukulam, T. C. Matise, K. B. McKusick, J. Morissette, A. Mungall, D. Muselet, H. C. Nusbaum, D. C. Page, A. Peck, S. Perkins, M. Piercy, F. Qin, J. Quackenbush, S. Ranby, T. Reif, S. Rozen, C. Sanders, X. She, J. Silva, D. K. Slonim, C. Soderlund, W.-L. Sun, P. Tabar, T. Thangarajah, N. Vega-Czarny, D. Vollrath, S. Voyticky, T. Wilmer, X. Wu, M. D. Adams, C. Auffray, N. A. R. Walter, R. Brandon, A. Dehejia, P. N. Goodfellow, R. Houlgatte, J. R. Hudson, Jr., S. E. Ilde, K. R. Iorio, W. Y. Lee, N. Seki, T. Nagase, K. Ishikawa, N. Nomura, C. Phillips, M. H. Polymeropoulos, M. Sandusky, K. Schmitt, R. Berry, K. Swanson, R. Torres, J. C. Venter, J. M. Sikela, J. S. Beckmann, J. Weissenbach, R. M. Meyers, D. R. Cox, M. R. James, D. Bentley, P. Deloukas, E. S. Lander, and T. J. Hudson, *Science* **274,** 540 (1996).

[17] A. Bird, M. Taggart, M. Frommer, O. J. Miller, and D. Macleod, *Cell* **40,** 91 (1986).

[18] F. Antequera and A. Bird, *Proc. Natl. Acad. Sci. U.S.A.* **90,** 11995 (1993).

[19] S. H. Cross, J. A. Charlton, X. Nan, and A. P. Bird, *Nature Genet.* **6,** 236 (1994).

splice sites.[20,21] Although sequence-based exon prediction methods can be useful for identifying genes and determining gene structure, it is currently unclear whether computer-based gene prediction algorithms will be reliable for comprehensive identification of genes in genomes of this complexity.[22]

Another set of methods relies on the biological selection of functional motifs in genomic DNA. Exon trapping and exon amplification are based on the selection of exons: RNA sequences that are flanked by functional 5′ and 3′ splice sites.[23–25] Fragments of genomic DNA are inserted into an intron that is flanked by 5′ and 3′ splice sites and exons of a reporter gene contained within a shuttle vector. These constructs are introduced into mammalian cells, facilitating transcription of the reporter gene and *in vivo* processing of the resulting RNA transcripts. Appropriately oriented splice sites contained within RNA derived from the inserted genomic fragment are paired with those of the reporter gene. The processed RNA contains the previously unidentified exon "trapped" between the reporter gene exons, and is cloned after conversion to cDNA. A similar method has also been developed to rescue 3′ terminal exons of genes by virtue of the presence of a flanking 3′ splice site and a poly(A) addition signal.[26] These methods (along with cDNA selection methods) have been widely and successfully used in numerous positional cloning studies. An important feature of these methods is that they do not require knowledge of the natural expression of a gene to gain an entry point into its cloning (i.e., "expression-independent cloning"). It must be noted, however, that intronless genes will not be detected by methods that search for functional splice sites. Thus the use of these methods alone will not suffice in the generation of comprehensive gene maps.

Evolution of Exon Amplification

Exon amplification is the most widely used of the functional selection methods. In the original version of the technique (see Fig. 1), genomic

[20] E. C. Uberbacher, Y. Xu, and R. J. Mural, *Methods Enzymol.* **266,** 259 (1996).

[21] K. A. Frazer, Y. Ueda, Y. Zhu, V. R. Gifford, M. R. Garofalo, N. Mohandas, C. H. Martin, M. J. Palazzolo, J. F. Cheng, and E. M. Rubin, *Genome Res.* **7,** 495 (1997).

[22] M. Burset and R. Guigo, *Genomics* **34,** 353 (1996).

[23] G. M. Duyk, S. W. Kim, R. M. Myers, and D. R. Cox, *Proc. Natl. Acad. Sci. U.S.A.* **87,** 8995 (1990).

[24] D. Auch and M. Reth, *Nucleic Acids Res.* **18,** 6743 (1990).

[25] A. J. Buckler, D. D. Chang, S. L. Graw, J. D. Brook, D. A. Haber, P. A. Sharp, and D. E. Housman, *Proc. Natl. Acad. Sci. U.S.A.* **88,** 4005 (1991).

[26] D. B. Krizman and S. M. Berget, *Nucleic Acids Res.* **21,** 5198 (1993).

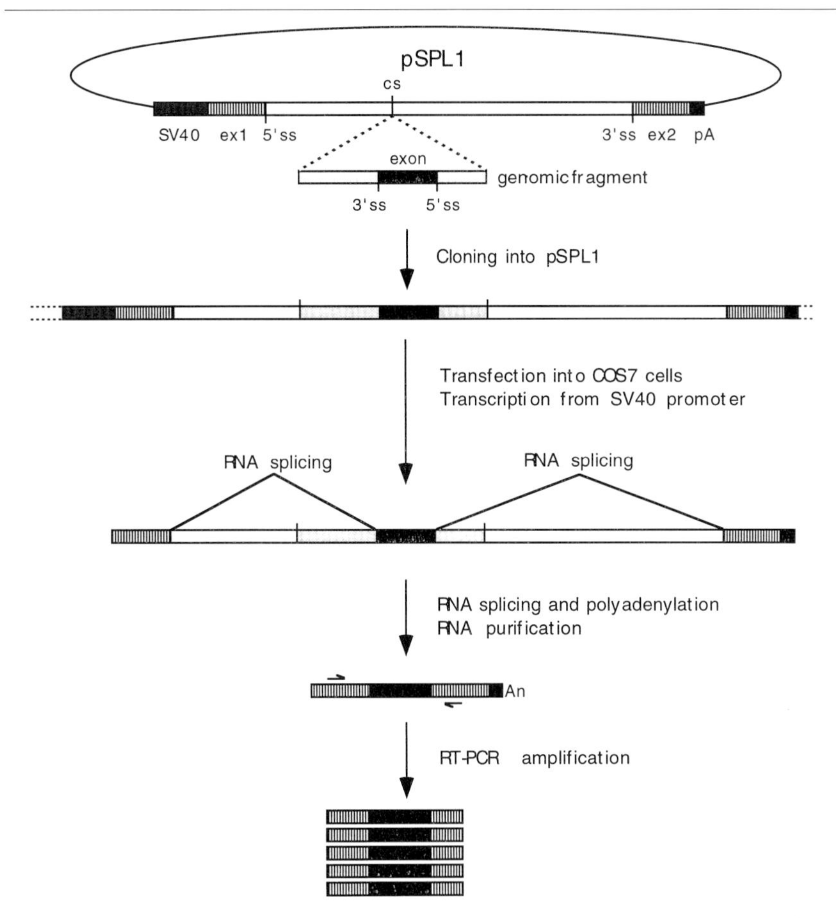

FIG. 1. Exon amplification. In this simplified initial scheme, a genomic fragment containing an exon is inserted into a unique cloning site (cs; in pSPL1, this is a *Bam*HI site) engineered into the intron of the reporter gene contained within pSPL1 (white bar). When the genomic fragment is properly oriented, two artificial introns are created from the intron sequences of the reporter gene and the genomic fragment (stippled bar). On transfection into COS7 cells, a high level of transcription is facilitated through the SV40 early promoter. RNA splice sites flanking the introduced exon will pair with the splice sites of the reporter gene (5′ss and 3′ss), thus joining the reporter gene exons (ex1 and ex2) with the novel exon. The spliced, polyadenylated RNA is then transported to the cytoplasm. RNA isolated from the transfectants is then converted to cDNA, using oligodeoxynucleotide primers corresponding to the reporter gene exons that now flank the unknown sequence of the "trapped" exon. This is followed by PCR amplification with nested primers, and cloning of the resultant product. If the genomic fragment was inserted in the opposite orientation or did not contain an exon, then the splice sites of the reporter gene will pair with each other. This results in the production of mature RNA with no novel sequence ("vector-only" splicing). SV40, SV40 origin of replication and early promoter; pA, SV40, polyadenylation signal.

DNA fragments were inserted into an intron present within the *in vivo* splicing plasmid pSPL1.[25] The insertion site is within an intron from the HIV-1 *tat* gene, whose flanking exons and splice sites were substituted for the second intron of the rabbit β-globin gene in the mammalian shuttle vector pSβ-IVS2.[27] On transfection of this plasmid construct into COS7 cells, transcription of the reporter gene is facilitated through the simian virus 40 (SV40) early promoter. RNA transcripts are then spliced to remove *tat* intron, polyadenylated, and transported to the cytoplasm. If the inserted genomic fragment contains an exon with flanking intron sequence in the sense orientation, heterologous pairing of vector and genomic fragment splice sites occurs. The exon is retained in the mature poly(A)$^+$ cytoplasmic RNA, and is flanked by reporter gene exons. Complementary DNA (cDNA) is produced via coupled reverse transcription and polymerase chain reaction (RT-PCR) amplification using oligodeoxynucleotide primers corresponding to reporter gene exon sequences, and the resulting products are cloned.

Preliminary experiments using this method demonstrated that exon amplification facilitated efficient isolation of exons from single or multiple genomic fragments (as pools of constructs) having a complexity as great as 175 kilobase pairs (kbp).[28,29] Although the method proved effective in gene isolation, further improvements were necessary to adapt the method for large-scale analysis. Enhancements were implemented to address certain weaknesses of the method. Two major drawbacks reduced its sensitivity and efficiency. First, in experiments in which large genomic fragments were screened as pools of constructs, the most abundant RT-PCR product contains only pSPL1 exon sequences without a trapped exon ("vector-only" products).[25,28,29] This is due to the presence of a large fraction of constructs in the pool that contain genomic fragments that are devoid of exons. As a result, competition among PCR templates will favor this smallest and most abundant template. The second problem was discovered on further characterization of a large sampling of cloned products. A specific class of readily identifiable artifacts was discovered, and comprises up to 50% of all RT-PCR clones.[28,29] These artifacts result from the activation of cryptic splice sites in the *tat* intron and in the genomic fragment, both near the 3′ end of the insertion. The general structure of these artifacts is presented in Fig. 2.

[27] A. R. Buchman and P. Berg, *Mol. Cell. Biol.* **8,** 4395 (1988).

[28] D. M. Church, A. C. Rogers, S. L. Graw, D. E. Housman, J. F. Gusella, and A. J. Buckler, *Hum. Mol. Genet.* **2,** 1915 (1993).

[29] M. A. North, P. Sanseau, A. J. Buckler, D. Church, A. Jackson, K. Patel, J. Trowsdale, and H. Lehrach, *Mammalian Genome* **4,** 466 (1993).

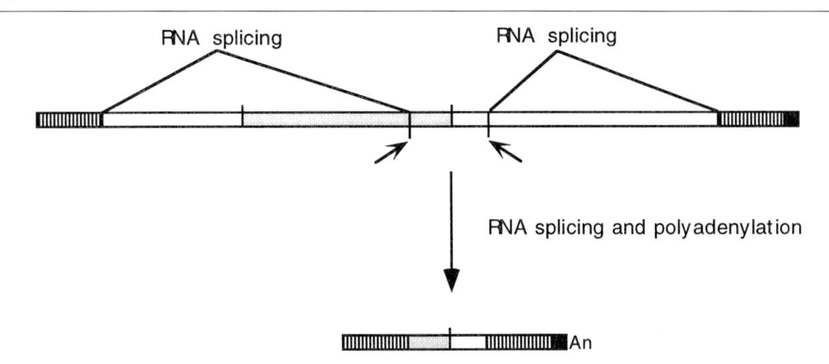

FIG. 2. A predominant class of artifact produced in exon amplification experiments using pSPL1. Activation of cryptic splice sites (arrows) in the genomic fragment and the reporter gene intron occurs when these sites are near the 3′ cloning junction. After transcription of the reporter gene, these cryptic splice sites are paired with the reporter gene splice sites, thus producing an RNA that contains sequences corresponding to the 3′ cloning junction of the original construct.

To enhance the effectiveness of exon amplification, pSPL1 and certain aspects of the protocol were modified to eliminate cloning of both vector-only products and the predominant class of artifacts.[30] Sequence modifications were made to pSPL1 that allowed for the introduction of a specific restriction endonuclease recognition site into either the vector-only or the artifact RNA sequences. When these RNA species are then converted to double-stranded cDNA, they are cleaved by the restriction endonuclease and will not amplify in the PCR.

First, the *tat* exon sequences that flank the vector intron were mutated such that a restriction endonuclease recognition site is formed when the *tat* exons are joined by splicing. Consideration was given to the fact that splice site function can be disrupted by mutations in nearby exon sequences.[31–33] In an attempt to avoid this problem, sequence alterations were made in pSPL1 that left intact the three *tat* exon nucleotides immediately adjacent to each splice site. This was accomplished by the introduction of half-sites for *Bst*XI, a restriction endonuclease specific for an interrupted palindrome sequence ($CCAN_6TGG$). The internal six nucleotides that correspond to the original *tat* exons comprise the interruption in the palindrome. Splicing joins the half-sites only in RNAs containing no trapped exons (i.e., vector-only), which are then cleaved by *Bst*XI after conversion

[30] D. M. Church, C. J. Stotler, J. L. Rutter, J. R. Murrell, J. A. Trofatter, and A. J. Buckler, *Nature Genet.* **6,** 98 (1994).

[31] H. J. Mardon, G. Sebastio, and F. Barelle, *Nucleic Acids Res.* **15,** 7725 (1987).

[32] T. A. Cooper and C. P. Ordahl, *Nucleic Acids Res.* **17,** 7905 (1989).

[33] P. E. Hodges and L. E. Rosenberg, *Proc. Natl. Acad. Sci. U.S.A.* **86,** 4142 (1989).

to double-stranded cDNA. The predominant class of artifact was similarly removed from the PCR substrate. A multiple cloning site that includes a *Bst*XI site at its 3′ end was inserted into the *tat* intron. As a result, any processed RNA molecule originating via the described cryptic splice site activation will include this *Bst*XI site.

Subsequent studies showed that sequences derived solely from the modified vector, pSPL3, were virtually eliminated from the final PCR product, as were those resulting from cryptic splice site activation. The findings clearly indicated that these modifications to exon amplification have dramatically increased its speed, sensitivity, and reliability. Indeed, in a single assay exon amplification can be used to effectively trap exons from up to 2–3 Mb of genomic DNA.[30,34] More recently, additional modifications have been made to pSPL3 to increase its efficiency,[35] with the removal of an alternatively spliced exon from the *tat* intron within pSPL3. This alternatively spliced product can represent anywhere from 2 to 50% of an RT-PCR product,[30,34,35] and this frequency may be related to the level of success in production of recombinant construct pools in pSPL3.

Exon amplification is now an effective means for isolation of coding sequences from genomic DNA of relatively high complexity. It has been an important tool in the successful identification of more than 30 genes involved in human and murine diseases. It is likely to remain a key component in positional cloning efforts and gene map construction for mammalian genomes, in combination with methods such as cDNA selection, EST mapping, and genomic sequencing.

Considerations for Exon Amplification Experiments

A diagram of the basic scheme of exon amplification with pSPL3 is shown in Fig. 3. The initial step in the protocol is the production of a genomic library within pSPL3. The construction of a high-quality library is critical to the success of exon amplification. The genomic DNA source for construction of this library can come from λ phage, cosmids, BACs, PACs (P1 phage-derived artificial chromosomes), YACs, pools of these clones, or total genomic DNA.[25,28–30,34,36] An important aspect of the input genomic DNA is its purity; contamination of mammalian DNA with either yeast or bacterial DNA can lead to the identification of artifactual exon

[34] J. A. Trofatter, K. R. Long, J. L. Rutter, C. J. Stotler, A. N. Turner, J. R. Murrell, and A. J. Buckler, *Genome Res.* **5,** 214 (1995).

[35] T. C. Burn, T. D. Connors, K. W. Klinger, and G. M. Landes, *Gene* **161,** 183 (1995).

[36] F. Gibson, H. Lehrach, A. J. Buckler, S. D. Brown, and M. A. North, *Biotechniques* **16,** 453 (1994).

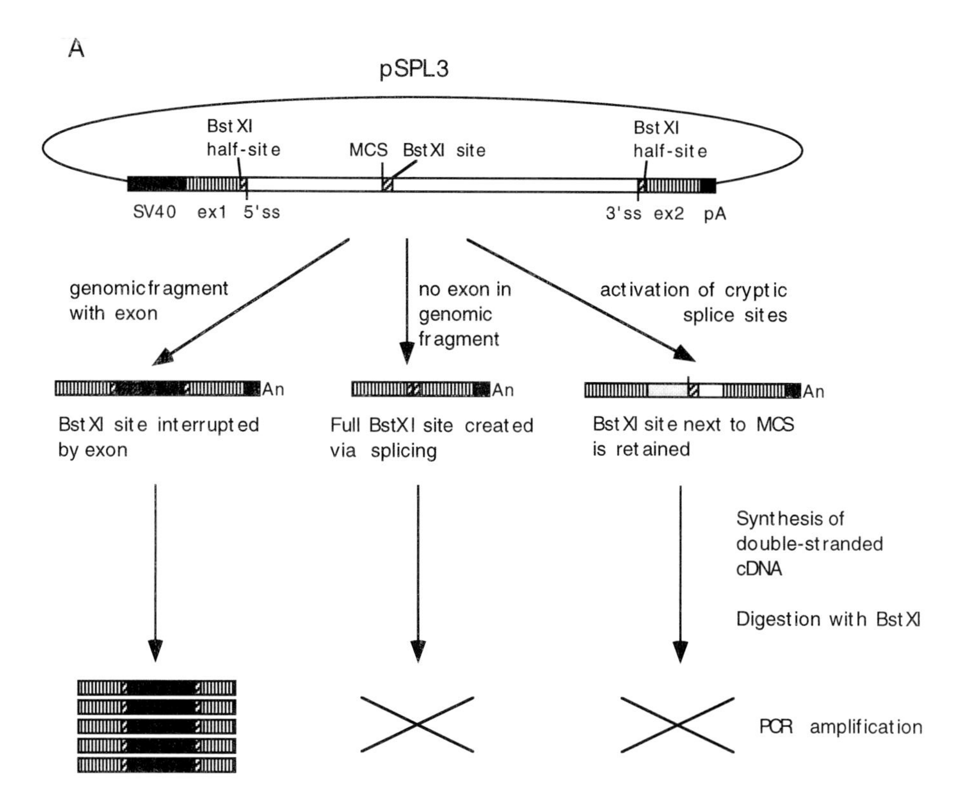

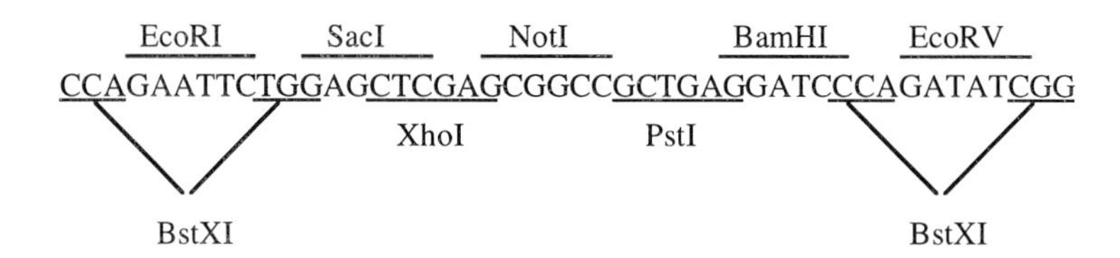

FIG. 3. pSPL3 and modifications to improve exon amplification. Half-sites for the restriction endonuclease *Bst*XI have been engineered into the reporter gene exon sequences immediately flanking the intron. In addition, a multiple cloning site (MCS) has been introduced, and is flanked by *Bst*XI sites on both sides. When "vector-only" RNA splicing occurs (genomic fragments lacking exons), the half-sites are joined to form what will be a full *Bst*XI site after synthesis of double-stranded cDNA. When the predominant class of artifact is produced by activation of cryptic splice sites (see Fig. 2), the 3′ cloning junction, including the *Bst*XI site, will be retained in the spliced RNA and subsequent cDNA. However, if an exon is "trapped" from the genomic fragment the *Bst*XI half-sites are interrupted and thus do not form a functional restriction site. Digestion with *Bst*XI separates the primer-binding sites, preventing PCR amplification. cDNAs containing a trapped exon are left intact, and will amplify in the PCR. (B) pSPL3-IV multiple cloning site (that of pSPL3-VI is in the reverse orientation).

trapping products (ETPs) derived from the yeast or bacterial genomes. While these clones are now easily identified via sequence analysis, it is likely that they will lower the production of desired ETPs.

Before the initiation of exon amplification, the complexity of the source DNA for each assay should be considered in the context of the experimental goal. Previous studies have shown that complex input DNA can be utilized to identify ETPs for a given region.[30] However, as the complexity of the input DNA increases, the yield of ETPs decreases. For example, experiments using shotgun subclone libraries of single cosmids (35–40 kb) yielded an average of approximately 3 exons per assay, or 1 ETP for every 12–14 kb. However, simultaneous analysis of 385–440 kb (11 cosmids in a single pool) yielded approximately 16 ETPs, or 1 ETP per 24–28 kb. Finally, analysis of pools of 1000 constructs generated directly from total genomic DNA (~3 Mb) yielded approximately 40 ETPs per pool, or an ETP for every 75 kb of target DNA. Although a significant decrease in the recovery of exons occurs as DNA complexity increases, this is compensated for by the advantage of analyzing significantly more DNA in a single experiment. The reasons for the decrease in exon recovery as the DNA sequence complexity increases are unclear. It is likely that there is competition at several stages in the process. For example, preferential growth of certain clones in the construct pools may lead to some representational biases. Preferential splicing of certain genomic splice sites that are more compatible with *tat* splice sites may play a role. In addition, differential amplification may occur in the RT-PCR. It should be noted that, in the results described above, although a larger number of ETPs were analyzed as the complexity increased, this number was not proportional to the increase in target DNA complexity. Therefore, it is entirely possible that analysis of a larger number of ETP clones would increase the yield in experiments using more complex DNA. Thus, if the goal of an exon amplification experiment is to isolate all possible ETPs from a given genomic region, one should perform assays with low complexity DNA, perhaps constructing multiple libraries using different restriction endonucleases. Alternatively, a greater number of ETP clones may be required for analysis when the complexity of genomic DNA is high.

Methodology

Vector Construction

A diagram of the vector pSPL3 is shown in Fig. 3A. The complete sequence of this vector has been deposited into GenBank (accession number XXU19867). Derivation of the vector is described briefly above, and in greater detail elsewhere.[25,30] A multiple cloning site has been introduced into the intron and contains six unique enzyme restriction endonuclease sites and is flanked by *Bst*XI sites (Fig. 3B). pSPL3-IV and pSPL3-VI contain the multiple cloning site in the *Eco*RI–*Eco*RV and *Eco*RV–*Eco*RI orientations, respectively (these plasmids, as well as pSPL3B,[35] are available from Life Technologies, Gaithersburg, MD). It should be noted that use of *Eco*RI or *Eco*RV will destroy one of the two flanking *Bst*XI sites. Constructs generated with *Eco*RI require the use of pSPL3-IV, and *Eco*RV requires the use of pSPL3-VI.

Protocol

Cloning into pSPL3. In general, most experiments will involve constructing shotgun subclone libraries in pSPL3 to allow for screening of complex DNA targets. Of course, this does not preclude the use of exon amplification in screening individual genomic fragments as single constructs. Genomic DNA is digested with a restriction endonuclease that produces ends compatible with sites within the pSPL3 polylinker. The resulting digestion fragments are prepared for ligation by inactivation or removal of the restriction enzyme. In a 20-μl reaction and under standard conditions,[37] 50–150 ng of the digested DNA is ligated to 50 ng of digested and dephosphorylated pSPL3. The resulting ligation is transformed into *Escherichia coli* HB101. We have found that the use of *E. coli* HB101 for propagation of pSPL3 and subclones in pSPL3 ensures stability of the construct. In our hands, propagation of pSPL3 in certain other *E. coli* strains (such as JM83 or DH5α) has led to instability of the vector. For individual cosmids, one-tenth of the transformation is plated to test for cloning efficiency, as compared with a self-ligation of the vector alone. The remainder of the transformation is inoculated into 10 ml of Luria-Bertani (LB) agar containing ampicillin (50 μg/ml), and incubated at 37° overnight in a shaking incubator. When shotgun cloning cosmid pools, BACs, PACs, or genomic DNA, the entire transformation is plated on 150-mm dishes of LB agar containing ampicillin (50 μg/ml) in parallel with a vector self-ligation control. This is done to minimize competition for nutrients that would result in differential growth (and, thus, unequal representation in the plasmid pool) of genomic subclones in liquid culture. The resultant colonies are subsequently pooled by flooding the plates with 10 ml of LB and resuspending the colonies with a plate spreader. Plasmid DNA is then isolated using standard methods; we typically use the alkaline lysis method.

[37] T. Maniatis, E. F. Fritsch, and J. Sambrook, "Molecular Cloning: A Laboratory Manual." Cold Spring Harbor Laboratory Press, Cold Spring Harbor, New York, 1989.

To ensure that a complex library has been constructed in pSPL3, a diagnostic digest can be performed. *Pvu*II restriction endonuclease sites flank the insertion site and produce an approximately 2-kb fragment on digestion, which will vary in length after insertion of a genomic fragment. An invariant fragment of approximately 4 kb will also be generated on digestion. Typically, one-tenth of the resulting plasmid DNA preparation is digested with *Pvu*II, size fractionated by electrophoresis in a 1% agarose gel [Tris–acetate–EDTA (TAE) buffer], and visualized by staining the gel with ethidium bromide. Cloning efficiency and library complexity can be ascertained by the diminished intensity of the 2-kb band relative to the 4-kb band. In a robust shotgun cloning, the 2-kb fragment is barely visible; this will minimize the yield of pSPL3-derived artifacts.

Electroporation into COS7 Cells. DNA from the resulting construct pools is then transfected into COS7 cells. The COS7 cell line was derived from the African green monkey kidney cell line, CV-1, by transformation with an origin-defective mutant of SV40 that codes for wild-type T antigen.[38] Thus, high levels of transcription can be driven from the SV40 early region promoter located in pSPL3 upstream of the reporter gene.

COS7 cells are grown in Dulbecco's modified Eagle's medium (DMEM) supplemented with 10% heat-inactivated fetal calf serum (500 ml at 56° for 40 min). In preparation for transfection, COS7 cells are grown to 75–85% confluency, trypsinized, collected by centrifugation, and washed in ice-cold phosphate-buffered saline (PBS) in the absence of divalent cations. The cells (4×10^6) are then resuspended in 0.7 ml of cold PBS and combined in a precooled electroporation cuvette (0.4-cm chamber; Bio-Rad, Hercules, CA) with 0.1 ml of PBS containing 1–15 μg of the plasmid pool DNA. After equilibrating on ice for 10 min, the cells are gently resuspended, electroporated [1.2 kV (3 kV/cm), 25 μF] in a Bio-Rad Gene Pulser, and immediately returned to ice. After recovering for 10 min the cells are transferred to a tissue culture dish (100 mm) containing 10 ml of prewarmed, preequilibrated culture medium. As an alternative to electroporation, COS7 cells are readily transfected by lipofection.[39]

RNA Isolation. RNA is isolated 48–72 hr posttransfection. Multiple standard RNA isolation procedures will suffice, but the method that we routinely use is a simple cytoplasmic RNA extraction (all solutions are RNase free). The cultures, still attached to the dish, are washed three times with ice-cold PBS (no divalent cations), and placed on a bed of ice. Ice-cold PBS (10 ml) is added to each plate, and the cells are removed with a

[38] Y. Gluzman, *Cell* **23,** 175 (1981).

[39] P. L. Felgner, T. R. Gadek, M. Holm, R. Roman, H. W. Chan, M. Wenz, J. P. Northrop, G. M. Ringold, and M. Danielsen, *Proc. Natl. Acad. Sci. U.S.A.* **84,** 7413 (1987).

cell scraper. The suspension is then transferred to a 15-ml conical tube and placed in ice. The cells are then pelleted by centrifugation [1200 rpm in a Sorvall (Norwalk, CT) RT6000B] at 4° for 7–8 min. The supernatant is removed completely and the cells are swelled by resuspension in 300 μl of a hypotonic solution of 10 m*M* Tris-HCl (pH 7.5), 10 m*M* KCl, and 1 m*M* $MgCl_2$, and incubate on ice for 5 min. Triton X-100 (15 μl of a 10% solution) is then gently mixed into the solution and incubation is continued for 5 min. This causes the cytoplasmic membrane to rupture, but leaves the nucleus intact. Nuclei and other cellular debris are sedimented by centrifugation (1500 rpm) at 4° for 7–8 min. The cytosolic supernatant is removed to a microcentrifuge tube that has been prechilled on ice. Twenty microliters of 5% sodium dodecyl sulfate and 300 μl of Tris-saturated phenol are then added, and the tube is vortexed vigorously. Phase separation is achieved by centrifugation for 5 min at 4° in a microcentrifuge. The aqueous phase is removed to a fresh microcentrifuge tube containing 300 μl of phenol–chloroform (1:1), and vortexed vigorously. Phase separation is achieved by centrifugation for 5 min at room temperature in a microcentrifuge. The aqueous phase is then removed to a prechilled microcentrifuge tube containing 12 μl of 5 *M* NaCl, 750 μl of ethanol is added, mixed, and the RNA is precipitated at −20° for ≥3 hr (or at −80° for 30 min). The RNA is sedimented by microcentrifugation at 4° for 30 min, and the resulting pellet is washed with 70% ethanol, air dried, and resuspended in 20 μl of H_2O.

Reverse Transcriptase-Polymerase Chain Reaction and Cloning. Although the overall number will depend on the complexity of the target genomic DNA, it is likely that multiple trapped sequences exist in each RNA sample, flanked by known exon sequences originating from the pSPL3 reporter gene. These known sequences are the basis for cloning the novel ETPs. RT-PCR amplification is performed using nested oligodeoxynucleotide primer pairs that correspond to the pSPL3-derived exon sequences. Subsequent cloning of the amplicons can be performed in multiple ways, and the method used may influence the choice of an internal primer pair. The products can be ligated to a T-A cloning vector,[40,41] a standard method for cloning products of PCR. Alternatively, the RT-PCR products can be digested with *Sal*I and *Bgl*II, for which restriction sites have been engineered into pSPL3 exons, and cloned into commonly used plasmids that contain compatible sites (e.g., the *Sal*I and *Bam*HI sites of pBluescript vectors).[30] These two methods, however, have certain limitations. In our experience,

[40] D. Marchuk, M. Drumm, A. Saulino, and F. S. Collins, *Nucleic Acids Res.* **19,** 1154 (1991).
[41] D. A. Mead, N. K. Pey, C. Herrnstadt, R. A. Marcil, and L. M. Smith, *Biotechnology* **9,** 657 (1991).

T-A cloning is inefficient and results in a high frequency of nonrecombinant clones. Cloning via restriction digestion is also problematic in that it requires multiple post-PCR steps, and RT-PCR product coligation occurs and results in a significant frequency of clone chimerism.[34]

We found that the problems of low efficiency cloning, post-PCR manipulations, and clone chimerism are essentially eliminated when the preceding methods were replaced by ligation-independent cloning using uracil DNA glycosylase (UDG).[42] This method entails the utilization of UDG to create long complementary single-stranded ends for both an insert and the cloning vector. This is achieved by synthesizing oligodeoxynucleotide primers for amplification of both vector and insert DNA. These primers are designed such that the 5′ end sequence of each vector primer is complementary to one of the insert primers, and the regions of complementarity for all primers contain dU residues rather than dT. After PCR amplification of the insert and the vector, the dU residues can be selectively removed from the products by digestion with UDG. This results in a fragmentation of the dU-containing region, and a destabilization of the remaining nucleotide hybrid that leaves the ends as single strands. The single-stranded vector and insert ends, which are complementary, can then form stable hybrids that can be transformed into an appropriate bacterial host. Nicks and gaps in the hybrid are then repaired by the host DNA repair enzymes.

Four oligodeoxynucleotide primers have been designed for cDNA synthesis and subsequent PCR amplification.[34] The sequences of these primers are as follows:

cDNA synthesis (first and second strand, respectively):
SA2: 5′-ATCTCAGTGGTATTTGTGAGC-3′
SD6: 5′-TCTGAGTCACCTGGACAACC-3′
PCR amplification of double-stranded cDNA for UDG cloning:
SDDU: 5′-AUAAGCUUGAUCUCACAAGCTGCACGCTCTAG-3′
SADU: 5′-UUCGAGUAGUACUTTCTATTCCTTCGGGCCTGT-3′

Complementary primers have been designed for the UDG-based cloning into the vector pBluescriptIIKS+ (Stratagene, La Jolla, CA), surrounding the *Eco*RV site. These are as follows:

BSDU: 5′-GAUCAAGCUUAUCGATACCGTCGACC-3′
BSAU: 5′-AGUACUACUCGAAUTCCTGCAGCC-3′

As an alternative to these primers designed specifically for cloning ETPs, a universal cloning system has been designed for UDG (Life Technologies).

[42] A. Rashtchian, G. W. Buchman, D. M. Schuster, and M. S. Berhinger, *Anal. Biochem.* **206,** 91 (1991).

For first-strand cDNA synthesis, approximately 2.5–5.0 μg of cytoplasmic RNA is added to a solution containing 1 × PCR buffer [1 × PCR buffer is 10 mM Tris-HCl (pH 8.3), 50 mM KCl, 1.5 mM $MgCl_2$, 0.001% gelatin], 4 mM dithiotreitol, 200 μM dNTPs, and 1 μM SA2. This mixture is heated to 65° for 5 min, then placed at 42°, followed by addition of an RNase inhibitor (RNasin, 3.5 U; Promega, Madison, WI) and mouse mammary leukemia virus (MMLV) reverse transcriptase (200 U; Life Technologies). The reaction (25 μl) is allowed to proceed at 42° for 90–120 min.

The entire first-strand cDNA is then converted to double-stranded cDNA using a limited number of PCR amplification cycles. The 25-μl first-strand reaction is added to 75 μl of a reaction mixture containing 1 × PCR buffer, 200 μM dNTPs, 1 μM SA2, 1.33 μM SD6, and 2.5 U of *Taq* polymerase. After the initial heat denaturation step (94°, 5 min), five or six amplification cycles are performed using the following conditions: 94° for 1 min, 62° for 1 min, 72° for 5 min. A 5-min polymerization time is used in an attempt to minimize bias toward smaller ETPs. To eliminate the main class of artifactual and vector-only spliced RNAs, 50 U of *Bst*XI is added to the amplification mixture and incubated overnight at 55°. To ensure complete digestion, the reaction should be supplemented the following morning with an additional 10–20 U of *Bst*XI and allowed to proceed for another 2 hr. A nested amplification of 10 μl of this digested DNA is carried out in a 100-μl reaction containing 1 × PCR buffer, 200 μM dNTPs, 1 μM SADU, 1 μM SDDU, and 2.5 U of *Taq* polymerase. After an initial heat denaturation step (94° for 1 min) 30 cycles are performed using the following conditions: 94° for 30 sec, 62° for 30 sec, 72° for 3 min.

The RT-PCR products (5–10 μl) are visualized by electrophoresis in 2% agarose gels [Tris–borate–EDTA (TBE) buffer] and staining with ethidium bromide. The RT-PCR products should appear as multiple discrete bands, or as a "smear," ranging in size from approximately 100 to several hundred base pairs, and the number of bands will be dependent on the complexity of the starting genomic DNA. The amplification products are directionally cloned into pBluescriptIIKS+ that has been prepared for ligation-independent cloning using UDG.[42] *Eco*RV-digested pBluescriptIIKS+ (10 ng) is PCR amplified with BSDU and BSAU, using the following reaction conditions: 1 × PCR buffer, 200 μM dNTPs, 1 μM BSDU, 1 μM BSAU, and 2.5 U of *Taq* polymerase. After an initial heat denaturation step (94° for 1 min), 25 amplification cycles are performed using the following conditions: 94° for 30 sec, 62° for 30 sec, 72° for 3 min. For cloning, 50 ng of the amplified plasmid is mixed with 50–100 ng of RT-PCR product (typically 2–3 μl), and the mixture is digested with 1 U of UDG (Life Technologies) for 30 min at 37° in 1 × PCR buffer. The total digestion/annealing reaction volume is 10 μl. The digested and annealed products are immediately transformed into *E. coli* DH5α or other appropriate host. It is important that the digestion/annealing reaction be immediately transformed after removal from 37°, because improper annealing of the long single-stranded ends may occur at lower temperatures and result in cloning artifacts. Clones from each sample can then be picked, propagated, frozen, and stored in 96- or 384-well microtiter plates.

Analysis of Exon-Trapping Products

Characterization of the cloned RT-PCR products is straightforward. The first step toward examining ETPs is sequence analysis, which can reveal functional and similarity information, as well as identify artifacts of the method. In addition to searches for matches or similarities to well-characterized genes, comparisons can be made with the several hundred thousand ESTs that reside in public databases. Identity with any of these sequences can provide rapid access to a cDNA corresponding to an ETP. However, it should be noted that the cloning and sequencing biases of the methods used to produce ETPs and ESTs may result in a low probability of obtaining overlapping sequences. ETPs are derived solely from internal exons, owing to the requirement of both 5′ and 3′ splice sites for exon amplification. Most ESTs within the databases are derived from 3′ enriched, oligo(dT)-primed cDNA libraries, and thus the percentage of ETPs identifying ESTs might not be great. Indeed, we have found that currently only 15–20% of ETP sequences match a portion of an EST.[34] Finally, database comparisons can reveal obvious artifacts of exon amplification. With pSPL3, we have found that approximately 15% of all ETPs are readily identifiable artifacts that can be eliminated from further analysis by database comparisons. These artifacts include pSPL3 sequences, known repetitive elements, *E. coli* sequences, and genomic clone vector sequences.[28,30,34]

For positional cloning studies, it is also important to localize the exons back to the region of interest. Mapping each ETP to the appropriate clone-based map or chromosomal region, either by PCR or hybridization, will allow for ordering of the ETPs with respect to each other and for refinement of the clone map. This may lead to an indication of which ETPs are derived from the same gene, and subsequently reduce the effort needed to obtain full-length representations of each gene corresponding to an ETP. Finally, this analysis may also help identify artifactual ETPs that are less obvious. For example, presently uncharacterized repetitive elements have been isolated using exon amplification.[28,43] The frequency with which these artifacts occur is low, but may depend on the genomic region being studied. In

[43] D. M. Church, J. Yang, M. Bocian, R. Shiang, and J. J. Wasmuth, *Genome Res.* **7,** 787 (1997).

larger scale exon amplification studies (e.g., whole chromosomes[34]), the single-copy nature of most ETPs makes them useful as probes in map construction.

ETPs are also amenable to mRNA expression analyses. Knowledge of tissue expression is quite useful in prioritizing clones for more detailed analysis in positional cloning. ETPs have been employed in numerous Northern blot experiments, but their utility as probes may be limited owing to their relatively short lengths (average length of 135 bp). ETPs have also been successfully used in RNase protection and RNA *in situ* hybridization studies.[44,45] Probes are readily developed from the ETPs, as they are directionally cloned into a vector that possesses bacteriophage promoter sequences flanking the cloning site. Also, oligodeoxynucleotide primers can be designed from the ETP sequence and can be utilized to analyze expression by RT-PCR.

Once the sequence, position, and/or expression analyses have been performed, interesting ETPs can be used as starting points for full-length mRNA (cDNA) cloning. ETPs can be used singly or in pools to screen cDNA libraries, using conventional methods.[8] Alternatively, the sequences of an ETP can be used to design oligodeoxynucleotide primers for 5′ and 3′ rapid amplification of cDNA ends (RACE),[46,47] for PCR-based cDNA library screening,[48] or for screening using selection methods such as GeneTrapper (Life Technologies).[49]

[44] D. M. Church, A. Carey, and A. J. Buckler, unpublished results (1997).

[45] S. Chu, K. R. Long, A. J. Buckler, and M. P. Duyao, unpublished results (1997).

[46] M. A. Frohman, M. K. Dush, and G. R. Martin, *Proc. Natl. Acad. Sci. U.S.A.* **85,** 8998 (1988).

[47] M. A. Frohman, *Methods Enzymol.* **218,** 340 (1993).

[48] D. J. Munroe, R. Loebbert, E. Bric, T. Whitton, D. Prawitt, D. Vu, A. Buckler, A. Winterpracht, B. Zabel, and D. E. Housman, *Proc. Natl. Acad. Sci. U.S.A.* **92,** 2209 (1995).

[49] T. D. Connors, T. C. Burn, J. M. Millholland, T. J. Van Raay, G. M. Landes, Q. Wang, J. Shen, J. Splawski, M. E. Curran, and M. T. Keating, *BRL Focus* **18,** 31 (1996).

[7] Cosmid-Based Exon Trapping

By JOHAN T. DEN DUNNEN

Introduction

Rationale

The number of mammalin genes that are currently identified is increasing at an incredible rate. Most of these sequences are originating from large-scale sequencing efforts of expressed sequences, cDNAs or expressed sequence tages (ESTs). It is unclear, however, what proportion of genes will be missed by such efforts. Up to 10–20% of all genes might be expressed at low levels or only during brief stages of development, making them difficult to track on the basis of their expression. Consequently, to complete the human gene catalog, efficient methods will be required to detect transcribed sequences independent of expression. We have established a system that provides such a possibility, i.e., large-insert (here cosmid based), multiple-exon trapping.[1]

Exon trapping is a widely used gene identification method.[1–4] In contrast to RNA-based methods, such as screening of cDNA libraries and cDNA selection, exon trapping is independent of gene expression. Exon trapping provides a functional assay in which cloned DNA sequences are selectively isolated on the basis of the presence of functional splice sites. Consequently, sequences are isolated directly from the clone under analysis without knowledge or availability of tissues expressing the unknown gene.

Large-insert exon trapping, here cosmid based (Fig. 1), was developed to bypass problems related to plasmid-based exon trapping (see [6] in this volume[5]). In the latter system, the genomic region of interest is subcloned in a special vector, yielding 1- to 2-kb inserts. Owing to the small insert size, it is unlikely that multiple exons will be cloned together and, consequently, most exon trap clones contain single exons, which are often small. Many putative exons are isolated from one clone and the intrinsic information of "order" is lost in the course of the trapping experiment. The single

[1] N. A. Datson, E. Van De Vosse, J. G. Dauwerse, M. Bout, G. J. B. Van Ommen, and J. T. Den Dunnen, *Nucleic Acids Res.* **24,** 1105 (1996).

[2] D. Auch and M. Reth, *Nucleic Acids Res.* **18,** 6743 (1990).

[3] G. M. Duyk, S. Kim, R. M. Myers, and D. R. Cox, *Proc. Natl. Acad. Sci. U.S.A.* **87,** 8995 (1990).

[4] D. M. Church, C. J. Stotler, J. L. Rutter, J. R. Murrell, J. A. Trofatter, and A. J. Buckler, *Nature Genet.* **6,** 98 (1994).

[5] D. M. Church and A. J. Buckler, *Methods Enzymol.* **303,** [6], 1999 (this volume).

Copyright © 1999 by Academic Press
All rights of reproduction in any form reserved.

0076-6879/99 $30.00

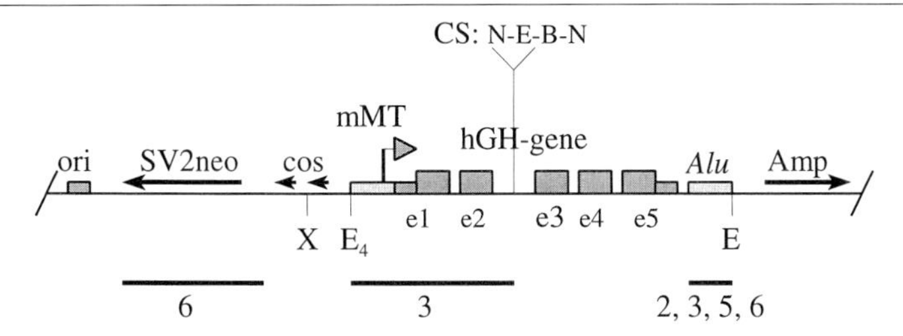

FIG. 1. The sCOGH-exon trap vectors. The basic figure shows sCOGH1; the thicker bars below represent the segments that are missing in the specific vectors, indicated by their number [e.g., 2 for the part missing in sCOGH2, i.e., the Alu repeat (*Alu*) directly downstream of the hGH gene]. Amp, Ampicillin resistance gene; cos, bacteriophage λ-derived cos sites; CS, *Bam*HI (B) cloning site flanked by *Not*I (N) sites; E, *Eco*RI site; e1–e5, exons 1 to 5 of the human growth hormone (hGH) gene; mMT, mouse metallothionein promoter; ori, origin of replication for propagation in *E. coli*; SV2neo, eukaryotic neomycin resistance marker; X, *Xba*I site (used for linearization of the vector).

trapped exons are usually used as probes (owing to their small size yielding weak signals) for cDNA library screening, losing the initial advantage of working with a system that is independent of expression. Finally, because large-insert exon trapping conserves the genetic context, i.e., the overall structural exon/intron information of the genes cloned, the background of false-positive clones, e.g., trapped purely intronic sequences, is reduced considerably.

Two consequences of large-insert exon trapping deserve special attention. First, when (part of) a gene is present in the insert, it can be expected that all exons are trapped in one product (Fig. 2). Consequently, reverse transcription-polymerase chain reaction (RT-PCR) amplification of the trapped products requires great care to ensure amplification of large products (>4 kb). Together the exons may span several kilobases, especially when 3′-terminal exons are present. Second, inserts may contain one or more 5′-first and/or 3′-terminal exons (Fig. 2) and it cannot be predicted how the splicing machinery recognizes all of these elements. 5′-First exons are probably not recognized because the promoter will be inactive in the cell line used for transformation. Transcription thus probably runs over 5′-first exons and they are skipped by splicing (Fig. 2B and D). However, 3′-terminal exons are probably recognized and produce a transcriptional stop, although the formation of a "fusion gene product," splice skipping the 3′-terminal exon, cannot be ruled out (Fig. 2D). In such cases, a 3′-RACE (rapid amplification of cDNA ends) protocol will be required to trap the exons between a 5′ vector-derived exon and the insert-derived 3′-terminal

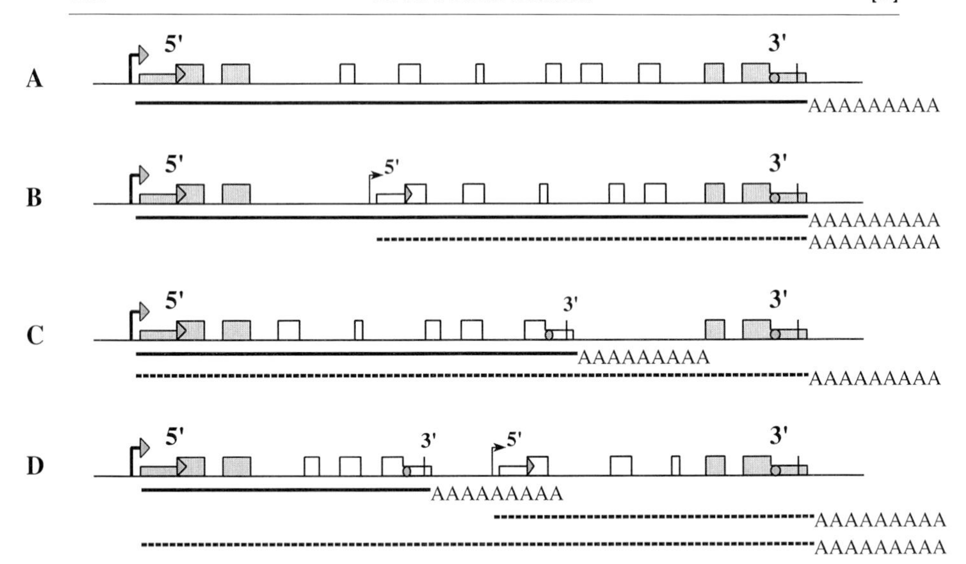

FIG. 2. Exon trapping with large-insert vectors. Example of cloned inserts (in white), flanked by the vector-derived gene (gray), containing (**A**) internal exons only, (**B**) a 5′-first and internal exons, (**C**) internal exons and a 3′-terminal exon, or (**D**) internal, a 5′-first, and a 3′-terminal exon. The transcripts that are most likely produced are drawn as a thick line. Transcripts that are probably not made, either because the promoter driving transcription of the 5′-first exon is likely to be inactive in the cell type used or because transcription terminates at a 3′-terminal exon, are drawn as broken lines. Large boxes represent internal exons and small boxes the 5′ or 3′ untranslated regions of the 5′-first and 3′-terminal exons, respectively. The prominent arrow represents a promoter; the As represent the poly(A) tail of a transcript.

exon.[6,7] Gene segments downstream will then be missed unless the promotor of the cloned gene is active and a 5′-RACE protocol is used for amplification.

Principle of Cosmid-Based Exon Trapping

The sCOGH cosmid vectors (Fig. 1) use a mouse metallothionein (mMT) promoter to drive expression of a human growth hormone gene (hGH) with a multiple cloning site in intron 2 (Fig. 2). The sCOGH vectors allow cloning and trapping of exons from inserts of up to 35 kb in one experiment. On transfer of the cloned DNA to mammalian cells, e.g., using electroporation, the hGH gene with insert is transiently transcribed from

[6] N. A. Datson, G. M. Duyk, G. J. B. Van Ommen, and J. T. Den Dunnen, *Nucleic Acids Res.* **22,** 4148 (1994).

[7] D. B. Krizman and S. M. Berget, *Nucleic Acids Res.* **21,** 5198 (1993).

the mMT promoter. After culturing for 48 hr, RNA is isolated and hGH-derived transcripts are amplified using reverse transcription and PCR (RT-PCR). If exonic sequences are present in the insert DNA they will be trapped between the insert-flanking hGH exons 2 and 3. On agarose gel electrophoresis, the trapped sequences will be detected as PCR products with increased sizes when compared with the products derived from the vector, i.e., hGH exon 2 spliced to exon 3.

The sCOGH system consists of several different vectors.[1] sCOGH2 is used for standard exon trapping. sCOGH5, missing the 3′ end of the hGH gene (exons 3, 4, and 5), was specifically designed to trap 3′-terminal exons although such exons can probably also be isolated using sCOGH2, applying a 3′-RACE protocol (see Fig. 2). sCOGH3 misses the mMT promoter and hGH exons 1 and 2 and can be used to trap 5′-first exons. To ensure trapping of such exons, sCOGH3 clones should be transformed into cells that express the gene to be trapped, because in these cells the promoter will be active and able to generate RNA from the cloned insert. Thus, sCOGH3 should facilitate the targeted isolation of tissue-specific genes.

The sCOGH system facilitates the scanning of large genomic regions, in one instance, for the presence of exons. Consequently, it would be advantageous to use sCOGH vectors at the stage of constructing contiguous sets of clones (contigs) in positional cloning projects because this facilitates a direct transition to the stage of gene identification. Cosmids can be generated, e.g., by subcloning a yeast artificial chromosome (YAC)- or P1 phage-derived artificial chromosome (PAC)-contig or by constructing a total genomic cosmid library and identifying the clones from the region of interest.

Protocols

Cosmid Cloning and DNA Isolation

The sCOGH1 vectors are propagated in *Escherichia coli*, e.g., strain 1046 or HB101. Construction of libraries in the sCOGH vectors uses standard cosmid cloning protocols.[8] sCOGH vector DNA is prepared for ligation by linearization with *Xba*I, dephosphorylation, and subsequent digestion with *Bam*HI. Input DNA (e.g., genomic DNA or YAC DNA–agarose plugs) is partially digested with *Mbo*I (size fractionation is optional) and ligated into the *Bam*HI site of the sCOGH vector. The ligated material is packaged using Gigapack II Plus packaging extract (Stratagene, La Jolla, CA) and used to transform *E. coli* 1046.

[8] T. Maniatis, E. F. Fritsch, and J. Sambrook, "Molecular Cloning: A Laboratory Manual." Cold Spring Harbor Laboratory Press, Cold Spring Harbor, New York, 1989.

The sCOGH vectors contain *Not*I sites flanking the cloning site. This facilitates the isolation of the insert and the construction of clones in which the orientation of the insert can be simply reversed if the insert is devoid of internal *Not*I sites (see Fig. 3). After *Not*I digestion, ligation (recircularization), and packaging, clones will be obtained with the insert in both possible orientations. The packaging step is essential to avoid obtaining clones that contain circularized cosmid vector only.

Preparing Cells for Transformation

Exon trapping with the sCOGH system can be performed using any mammalian cell line that grows rapidly and that can be transformed efficiently using cosmid-sized DNA. To facilitate RT-PCR amplification of sCOGH-derived exon trap products we have used rodent cell lines. These cells enable the design of human-specific growth hormone primers that will not amplify endogenous transcripts (see below), thereby reducing undesired background. We have worked most frequently with V79 Chinese hamster lung cells, but rat and murine cells also give acceptable results.

Cell culturing should start about 1 week before the transformation step of the exon trap procedure will be performed. The exact preparation time depends on the growth rate of the cells and the number of cells required for transformation. Always start with a fresh inoculate derived from a batch of cells stored in liquid nitrogen. In our hands, culturing cells longer than 3 weeks before transformation will have a negative effect on exon trap efficiency. During growth, take care that the cells are nicely spread and

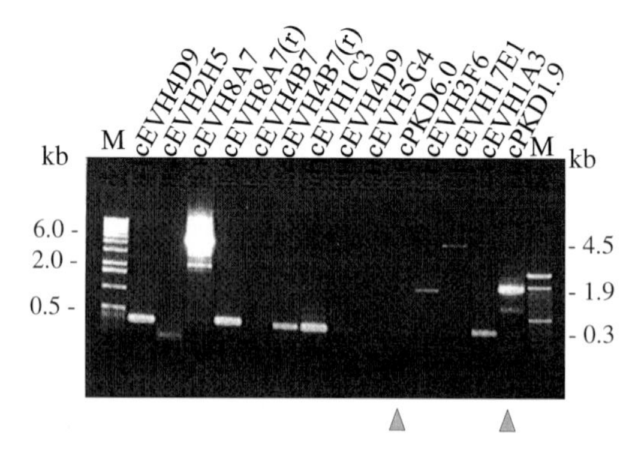

FIG. 3. Agarose gel analysis of internal exon trap products after RT-PCR analysis. The trapped products vary from 0.2 kb (the empty hGH exon 2/3 splice product) to 4.5 kb. For clones cEVH8A7 and cEVH4B7 inserts were also tested in the opposite transcriptional orientation (r). Sizes (in kilobase pairs) are indicated on the left and right. M, Molecular size marker.

that the flasks are not too full. Usually, we split one 650-ml flask into three new flasks. For 10–15 transformations, we use nine to twelve 650-ml flasks.

Harvesting Cells for Transformation

V79 cells are cultured in DMEM [Dulbecco's modified Eagle's medium with GlutaMAX 1, 25 m*M* HEPES, D-glucose (4500 mg/liter), and without sodium pyruvate (GIBCO-BRL, Gaithersburg, MD)] with 10% inactivated fetal calf serum (FCS, mycoplasma and virus screened; GIBCO-BRL).

1. Carefully remove the culture medium.
2. Wash the flask with 15 ml of Earle's balanced salt solution without calcium, magnesium, and phenol red (EBSS; GIBCO-BRL).
3. Remove the EBSS and add 3 ml of trypsin [5 ml of 10× trypsin–EDTA (GIBCO-BRL) in 45 ml of EBSS]. Incubate the flask for 2 min at 37°. Some cell types, e.g., V79 cells, tend to stick to each other. To prevent this, carefullly pipette the cells up and down a few times, using a 2-ml pipette.
4. Add 10 ml of DMEM$^+$ [500 ml of DMEM supplemented with 50 ml of inactivated FCS, 10 ml of penicillin–streptomycin (10× stock: 10,000 units of penicillin per milliliter and 10,000 μg of streptomycin per milliliter; GIBCO-BRL)] and collect the cells in a 50-ml tube.
5. Centrifuge the cells for 5 min at 1200 rpm, at room temperature.
6. Remove the DMEM$^+$ and resuspend the cells in 5 ml of ice-cold phosphate-buffered saline (Dulbecco's PBS without calcium, magnesium, and sodium bicarbonate; GIBCO-BRL).
7. Fill the tube up to 50 ml with ice-cold PBS.

Note: From this point on, work on ice.

8. Count the cells.
9. Centrifuge the cells for 5 min at 1200 rpm at 4°; remove the PBS and store on ice.

Transformation (Electroporation)

Most protocols that yield DNA that can be digested with restriction enzymes, without recognizable degradation, can be used for transformation. Transformation can be performed using many protocols, e.g., lipofection, electroporation, and calcium phosphate precipitation. In general, the standard laboratory transformation protocol is most effective. However, a test is recommended to determine the transformation efficiency of the cosmid DNA in combination with the cell line used. The hGH gene facilitates testing of the transformation efficiency; 100 μl of cell culture medium may be assayed for hGH concentration, e.g., using the Allégro hGH transient gene assay kit (Nichols Institute, San Juan Capistrano, CA.[1] It should be noted, however, that using an empty sCOGH vector (measuring ~10 kb) to determine the optimal transformation protocol is probably not fully conclusive about the transformation efficiency obtained with an ~50-kb cosmid clone.

1. Fill the required number of 100-mm tissue culture dishes with 10 ml of DMEM$^+$. Prewarm the dishes by incubation at 37°.
2. Gently resuspend the cell suspension in PBS at 2×10^7 cells/ml.
3. Resuspend the ethanol-precipitated DNA in TE (10 m*M* Tris-HCl, 0.1 m*M* EDTA, pH 8.0) at a concentration of 0.5 μg/μl or higher (high DNA concentrations facilitate efficient transformation).
4. Transfer 20 μl of DNA solution, containing at least 10 μg of DNA, to a fresh tube.
5. Transfer 0.5 ml of cell suspension to the tube containing the DNA solution.
6. Mix carefully and transfer to a prechilled electroporation cuvette (0.4-cm chamber; Bio-Rad, Hercules, CA).
7. Set the Bio-Rad Gene Pulser to 300 V (750 V/cm) and 960 μF and perform the electroporation.
8. Place the cuvette for exactly 5 min. on ice.
9. Transfer the cells gently to a 100-mm tissue culture dish and incubate for 48–72 hr at 37°.

RNA Isolation

The quality of the RNA preparation is one of the most critical elements of the entire procedure, especially for the reverse transcription reaction.

1. Gently wash the tissue culture dishes twice with 5 ml of ice-cold PBS. Handle them carefully; the cells easily detach from the dishes.
2. Remove all PBS and add 2.5 ml of RNAzolB (Campro Scientific, Veenendaal, the Netherlands).
3. Allow the cells to lyse. This may take some time. If necessary, check lysis under a microscope.
4. Divide the solution into two vials (i.e., 1.25 ml each).
5. Add 125 μl of chloroform, shake vigorously for 15 sec, and leave on ice for 15 min.
6. Pellet the suspension by centrifugation at 14,000 rpm, 4°, for 15 min.
7. Transfer the upper, aqueous phase to a fresh tube and add an equal volume of 2-propanol. Incubate the tube for 15 min on ice.
8. Centrifuge the tube at 14,000 rpm, for 15 min.

9. Remove the supernatant and wash the pellet once with 1 ml of 75% ethanol (750 ml of pure ethanol with 250 ml of H_2O) by vortexing.
10. Centrifuge the tube at 14,000 rpm, 4°, for 10 min.
11. Remove all supernatant and dissolve the pellet in 50 μl of MilliQ/DEPC [DEPC treatment: add 0.05–0.1% (v/v) diethyl pyrocarbonate (Sigma, St. Louis, MO), leave overnight, and autoclave]. Incubate for 10 min at 65°.
12. Take 1 μl and analyze the RNA on a 2% TBE agarose gel [2 g of agarose (SeaKem LE; FMC, Philadelphia, PA) in 100 ml of 90 m*M* Tris–90 m*M* boric acid–1 m*M* EDTA (pH 8.3)]. Use autoclaved RNase-free solutions and a clean electrophoresis tank.

RT-PCR analysis can be performed using total RNA but, especially when long products must be generated, we have obtained improved results using poly(A)$^+$ RNA. For poly(A)$^+$ RNA isolation, several excellent techniques/kits are available. We have applied mostly the magnetic mRNA isolation kit (HT Biotechnology, Cambridge, United Kingdom), using the manufacturer's protocol.

13. Add again 50 μl of MilliQ/DEPC, heat the RNA samples for 2 min at 65° (disrupting secondary structures), and put directly on ice.
14. Wash the magnetic beads two or three times with BE buffer (binding/elution buffer: PBS with 1.0 *M* NaCl, RNase free).
15. Resuspend the beads in BE2 buffer (BE buffer with 0.1% Tween 20, RNase free).
16. Add 100 μl of magnetic beads solution to the tubes containing the RNA samples. Incubate at 18–25° for at least 10 min. Gently mix the beads during incubation.
17. Capture the beads, remove the supernatant, and wash twice with binding/elution buffer.
18. Wash the beads once or twice with twice-diluted binding/elution buffer.
19. Resuspend the beads in 20 μl of MilliQ/DEPC.
20. Elute the mRNA by heating to 65° for 10 min.

RNA quality and quantity can be checked on a gel (optional). In general, 2 μl of the final RNA sample is sufficient for one RT-PCR analysis.

Reverse Transcription-Polymerase Chain Reaction Amplification

Reverse transcription (RT) and amplification by polymerase chain reaction (PCR) is the most critical step of the entire procedure. The primers used during amplification determine whether internal or 3′-terminal exons are isolated. Furthermore, when (part of) a gene is present on the large, cosmid-sized inserts used, it can be expected that large exon-containing fragments need to be amplified, especially when 3′-terminal exons are present (see above). Special care must thus be taken during RT-PCR to facilitate the amplification of cDNA products in excess of 4 kb. The use of long-range PCR protocols is thus highly recommended.

Amplification of long RT-PCR products also critically depends on the absence of other, small target sequences, because these will be highly favored during the two-step amplification protocol applied. In this respect it is important that no endogenous growth hormone transcripts be amplified, because these would yield RT-PCR products of 0.2 kb. In the protocol we achieve this by using rodent cell lines, in which, owing to the many sequence differences, the human growth hormone primers find no endogenous target. Background is further reduced by the use of primers that are not able to amplify genomic or vector DNA; primers hGHex 1/2 and hGHex4/5 span two flanking exons and can bind only on correctly spliced RNA.

Another consequence of these considerations is that cosmids should not be mixed in an exon trap experiment, because this will favor amplification of the smallest, i.e., the least interesting, exon trap products. In this respect, note that 50% of the inserts will be in the wrong transcriptional orientation when the vector-derived hGH gene and the insert gene are compared. Consequently, in 50% of the clones the splicing machinery will not recognize any exons and an "empty" (only 0.2 kb; Fig. 3) hGH exon 2 to 3 splice product will be obtained.

In general, two separate amplifications are performed with each cosmid: one to trap internal exons and one to trap 3′-terminal exons (with coamplified upstream internal exons). Reverse transcription and first- and second-round PCR can be performed using standard laboratory protocols or according to the recommendations of the manufacturers of the RT, RT-PCR, and PCR systems. We have frequently used the Access RT-PCR system (Promega, Madison, WI), the Gene Amp XL RNA PCR kit (Perkin-Elmer, Norwalk, CT), and the Expand long template PCR system (Boehringer Mannheim, Indianapolis, IN). After reverse transcription, we perform an RNase A treatment, which significantly increases yields.

1. Use 2 μl of RNA sample per reaction.

2a. *Trapping internal exons:* Prime reverse transcription with primer hGHex4/5. After reverse transcription, degrade the RNA using RNase A. The first-round PCR is performed by adding primer hGHex1/2. The nested PCR is performed with primers hGHa and hGHb.

2b. 3′ *Exon trapping:* Prime reverse transcription with primer PolyT-REP/5. After reverse transcription, degrade the RNA using RNase

A. The first-round PCR is performed by adding primer hGHex1/2. The nested PCR is performed with primers hGHa and REP.

3. PCR cycling conditions are as follows: one cycle of 5 min at 94°; 30 cycles of 1 min at 94°, 1 min at 60°, and 2 min at 72°; and one cycle of 10 min at 72°. Products are stored at 4°.

When RT-PCR products are to be used for *in vitro* transcription/translation (see below), three independent nested PCRs are performed, exchanging primer hGHa for primers hGHaORF1, hGHaORF2, and hGHaORF3, respectively.

Primers

All of the following primers are 5′ to 3′:
- hGHa: [T7]-CGTCTGCACCAGCTGGCCTTTGAC
- hGHex1/2: ATGGCTACAGGCTCCCGGA
- hGHaORF1: [TT]-CAGCTGGCCTTTGACACCTACCAGGAG
- hGHaORF2: [TT]-GCAGCTGGCCTTT_C_ACACCTACCAGGAG (*Note:* _C_ differs from hGH sequence to provide an open reading frame)
- GHaORF3: [TT]-GGCAGCTGGCCTTTGACACCTACCAGGAG
- hGHb: cgggatccCGTCTAGAGGGTTCTGCAGGAATGAATACTT (with *Bam*HI site facilitating cloning)
- hGHex4/5: TTCCAGCCTCCCCCATCAGCG
- PolyT-REP: GGATCCGTCGACATCGATGAATTC$(T)_{25}$
- REP: GGATCCGTCGACATCGATGAATTC
- T7-tail: CGGGATCCTAATACGACTCACTATAGG (with *Bam*HI site facilitating cloning)
- TT-tail: CGGGATCCTAATACGACTCACTATAGGACAGACCACCATG (containing a *Bam*HI site for cloning, a T7 RNA polymerase promoter, and a Kozak translation initiation sequence)

Analysis of Reverse Transcription-Polymerase Chain Reaction Products

RT-PCR products are analyzed on a 0.8% TBE–agarose gel. An example is shown in Fig. 3. The trapped products vary from 0.2 kb (the empty product) to 4.5 kb (clones cEVH8A7 and cEVH17E1). Control clones cPKD1.9 and cPKD6.0 contain inserts from the human PKD1 (polycystin) gene, which should yield 1.9- and 6.0-kb products, respectively. The 1.9-kb fragment is visible; amplification of the 6.0-kb fragment failed. RT-PCR using internal polycystin primers showed that the expected RNA was present, thus indicating that, in this experiment, RT-PCR failed to amplify products with sizes of 6 kb and larger. Clones 4B7, 4D9, and 5G4 gave no product, indicating that the trapped sequences might also be too large to ampify. Clones 4D9, 8A7r,

4B7r, 1C3, and 1A3 yield the 0.2-kb "empty" product, i.e., the vector-derived hGH exon 2 directly spliced to exon 3, indicating that the insert, in the orientation cloned, does not contain exonic sequences. The inserts of clones 8A7 and 4B7 were reversed (using *Not*I digestion; see above) yielding 8A7(r) and 4B7(r), respectively. Interestingly, one orientation revealed the "empty" product while the other orientation yielded a 4.5-kb (8A7) or a "too-large-to-amplify" product (4B7). Such a result would be the ideal from clones, containing exonic segments of one gene, when the exon trap system would be efficient and with low background.

Depending on the size and number of fragments obtained after RT-PCR, a number of further experiments can be performed (see also [6] in this volume[5]). Sequencing, followed by database searches, is the most direct way to obtain detailed information from the trapped sequences. Trapped products are putative cDNAs and can also be used as such, e.g., to probe cDNA libraries, Northern blots, or tissue sections (see [2] in this volume[9]). Because they provide an excellent marker by which to identify genuine coding gene sequences, we have designed an *in vitro* transcription/translation protocol to scan the trapped products for the presence of large open reading frames.[1] For this purpose, three second-round PCRs were performed using three alternative forward primers, one for each reading frame to test.

As soon as sequences become available from the ends of the trapped products, primers can be designed for further analysis. Two types of analysis are most obvious: amplification of genomic DNA to determine if the amplified segment is interrupted by introns, and RT-PCR amplification of RNA to verify expression and to determine the expression profile. If in these analyses patient-derived material is used, one can directly proceed to mutation detection without waiting until the entire sequence or gene structure has been determined. In this respect, the combination with *in vitro* transcription/translation is especially attractive as a possibility to scan quickly a large set of patient genomes for truncating mutations.[10]

Acknowledgments

Gert-Jan Van Ommen, Hans Dauwerse, Nicole Datson, Esther van de Vosse, and Paola van de Bent-Klootwijk are acknowledged for technical assistance and scientific input during development of the sCOGH exon trap system. The exon trap system was developed with financial support from the Netherlands Organization for Scientific Research (NWO) and the European Union (BIOMED-2 Grant PL962674).

[9] P. Carninci and Y. Hayashizaki *Methods Enzymol.* **303,** [2], 1999 (this volume).

[10] F. Petrij, R. H. Giles, J. G. Dauwerse, J. J. Saris, R. C. M. Hennekam, M. Masuno, N. Tommerup, G. J. B. Van Ommen, R. H. Goodman, D. J. M. Peters, and M. H. Breuning, *Nature* (*London*) **376,** 348 (1995).

[8] Direct cDNA Selection Using Large Genomic DNA Targets

By ANDREW D. SIMMONS and MICHAEL LOVETT

Introduction

One goal of the human genome project is to determine the DNA sequences of the estimated 50,000 to 100,000 human genes. This information will have relevance and utility for all those searching for the genetic basis of human diseases. An initiative that has accelerated gene discovery has been the large-scale random sequencing of expressed sequence tags (ESTs), in which academic and corporate groups have sequenced one or both ends of approximately 10^6 random cDNAs from conventional and normalized cDNA libraries.[1,2] These ESTs have proved useful in many areas of biology, including the identification of new members of gene families and the identification of evolutionary conserved functional motifs.[3–5] In human genetics and positional cloning projects, ESTs have been useful as positional candidate genes (when they have been mapped within the genome) and as one means for partially extending shorter cDNA fragments.[6–10] Unfortunately,

[1] D. Gerhold and C. T. Caskey, *Bioessays* **18,** 973 (1996).

[2] G. Lennon, C. Auffray, M. Polymeropoulos, and M. B. Soares, *Genomics* **33,** 151 (1996).

[3] M. Nangaku, S. J. Shankland, K. Kurokawa, K. Bomsztyk, R. J. Johnson, and W. G. Couser, *Immunogenetics* **46,** 99 (1997).

[4] C. A. Wise, G. A. Clines, H. Massa, B. J. Trask, and M. Lovett, *Genome Res.* **7,** 10 (1997).

[5] Z. Wu, J. Wu, E. Jacinto, and M. Karin, *Mol. Cell. Biol.* **17,** 7407 (1997).

[6] E. Levy-Lahad, W. Wasco, P. Poorkaj, D. M. Romano, J. Oshima, W. H. Pettingell, C. E. Yu, P. D. Jondro, S. D. Schmidt, K. Wang, A. C. Crowley, Y. H. Fu, S. Y. Guenette, D. Galas, E. Nemens, E. M. Wijsman, T. D. Bird, G. D. Schellenberg, and R. E. Tanzi, *Science* **269,** 973 (1995).

[7] E. I. Rogaev, R. Sherrington, E. A. Rogaeva, G. Levesque, M. Ikeda, Y. Liang, H. Chi, C. Lin, K. Holman, T. Tsuda, L. Mar, S. Sorbi, B. Nacmias, S. Piacentini, L. Amaducci, I. Chumakov, D. Cohen, L. Lannfelt, P. E. Fraser, J. M. Rommens, and P. H. St George-Hyslop, *Nature* (*London*) **376,** 775 (1995).

[8] N. Papadopoulos, N. C. Nicolaides, Y. F. Wei, S. M. Ruben, K. C. Carter, C. A. Rosen, W. A. Haseltine, R. D. Fleischmann, C. M. Fraser, M. D. Adams, J. C. Venter, S. R. Hamilton, G. M. Petersen, P. Watson, H. T. Lynch, P. Peltomaki, J. P. Mecklin, A. de la Chapelle, K. W. Kinzler, and B. Vogelstein, *Science* **263,** 1625 (1994).

[9] N. C. Nicolaides, N. Papadopoulos, B. Liu, Y. F. Wei, K. C. Carter, S. M. Ruben, C. A. Rosen, W. A. Haseltine, R. D. Fleischmann, C. M. Fraser, M. D. Adams, J. C. Venter, M. G. Dunlop, S. R. Hamilton, G. M. Petersen, A. de la Chapelle, B. Vogelstein, and K. W. Kinzler, *Nature* (*London*) **371,** 75 (1994).

Copyright © 1999 by Academic Press
All rights of reproduction in any form reserved.

0076-6879/99 $30.00

the EST database does not yet describe all human genes, and until that goal is reached all positional cloning projects will be confronted with the choice between relying on positional candidate genes from EST/gene databases and embarking on a targeted gene identification project within a specific genomic interval. Currently, the EST databases are estimated to contain pieces from more than 80% of all human cDNA sequences,[11] although no definite algorithm yet exists for determining the total human gene number. The lack of completeness in the EST database has occurred for two reasons, neither of which is simple to circumvent technically: (1) the method by which ESTs are currently derived inherently skews them toward more abundant transcripts in the sets of tissues that have been screened to date; and (2) conventionally constructed cDNA libaries, even when quasinormalized, do not adequately recapitulate the mRNA populations that are present in a complex organ such as the brain.

Direct cDNA selection[12,13] is one positional cloning methodology that can be used to conduct targeted and detailed gene identification in a large genomic interval. It can be used with high efficiencies on contigs several megabases in length, and can thus be employed when the genetic target interval is only broadly defined. However, in this chapter we present protocols for scaling this process up to an entire human chromosome. As one method of deriving the remaining genes in the human genome, direct cDNA selection with an entire human chromosome can be used to generate chromosome-specific cDNA libraries. In contrast to conventional cDNA libraries, these resources are greatly enriched for the gene sequences encoded by one specific chromosome. Furthermore, rare transcripts are greatly enriched in these libraries owing to the use of high-complexity cDNA pools (rather than cDNA libraries) as a starting source, and because direct selection can be designed to yield quasinormalized cDNA populations. In a previous report from our group,[14] prototype chromosome 5-specific cDNA libraries were constructed from several different tissue sources. Approximately 25% of the novel 3′-end cDNAs within these librar-

[10] S. Osborne-Lawrence, P. L. Welcsh, M. Spillman, S. C. Chandrasekharappa, T. D. Gallardo, M. Lovett, and A. M. Bowcock, *Genomics* **25,** 248 (1995).

[11] L. Hillier, G. Lennon, M. Becker, M. F. Bonaldo, B. Chiapelli, S. Chissoe, N. Dietrich, T. Dubuque, A. Favello, W. Gish, M. Hawkins, M. Hultman, T. Kucaba, M. Lacy, M. Le, N. Le, E. Mardis, B. Moore, M. Morris, C. Prange, L. Rifkin, T. Rohlfing, K. Schellenberg, M. Soares, F. Tan, E. Trevaskis, K. Underwood, P. Wohldmann, R. Waterston, R. Wilson, and M. Marra, *Genome Res.* **6,** 807 (1996).

[12] M. Lovett, J. Kere, and L. M. Hinton, *Proc. Natl. Acad. Sci. U.S.A.* **88,** 9628 (1991).

[13] S. Parimoo, S. R. Patanjali, H. Shukla, D. D. Chaplin, and S. M. Weissman, *Proc. Natl. Acad. Sci. U.S.A.* **88,** 9623 (1991).

[14] R. G. Del Mastro, L. P. Wang, A. D. Simmons, T. D. Gallardo, G. A. Clines, J. A. Ashley, C. J. Hilliard, J. J. Wasmuth, J. D. McPherson, and M. Lovett, *Genome Res.* **5,** 185 (1995).

ies do not overlap with ESTs in the database.[14,15] In addition to being sources for new ESTs, these resources have applications in positional cloning projects and in the identification of new gene families. Examples of these applications are gene identification by hybridization-based screening with genomic configs or motifs.[16] For example, direct screening by our group of our chromosome 11-specific fetal brain cDNA library led to the identification and positional cloning of the chromosome 11-linked, hereditary multiple exostoses type 2 gene.[17]

The average insert size of cDNA clones derived by most gene discovery techniques, including exon amplification[18] and direct cDNA selection, is less than 1 kb. Extension of short fragments to full length is a major bottleneck in positional cloning. Currently available ESTs do not solve this problem because they are on average only 1.5 kb in length. The two most common methods of extending cDNAs are direct screening of conventional cDNA libraries and the rapid amplification of cDNA ends.[17,19,20] Unfortunately, neither of these techniques is amenable to the high-throughput extension of a large number of cDNA sequences. Here we present a technique for extending a large number of direct-selected cDNAs *en masse* by capturing single-stranded cDNA circles after the selection, but prior to cloning.

Chromosome-Specific cDNA Libraries: Principle of Method and Considerations

Direct cDNA selection is an expression-based technique used to derive transcription maps and candidate genes from large genomic regions. It is based on the solution hybridization and subsequent capture of polymerase chain reaction (PCR)-amplifiable cDNAs using biotinylated genomic templates.[21] The starting template for most direct cDNA selections is a contig

[15] J. W. Touchman, G. G. Bouffard, L. A. Weintraub, J. R. Idol, L. P. Wang, C. M. Robbins, J. C. Nussbaum, M. Lovett, and E. D. Creen, *Genome Res.* **7,** 281 (1997).

[16] A. D. Simmons, J. Overhauser, and M. Lovett, *Genome Res.* **7,** 118 (1997).

[17] D. Stickens, G. Clines, D. Burbee, P. Ramos, S. Thomas, D. Hogue, J. T. Hecht, M. Lovett, and G. A. Evans, *Nature Genet.* **14,** 25 (1996).

[18] A. J. Buckler, D. D. Chang, S. L. Graw, J. D. Brook, D. A. Haber, P. A. Sharp, and D. E. Housman, *Proc. Natl. Acad. Sci. U.S.A.* **88,** 4005 (1991).

[19] G. Joslyn, M. Carlson, A. Thliveris, H. Albertsen, L. Gelbert, W. Samowitz, J. Groden, J. Stevens, L. Spirio, M. Robertson, L. Sargeant, K. Krapcho, E. Wolff, R. Burt, J. P. Hughes, J. Warrington, J. McPherson, J. Wasmuth, D. Le Paslier, H. Abderrahim, D. Cohen, M. Leppert, and R. White, *Cell* **66,** 601 (1991).

[20] M. R. Wallace, D. A. Marchuk, L. B. Andersen, R. Letcher, H. M. Odeh, A. M. Saulino, J. W. Fountain, A. Brereton, J. Nicholson, A. L. Mitchell, B. H. Brownstein, and F. S. Collins, *Science* **249,** 181 (1990).

[21] J. G. Morgan, G. M. Dolganov, S. E. Robbins, L. M. Hinton, and M. Lovett, *Nucleic Acids Res.* **20,** 5173 (1992).

derived from bacterial [cosmid, P1, plasmid artificial chromosome (PAC), or bacterial artificial chromosome (BAC)] and/or yeast (yeast artificial chromosome, YAC) clones that encompass a specific region of the human genome. However, this technique can also be scaled up to clone positionally on a chromosomal scale. As shown in Fig. 1, cosmid clones representing

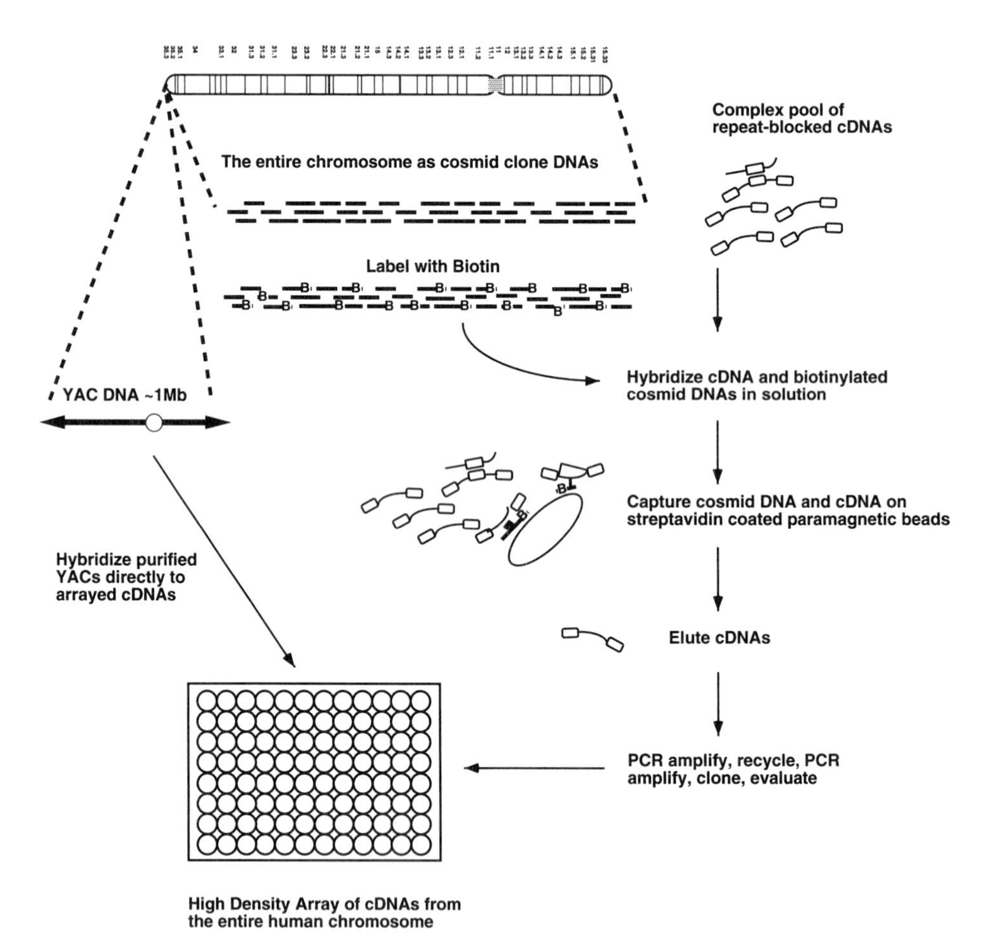

FIG. 1. Chromosome-specific cDNA library construction and screening scheme. Cosmid clones representing an entire human chromosome are biotinylated *en masse* and hybridized in solution to uncloned, repeat-suppressed, PCR-amplifiable cDNA pools. After washing, the genomic target and hybridizing cDNAs are captured using streptavidin-coated paramagnetic beads. Hybridizing cDNAs are eluted, PCR amplified, and recycled through a second round of direct cDNA selection. The cDNAs eluted from the second round of selection are PCR amplified, cloned, and arrayed for further analysis. YACs containing fragments of a specific human chromosome can then be directly hybridized to high-density arrays of these cDNA clones.

an entire human chromosome are biotinylated *en masse* and hybridized in solution to uncloned, PCR-amplifiable cDNA pools. After washing, the genomic target and hybridized cDNAs are captured using streptavidin-coated paramagnetic beads. Hybridizing cDNAs are eluted, PCR amplified, and recycled through a second round of direct cDNA selection. The resulting cDNAs are PCR amplified, cloned, and arrayed for further analysis. The left side of Fig. 1 illustrates one use of these resources, in which YACs containing fragments of a specific human chromosome are directly hybridized to high-density arrays of chromosome-specific cDNA libraries to identify candidate genes within a particular interval.[16]

Purification and Biotin Labeling of Genomic Template

Preparation of the genomic template is one of the critical steps in direct cDNA selection. For these experiments the genomic template is a pool of purified DNA from the cosmids representing an entire human chromosome.[22] However, this technique can be used with all bacterial-based cloning vectors, including cosmids, PACs, and BACs.[22–25] The clones within these libraries must be independently amplified to retain the initial complexity of the starting library. An efficient method by which to amplify these libraries is to stamp the clones on large cookie sheets containing LB agar with the appropriate antibiotic. A rubber policeman and additional medium are then used to scrape the library into one pool prior to the purification of the cosmid DNAs. An alternative method by which to amplify the library is to replicate the clones in 96- or 384-well microtiter plates and then combine the cultures after an overnight incubation. Clones containing the genes that encode ribosomal RNAs (present on chromosomes 13, 14, 15, 21, and 22) can give rise to a significant contamination problem when used in direct selection. These clones can either be removed prior to the selection procedure (which is time consuming but avoids most ribosomal contamination) or must be extensively blocked prior to the selection (see step 1 of Direct cDNA Selection). Efficient labeling of the genomic template is another critical step in the selection procedure and is indirectly monitored using trace amounts of [α-^{32}P]dCTP in addition to the biotin-labeled nucleotide triphosphate. A portion of the labeling reaction is then tested for its ability to bind streptavidin-coated magnetic beads. The ratio of bound to

[22] J. L. Longmire, N. C. Brown, L. J. Meincke, M. L. Campbell, K. L. Albright, J. J. Fawcett, E. W. Campbell, R. K. Moyzis, C. E. Hildebrand, G. A. Evans, and L. L. Deaven, *Genet. Anal. Tech. Appl.* **10,** 69 (1993).

[23] G. A. Evans, K. Lewis, and B. E. Rothenberg, *Gene* **79,** 9 (1989).

[24] P. A. Ioannou, C. T. Amemiya, J. Garnes, P. M. Kroisel, H. Shizuya, C. Chen, M. A. Batzer, and P. J. de Jong, *Nature Genet.* **6,** 84 (1994).

[25] H. Shizuya, B. Birren, U. J. Kim, V. Mancino, T. Slepak, Y. Tachiiri, and M. Simon, *Proc. Natl. Acad. Sci. U.S.A.* **89,** 8794 (1992).

unbound counts on the beads provides a rough estimate of the biotin incorporation. To increase the probability that biotin is uniformly incorporated within the genomic template, purified DNA from the amplified library is labeled using both random priming and nick translation. The labeling reactions are also performed using a 1:10 molar ratio of biotin-16-dUTP to dTTP to reduce steric hindrance during the hybridization and capture reactions.

Protocol

1. Stamp the bacterial library (n = 5000–25,000) on 25 × 37 cm cookie sheets containing LB agar with the appropriate antibiotic. After an overnight incubation at 37°, scrape the bacteria into a beaker, using the appropriate medium and a rubber policeman. Prepare plasmid DNA from the amplified library using a kit (e.g., Qiagen, Chatsworth, CA) or a standard alkaline lysis procedure.[26] Confirm that the purified plasmid DNA is free of contaminating bacterial genomic DNA by agarose gel electrophoresis. Test the purified genomic DNA clones for the presence of reporter genes, using the PCR.

2. Incorporate biotin into the cosmid DNAs using both nick translation and random primed labeling. The nick translation contains 1 μl of purified plasmid DNA (2 μg), 1 μl of biotin-16-dUTP [0.04 mmol/liter; Boehringer Mannheim (BM), Indianapolis, IN], 1 μl each of dGTP, dCTP, dATP, and dTTP (each at 0.4 mmol/liter), 2 μl of 10× reaction buffer (BM), 1 μl of [α-^{32}P]dCTP (3000 Ci/mmol; Amersham, Arlington Heights, IL), 9 μl of sterile water, and 2 μl of DNA polymerase I/DNase I (BM). The random primed labeling reaction is performed under the same conditions using 10× random primed labeling buffer (BM) and the Klenow fragment of *Escherichia coli* DNA polymerase (BM).

3. After a 2-hr incubation at the appropriate temperature (nick translation, 4°; random primed labeling, 37°), combine reactions, purify by Sephadex G-50 chromatography, and ethanol precipitate. Resuspend the pellet in 10 μl of sterile water and store at −20° until use.

4. An aliquot of the labeling reaction (~10%) is tested for its ability to bind streptavidin-coated magnetic beads (Dynal, Chantilly, VA). The conditions for this step are identical to those described in direct cDNA selection steps 3 and 4. Efficient biotin labeling is observed when the ratio of [α-^{32}P]dCTP bound to the beads rather than remaining in the supernatant is greater than 10:1. Do not proceed with the selection if this criterion is not met.

[26] J. Sambrook, E. F. Fritsch, and T. Maniatis (eds.), "Molecular Cloning: A Laboratory Manual," 2nd Ed. Cold Spring Harbor Laboratory Press, Cold Spring Harbor, New York, 1989.

Preparation of Starting cDNA Pools

The optimal cDNA sources for direct selections are uncloned PCR-amplifiable cDNA pools.[27] Whereas conventional cDNA libraries contain only a subset of the transcripts expressed in a complex tissue source, cDNA pools effectively recapitulate the mRNA complexity of the source. Cytoplasmic RNA is directly purified from mammalian cells or tissues to reduce the possibility of selecting contaminating heteronuclear RNA (hnRNA) and genomic DNA.[26,28] Polyadenylated RNA is purified by two rounds of oligo(dT)-cellulose chromatography. Double-stranded cDNA synthesis can be performed using oligo(dT) or random primers. The advantage to using random primed cDNA is that it yields fragments from essentially all parts of the mRNA. Thus, if a repeat-containing cDNA is identified from the candidate interval it can be discarded on the basis of the assumption that random primed, nonrepetitive fragments of the same cDNA will also be present within the selected material. The disadvantage to using random primed cDNAs is that they are not necessarily directly comparable to sequences that are in the EST databases and may be internal parts of previously identified cDNAs. It is also advisable to size-select cDNA products that are at least 500 bp in length by either column chromatography or gel electrophoresis prior to proceeding to the linkering steps. The aforementioned steps in cDNA construction have been described in detail elsewhere.[26] However, the addition of PCR-amplifiable linkers is a more specialized technique and is presented below. The most important feature of the linker cassette is that it does not significantly overlap with DNA sequences in vectors or the human genome. A 2-base pair (bp) overhang is present at the 3′ end of one of the linker primers, thus forcing the cassette to be ligated in only one orientation and also stopping the formation of oligomers of the cassette. Although multiplexing is not used for chromosome-specific direct cDNA selections owing to the complexity of the initial starting pools, direct cDNA selections are frequently performed using multiplexed cDNA pools, which can be separately PCR amplified and cloned for further analysis.[29]

Protocol

1. The appropriate oligonucleotide pairs [e.g., 5′-CTCGAGAATTCT GGATCCTC-3′ (oligo 3) and 5′-GAGGATCCAGAATTCTCGAGTT-3′

[27] G. R. Reyes and J. P. Kim, *Mol. Cell. Probes* **5,** 473 (1991).

[28] M. J. Clemens, *in* "Transcription and Translation: A Practical Approach," 2nd Ed. (B. M. Haines and S. Y. Higgins, eds.), p. 211. IRL Press, Washington, D.C., 1984.

[29] A. D. Simmons, S. A. Goodart, T. D. Gallardo, J. Overhauser, and M. Lovett, *Hum. Mol. Genet.* **4,** 295 (1995).

(oligo 4)] are separately phosphorylated at their 5′ terminus using polynucleotide kinase. Mix complementary primers in equal molar ratios (~2 μg each), denature for 10 min at 100°, and slowly cool to room temperature. Ligate the PCR amplification cassette to 3 μg of double-stranded, blunt-ended cDNA under the following conditions: 22 μl of cDNA (3 μg), 2 μl of PCR amplification cassette (1 μg/μl), 3 μl of 10× T4 DNA ligase buffer [0.66 *M* Tris-HCl (pH 7.6), 50 m*M* $MgCl_2$, 50 m*M* dithiothreitol, bovine serum albumin (fraction V, 1 mg/ml), 10 m*M* hexamminecobalt chloride, 2 m*M* ATP, 5 m*M* spermidine-HCl], and 3 μl of T4 DNA ligase (1 unit/μl).

2. Incubate the ligation overnight at 4° and inactivate the T4 DNA ligase by a 10-min incubation at 65°. Purify the ligation reaction by phenol–chloroform extraction, Sephadex G-50 chromatography, and ethanol precipitation.

3. Amplify cDNA pools in a Perkin-Elmer Cetus (Emeryville, CA) GeneAmp 9600 thermal cycler, using 30 cycles of amplification under the following conditions: 30 sec at 94°, 30 sec at 55°, and 1 min at 72°. The reactions are performed using 1.0 μl of DNA (10 ng), 5.0 μl of linker primer (oligo 3, 10 μ*M*), 2.5 μl of 10× PCR buffer [100 m*M* Tris-HCl (pH 8.3), 500 m*M* KCl, and 0.01% (w/v) gelatin], 2.5 μl of dNTP mixture (2.5 m*M* concentration of each nucleotide), 1.5 μl of 25 m*M* $MgCl_2$, 12.3 μl of sterile water, and 0.2 μl of *Taq* polymerase (5 units/μl; Perkin-Elmer Cetus). Approximately 2 μg of each PCR-amplified cDNA source should be purified by three phenol extractions and ethanol precipitation. The amplified cDNAs should be visually compared with the input linkered cDNA source by gel electrophoresis to ensure that the PCR products reflect the correct size distribution of cDNAs.

Direct cDNA Selection

Direct cDNA selection has been described in some detail.[30] However, some modifications were required to optimize the technique for a chromosome-sized target. The first round of selection is usually conducted under conditions of genomic DNA excess relative to rare transcripts, whereas the genomic DNA is limiting in concentration in the secondary round. This results in very large increases in the relative abundance of some (more abundant) cDNA species during the first round of selection, but during the second round insufficient genomic target is present to capture all of these abundant cDNAs. The high abundance-selected cDNAs are thus reduced in abundance, and lower abundance-selected cDNAs are increased in relative

[30] M. Lovett, *in* "Current Protocols in Human Genetics" (J. Seidman, ed.), p. 6.3.1. Wiley Interscience, New York, 1994.

abundance. To establish these same quasinormalization conditions for chromosome-specific direct cDNA selections, the genomic target (biotin-labeled cosmid DNA) is increased from 100 ng to 1 μg and the cDNA pool is decreased from 1 μg to 100 ng for the primary round of the selection procedure. The second round of direct cDNA selection is performed using 100 ng of the genomic template and 1 μg of the PCR-amplified primary selected cDNAs.

The efficiency of each round of the selection procedure is directly monitored using reporter genes from the appropriate chromosome. This is performed by hybridizing gene-specific DNA fragments to a Southern blot containing PCR-amplified starting cDNA, cDNAs from the first round of selection (the primary selected cDNA), and cDNAs from the second round of selection (the secondary selected cDNA). Both low- and high-abundance transcripts randomly spaced throughout the chromosome should be monitored. Rather than the 10^3- to 10^4-fold enrichments observed in standard direct cDNA selections, only 10^2- to 10^3-fold enrichment of most reporter genes is observed. However, low-abundance cDNAs should be enriched in both rounds of the direct cDNA selection. A more expensive, but more quantitative, method of evaluating the selection is to hybridize reporter genes to arrays of clones derived from the secondary selected cDNAs.[14]

Incorporation of restriction enzyme sites into the cDNA cassette facilitates the cloning of cDNAs into phage or plasmid vectors. However, we recommend cloning the secondary selected cDNAs into the uracil DNA glycosylase plasmid vector pAMP10 (GIBCO-BRL, Gaithersburg, MD). Briefly, the secondary selected cDNAs are PCR amplified with the appropriate linker primer (e.g., oligo 3) containing four copies of the trinucleotide repeat CUA at its 5′ end. Treatment with uracil DNA glycosylase then removes the uracils present at both ends of the PCR product as well as those in the cloning region of the vector. This disrupts base pairing and exposes 3′ overhangs, which then anneal to the complementary pAMP10 vector ends. This cloning technique is rapid, efficient, and eliminates the possibility that cDNAs are fragmented by restriction enzyme digestion.

Protocol

1. Repeat suppress 100 ng of the PCR-amplified cDNA pool using 2 μg of C_0t-1 DNA (GIBCO-BRL) and 1 μg of sheared sCos-1 plasmid DNA (Stratagene, La Jolla, CA). If ribosomal clones are present in the genomic target, at least 1 μg of cDNA derived from rRNA will also be necessary at this step. Combine these DNAs in a final volume of 10 μl, heat denature for 10 min at 100°, add 10 μl of 2× hybridization solution [1.5 M NaCl, 40 mM sodium phosphate buffer (pH 7.2), 10 mM EDTA, 10× Denhardt's

solution, and 0.2% sodium dodecyl sulfate (SDS)], and incubate for 4 hr at 65°. Perform the blocking reaction in a thermal cycler and overlay with mineral oil to prevent evaporation.

2. After the repeat suppression of the cDNA pools is complete, heat denature 5 μl of the biotinylated cosmid DNAs (1 μg) for 10 min at 100°. Combine the denatured cosmid DNAs with the repeat suppressed cDNA, add 5 μl of 2× hybridization solution, and hybridize for >54 hr at 65°. For the first round of chromosome-specific direct cDNA selections, use 100 ng of the PCR-amplified cDNA pool and 1 μg of the genomic target (biotin-labeled cosmid DNA). However, this ratio is reversed for the second round of selection, which is performed using 100 ng of the genomic template and 1 μg of the PCR-amplified primary selected cDNAs.

3. Genomic DNA and hybridizing cDNAs are captured using streptavidin-coated magnetic beads (Dynal). Magnetic beads (2 mg, 200 μl) are prepared by three 1-ml washes in binding buffer [10 m*M* Tris-HCl (pH 7.5), 1 m*M* EDTA (pH 8.0), 1 *M* NaCl]. After each wash, beads are removed from the binding buffer using a magnetic separator (Dynal). The washed beads are resuspended in 1× binding buffer at a final concentration of 10 mg/ml (200 μl). The genomic template and hybridizing cDNAs are captured by a 15-min incubation at room temperature with the prepared beads. During the incubation period, the solution is gently mixed to prevent the beads from settling.

4. The beads are removed from the solution using a magnetic separator and washed two times, 15 min each, in 1 ml of 1× SSC–0.1% SDS at room temperature followed by three washes, 15 min each, in 1 ml of 0.1× SSC–0.1% SDS at 65° (1× SSC is 0.15 *M* NaCl plus 0.015 *M* sodium citrate.)

5. Hybridizing cDNAs are eluted by a 10-min incubation at room temperature in 100 μl of 0.1 *M* NaOH. Eluted cDNAs are neutralized with an equal volume of 1 *M* Tris-HCl, pH 7.5, and purified by Sephadex G-50 chromatography.

6. The primary selected cDNAs are PCR amplified, phenol–chloroform extracted, ethanol precipitated, and then recycled through an additional round of direct cDNA selection.

7. Secondary selected cDNAs are PCR amplified with oligo 3 containing four copies of the trinucleotide repeat CUA at its 5′ end. PCR amplification reactions are purified by Sephadex G-50 chromatography, annealed to the vector pAMP10 (GIBCO-BRL), and transformed into DH5α competent cells (GIBCO-BRL).

Discussion

Preliminary analysis of the selection, as described above, is the hybridization of reporter genes to Southern blots containing PCR-amplified cDNAs.

A similar Southern blot should also be hydridized with a ribosomal RNA probe to monitor enrichment or depletion of this contaminant. A third Southern blot should be hybridized with C_0t-1 DNA to monitor repeat enrichment or depletion. If the reporter genes are enriched in the selections and the ribosomal and repeat elements are not, then the secondary selected cDNAs are cloned and arrayed into 96-well microtiter plates for further analysis. One of the advantages of chromosome-specific direct cDNA selections is that a relatively small number of recombinant clones is required to comprise a good sampling of essentially all of the genes expressed in a specific tissue source. For example, chromosome 5 is thought to encode approximately 5000 genes, based on the assumption that it is about 190 Mb in size and that the human genome contains about 100,000 randomly distributed genes. If a very high percentage (25%) of all genes are expressed from a specific tissue source (and current estimates indicate that this number is much lower), this corresponds to only 1250 genes per tissue for the entire chromosome. We usually array 4608 recombinant clones (48 × 96-well plates or 12 × 384-well plates) for subsequent analysis and find an average of four clones per gene when a tissue source such as placenta is used for cDNA synthesis.[14]

One of the first steps in the analysis of chromosome-specific cDNA selections is removal of any repetitive or ribosomal clones. This is usually performed by hybridization of high-density arrays of the cDNAs with radiolabeled fragments. As briefly mentioned above, a good source of ribosomal DNA can be generated by first-strand cDNA synthesis of the poly(A)$^-$ fraction obtained during construction of the cDNA pools. Approximately 10% of mRNAs contain repetitive elements and although repeat-containing cDNAs are usually not analyzed owing to the technical problems in manipulating these clones, they may in fact be bona fide chromosome-specific cDNAs. Preliminary analysis of the chromosome 5-specific cDNA libraries suggests that at least 70% of the Alu-containing cDNAs map back to chromosome 5 (based on PCR mapping of unique regions flanking the Alu repeat[31]). Initial analysis of chromosome-specific cDNA libraries includes hybridization of reporter genes to high-density arrays, which provides quantitative values for gene enrichment and some idea of normalization. The arrays of cloned cDNAs can then be directly used for hybridization of oligonucleotides, YACs, or for EST generation by random cDNA sequencing.

Single-Stranded cDNA Capture: Principle of Method and Considerations

One of the most labor-intensive and time-consuming aspects of positional cloning is gene extension. No currently available gene identification

[31] T. D. Gallardo and M. Lovett, in preparation (1999).

technique, whether it is direct selection, exon amplification,[18] or computational analysis of genomic DNA sequences,[32,33] can identify full-length protein-coding regions, far less full-length transcription units that include 5′ and 3′ untranslated regions. The current range of cDNA libraries used for EST generation are also less than full length. The average length of an EST-based cDNA is 1.5 kb, a selected cDNA is ~0.5 kb, and a trapped exon is about 0.15 kb. Here we describe a simple technique to rapidly extend a pool of cDNAs *en masse.* A pool of cDNA fragments is biotinylated *en masse* and used to capture single-stranded (ss) cDNA circles prepared *in vivo* from phagemid cDNA libraries (Fig. 2). After hybridization, the biotinylated cDNAs and annealing phagemid clones are captured using streptavidin-coated magnetic beads. After posthybridization washes the annealed phagemid clones are eluted, partially repaired by primer extension, and transformed into competent cells. This is nothing more than an additional selection step at the end of two PCR-based rounds of selection; in this case the biotinylated substrate is the selected cDNAs and these are used to capture longer length cDNA circles. This type of single-strand capture was initially proposed as one method of gene isolation from large genomic regions by Weissman.[34] In the protocol described here we show its application to converting a low-complexity set of cDNA fragments to larger cDNAs in a batch process.

Preparation of Single-Stranded cDNA

A critical factor of this technique is the quality and complexity of the starting cDNA libraries. To standardize these experiments, we used commercially available human adult brain and testis cDNA libraries cloned into the phagemid vector pCMV · SPORT (GIBCO-BRL). These libraries can also be used in a related type of single cDNA capture system marketed by GIBCO-BRL called the Genetrapper cDNA positive selection system. As with most conventional cDNA libraries, these contain a limited number of full-length cDNAs. However, this scheme is readily compatible with "true full-length" cDNA libraries.[35,36] Single-stranded (ss) cDNA is prepared *in vivo* from the cDNA libraries by coinfection with the helper phage M13KO7. Other vectors, including the filamentous bacteriophage M13, can

[32] T. M. Smith, M. K. Lee, C. I. Szabo, N. Jerome, M. McEuen, M. Taylor, L. Hood, and M. C. King, *Genome Res.* **6,** 1029 (1996).

[33] J. Flint, K. Thomas, G. Micklem, H. Raynham, K. Clark, N. A. Doggett, A. King, and D. R. Higgs, *Nature Genet.* **15,** 252 (1997).

[34] S. M. Weissman, *Mol. Biol. Med.* **4,** 133 (1987).

[35] P. Carninci, C. Kvam, A. Kitamura, T. Ohsumi, Y. Okazaki, M. Itoh, M. Kamiya, K. Shibata, N. Sasaki, M. Izawa, M. Muramatsu, Y. Hayashizaki, and C. Schneider, *Genomics* **37,** 327 (1996).

[36] I. Edery, L. L. Chu, N. Sonenberg, and J. Pelletier, *Mol. Cell. Biol.* **15,** 3363 (1995).

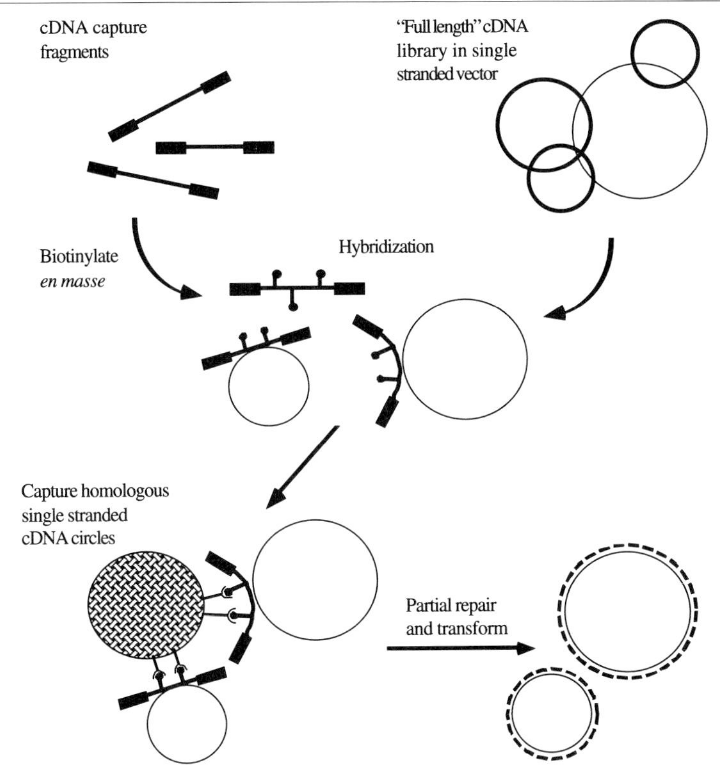

FIG. 2. Single-stranded cDNA capture scheme. A pool of cDNA fragments is biotinylated *en masse* and used to capture single-stranded cDNA circles prepared *in vivo* from phagemid cDNA libraries. After the hybridization, the biotinylated cDNAs and annealing phagemid clones are captured using streptavidin-coated paramagnetic beads. The beads are washed and the captured phagemid clones are eluted, partially repaired by primer extension, and transformed into a bacterial host.

also be used to prepare ss cDNA libraries. However, it is preferable that the cDNA library be directionally cloned. Preparation of ss cDNA from directionally cloned cDNA libraries yields either sense or antisense fragments (but not both) that do not reanneal during subsequent hybridization steps.

Protocol

1. Inoculate 125 ml of LB (with the appropriate antibiotic) with $>1 \times 10^6$ CFU/ml of the primary cDNA library. Incubate for 3 hr at 37° with strong agitation (275 rpm). Add M13KO7 helper phage (GIBCO-BRL) to a final concentration of $>2 \times 10^7$ PFU/ml and shake the culture for an

additional 1–2 hr at 37°. Select infected cells by the addition of 1.5 ml of 1% (w/v) kanamycin and incubate overnight at 37° with shaking (275 rpm).

2. Remove bacteria by centrifugation for 15 min at 15,000*g* (4°). Filter the resulting supernatant through a 0.2-μm pore size sterile filter.

3. Degrade any contaminating double-stranded DNA by the addition of 200 μl of DNase I (10 units/μl) (BM) to the filtered supernatant. Incubate for 3 hr at room temperature.

4. Transfer 100 ml of this solution to a sterile centrifuge bottle containing 25 ml of 40% polyethylene glycol (PEG) 3250, vortex well, and incubate on ice for 1 hr.

5. Recover bacteriophage particles by centrifugation for 20 min at 15,000*g* (4°). Resuspend the pellet in 2 ml of 1× TE and add 20 μl of proteinase K (10 mg/ml) and 10 μl of 20% SDS. Incubate the lysis reaction for 1 hr at 45°.

6. Purify the recovered ss DNA circles by four phenol–chloroform–isoamyl alcohol extractions and ethanol precipitation. Resuspend the resulting pellet in 100 μl of 1× TE and incubate for 1 hr at −20° to precipitate polysaccharides, which are then removed by a 15-min centrifugation at 14,000*g* (4°).

7. The ss DNA-containing supernatant is further purified by four phenol–chloroform–isoamyl alcohol extractions and ethanol precipitation. Approximately 20 μg of ss DNA is recovered from 100 ml of cells.

Capture Reaction

The amplified ss cDNA library is hybridized in solution with the biotinylated cDNA fragments to be extended. The cDNA fragments have been directly biotinylated using nick translation. The biotinylated cDNA fragments and hybridizing ss cDNA circles are then captured using streptavidin-coated paramagnetic beads. Captured single-stranded circular cDNAs are eluted and partially repaired for transformation.

Protocol

1. The cDNA capture fragments are prepared by PCR amplification using flanking vector primers. Capture fragments can also be excised from recombinant clones with the appropriate restriction endonuclease(s). The capture fragments are separated by agarose gel electrophoresis and purified from the agarose using QIAquick spin columns (Qiagen). Purified DNAs from the capture fragments are combined, purified by phenol–chloroform extraction, and concentrated by ethanol precipitation.

2. Biotin label approximately 200 ng of the pooled capture fragments by nick translation as described above. Purify biotin-labeled cDNAs by

Sephadex G-50 chromatography, phenol–chloroform extraction, and ethanol precipitation.

3. Denature the labeled cDNA fragments (100 ng) for 10 min at 100° under mineral oil. After 9 min of incubation, add 5 μl of the purified single-stranded cDNA library (10 μg) to the denatured cDNAs and incubate for an additional minute at 100°. Add 10 μl of 2× hybridization solution [1.5 *M* NaCl, 40 m*M* sodium phosphate buffer (pH 7.2), 20 m*M* EDTA, 10× Denhardt's solution, 0.4% SDS] and hybridize overnight at 65°.

4. The biotin-labeled cDNAs and hybridizing single-stranded cDNA circles are captured using streptavidin-coated magnetic beads as described in the direct cDNA selection steps 3 and 4. However, the washing and elution conditions are slightly modified. The beads are washed once for 15 min in 1 ml of 2× SSC–0.1% SDS at room temperature, once for 15 min in 1 ml of 2× SSC–0.1% SDS at 65°, and twice for 10 min each in 1 ml of 0.1× SSC–0.1% SDS at 65°.

5. Elute hybridizing single-stranded cDNA circles for 10 min at 65° in 50 μl of 50% redistilled formamide (GIBCO-BRL) and neutralize by the addition of 50 μl of 1 *M* Tris-HCl, pH 7.5. This differs slightly from the usual direct selection capture reaction in avoiding the use of NaOH, which can introduce nicks into the single-stranded circles. Purify the eluted single-stranded cDNA circles by phenol–chloroform–isoamyl alcohol extraction and ethanol precipitation after the addition of 1 μl of glycogen (20 μg/μl; BM). Resuspend the recovered ss DNA in 10 μl of 1× TE.

Repair Reaction

The recovered ss DNA circles are partially repaired by primer extension to increase the efficiency of transformation. The repair reaction is performed using two vector-specific oligonucleotides of the appropriate polarity derived from the polylinker region of the vector (e.g., pCMV · SPORT). One of the primers (e.g., T7) is designed to repair across the insert whereas the second primer (e.g., the reverse complement of SP6) repairs across the vector sequence. Two primers are used to ensure that a substantial proportion of the ss cDNA circle is repaired. The repaired cDNAs are propagated by transformation into bacterial cells. Initial assessment of the cDNAs is performed by hybridization of bacterial lifts of the recovered phagemid clones with the capture fragments. On the basis of these results, either random sequencing or additional screening strategies can be used to analyze the recovered clones.

Protocol

1. The repair reaction is composed of 3 μl of 10× Expand PCR buffer (BM), 0.5 μl of dNTPs (2.5 m*M* each), 5.0 μl of captured ss DNA, 1.0 μl

of each forward vector primer (10 mM each), 19.25 μl of sterile water, and 0.25 μl of Expand enzyme mixture (BM). This enzyme mixture contains both thermostable *Taq* and *Pwo* DNA polymerases from the Expand High Fidelity PCR system (BM).

2. Incubate the repair reaction for 2 min at 95°. Transfer the repair reaction to 65° and incubate for an additional hour.

3. Precipitate the repair reaction for 1 hr on dry ice following the addition of 1 μl of glycogen (20 μg/μl), 16 μl of 7.5 M ammonium acetate, and 125 μl of −20° ethanol. Resuspend the repaired DNA in 10 μl of 1× TE and transform 7 μl of the repair reaction into MAX efficiency DH5α competent cells (GIBCO-BRL).

Discussion

We applied this single-strand capture technique to 28 cDNA fragments (average length of 400 bp) from the Cri-du-chat critical region of chromosome 5p. Using the preceding protocol we recovered 1×10^4 recombinant clones. Positive clones were identified by hybridization of purified DNA from the pool of 28 cDNA clones to colony lifts of the recovered phagemid clones. This resulted in the extension of these cDNAs to an average length of 1.2 kb per cDNA. In other pilot experiments we have used pools of 48 cDNA fragments in separate extensions with similar success rates. While we have not yet explored larger numbers than these, we anticipate that the upper limit will be dependent on the transformation efficiencies that can be achieved with repair circular cDNA clones, and on the relative abundance of target clones in the cDNA capture library. As currently configured, this protocol is designed to be driven with an excess concentration of biotinylated fragments in order to capture as many low-abundance species as possible. This reverses the abundance normalization that occurs during selection, but extended clones can be easily identified by hybridization even if they compose only a few percent of the extended material. There is an unfortunate and inherent conundrum in first selecting cDNA fragments that may be derived from very low-abundance mRNAs, and then trying to extend them within a conventionally cloned cDNA library that is limited in its complexity and may therefore not contain them. Some selected cDNAs will be refractory to extension by this method and will have to be painstakingly extended by RACE (rapid amplification of cDNA ends) or similar targeted single-gene methods. Obviously, a critical parameter in this technique is the size and complexity of the starting phagemid cDNA library. Although to date, we have used only conventional cDNA libraries in these types of enrichment, this scheme is readily compatible with techniques that select for longer length cDNAs or with schemes in which size-selected full-length cDNA libraries[35,36] are subpartitioned for cDNA clone extensions.

[9] cDNA Selection: An Approach for Isolation of Chromosome-Specific cDNAs

By SATISH PARIMOO and SHERMAN M. WEISSMAN

Introduction

Major efforts to generate expressed sequence tags (ESTs) from 3′ and 5′ ends of random cDNAs have led to the establishment of a public domain database—dbEST.[1] The latest Internet release (August 22, 1997) from the NCBI (*http://www.ncbi.nlm.nih.gov/dbEST/*) indicates more than 1 million ESTs from all species, with human and mouse contributing 794,650 and 218,441 ESTs, respectively. As of 1997, 51 of 62 disease genes (82%), cloned by positional cloning approaches,[2] had corresponding ESTs in the EST database. This suggests that although the EST database has a huge collection of entries, it may not necessarily represent expressed sequences from all the genes. Moreover, despite international consortium efforts, only a fraction of unique ESTs have been mapped on the individual human chromosomes.[3] Although tremendous progress has been made in sequencing technology, as reflected by complete sequencing of the yeast genome,[4] sequencing of the entire human genome is expected to be completed only by the year 2005. In the interim, individual investigators interested in positional cloning of disease genes or generation of chromosome-specific transcript

[1] M. S. Boguski and G. D. Schuler, *Nature Genet.* **4,** 332 (1993).
[2] F. S. Collins, *Nature Genet.* **1,** 3 (1992).
[3] G. D. Schuler, M. S. Boguski, E. A. Stewart, L. D. Stein, G. Gyapay, K. Rice, R. E. White, P. Rodriquez-Tome, A. Aggarwal, E. Bajorek, S. Bentolilia, B. B. Birren, A. Butler, A. B. Castle, N. Chiannikulchai, A. Chu, C. Clee, S. Cowles, P. J. R. Day, T. Dibling, N. Drouot, I. Dunham, S. Duprat, C. East, C. Edwards, J.-B. Fan, N. Fang, C. Fizames, C. Garret, L. O. Green, D. Hadley, M. Marris, P. Jarrison, S. Brady, A. Hicks, E. Holloway, L. Hui, S. Hussain, C. Louis-Dit-Sully, J. Ma, A. MacGilvery, C. Madder, A. Maratukulam, T. C. Matise, K. B. McKusick, J. Morissette, A. Mungall, D. Muselet, H. C. Nusbaum, D. C. Page, A. Peck, S. Perkins, M. Piercy, F. Qin, J. Quackenbush, S. Ranby, T. Reif, S. Rozen, C. Sanders, X. She, J. Silva, D. K. Slonim, C. Soderlund, W. L. Sun, P. Tabar, T. Thangarajah, N. Nega-Czarny, D. Vollrath, S. Voyticky, T. Wilmer, X. Wu, M. D. Adams, C. Auffray, N. A. R. Walter, R. Brandon, A. Dehejia, P. N. Goodfellow, R. Houlgatte, J. R. Hudson, Jr., S. E. Ide, K. R. Iorio, W. Y. Lee, N. Seki, T. Nagase, K. Ishikawa, N. Nomura, C. Phillips, M. H. Polymeropoulos, M. Sandusky, K. Schmitt, R. Berry, K. Swanson, R. Torres, J. C. Venter, J. M. Sikela, J. S. Beckman, J. Weissenbach, R. M. Myers, D. R. Cox, M. R. James, D. Bentley, P. Deloukas, E. S. Lander, and T. J. Hudson, *Science* **274,** 540 (1996).
[4] A. Goffeau, R. Aert, M. L. Agostini-Carbone, *et al., Nature* (*London*) **387**(Suppl.), 5 (1997).

Copyright © 1999 by Academic Press
All rights of reproduction in any form reserved.

0076-6879/99 $30.00

maps have the option to choose from a variety of methodologies to isolate coding sequences from large genomic segments such as yeast artificial chromosomes (YACs), P1 phage artificial chromosomes (PACs), bacterial artificial chromosomes (BACs), or cosmids.[5] Identification of coding sequences is a key element in positional cloning before one embarks on a systematic search for mutations in the candidate genes that are identified in a particular segment defined by linkage studies. Remarkable progress has been made in mapping, gene identification, and mutational analysis of several disease genes.

One of the approaches that we developed for the identification of coding sequences in large genomic segments is a polymerase chain reaction (PCR)-based cDNA hybridization selection method involving the use of nylon membrane.[6,7] Other investigators have also described cDNA selection, using either membrane- or solution-based hybridization methodology, with minor differences in technical details.[8–14] This chapter details the simple membrane-based cDNA selection approach with random primed cDNA libraries and the results obtained in its application in the human *HLA* locus.

Principle of cDNA Selection Approach

All cDNA-based selection approaches share the basic underlying principle of cDNA–genomic DNA reassociation kinetics under well-defined conditions and recovery of selected (specific) cDNAs. Our approach to PCR-based cDNA selection consists of immobilizing the genomic target DNA fragments (YAC or cosmid DNA) on a small piece of nylon paper disk followed by hybridization selection of cDNA inserts after blocking the

[5] S. Parimoo, S. R. Patanjali, R. Kolluri, H. Xu, H. Wei, and S. M. Weissman, *Anal. Biochem.* **228,** 1 (1995).

[6] S. Parimoo, S. R. Patanjali, H. Shukla, D. Chaplin, and S. M. Weissman, *Proc. Natl. Acad. Sci. U.S.A.* **88,** 9623 (1991).

[7] S. Parimoo, R. Kolluri, and S. M. Weissman, *Nucleic Acids Res.* **18,** 4422 (1993).

[8] M. Lovett, J. Kere, and L. Hinton, *Proc. Natl. Acad. Sci. U.S.A.* **88,** 9628 (1991).

[9] J. G. Morgan, G. M. Dolganov, S. E. Robbins, L. M. Hinton, and M. Lovett, *Nucleic Acids Res.* **20,** 5173 (1992).

[10] D. Vetrie, I, Vorechovsky, P. Sideras, J. Holland, A. Davies, F. Flinter, L. Hammarstrom, C. Kinnon, R. Levinsky, M. Bobrow, C. I. E. Smith, and D. R. Bentley, *Nature* (*London*) **361,** 226 (1993).

[11] J. Gecz, L. Villard, A. M. Lossi, P. Millasseau, M. Djabali, and M. Fontes, *Hum. Mol. Genet.* **2,** 1389 (1993).

[12] J. M. J. Derry, H. D. Ochs, and U. Franke, *Cell* **78,** 635 (1994).

[13] B. Korn, Z. Sedlacek, A. Manca, P. Kioschis, D. Konecki, H. Lehrach, and A. Poustka, *Hum. Mol. Genet.* **1,** 235 (1992).

[14] M. F. Bonaldo, M.-T. Yu, P. Jelenc, S. Brown, L. Su, L. Lawton, L. Deavon, A. Efstratiadis, D. Warburton, and M. B. Soares, *Hum. Mol. Genet.* **3,** 1663 (1994).

genomic repeat and GC-rich sequences, and PCR amplification of the specific hybrid bound cDNA molecules using flanking vector primers as depicted in Fig. 1. Two cycles of cDNA selection give better enrichment than samples selected once. Our original report of cDNA selection with random primed cDNA libraries and purified YACs and cosmids as genomic targets[6] was followed by another, modified approach using total yeast DNA containing YACs as a target for selection of YAC-specific cDNAs.[7] These reports detailed qualitative and quantitative data to evaluate the sensitivity and reliability of the method. The original nylon membrane-based selection method has been used extensively with several YACs, and a number of novel genes have been discovered as a result of this approach.[15–21] The cDNA selection process is carried out with small amounts of cDNA and target genomic DNA (YAC or cosmid) because it involves the use of PCR for the recovery of selected material. The selection can be carried out with either a single cDNA library or with a mixture of cDNA libraries in order to obtain a more comprehensive coverage of cDNAs. The advantage of using a short-fragment random primed cDNA library rather than an oligo$(dT)_{12-18}$-primed cDNA library in the selection process is that multiple fragments of the appropriate size corresponding to a transcript are present in a cDNA library. Therefore, chances of recovering even a rare cDNA in a selected material are better because any bias introduced by PCR, as a result of the size of cDNAs or GC-rich regions, is unlikely to eliminate all of the fragments corresponding to a message.

Genomic Target DNA for cDNA Selection

We have tested the method with cosmids, YACs, and total yeast DNA containing a YAC; hence it should also work with BACs or PACs. The optimum target DNA and cDNA concentrations for selection with cosmids,

[15] W.-F. Fan, X. Wei, H. Shukla, S. Parimoo, H. Xu, P. Sankhavaram, Z. Li, and S. M. Weissman, *Genomics* **17,** 575 (1993).

[16] H. Wei, W.-F. Fan, H. Xu, S. Parimoo, H. Shukla, D. Chaplin, and S. M. Weissman, *Proc. Natl. Acad. Sci. U.S.A.* **90,** 11870 (1993).

[17] V. L. Goei, S. Parimoo, A. Caposscla, T. W. Chu, and J. R. Gruen, *Am. J. Hum. Genet.* **54,** 244 (1994).

[18] W. Fan, Y. C. Liu, S. Parimoo, and S. M. Weissman, *Genomics* **27,** 119 (1995).

[19] S. R. Nalabolu, H. Shukla, G. Nallur, S. Parimoo, and S. M. Weissman, *Genomics* **31,** 215 (1996).

[20] W. Fan, W. Cai, S. Parimoo, G. G. Lennon, and S. M. Weissman, *Immunogenetics* **44,** 97 (1996).

[21] J. R. Gruen, S. R. Nalabolu, T. W. Chu, C. Bowlus, W.-F. Fan, V. L. Goei, H. Wei, R. Sivakamasundari, Y.-C. Liu, H. X. Xu, S. Parimoo, G. Nallur, R. Ajioka, H. Shukla, P. Bray-Ward, J. Pan, and S. M. Weissman, *Genomics* **36,** 70 (1996).

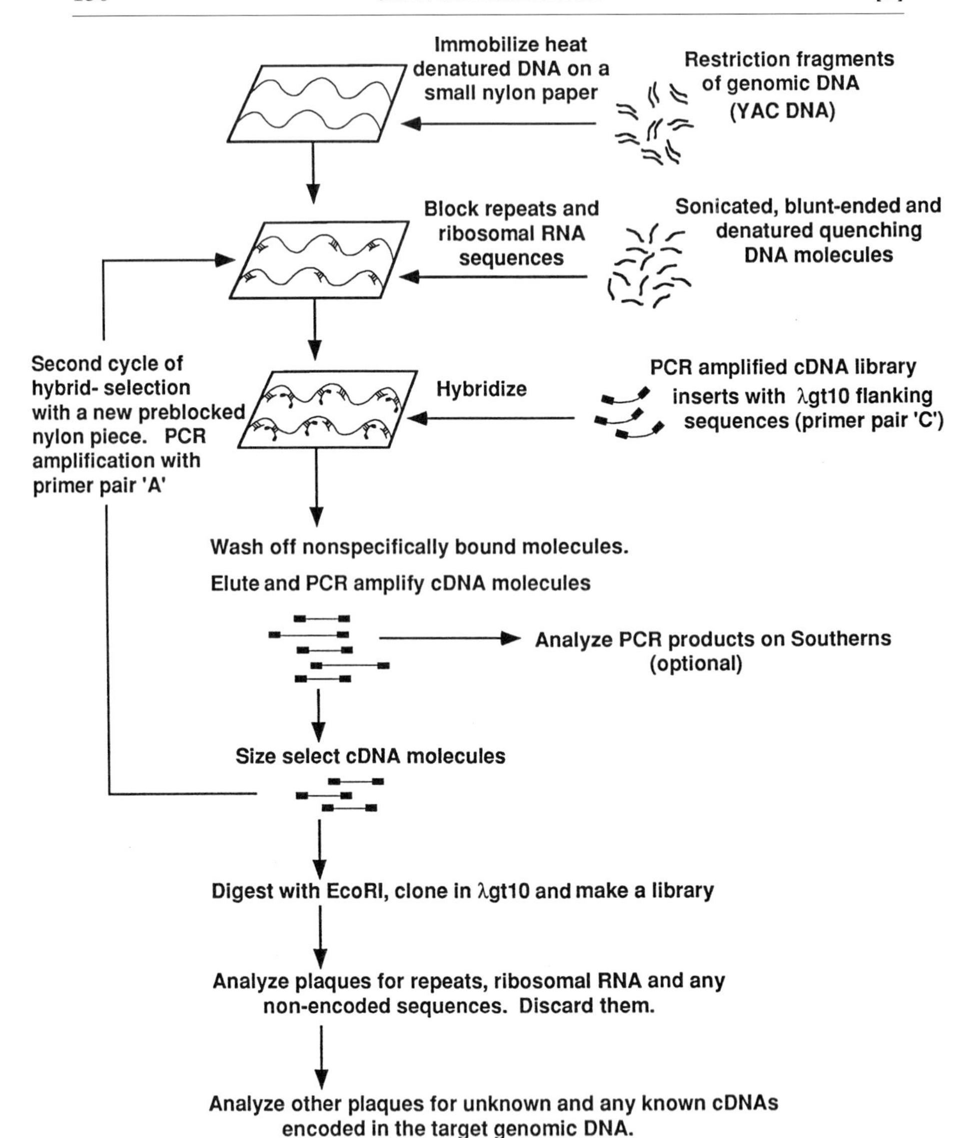

FIG. 1. Flow chart representation of the major steps in cDNA selection. Primer pair C is the outer primer I set of the λgt10 vector, and primer pair A is the inner primer II set of the λgt10 vector, as described in text.

purified YACs, or total yeast DNA have been determined, taking into account the sensitivity and specificity of the process.[6,7] Hence, whereas the amount of DNA required is minimum for a cosmid, it is maximum for total yeast DNA containing a YAC. For bigger or smaller YACs one must adjust the concentration of immobilized DNA accordingly. Because the method is PCR based, small amounts of target DNA and hybridizing cDNA species are used so as to keep the level of coselected undesirable sequences to a minimum in the final selected cDNA.

Preparation of Reagents

Genomic Target DNA

YAC DNA: Yeast strains containing YACs (YAC protocols)[22] are grown in AHC medium and their chromosomes and YACs are separated by pulsed-field gel electrophoresis (PFGE) of agarose plugs in 1% SeaKem GTG agarose (FMC, Philadelphia, PA) on a contour-clamped homogeneous electric field (CHEF) electrophoresis apparatus (Bio-Rad, Hercules, CA) by standard methods.[22] After identification of the YAC by either ethidium bromide staining or probing of the Southern blot of the PFG with pBR322 or human Cot1 DNA (BRL, Gaithersburg, MD), the YAC band of interest is cut out from the gel and the agarose slice (~8 cm) containing the DNA is equilibrated with a restriction enzyme (RE) buffer for at least 2 hr at 4° with a change of buffer in between. Subsequently the agarose slice is immersed in 3–4 ml of RE buffer with an enzyme such as *Hin*dIII or *Bam*HI (0.25–0.5 units/μl) and incubated at 37° for at least 6–8 hr. EDTA, pH 8, is added to a final concentration of 20 m*M* to stop the reaction. The DNA from the agarose slices is extracted by electroelution in 0.5× TBE (Tris–borate–EDTA) in dialysis tubing or by agarase (New England BioLabs, Beverly, MA) treatment. The YAC DNA is adjusted to a concentration of 40–60 ng/μl

Total yeast DNA containing YAC: Total yeast DNA is prepared using published protocols[23]

Cosmid/BAC or PAC DNA: The DNAs are prepared according to standard protocols[24]

[22] J. Ragoussis, *Methods Mol. Biol.* **54,** 69 (1996).

[23] M. D. Rose, F. Winston, and P. Hieter, "Methods in Yeast Genetics: A Laboratory Course Manual," p. 198. Cold Spring Harbor Laboratory Press, Cold Spring Harbor, New York, 1990.

[24] J. Sambrook, E. F. Fritsch, and T. Maniatis, "Molecular Cloning: A Laboratory Manual." Cold Spring Harbor Laboratory Press, Cold Spring Harbor, New York, 1990.

Short-Fragment Random Primed cDNA Libraries

Total RNA is isolated from a particular tissue or cell line by a combination of guanidine thiocyanate–phenol extraction,[25] and selective precipitation of RNA by lithium chloride.[26] After treating total RNA with RNase-free DNase (Boehringer Mannheim, Indianapolis, IN), poly(A)$^+$ RNA is isolated using oligo(dT)$_{12-18}$ cellulose columns by standard methods.[24] Short-fragment (sf), double-stranded cDNA is synthesized from 5 μg of poly(A)$^+$ RNA according to the Gubler and Hoffman method,[27] using one of the commercial cDNA synthesis kits (e.g., kit 18267-013); BRL). The reaction conditions are essentially the same as recommended by the manufacturer, except that the first-strand cDNA synthesis is 40 μl is carried out with 6.75 ng of random hexanucleotide [p(dN)$_6$] instead of oligo(dT)$_{12-18}$. The cDNA is made blunt ended by treatment with mung bean nuclease (0.5 units/μg cDNA at 29°). *Eco*RI adapters are ligated to the double-stranded cDNA by standard procedures.[24] cDNA of the size 400–1500 bp, fractionated in a 1% low melting point agarose gel, is cloned in λgt10 vector at the *Eco*RI site. Three million or more recombinant plaques (in complex tissue libraries) are amplified by plating 75,000–100,000 per 150-mm petri dish and pooling phage lysate in SM medium.[24] Phage DNA is purified by standard methods.[24]

Polymerase Chain Reaction and Release of cDNA Inserts

DNA inserts from the sf cDNA library are amplified in at least 10 tubes (100 μl each) by PCR, using vector λgt10 outer primer pair (set I) 5′ CCACCTTTTGAGCAAGTTCAG 3′ and 5′ GAGGTGGCTTAT GAGTATTTC 3′.

The reaction conditions in the final 100 μl of PCR mixture are as follows: 10 ng of recombinant λgt10 phage DNA, a 0.6 μM concentration of each primer, 2.5 units of AmpliTaq DNA polymerase (Perkin-Elmer, Norwalk, CT), with the recommended buffer except that $MgCl_2$ is at 2 m*M*. The hot start PCR is initiated by first denaturing the DNA along with all of the components of the PCR mixture except the dNTPs in a volume of 90 μl (covered with mineral oil) at 94° for 3 min and then setting the machine at pause mode at 80° and underlayering 10 μl of 2 m*M* dNTP solution in each of the tubes at 80°. Hot start PCR can also be initiated using commercially available Ampliwax beads (Perkin-Elmer). The PCR cycling conditions for

[25] P. Chomczynski and N. Sacchi, *Anal. Biochem.* **162,** 156 (1987).

[26] G. Cathala, J.-F. Sayouret, B. Mendez, B. L. West, M. Karin, J. A. Marctial, and J. D. Baxter, *DNA* **2,** 329 (1983).

[27] U. Gubler and B. J. Hoffman, *Gene* **25,** 263 (1983).

a Perkin-Elmer 9600 machine are 94° for 45 sec/50° for 45 sec/72° for 2 min for 30 cycles with a final extension of 5 min at 72° at the end of 30 cycles. The cDNA (400 and 2000 bp) is size fractionated on a low melting point agarose gel.

An alternative to cDNA libraries is to use size-fractionated random primed cDNA with ligated amplifying primers directly for cDNA selection if a good supply of RNA is at hand.

Blocking/Quenching Reagents

Ribosomal RNA clones: Human ribosomal RNA 45S precursor-coding region *Eco*RI fragments of 7.3 kb (pR7.3) and 5.8 kb (pR5.8) cloned in the plasmid pBR322[28]

Repeat libraries, RL-I: A high-copy genomic repetitive sequence library is prepared from a human partial digest genomic library by pooling DNA from 500 or more plaques that hybridize to ^{32}P-labeled total genomic DNA. Alternatively, commercially available Cot1 DNA (BRL) supplied as sonicated DNA can be used

Poly(dI·dC): From Pharmacia (27-7880; Pharmacia, Piscataway, NJ); it is dissolved in water and used as such without any further treatment

RL-II: Single human chromosome-specific genomic library subcloned in a high-copy plasmid vector such as Bluescript (Stratagene, La Jolla, CA). This is an optional blocking reagent to quench some of the medium-abundance repeats. The chromosome library to be used for this purpose should be derived from a chromosome different from the one from which the YAC is derived

Yeast DNA: DNA from yeast host alone (without any YACs) such as *Saccharomyces cerevisiae* AB1380

Yeast ribosomal RNA: The following two additional blocking reagents derived from yeast are needed when using yeast total DNA containing a YAC as the genomic target for selection. RibH15 and RibH7 clones contain the entire yeast ribosomal RNA precursor sequence as 7.3-kb and 2.5-kb *Hin*dIII fragments, respectively, in pBR322[7,29]

Processing of Blocking Reagents

DNA samples (yeast or plasmid) are prepared by standard methods,[23,24] digested with *Eco*RI, ethanol precipitated after phenol–chloroform extrac-

[28] J. M. Erickson, C. L. Rushford, D. J. Dorney, G. N. Wilson, and R. D. Schmickel, *Gene* **16,** 1 (1981).

[29] P. Philippsen, M. Thomas, R. A. Kramer, and R. W. Davis, *J. Mol. Biol.* **123,** 387 (1978).

tions, and sonicated in 10 mM Tris-HCl (pH 8.0)–10 mM EDTA to generate DNA of ~0.2–0.8 kb in size. After treatment with mung bean nuclease (New England BioLabs), 1 unit/μg DNA for 30 min at room temperature to make them blunt ended, the DNA samples are extracted with phenol–chloroform and ethanol precipitated. Concentrated DNA stocks (20×) of each quenching reagent are prepared as follows: RL-I (or Cot1 DNA), 0.5 μg/μl; RL-II, 1 μg/μl; pR7.3 and pR5.8 (1:0.8 ratio) mixture, 0.8 μg/μl; poly(dI·dC), 0.4 μg/μl; RibH15, 4 μg/μl; RibH7, 2 μg/μl; yeast host DNA (AB1380), 0.5 μg/μl.

Hybridization Selection on Nylon Paper Disks

Immobilization of Target Genomic DNA onto Nylon Disks

Hybond-N nylon (Amersham, Arlington Heights, IL) is cut into small (2.5 × 2.5 mm) squares (3 × 3 mm if using total yeast DNA containing a YAC). A batch of Hybond-N that gives a clean background is the obvious choice. The nylon disks are marked on the back (non-DNA side) with a pencil. The genomic target DNA (cosmid, YAC, etc.) along with carrier DNA (*Hae*III-digested ϕ × 174) is denatured at 95–98° for 3 min and then chilled on ice immediately, and an equal volume of 20× SSC, pH 7, is added and mixed (1× SSC is 0.15 M NaCl plus 0.015 M sodium citrate). From this mixture, a total of 0.2 ng of cosmid DNA or 10–15 ng of YAC DNA (for a 300-kb YAC) along with 50 ng of carrier DNA is spotted onto nylon disks. In the case of the total yeast DNA containing a 300-kb YAC, we use 50–200 ng of total yeast DNA without any carrier DNA. Other genomic targets such as P1 or BAC DNA (~5 ng of DNA) can also be used similarly. The dried disks are placed on a small sheet of Whatman (Clifton, NJ) paper and placed sequentially onto several sheets of Whatman 3MM presoaked with 0.5 M NaOH–1.5 M NaCl (5 min), 0.5 M Tris (pH 7.5)–1.5 M NaCl (1 min, and another 2–3 min on a new stack with the same solution); and finally to 10× SSC for 1 min. The disks are transferred onto a dry Whatman filter and UV cross-linked in a Stratalinker (Stratagene) set on Autocrosslink mode for 1 min while they are damp. The disks are then transferred into 0.5-ml Eppendorf tubes and the tubes are baked at 80° with lids open in a vacuum oven for 30 min. The disks are stored dry at 4°.

Quenching/Blocking Step

The quenching/blocking step blocks repeats, ribosomal RNA, and other GC-rich sequences. Prehybridization of the target DNA on membrane disks is set up in 0.5-ml Eppendorf tubes in a total volume of 50 μl (overlaid with mineral oil) with a final concentration of 5× SSPE [1× SSPE is 0.18 *M* NaCl, 10 m*M* $NaPO_4$, and 1 m*M* EDTA (pH 7.7)],[24] 5× Denhardt's,[24] 0.5% sodium dodecyl sulfate (SDS), and 1× heat-denatured (5 min at 98°) quenching agents prepared from 20× concentrated solution (see Processing of Blocking Reagents). After completion of prehybridization (65° for 24 hr), the oil phase is removed carefully and the nylon disks are washed twice briefly with 5× SSPE containing 0.1% SDS at room temperature. The disks are kept in this solution at room temperature until ready for hybridization.

Hybridization of cDNA Inserts to Target Genomic DNA (YACs or Cosmids)

Aliquots (40–50 μl) of hybridization mix [same composition as prehybridization mix, excluding yeast DNA and poly(dI:dC)] containing heat-denatured PCR-amplified cDNA library are distributed into 0.5-ml Eppendorf tubes, and the wet nylon disks from the prehybridization tubes are transferred on Pipetman tips into these tubes. The concentration of cDNA (prepared by PCR with the flanking vector primers) is 10 μg/ml in the final hybridization mix. If a mixture of several cDNA libraries (multiplex cDNA) is desired, then the concentration of each cDNA library is 5 μg/ml in the final hybridization mix. The hybridization samples in the tubes are covered with mineral oil and incubated at 65° for 36–40 hr with gentle rocking. Other buffer systems, e.g., Church-Gilbert buffer, have also given good results.[30]

Removal of Nonspecific Bound cDNAs

The top oil phase and hybridization solution are removed from the tubes after completion of hybridization, without letting the disks dry at any stage. The nylon disks are washed briefly twice with 300 μl of wash solution I (2× SCC–0.1% SDS) at room temperature. Using Pipetman tips, transfer the wet disks to fresh 1.5-ml Eppendorf tubes, each containing 600 μl of prewarmed solution I; do not allow the disks to make contact with the oily walls of the tubes. The disks are washed with 600 μl of solution I at 65°, three times for 20 min each in a shaking water bath; remove the wash solution in between washes as before and vortex intermittently. Next, the disks are transferred to fresh tubes again and washed with 600 μl of solution II (1× SSC–0.1% SDS) at 65° once for 20 min and with solution III (0.2× SSC–0.1% SDS) once at 65° for 15 min. The disks are transferred to fresh

[30] S. R. Patanjali, H. X. Xu, S. Parimoo, and S. M. Weissman, *in* "Identification of Transcribed Sequences" (U. Hochgeschwender and K. Gardiner, eds.), p. 29. Plenum Press, New York, 1994.

tubes and washed with 600 μl solution IV (0.1× SSC–0.1% SDS) twice at 65° for 20 min each. Subsequently the disks are rinsed with solution V (0.1× SSC) twice at room temperature, and transferred to 0.5-ml siliconized eppendorf tubes containing 40 μl of autoclaved water without carrying over the salt solution from the previous tube. The tubes can be stored overnight at 4° or processed immediately for PCR amplification.

Elution of Target-Bound cDNAs and Their Amplification

The tubes containing the disks in 40 μl of water are heated in the PCR machine at 98° for 5 min, and briefly centrifuged before being chilled. From each tube 20 μl is transferred to fresh PCR tubes for PCR amplification, and the rest is stored as backup samples at −20°. Hot start PCR is carried out with 20 μl of eluate from the nylon disks in a 50-μl volume, using the same primer set I for 30 cycles with the same cycling conditions as described above, except that the concentration of primers is 0.3 μM in the final 50-μl PCR mixture. To generate enough material for subsequent steps, a second round of PCR is performed by taking out 3 μl from the first round of PCR for reamplification by PCR in 100-μl aliquots, with a 0.6 μM final concentration of primer set I as described earlier. This PCR product cDNA is size fractionated on a 1% low melting point agarose gel and DNA greater than 0.35 kb is cut out and processed for DNA extraction and ethanol precipitated as described above. Centricon-30 or Centricon-100 (Amicon, Danvers, MA) can be used to reduce the volumes. The DNA is dissolved in 50 μl of TE (10 mM Tris-HCl, pH 8.0, 1 mM EDTA) and an aliquot is taken for estimation of DNA concentration.

Second Cycle of cDNA Selection

To further enhance the enrichment, an additonal cycle of cDNA selection with two rounds of PCR can be performed with once-selected material and new disks bearing the target genomic DNA as before, but with a few changes. First, during hybridization, a lower concentration of cDNA (from once-selected material) is used; the final concentration of cDNA is 0.25–0.5 μg/ml for single cDNA library-derived material or 0.5–1.0 μg/ml if multiplex cDNA libraries are used during the first cycle of cDNA selection. Second, the PCR amplification is carried out under the same conditions as for the first cycle except that the vector inner primers (set II: 5′ AGCCTGG TTAAGTCCAAGCTG 3′ and 5′ CTTCCAGGGTAAAAAGCAAAAG 3′) are used for two rounds of PCR amplification as before. The twice-selected cDNA is extracted sequentially with phenol, phenol–chloroform, and chloroform alone, and ethanol precipitated after addition of 1/2 vol of 7.5 M ammonium acetate and 2 μg of glycogen as carrier.

Southern Blot Analysis of Selected cDNA

The Southern blot analysis step is optional. An aliquot (0.5 μl) of cDNA from the PCR mix of the first and second selection is run along with the total PCR-amplified cDNA from the cDNA library on an agarose gel and transferred to a Southern blot. The blots are analyzed with various specific probes (single-copy sequences encoded by the YAC) if available and nonspecific probes such as total genomic DNA (for repeats) and ribosomal RNA probes. An example of representative results of selection with a YAC is shown in Fig. 2. The Southern blot shows enrichment of MHC class I genes and an anonymous sequence (B30.7) encoded by the YAC B30H3. The gene B30.7 codes for a rare-abundance RNA species (hence hardly visible in total cDNA before selection). The signal from the selected cDNA lanes with the probes for repeats and ribosomal probes (nonspecific se-

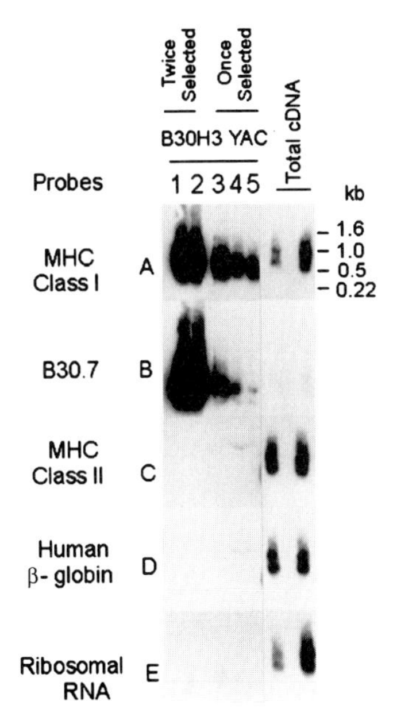

FIG. 2. Southern blot of cDNAs before selection (total cDNA lanes) and after one or two cycles of selection, as indicated, with YAC B30H3 from the human *HLA-A* region. Lanes 1–4 contain selected material obtained after selection with 5 ng of purified YAC DNA as the target genomic DNA; lane 5 carries selected material obtained by hybridization selection with a nylon strip carrying YAC DNA transferred directly from a CHEF gel. Identical blots were probed with various YAC sequence-specific probes: (A) *HLA-A;* (B) anonymous clone B30.7, or nonspecific probes; (C) human MHC class II; (D) human β-globin; (E) human ribosomal RNA probe. The final concentration of spleen short-fragment cDNA was 10 and 0.25 μg/ml during the first and second cycle selection, respectively.

quences) is invisible in the selected samples, in contrast to the total cDNA sample before selection.

Cloning and Characterization of Selected Library

The twice-selected cDNA is digested with *Eco*RI and size fractionated on a 1% low melting point agarose gel. The cDNA (>0.3 kb) is extracted and ethanol precipitated in the presence of ammonium acetate and glycogen as described earlier, and dissolved in 10 m*M* Tris-HCl (pH 7.5)–0.1 m*M* EDTA (TLE). A ligation reaction with 200 units of T_4 DNA ligase (New England BioLabs) is set up in 5 μl with 0.5 μg of *Eco*RI-cut and dephosphorylated phage in λgt10 arms [Promega (Madison, WI) or Stratagene] and 5 ng of twice-selected cDNA in the presence of 50 m*M* Tris-HCl (pH 7.8)–10 m*M* $MgCl_2$–1 m*M* ATP–5 m*M* dithiothreitol (DTT). A 4-μl aliquot of the ligation mix is used for *in vitro* packaging with a Gigapack Plus packaging kit (Stratagene) and subsequent infection into *Escherichia coli* c600 hfl cells as per the manufacturer instructions. A typical single packaging reaction yields 0.1–3 million phage plaques. Phage lysate is harvested with 15 ml of SM from one confluent plate. This represents a high-titer selected library (titer > ~10). One can use 1 μl out of 100 μl from a single-phage suspension for a 50-μl PCR. Distinct single bands ranging in size from 300 to 600 bp or more are visible on analyzing 5–10 μl on agarose gels. At least 90% or more of the plaques are cDNA insert-bearing recombinant plaques.

Screening of Selected cDNA Library for Novel Clones

At least 50 plaques for every 100 kb of YAC are picked at random, resuspended individually in 100 μl of SM, and amplified by PCR of 1 μl in 50 μl as described earlier. A small aliquot (1 μl) of the PCR product is dot blotted from each sample in an array of 96 dots and multiple blots are made depending on the number of the probes to be used. The blots are probed with ribosomal RNA probes and total genomic DNA (such as human DNA) to assess the level of coselection of undesired sequences. The remaining clones are potentially novel clones. The level of enrichment of individual cDNAs in the selected library in relation to the original library is a function of both genome complexity of individual YACs (such as the presence of pseudogenes and multigene family members) and the extent of coselection of undesirable sequences such as ribosomal RNA sequences, repeats, etc. Pools of the PCR products from rows and columns of the 96-grid plate can be used as probes to identify overlapping clones. Single-pass sequencing of unique clones can be achieved by cycle sequencing (BRL PCR-based cycling sequence kit). For sequencing, 2–3 μl is taken out of 50 μl of the PCR-amplified product (amplified with outer primer set

I) and subjected to PCR-based cycling sequencing, without any intermediate purification, using one of the inner primers of set II as ^{32}P end-labeled sequencing oligonucleotide. These sequences can be used for any sequence matches in the GENEMBL databank. Any unique (novel) clones from a selected library can be confirmed by Southern blots using total genomic DNA, YAC DNA, and DNA from a somatic hybrid cell line that is known to harbor a chromosome region corresponding to the YAC of interest. A full-length cDNA library can be screened subsequently, using a selected cDNA clone as a probe.

Controlling Parameters for cDNA Selection

Since cDNA selection is a PCR-based method, all of the necessary precautions and good laboratory practices for PCR that are necessary to avoid contamination problems should be followed, such as use of aerosol-resistant Pipetman tips, good-quality reagents, and autoclaved plasticware and solutions (whenever possible). An excessive number of PCR cycles (>30) should be avoided as this generates high molecular weight products after primers are exhausted. Nylon membranes used for immobilization of target DNA should be prescreened for high signal-to-noise ratio in a typical Southern blot experiment. Care should be exercised not to allow disks to be covered with mineral oil during the experimentation. The immobilized target DNA (e.g., YAC DNA) should be free of yeast RNA or any oligonucleotides. All of the blocking reagents (plasmid and DNA) should be completely digested with *Eco*RI and sonicated and made blunt ended by treatment with mung bean nuclease as described above. The presence of a large number of clones of repeats or ribosomal RNA sequences in the selected library could reflect either a YAC that is deficient in genes or improper preparation of the blocking reagents. If a YAC comigrates with a yeast chromosome in the 1- to 2-Mb region, one should include yeast ribosomal RNA quenching agents as additional blocking reagents as in the total yeast DNA selection.

Applications of cDNA Selection Approach

The obvious application of cDNA selection is in the positional cloning of disease genes, which involves the identification of candidate genes in a chromosomal segment defined by genetic linkage studies. Although in many cases a good judgment can be made as to which cDNA libraries should be used for searching candidate genes, many times it may be necessary to use more than one library if one is unsure about the tissues that most likely express the disease gene.

With the availability of chromosome-specific cosmid libraries and

YAC-based physical maps, another application of cDNA selection is in the generation of transcript maps, which can be subchromosome specific or chromosome specific depending on the range of YAC contigs (overlapping clones) used. In fact, chromosome 7-specific cDNAs, isolated by cDNA selection, have been reported.[31]

The method can be used to isolate chromosome-specific or subchromosome-specific, tissue-specific cDNAs if the cDNA selection is carried out using two different cDNA libraries followed by subtractive hybridization.[31a] Alternatively, differential display of the selected cDNA libraries can identify differentially regulated cDNAs.[31b]

The selection procedure has been used to isolate polymorphic clones containing CA dinucleotide repeats.[32] A selection approach has also been described for enrichment of sequences conserved between species.[33] The method has a potential to select jumping and linking clones.[33a,33b]

cDNA selection allows handling of multiple YACs and cDNA libraries simultaneously. Hence the method has the potential for automation for large-scale transcript map projects.

Human Major Histocompatibility Complex Region as a Model System for cDNA Selection

The cDNA selection approach with the multiplex cDNA libraries has been used extensively in the human major histocompatibility (MHC) region.[15–21] In the 3-Mb region, spanning *HLA-E* to the telomeric histone cluster, more than 90 distinct genes, pseudogenes, or ESTs have been identified in addition to approximately 15 class I genes and their pseudogenes, making this region gene dense, with an average density of a gene every ~20 kb.[21] Of the 90 potential coding sequences, 63 have been characterized as distinct genes with homologies to known gene/gene families (35 of 63) or of unknown function (28 of 63). Each of the 63 nonredundant genes was classified as distinct on the basis of several criteria such as

[31] J. W. Touchman, G. G. Bouffard, L. A. Weintraub, J. R. Idol, L. Wang, C. M. Robbins, J. C. Nussbaum, M. Lovett, and E. D. Green, *Genome Res.* **7,** 281 (1997).

[31a] A. Swaroop, J. Xu, N. Agarwal, and S. M. Weissman, *Nucleic Acids Res.* **19,** 1954 (1991).

[31b] J. R. Pardinas, N. J. Combates, S. M. Prouty, K. S. Stenn, and S. Parimoo, *Anal. Biochem.* **257,** 161 (1998).

[32] R. P. Kandpal, G. Kandpal, and S. M. Weissman, *Proc. Natl. Acad. Sci. U.S.A.* **91,** 88 (1994).

[33] Z. Sedlacek, D. S. Konecki, R. Siebenhaar, P. Kioschis, and A. Poustka, *Nucleic Acids Res.* **21,** 3419 (1993).

[33a] F. S. Collins and S. M. Weissman, *Proc. Natl. Acad. Sci. U.S.A.* **81,** 6812 (1984).

[33b] H. Arenstorf, R. P. Kandpal, N. Baskaran, S. Parimoo, Y. Tanaka, S. Kitajima, Y. Yasukochi, and S. M. Weissman, *Genomics* **11,** 115 (1991).

hybridization to RNAs from several cell lines/tissues on Northern blots, isolation of their cognate full-length/near full-length cDNA clones from oligo$(dT)_{12-18}$ libraries, and sequence analysis. Twenty-one of the 63 genes had been described by others also. Of the remaining 42 genes, only half had homologous sequences in the dbEST.[21] Detailed analyses of expression studies with genes have indicated some genes to be restricted in expression to cells of the immune system.[19,20] Of all the known genes in this region only a small fraction were absent in our selected libraries.[21] However, it is not clear whether it is because of the absence of these clones in our starting cDNA libraries or the use of YACs in certain regions that may represent minor rearrangements in different YACs. On the other hand, the region telomeric to the class II region, including the tumor necrosis factor (TNF) locus, is one of the well-characterized MHC regions, yet cDNA selection, besides identifying all of the known genes of the region, identified two novel genes whose expression was limited to cells of the immune system.[19] In addition, cDNA selection was able to identify the human homolog of mouse olfactory receptor gene in the *HLA-F* region. The abundance of this cDNA was less than 1 in a million in the starting cDNA library. Overall, extensive experimentation with cDNA selection in the MHC region has demonstrated it to be an efficient and rapid method of identifying genes within large genomic segment(s) and sometimes proved more fruitful than sequence scanning for exons by GRAIL[21,34]; this is particularly important because some coding sequences are embedded within introns of other genes.[35] Detailed analysis of several randomly picked twice selected cDNA clones from several YAC selected libraries showed that essentially all of the selected clones, after eliminating ribosomal RNA and repeat clones, did hybridize back to their cognate YACs.[16,21] Additional cycles of selection do not seem to enhance enrichment.[11]

cDNA Selection with Other Chromosomal Regions

In collaborations with other investigators, the method has been applied to YACs/YAC contigs from the long arm of chromosome 21,[36] the DiGeorge syndrome region in chromosome 22q11,[37] and the Werner syn-

[34] E. C. Uberbacher and R. J. Mural, *Proc. Natl. Acad. Sci. U.S.A.* **88,** 11261 (1991).

[35] R. D. Campbell and J. Trowsdale, *Immunol. Today* **14,** 49 (1993).

[36] H. Xu. H. Wei, F. Tassone, S. Graw, K. Gardiner, and S. M. Weissman, *Genomics* **27,** 1 (1995).

[37] H. Sirotkin, H. O'Donnell, R. DasGupta, S. Halford, B. St-Jore, A. Puech, S. Parimoo, B. Morrow, A. Skoultchi, S. M. Weissman, P. Scambler, and R. Kucherlapati, *Genomics* **41,** 75 (1997).

drome region of chromosome 8p.[38] cDNA clones of rare abundance in the starting cDNA library have been selected from other YACs besides those from the MHC region.[30] The level of coselected undesirable sequences as examined with three different YACs from three different chromosomes is less than 10%. Although a strict comparison between cDNA selection carried out with a purified YAC versus total yeast DNA containing the same YAC has not been done, it is possible that purified YACs in some instances may result in better enrichments. Nevertheless, selection with total yeast DNA containing a YAC is a rapid approach and can identify a large number of cDNAs as a first step in identification of transcribed sequences.[21]

Critical Analysis of cDNA Selection

The method described here uses short-fragment random primed cDNA libraries. Their use is essential in this PCR-based selection approach because PCR is biased against long cDNAs and GC-rich regions and can eliminate some rare cDNA species if oligo(dT)-primed cDNA libraries are used. The downside of short-fragment cDNA libraries is that eventually full-length cDNA clones must be isolated by traditional or other approaches.[39] Rare cDNAs that may contain multiple internal *Eco*RI sites, or that are very short (less than 200 bp) may be missed. While the former problem can be solved by direct cloning of PCR products, the latter problem may be dealt with by carrying out selection separately with oligo$(dT)_{12-18}$-primed cDNA libraries. Although long-range PCR kits are available now, their use in selection process has not been explored. Being a cDNA-based approach, an ideal situation requires a cDNA library representing all possible sequences. Although there are serious practical obstacles to obtaining such libraries for humans, multiplexing cDNA libraries from different tissues and/or the use of pooled and normalized libraries makes it possible to approach such a complete library.[40–42] One group of investigators has used a normalized cDNA library for cDNA selection.[14]

The membrane-based method, in contrast to the solution-based method, is a relatively simple approach because it does not involve amplification and subsequent manipulation of the genomic target DNA. Whereas the

[38] C. E. Yu, J. Oshima, Y.-H. Fu, E. M. Wijsman, F. Hisama, R. Alisch, S. Matthews, J. Nakura, T. Miki, S. Ouais, G. M. Martin, J. Mulligan, and G. D. Schellenberg, *Science* **272,** 258 (1996).

[39] H. Xu, H. Wei, R. Kolluri, and S. M. Weissman, *Nucleic Acids Res.* **23,** 4528 (1995).

[40] S. R. Patanjali, S. Parimoo, and S. M. Weissman, *Proc. Natl. Acad. Sci. U.S.A.* **88,** 1943 (1991).

[41] M. S. H. Ko, *Nucleic Acids Res.* **18,** 5705 (1990).

[42] M. B. Soares, M. F. Bonaldo, P. Jelene, L. Su, L. Lawton, and A. Efstratiadis, *Proc. Natl. Acad. Sci. U.S.A.* **91,** 9228 (1994).

exon-trapping methods[43,44] are sensitive to cryptic splice sites, cDNA selection, being a hybridization selection approach, is independent of splicing mechanisms and can identify genes that lack introns. The cDNA selection method described in this chapter is a PCR-based method, and hence point mutations induced by PCR in the selected cDNAs are possible. It is not a serious problem because one must eventually isolate full-length cDNA clones and the selected cDNA are good probes for that purpose. Nevertheless, with the availability of proofreading *Taq* polymerases, it is possible to overcome this problem.

Some YACs may contain region-specific, low-level repeats that may not be competed out by the quenching cocktail. In such cases, a significant number of cDNAs with homologies to such repeats may be selected. One way of overcoming this problem is to hybridize the selected library plaques to such repeats and to remove them physically. Alternatively, one can make additional quenching/blocking reagents from several plaques that hybridize to such repeats. YACs encoding pseudogenes or sequences related to multigene families may select sequences encoded by chromosomes different from the one related to the YAC of interest. Such clones can be discriminated by hybridization to Southern blots of somatic cell hybrids containing the individual chromosomes (available commercially). If, on the other hand, these multigene family members are resident on a single chromosome, sequencing and/or oligonucleotide hybridization should be able to discriminate such clones.

Conclusion

The cDNA selection approach is a relatively simple, efficient, and rapid method of identification of chromosome- or subchromosome-specific cDNAs. The approach is certainly more efficient and sensitive in detecting rare species of cDNAs than is direct screening of full-length cDNA libraries with a series of YACs. Many disease genes have been identified using either cDNA selection or exon amplification methods.[5] Although any particular method is not without limitations, the decision to use a particular methodology for identifying genes is largely dependent on the availability of starting materials and the ultimate goal. Hence, a particular situation may even demand the use of more than one approach to complement the limitations of the others. Until the time when the human genome is completely sequenced, cDNA selection will remain one of the simple and reliable methods for identification of cDNAs from large genomic segments.

[43] G. M. Dyuk, S. Kim, R. M. Myers, and D. R. Cox, *Proc. Natl. Acad. Sci. U.S.A.* **87,** 8995 (1990).

[44] A. J. Buckler, D. D. Chang, S. L. Graw, J. D. Brook, D. A. Haber, P. A. Sharp, and D. E. Housman, *Proc. Natl. Acad. Sci. U.S.A.* **88,** 4005 (1991).

[10] Saturation Identification of Coding Sequences in Genomic DNA

By KATHELEEN GARDINER

Introduction

The identification of coding sequences within genomic DNA of higher organisms is complicated by the presence of introns, large intergenic distances, and transcribed repetitive sequences. Genes within introns, overlapping genes, and small exons buried within large introns increase the difficulty. When identification of all genes within a region is the aim, the problem is further complicated by issues of tissue specificity and developmental timing of expression.

Several chapters in this volume deal with specific aspects of and individual approaches to gene identification (e.g., construction and analysis of cDNA libraries, cDNA selection). This chapter is designed to provide an overview of comprehensive transcriptional mapping as it would be applied to the genomes of mammals or other higher vertebrates. In this context, we have chosen to discuss four techniques for gene identification that have demonstrated their utility for reliable analysis of large genomic regions. These are as follows.

dbEST analysis[1]: The dbEST (database of expressed sequence tags) is an NCBI (National Center for Biotechnology Information) database

[1] G. D. Schuler, M. S. Boguski, E. A. Stewart, L. D. Stein, G. Gyapay, K. Rice, R. E. White, P. Rodriquez-Tome, A. Aggarwal, E. Bajorek, S. Bentolilia, B. B. Birren, A. Butler, A. B. Castle, N. Chiannikulchai, A. Chu, C. Clee, S. Cowles, P. J. R. Day, T. Dibling, N. Drouot, I. Dunham, S. Duprat, C. East, C. Edwards, J.-B. Fan, N. Fang, C. Fizames, C. Garret, L. O. Green, D. Hadley, M. Marris, P. Jarrison, S. Brady, A. Hicks, E. Holloway, L. Hui, S. Hussain, C. Louis-Dit-Sully, J. Ma, A. MacGilvery, C. Mader, A. Maratukulam, T. C. Matise, K. B. McKusick, J. Morissette, A. Mungall, D. Muselet, H. C. Nusbaum, D. C. Page, A. Peck, S. Perkins, M. Piercy, F. Qin, J. Quackenbush, S. Ranby, T. Reif, S. Rozen, C. Sanders, X. She, J. Silva, D. K. Slonim, C. Soderlund, W. L. Sun, P. Tabar, T. Thangarajah, N. Nega-Czarny, D. Vollrath, S. Voyticky, T. Wilmer, X. Wu, M. D. Adams, C. Auffray, N. A. R. Walter, R. Brandon, A. Dehejia, P. N. Goodfellow, R. Houlgatte, J. R. Hudson, Jr., S. E. Ide, K. R. Iorio, W. Y. Lee, N. Seki, T. Nagase, K. Ishikawa, N. Nomura, C. Phillips, M. H. Polymeropoulos, M. Sandusky, K. Schmitt, R. Berry, K. Swanson, R. Torres, J. C. Venter, J. M. Sikela, J. S. Beckmann, J. Weissenbach, R. M. Myers, D. R. Cox, M. R. James, D. Bentley, P. Deloukas, E. S. Lander, and T. J. Hudson, *Science* **274,** 547 (1967).

Copyright © 1999 by Academic Press
All rights of reproduction in any form reserved.

0076-6879/99 $30.00

containing end sequences of random, arrayed cDNA clones from a large number of tissue sources. These cDNA clones are now being mapped to chromosomal regions, using a panel of radiation hybrid cell lines.

cDNA selection: cDNAs from any tissue source are isolated on the basis of their hybridization to genomic clones from the region of interest.[2,3]

Exon trapping: Genomic fragments are subcloned into a "splicing vector" and transfected into cos cells. Cellular machinery is used to splice exons *in vivo* from the fragment.[4]

Genomic sequence analysis: A growing number of increasingly sophisticated programs provide predictions of exon locations and gene models within DNA sequence.[5–8]

Generally developed for analysis of mammalian genomes, these techniques are applicable to any genome where the presence of introns, large intergenic distances, and/or repetitive DNA significantly reduces the proportion of coding to noncoding sequence.

More traditional techniques, including CpG island identification and detection of evolutionary conserved sequences combined with direct screening of cDNA libraries,[9,10] are inappropriately laborious for large chromosomal regions. They are also by no means exhaustive, even for cosmid-sized segments. This is true, in the first case, because CpG islands are associated with only 40–60% of genes.[11] In the second case, while exons are indeed generally highly conserved among species and, therefore, can sometimes be identified by cross-hybridization on "zoo blot" Southerns, in many cases small exons are buried within large introns and make ineffective hybridization probes.[12]

Additional procedures include screening of genomic libraries directly

[2] S. Parimoo, S. R. Patanjali, H. Shukla, D. D. Chaplin, and S. M. Weissman, *Proc. Natl. Acad. Sci. U.S.A.* **88,** 9623 (1991).

[3] J. G. Morgan, G. M. Dolganov, S. E. Robbins, L. M. Hinton, and M. Lovett, *Nucleic Acids Res.* **20,** 5173 (1992).

[4] A. J. Buckler, D. D. Chang, S. L. Graw, J. D. Brook, D. A. Haber, P. A. Sharp, and D. E. Housman, *Proc. Natl. Acad. Sci. U.S.A.* **88,** 4005 (1991).

[5] E. C. Uberbacher and R. J. Mural, *Proc. Natl. Acad. Sci. U.S.A.* **88,** 11261 (1991).

[6] E. E. Snyder and G. D. Stormo, *J. Mol. Biol.* **248,** 1 (1995).

[7] N. L. Harris, *Genome Res.* **7,** 754 (1997).

[8] C. Burge and S. Karlin, *J. Mol. Biol.* **268,** 78 (1997).

[9] A. P. Bird, *Nature (London)* **321,** 209 (1986).

[10] A. P. Monaco, R. L. Neve, C. Colletti-Feener, C. J. Bertelson, D. M. Kurnit, and L. M. Kunkel, *Nature (London)* **323,** 646 (1986).

[11] F. Larsen, G. Gunderson, R. Lopez, and H. Prydz, *Genomics* **13,** 1095 (1992).

[12] I.-T. Chen and L. A. Chasin, *Mol. Cell. Biol.* **14,** 2140 (1994).

TABLE I
LIMITATIONS OF INDIVIDUAL GENE IDENTIFICATION TECHNIQUES

dbEST
Clones are 3′-end specific, i.e., largely 3′ UTRs
No coding sequence for larger transcripts
Dependent on expression
Dependent on the efficiency of reverse transcriptase
Dependent on cloning efficiency
Presence of priming from internal poly(A)$^+$ tracts; possible intron sequences
cDNA selection
Dependent on expression
Intronic material selected from incompletely processed mRNA
Occurrence of PCR artifacts increases in gene-poor or expression-poor regions
Exon trapping
Dependent on the presence of introns
Variable efficiency with which exons are trapped
Biologically irrelevant sequences trapped
Small size of single coding exons
Genomic sequence analysis
Generation of genomic sequence not universally accessible
Genes within introns, overlapping genes, very large genes confound gene model predictions
AT-rich sequences produce higher false-positive rates

with cDNA and reciprocal screening of genomic and cDNA libraries.[13,14] These techniques have been productive,[14–16] although they have not generally received the attention that they deserve. Their particular strengths lie in identifying genes expressed in specific tissues and in determining differential expression. Although they are not discussed further here, they are worth considering if specificity of expression is a critical issue.

All four techniques listed above are powerful, but as summarized in Table I and discussed below, each has limitations and artifacts associated with it. For effective gene searches, it must be emphasized that no single approach will be adequate. For truly comprehensive gene identification, all four techniques must be applied. While this cannot guarantee identification of all coding regions, the complementarities among approaches will go a long way toward this goal. Furthermore, identification of the same

[13] U. Hochgeschwender, J. G. Sutcliffe, and M. B. Brennan, *Proc. Natl. Acad. Sci. U.S.A.* **86,** 8482 (1989).

[14] C. C. Lee, A. Yazdani, M. Wehnert, Z. Y. Zhao, E. A. Lindsay, J. Bailey, M. I. Coolbaugh, L. Couch, M. Xiong, A. C. Chinault, A. Baldini, and C. T. Caskey, *Hum. Mol. Genet.* **4,** 1373 (1995).

[15] M. B. Brennan and U. Hochgeschwender, *Hum. Mol. Genet.* **3,** 2019 (1994).

[16] J. F. Coy, P. Kioschis, Z. Sedlacek, and A. Poustka, *Mammalian Genome* **5,** 131 (1994).

sequence using more than one technique decreases the probability of having an artifact and increases the likelihood of a bona fide coding sequence. A second point requires emphasis: at every step of transcriptional mapping, verification is required: verification that the identified sequences map to the correct genomic clones, and that they are not only transcribed, but represent part of the mature message (i.e., nonintronic). Neglecting to verify constantly can lead one down exciting alleys that are, nevertheless, blind.

Before describing the general advantages and disadvantages of each technique, two points will be discussed: (1) the preliminary considerations for defining the magnitude and scope of the gene search, and (2) necessary resources for beginning the gene search.

Preliminary Considerations

Once the genomic region of interest has been defined, there are several considerations that are of primary importance for determining how much emphasis should be given to each technique and how success of the gene search will be assessed.

Size of Genomic Region

Regions that are cosmid sized (to 100–200 kb) are ideally suited for exon trapping and genomic sequencing. Exon trapping can be applied exhaustively, using multiple cloning enzymes and individual cosmids. With access to automated sequencing, genomic sequencing in a region this size remains practical for essentially any laboratory. cDNA selection and dbEST analysis can be included with limited extra effort. In contrast, regions that range in size from megabases to entire chromosomal bands are not suitable for exon trapping and genomic sequencing as a first line of attack. In these cases, cDNA selection is most productive and dbEST has its greatest potential.

Type of Genomic Region

Although rarely given much consideration, the characteristics of the chromosomal band in which the genomic region lies are highly relevant, particularly in assessing the results of a gene search.

Isochore analysis has shown quite clearly that gene density varies dramatically among mammalian chromosomal bands, with a variation that is related to base composition.[17] Thus, the subclass of R bands known as T

[17] G. Bernardi, *Annu. Rev. Genet.* **29,** 445 (1995).

(for telomeric) bands are high in GC content and show the highest gene densities, estimated to be as high as one gene per 5–10 kb.[17,18] T bands are located at most human chromosome telomeres (exceptions include 3p and 18p) and at a number of internal locations. The best and most recent cataloguing of T bands in the human karyotype is in Saccone *et al.*[19] Gene searches in these regions should be easily productive, but care must be taken. While a perceived gene density of one per 30–50 kb may be satisfying, it is likely to be incomplete.

In contrast, G bands present opposite characteristics. Isochore analysis, mapping of human genes, and the in-depth analysis of a limited number of such bands, all indicate that G bands are gene poor.[17,20–23] Isochore analysis here predicts gene densities averaging perhaps one per 200–300 kb, including many genes from the largest size class and containing large (tens of kilobases) introns. The problems and cautions associated with these regions are obviously different from those of T bands. Here, exon trapping and cDNA selection efforts produce few bona fide coding sequences and generally also an increased proportion of nonspecific products and cloning artifacts. Exon prediction from sequence analysis suffers from a higher false-positive rate, leading to an increased proportion of failures to verify exon predictions in reverse transcriptase-polymerase chain reaction (RT-PCR) and/or Northern experiments. It is easy to blame such negative results unfairly on technical errors.

The remaining chromosomal regions are admittedly less straightforward to classify. The non-T-band R bands are of two types[24]: those that are moderately GC and gene rich, and those that are poor in both GC and gene content. The former are predicted to have gene densities of one per 30–50 kb; the latter resemble G bands.[17] Without genomic sequence data, these regular R-band types are indistinguishable and in such cases, band location is of little predictive value.

An additional problem, of course, lies with determining band location. This is especially equivocal with smaller internal bands and regions lying at band borders. Band location is, however, unambiguous with larger bands, especially telomeric bands and larger G bands, and in these cases the information is valuable and should be taken into account.

[18] B. Dutrillaux, *Chromosoma* **41,** 395 (1973).

[19] S. Saccone, S. Cacciò, J. Kusuda, L. Andreozzi, and G. Bernardi, *Gene* **174,** 85 (1996).

[20] K. K. Kidd, A. M. Bowcock, J. Schmidtke, R. K. Track, F. Ricciuti, G. Hutchings, A. Bale, P. Pearson, and H. F. Willard, *Cytogenet. Cell Genet.* **51,** 622 (1989).

[21] H. Xu, H. Wei, F. Tassone, S. Graw, K. Gardiner, and S. M. Weissman, *Genomics* **27,** 1 (1995).

[22] K. Gardiner, *Gene* **205,** 39 (1997).

[23] G. Pilia, R. D. Little, B. Aïssani, G. Bernardi, and D. Schlessinger, *Genomics* **17,** 456 (1993).

[24] G. P. Holmquist, *Am. J. Hum. Genet.* **51,** 17 (1992).

Assessing Success

The precise requirements of individual gene searches vary. For example, a search for developmentally relevant genes may be defined by their timing and tissue specificity of expression. In such cases, cDNA selection is of primary importance. Searches for disease genes may also be defined by the tissue affected. However, precisely how is less certain because a disease gene may be associated with expression that is inappropriate in place or level, either expressed in a tissue where it is normally silent, or expressed in the appropriate tissue but at levels greater or less than normal. Such ambiguity requires more thorough gene searches, in which all four techniques are applied.

Another point may be the general expression activity of a region. A DNA segment with genes of specific and/or limited expression may respond poorly to cDNA selection, but well to exon trapping.

Starting Materials

Yeast Artificial Chromosome Contiguous Genomic Clones

The most valuable resources for initiating a comprehensive gene search are a yeast artificial chromosome (YAC) contiguous genomic clone set (contig), a chromosome-specific arrayed cosmid library, and dbEST/RH mapped cDNAs.

One must, of course, have the genomic clones available, or it is not yet a transcriptional mapping project. As of this writing, YAC clones are the most prevalent of genomic clones, with the CEPH map and the developing MIT map being the two major sources of whole genome YAC sets.[25,26]

[25] I. M. Chumakov, P. Rigault, I. Le Gall, C. Bellané-Chantelot, A. Billault, S. Guillou, P. Soularue, G. Guasconi, E. Poullier, I. Gros, M. Belova, J.-L. Sambucy, L. Susini, P. Gervy, F. Gilbert, S. Beaufils, H. Bui, V. Perrot, M. Saumier, C. Soravito, R. Bahouayila, A. Cohen-Akenine, E. Barillot, S. Bertrand, J.-J. Codani, D. Caterina, I. Georges, B. Lacroix, G. Lucotte, M. Sahbatou, C. Schmit, M. Sangouard, E. Tubacher, C. Dib, S. Fauré, C. Fizames, G. Gyapay, P. Millasseua, S. Nguyen, D. Muselet, A. Vignal, J. Morissette, J. Menninger, J. Lieman, T. Desai, A. Banks, P. Bray-Ward, D. Ward, T. Hudson, S. Gerety, S. Foote, L. Stein, D. C. Page, E. S. Lander, J. Weissenbach, D. Le Paslier, and D. Cohen, *Nature (London)* **377,** 175 (1995).

[26] T. J. Hudson, L. D. Stein, S. S. Gerety, J. Ma, A. B. Castle, J. Silva, D. K. Slonim, R. Baptista, L. Kruglyak, S.-H. Xu, X. Hu, A. M. E. Colbert, C. Rosenberg, M. P. Reeve-Daly, S. Rozen, L. Hui, X. Wu, C. Vestergaard, K. M. Wilson, J. S. Bae, S. Maitra, S. Ganiatsas, C. A. Evans, M. A. DeAngelis, K. A. Ingalls, R. W. Nahf, L. T. Horton, Jr., M. O. Anderson, A. J. Collymore, W. Ye, V. Kouyoumjian, I. S. Zemsteva, J. Tam, R. Devine, D. F. Courtney, M. R. Renaud, H. Nguyen, T. J. O'Connor, C. Fizames, S. Fauré, G. Gyapay, C. Dib, J. Morissette, J. B. Orlin, B. W. Birren, N. Goodman, J. Weissenbach, T. L. Hawkins, S. Foote, D. C. Page, and E. S. Lander, *Science* **270,** 1945 (1995).

Many additional chromosome-specific or region-specific contigs have been and continue to be reported. However, even if the relevant contigs through the region of interest are not established, there are certainly individual clones to serve as starting points either in or near the region from which contig construction can then begin.

It is arguable whether YACs are the best source of genomic coverage, and undeniably, they have limitations.[27–30] These include deletions and rearrangements, which seem to be particularly prevalent in YACs deriving from telomeric bands, and possibly are prevalent in internal T bands as well.

The problem of deletions and rearrangements can be overcome by using the individual YAC clones as probes to screen an arrayed chromosome-specific cosmid library. Pulsed-field gel-purified YAC DNA can be labeled and repetitive sequences efficiently blocked with total human or Cot1 DNA. YACs up to 500–800 kb can be used in this fashion to identify a set of cosmids that can then be analyzed to define a minimal tiling path spanning the YAC. The advantages of this approach are that deletions in a YAC or gaps in a YAC contig are identified as holes in the cosmid contig; these can be closed by using cosmid end probes to rescreen the cosmid library and thus "walk" across the missing region. Rearrangements in YACs become irrelevant as long as cosmids can be obtained. Chimeric segments of the YACs also disappear because they are unlikely to derive from the same chromosome, and therefore will not occur in the chromosome-specific cosmid library.

Alternatively, as large-scale genomic sequencing of the human genome progresses, more and more chromosomes and chromosome segments will be converted to "sequence ready" maps, i.e., contigs of genomic clones that can be sequenced directly. These will most likely not be YACs, but instead, bacterial artificial chromosomes (BACs), phage artificial chromosomes (PACs), or P1 clones, which are believed to be more stable and reliable than YACs and that can also be more easily purified to a point suitable for subcloning for sequencing efforts.

Whatever the source of the genomic clones, it is worthwhile to spend some time to verify that the contig is complete, lacks holes, and faithfully represents the human genomic segment. The ideal way to do this is to

[27] W. R. Dackowski, T. D. Connors, A. E. Bowe, V. Stanton, Jr., D. Housman, N. A. Doggett, G. M. Landes, and K. W. Klinger, *Genome Res.* **6,** 515 (1996).

[28] A. De Sario, E.-M. Geigl, G. Palmieri, M. D'Urso, and G. Bernardi, *Proc. Natl. Acad. Sci. U.S.A.* **93,** 1298 (1996).

[29] K. Gardiner, S. Graw, H. Ichikawa, M. Ohki, A. Joetham, P. Gervy, I. Chumakov, and D. Patterson, *Somat. Cell Mol. Genet.* **21,** 399 (1995).

[30] P. R. Cooper, N. J. Nowak, M. J. Higgins, S. A. Simpson, A. Marquardt, H. Stoehr, B. H. Weber, D. S. Gerhard, P. J. de Jong, and T. B. Shows, *Genomics* **41,** 185 (1997).

physically map (generally by pulsed-field analysis) directly in human genomic DNA (DNA from an established cell line or from fresh lymphocytes) and to compare this map with that of the cloned contigs. This is extremely laborious and unlikely to be practical. A reasonable alternative is to develop a pair of contigs, such as YACs and cosmids, in two different cloning vectors. The hope is that segments that are unclonable or unstable in one vector–host combination will be found reliably in the second.

Identification of Coding Sequences

Database of Expressed Sequence Tags and Radiation Hybrid Mapping

This approach is discussed first, not because it is currently the most powerful or productive, but because it is the easiest to carry out.

The starting point is the database of expressed sequence tags (dbEST). This was developed from a series of cDNA libraries originally constructed by M. B. Soares *et al.*,[31] but now also coming from some additional sources. These libraries have been constructed using a wide variety of tissue sources, fetal and adult, including some unusual tissues such as cochlea, melanoma, and pregnant uterus. The libraries are oligo(dT) primed and directionally cloned, with average insert sizes of 1–2 kb. The cDNA clones from each library have been arrayed in microtiter plates, and groups largely from MERCK and TIGR are end sequencing thousands of clones from each.[32,33] The single run of sequence generated from each end of each clone provides approximately 300 nucleotides of sequence from the poly(A)$^+$ tail (the 3′ sequence) and from a point 1–2 kb upstream of the poly(A)$^+$ tail (the so-called 5′ sequence). This last point is particularly important. Most mRNAs are longer than 1–2 kb in length; thus the 5′ end sequence from these cDNAs will not represent the 5′ end of the message. Furthermore, a significant percentage of messages are longer than 4–5 kb and contain significant stretches of 3′ untranslated regions (UTRs). Thus, for many of these cases, the "5′" sequence in dbEST will likely not even be located within coding sequence.

[31] M. B. Soares, M. F. Bonaldo, P. Jelene, L. Su, L. Lawton, and A. Efstratiadis, *Proc. Natl. Acad. Sci. U.S.A.* **91,** 9228 (1994).

[32] L. Hillier, G. Lennon, M. Becker, M. F. Bonaldo, B. Chiapelli, S. Chissoe, N. Dietrich, T. DuBuque, A. Favello, W. Gish, M. Hawkins, M. Hultman, T. Kucaba, M. Lacy, M. Le, N. Le, E. Mardis, B. Moore, M. Morris, C. Prange, L. Rifkin, T. Rohlfing, K. Schellenberg, M. Soares, F. Tan, E. Trevaskis, K. Underwood, P. Wohldmann, R. Waterston, R. Wilson, and M. Marra, *Genome Res.* **6,** 807 (1996).

[33] M. D. Adams, M. B. Soares, A. R. Kerlavage, C. Fields, and J. C. Venter, *Nature Genet.* **1,** 373 (1993).

There are two additional cautions regarding the use of dbEST information. First, in the construction of any cDNA library with oligo(dT) priming, some priming from internal poly(A) stretches will occur. Thus, there will be clones containing apparent poly$(A)^+$ tails that do not represent bona fide 3′ ends of messages; some of these may even represent priming from sites within introns that had not yet been spliced from the message population. Therefore, not every sequence in dbEST can be assumed to be bona fide. Second, it is unlikely that all messages in a tissue source will be represented, no matter how exhaustive the sequencing efforts become. This is because reverse transcriptase will be rendered ineffective by secondary structures in some proportion of messages, and these and other messages, for other reasons, will clone poorly or not at all.

A powerful addition to the dbEST data, and what makes it valuable here, has been provided by the mapping of more than 16,000 of the sequence entries on the sequence-tagged side (STS)-based radiation hybrid (RH) map of the human genome.[1] At *http:/www.ncbi.nlm.nih.gov/Science96/* can be found ESTs localized by interval to chromosomal regions. The STSs used to define the intervals can, in theory, be related to physical and clone maps. Also in dbEST are some known, well-studied genes, whose map locations are probably unambiguous to researchers involved with a particular chromosome or region. These can also be used to orient oneself and decide which cDNAs are most likely to lie within the region for which the comprehensive transcriptional map is desired. Each dbEST cDNA is archived at several commercial sources and can be obtained for a nominal fee.

How many genes will be defined in this manner? If the current 16,000 ESTs were uniformly distributed throughout the 3000 Mb of the human genome, a typical 3-Mb genomic region would contain 16 cDNAs—not a bad start. On the other hand, the genes are clearly not uniformly distributed. Of the 16,000 ESTs, 141 mapped to chromosome 21q (already lower than the 160 that would have been estimated from the 1% of the genome estimated to be contained in chromosome 21q). Approximately 15% of these mapped within the proximal half of the long arm that contains 50% of the DNA; and 40% mapped within the distal 20%.[22]

Once the EST cDNA clones are in hand, the next steps are straightforward. First, the clones should be verified by end sequence to be those desired and verified to map within the appropriate chromosomal region, i.e., to one or more of the clones defining the clone contig. Errors in one or both of these are not common, but have been found. Second, the EST should be completely sequenced and analyzed on a Northern blot to determine whether or not it represents close to a complete transcript. If necessary, the EST can then be used to screen additional cDNA libraries—random hexamer primed would be particularly useful—to obtain a complete cDNA, or to obtain coding sequence if the transcript is large.

One last point: the developing mouse EST database should not be ignored.[34] As mouse ESTs are placed on the mouse genetic map, their positions within regions of conserved synteny with the human map will be defined. The sources of mouse cDNAs that are being emphasized are not the same as the human sources, e.g., staged mouse embryos are being strongly represented and these cDNA sources are clearly underrepresented in the human dbEST. Thus, developmentally regulated genes can be expected to appear with higher frequency in the mouse dbEST. Those with specific developmental regulation may be unique to the mouse database. As appropriate, these can be added to the human transcriptional map.

cDNA Selection

cDNA Sources. In using cDNA selection for comprehensive transcriptional mapping, choosing the sources of the cDNAs is a key issue. Because cDNA selection requires expression for a gene to be identified, failure to choose the appropriate tissues will result in failure to identify associated genes. In general, therefore, it is important to obtain the most complex set of mRNAs possible. Brain expresses a complex population of messages,[35] and therefore inclusion of fetal and adult brain samples is important. Use of whole fetus adds potentially developmentally specific genes (although obviously restricted to the fetal ages obtainable). Anecdotally, testis is also a good source of complex message populations. This has been shown explicitly for the HeLa cell line,[36] and is reported to be characteristic of several brain tumor cell lines as well.[37]

Once tissues/cell lines are chosen, poly(A)$^+$ RNA is prepared and reverse transcribed. Using random primers in the reverse transcription step ensures that a higher proportion of coding sequence will be represented, and this aids significantly in the analysis stage, both in putting together complete cDNAs and in identifying functional homologies. cDNA is either cloned or ligated to linkers. The advantage of cloning is a stable, reliable, and renewable source of cDNA. The disadvantage lies in possibly losing clones with poorer relative growth characteristics.

The advantage of storing cDNA simply with linkers ligated is that

[34] M. F. Bonaldo, G. Lennon, and M. B. Soares, *Genome Res.* **6,** 791 (1996).
[35] N. D. Hastie and J. O. Bishop, *Cell* **9,** 761 (1976).
[36] J. O. Bishop, J. G. Morton, M. Rosbash, and M. Richardson, *Nature (London)* **250,** 199 (1974).
[37] L. Carlock, D. Wisniewski, M. Lorincz, A. Pandrangi, and T. Vo, *Genomics* **13,** 1108 (1992).

slightly different linker sequences can be used to tag cDNA from each tissue, permanently preserving this information after pooling steps.[38] This is not always useful because it requires sequencing of large numbers of cDNAs to define all sources and proportions, and often similar clones are screened out by hybridization, so this information is lost anyway. The second advantage of using linkers, however, lies in not cloning; this may help to preserve a more complete representation of the message population.

In either scenario, cDNAs are then separately amplified by PCR. At this point, they may be pooled for the hybridization step, or kept separate and handled in parallel.

Genomic DNA. Several factors are important here regarding the type and preparation of the genomic DNA, and the use of blocking reagents.

As discussed above, YAC clones may be the most readily available. Both whole yeast DNA and pulsed-field gel-purified YAC-specific DNA have been used successfully.[2,21] In both cases, the presence of ribosomal RNA-encoding DNA contaminants can cause problems. Even though it is yeast, it can effectively "select" rRNA sequences remaining in the prepared mammalian cDNA, and can result in quite significant contamination of the resultant selected cDNA libraries. BAC, PAC, P1, and cosmid clones can be purified through CsCl, largely eliminating this problem.

Regardless of type of genomic clone, the DNA is digested with an appropriate frequent cutter and ligated to compatible linkers (different in sequence from those used for the cDNA).

Again, as with the cDNA preparation, the question of pooling of sequences arises. The answer to this will depend in part on the size of the genomic region being analyzed. For a 2- to 3-Mb segment, pools of genomic clones totaling 400–500 kb are appropriate. This would correspond to a single YAC clone, 3 or 4 BAC clones, and 10–15 cosmids, and would imply 5 or 6 separate genomic pools to cover the region. Decreasing the size of the genomic pools too much vastly increases the amount of work necessary to analyze the products. Pools that are too large can cause losses and omissions.

Hybridizations. Hybridizations can be set up in two ways.[2,3] The genomic DNA can be bound directly to a small nylon disk filter, or the DNA can be biotinylated in a PCR reaction using a labeled nucleotide or a labeled primer to the linker. For the biotinylated DNA, hybridization occurs in solution, followed by binding of the biotinylated DNA to streptavidin-coated magnetic beads. In either case, thorough blocking of the DNA in

[38] J. M. Rommens, B. Lin, G. B. Hutchinson, S. E. Andrew, Y. P. Goldberg, M. L. Glaves, R. Graham, V. Lai, J. McArthur, J. Nasir, J. Thelmann, H. McDonald, M. Kalchman, L. A. Clarke, K. Schappert, and M. R. Hayden, *Hum. Mol. Genet.* **2,** 901 (1993).

a prehybridization step is required, generally using a cocktail containing human ribosomal RNA sequences, human Cot1 DNA, and poly(dI:dC) (to block nonspecific hybridization of high-GC sequences).

After prehybridization, PCR-amplified cDNA is added and standard hybridization proceeds for 24–48 hr. Nonhybridizing cDNA is then washed from the DNA that is bound either to the nylon filter or to the magnetic beads. cDNAs that have hybridized to their complement in the genomic DNA remain bound to the filter or the beads and can then be eluted.

Generally, this selection procedure is repeated at least once, to decrease the nonspecific contaminants. After the second round of selection, the eluted cDNAs are cloned.

Several refinements to the basic procedure have been reported. Using limiting quantities of input, genomic DNA will produce some normalization of the resultant cDNAs. This helps in detecting those messages that may be expressed at relatively low levels, although information on relative expression levels among genes is lost. An alternative approach to achieve the same thing is to use normalized cDNA libraries[21] as the cDNA source. A drawback to cDNA selection is often the small size of the final cDNA fragments (averaging <200–400 bp). This can be partially alleviated by not amplifying the cDNA between the first and second round of selection. Size selecting both the input cDNA and the product of the first round of selection also can help to increase final product size. Such methods have been reported to yield products averaging 700 bp.[38,39]

Cloning and Analysis. The cDNA eluted from the final round of selection is amplified by PCR and cloned; clones are picked to microtiter plates and stored in duplicate. How many clones should be picked? Again, regional variations in gene density and expressivity will dictate requirements. A rule of thumb has been 100–200 clones per 100 kb of genomic DNA. Exact numbers can reflect the number of tissues used/pooled and the general transcriptional activity of the region, once a feel for this has been obtained.

Analysis of the clones can be laborious, but must be thorough for saturation transcriptional mapping. Analysis includes checking the size of inserts and the proportion of clones with inserts, verifying that individual clones map to at least one of the genomic clones used in the selection and/or the appropriate chromosomal region if somatic cell hybrids are available, hybridizing individual clones to the entire selected library to identify identical or overlapping clones, and sequencing nonidentical inserts.

In this way, a set of nonredundant cDNA fragments is identified. It is

[39] P. Koischis, U. C. Rogner, E. Pick, S. M. Klauck, N. Heiss, R. Siebenhaar, B. Korn, J. F. Coy, J. Laporte, S. Liechti-Gallati, and A. Poustka, *Genomics* **33,** 365 (1996).

important to remember here that this does not necessarily translate into information on the number of nonredundant genes present or identified.

Artifacts. Common artifacts include ribosomal RNA sequences, mitochondrial DNA sequences, pseudogenes, and intronic material. Intronic material presumably finds its way in because of the presence of slowly and incompletely processed transcripts. Pseudogenes can be difficult to distinguish without sequencing the genomic DNA, and this may be necessary if a gene family member is found in an unexpected location.

Cloning artifacts include vector sequences and concatamers of primer sequences used in PCR amplifications. These two types of artifacts are found particularly in regions of apparently low transcriptional activity. For example, they were observed at a frequency of <5% in a gene-rich YAC, and, in a parallel experiment, at >95% in a gene-poor YAC.[21] Therefore, if a plethora of apparent artifacts appears, it is important to ferret out the cause, e.g., poor blocking reagents, poor preparation of genomic DNA, or a genuinely transcriptionally inactive region. If it is the latter, using additional tissues may be helpful, although application of exon trapping may be the more productive approach at such a point.

It should be noted that regions of low transcriptional activity, i.e., those regions with either few genes, or genes of generally low levels of expression, are not necessarily uninteresting biologically. For example, the 2.4 Mb spanned by the Dystrophin gene yields only an 11-kb mature message; the 300 kb spanned by the amyloid precursor protein (APP) gene, a 4-kb message.[40,41] In both these regions there will be large amounts of transcribed, noncoding sequences, spanning possibly several YACs.

Complete Coding Regions. Without complete coding sequence, there may be little hope for functional assessment of a novel gene, unless a strong homology is found to a mammalian gene. Two hundred to 300 nucleotides of a human sequence may produce unconvincing matches with evolutionarily distant organisms, whereas 2–3 kb may be extremely informative.

Screening of a large-insert cDNA library with the smaller cDNA selected fragment can be effective. It helps here to know the tissues and levels of expression, determined by RT-PCR of input, nonpooled cDNA. With longer cDNAs, Northern blots can be efficient.

Exon Trapping

Basic Procedures. The various exon-trapping vectors share similar features. The vector contains a promoter plus an intron flanked by exons of

[40] J. Chelly, J. C. Kaplan, P. Maire, S. Gautron, and A. Kahn, *Nature (London)* **333,** 858 (1988).

[41] M. Hattori, F. Tsukahara, Y. Furuhata, H. Tanahashi, M. Hirose, M. Saito, S. Tsukuni, and Y. Sakaki, *Nucleic Acid Res.* **25,** 1802 (1997).

a known gene. The genomic fragment of interest is cloned into the intron sequence, and the clone is then transfected into cos cells. There, the insert will be transcribed from the vector promoter. If the insert contains an exon in the correct orientation, the transcript will be spliced to produce a "mature" mRNA containing the two vector-encoded exons plus the clone-derived exon(s) present between them. The presence of an included exon is detected by RT-PCR using vector-specific primers and cytoplasmic RNA.

The major advantage of exon trapping is its independence of expression. This makes it an excellent complement to cDNA selection, especially when tissues are limited or suspected of being inappropriate.

Vectors. There are three vector choices: a plasmid that will detect internal exons, a plasmid designed to detect the 3′ terminal exon, and a cosmid vector for larger inserts.[42–44]

The advantage of the 3′ exon trap system is the large average size of 3′ terminal exons (>1000 nucleotides versus 125 nucleotides for internal exons).[12] The disadvantage is that 3′ exons are largely, if not entirely, noncoding, and therefore produce little possibility of protein homology. A significant proportion also contains repetitive sequences, which can make their use as hybridization probes difficult. Such drawbacks may, however, be outweighed by the increased efficiency of longer probes in hybridizations to Northern blots, Southern blots, and cDNA library screenings.

The major and important advantage of the cosmid exon trap vector is that it allows cloning of 30- to 40-kb genomic fragments, potentially containing multiple exons. If several exons are present, these can be spliced together, yielding a larger, more informative, more functional product.

How can the cosmid vector be used most efficiently? As discussed above, the deletions and rearrangements possibly present in YAC clones make it unattractive to subclone them directly as cosmids. However, this can be done, as long as one remains aware that genes might be missed, or that weirdnesses such as not finding parts of apparent genes may occur.

One alternative is to reclone *en masse* a chromosome-specific cosmid library into the cosmid trap vector (C. Lau, University of San Francisco; personal communication, 1998). Appropriate restriction sites, e.g., *Bss*HII, *Not*I, and *Eag*I, exist in the Lawrist vector to make this reasonable, at least in regions where such sites are rare in the genomic DNA. For human DNA, these sites are actually quite frequent (one per 20–50 kb) in the most GC/gene-rich regions.

[42] T. C. Burn, T. D. Connors, K. W. Klinger, and G. M. Landes, *Gene* **161,** 183 (1995).

[43] D. B. Krizman and S. M. Berget, *Nucleic Acids Res.* **21,** 5198 (1993).

[44] N. A. Datson, E. van de Vosse, H. G. Dauwerse, M. Bout, G.-J. B. van Ommen, and J. T. den Dunnen, *Nucleic Acids Res.* **24,** 1105 (1996).

A variation on this procedure is possible if a cosmid contig has already been constructed from YACs by screening of the chromosome-specific library. In this case, individual cosmid inserts could then be excised and recloned into the trap cosmid vector. Efficiency and fidelity can then be monitored directly.

Regardless of the vector choice and genomic starting material, it is worthwhile to consider use of several restriction enzymes in the genomic subcloning. *Pst*I, partial *Mbo*I (into *Bam*HI sites), and partial *Tsp*509 (into *Eco*RI sites) are reasonable choices for the plasmid vector, and for the partial digests, also in the cosmid vector. Multiple cloning sites increase the chances that exons will be properly contained.

Most reports of large-scale exon trapping have subcloned pooled cosmids into the plasmid vector.[45] Undoubtedly, the larger the pool of clones, the more exons (and therefore potential genes) are missed. Averages of one exon per 20–40 kb are typical and clearly underrepresent the genome in many regions. If only one exon per gene is sought or required, this may be sufficient; however, single exons, because of their small size, make poor probes and give poor homology matches. Multiple exons from a single gene are required if nothing is to be overlooked.

Analysis. Successful exon trapping is detected as an RT-PCR product that is larger than the vector lacking insert control. These larger bands are cloned, verified to map to a parent genomic clone, and sequenced. Common artifacts include vector sequences, nonbiologically relevant splicing, and splicing on the antisense strand.[46] Sequencing or PCR with appropriate vector-related primers will eliminate the common vector problems. Direct sequencing of the parent genomic clone using primers to the putative exon will detect consensus splice sites if the exon is bona fide. This may be the fastest, most efficient method of exon verification, in particular when single exons have been trapped.

Genomic Sequence Analysis

Genomic sequencing of significantly large DNA segments is impractical for the independent laboratory. However, large-scale sequencing of the human genome has begun in several centers worldwide (in particular, for the X chromosome and chromosomes 4, 7, 12, 19, 21, and 22), and completion of the entire human genome sequence is still scheduled for 2005.

To handle these data, there are increasing numbers of programs de-

[45] M.-L. Yaspo, L. Gellen, R. Mott, B. Korn, D. Nizetic, A. Poustka, and H. Lehrach, *Hum. Mol. Genet.* **8,** 1291 (1995).

[46] H. Gen, R. Chrast, C. Rossier, M. A. Morris, M. D. Lalioti, and S. E. Antonarakis, *Genome Res.* **6,** 747 (1996).

signed to analyze and annotate DNA sequence.[47] They will predict CpG islands, promoters, and polyadenylation sites and identify various classes of high, moderate, and simple repeats. Many also predict the locations of exons, and knit these together to predict gene models. Some readily accessible and user-friendly examples include

XGRAIL: Available from Oak Ridge National Laboratory[5]

GRAIL, Fxgene, and splite site prediction: Available at the Baylor web site[48]

Grail, Xpound, Genie, and Genefinder: Available from Lawrence Berkeley Laboratory as the Genotator web server[7]

Genscan[8]

The problem each of these programs aims to solve is how to identify a high proportion of true exons while maintaining a low false-positive rate. In general, each program provides a different picture of the same sequence; some will differ wildly, others only slightly in the number and placement of putative exons. Boundaries of exons will most often differ, and should not be regarded as precise.

A productive approach is to run several exon prediction programs on the same sequence, and follow up on "exons" that are identified by two or more progams. Exons predicted by three or more programs can be considered highly reliable. Genotator is particularly nice in this respect; the output aligns the exon predictions of four programs, plus GenPept and dbEST matches, all in one picture. Commonalities are easy to see and note.

Exon and gene model predictions easily divide themselves into different levels of credibility.

Obviously the strongest cases are those with matches in BlastX searches to successive segments of a single protein. Detailed examination of the DNA sequence would most likely reveal the corresponding exon structure of the gene, or a closely related family member. Nonprocessed pseudogenes will usually be easy to detect from the patterns of stop codons with "exons."

Similarly high-confidence results include those showing clusters of exons predicted by two or more programs and associated with dbEST matches of >95% identity. (Note that because of the high proportion of 3′ UTR sequences in dbEST, matches of <95% need to be inspected carefully to ensure that they are relevant.) Such results would suggest identification of a novel protein.

Problems will arise with results showing no protein matches, and dispersed exon predictions. This is a more frequent occurrence in AT-rich regions, where genes tend to be larger and have larger introns. Genscan

[47] J. W. Fickett, *Trends Genet.* **12,** 316 (1996).

[48] K. C. Worley, B. A. Wiese, and R. F. Smith, *Genome Res.* **5,** 173 (1995).

is reported to function better in AT-rich DNA than earlier programs, most of which were more successful at analyzing GC-rich sequences, of the kind that they were trained on.

Problems can also be expected in cases of genes within introns of other genes, and where genes may overlap. It is also to be noted that these programs tend to identify coding sequences best. Thus, the end point of genes, the extents of 5′ and 3′ UTRs, will not be predicted with confidence, although inferences can be made from the presence of CpG islands and polyadenylation sites.

A critical issue with the use of "software trapping" is that these programs provide predictions only. As with cDNA selection and exon trapping, the putative exons must be verified, again using RT-PCR and Northern analysis. By these means, boundaries of coding regions, internal exons that may have been missed, alternative splicing sites, and transcript sizes are determined.

Data Consolidation

Each of the four gene identification approaches has been considered separately. The strength, however, of the multiple approaches lies in compiling results obtained for all four. This is most important when novel genes are found, genes with weak or no significant homologies to known protein sequences of any organism or only distantly related organisms. In these cases, obvious important correlations include an exon-trapping product whose sequence is found within a selected cDNA fragment or a dbEST entry; or a cDNA whose sequence is found in a genomic fragment as a series of Genotator-predicted exons, each bounded by consensus splice sites. The more such indications that a sequence is expressed and is coding, the better. One should not, however, ignore expression analysis. Northern analysis provides important information about transcript size. This verifies predictions of coding regions and helps to define the extents of 5′ and 3′ UTRs. These latter data will help in the identification of promoter and other regulatory sequences. Products of RT-PCR with primers at the limits of the predicted coding region can be used as probes for these Northern blots. RT-PCR with primers to various exons can be used to identify alternatively spliced rare exons, although this needs to be used with some caution. Limiting the number of cycles in PCR is crucial to avoiding amplification of introns that may be present in only a minor proportion of messages.

It remains difficult to assess when a transcript map is complete. Genes with low levels of expression that may also be restricted to uncommon times or tissues may be abandoned or overlooked because they cannot be verified by Northern blots or RT-PCR. Worse, given the dramatic differences in gene density so far apparent among different genomic regions, it is not possible to use the number of genes identified per megabase as a measure of completeness. It can be assumed that as more mammalian genomic sequence accumulates, and is analyzed both by software and by experiment, our ability to interpret transcriptional mapping data will improve. It may still evolve, however, that in practice a transcript map is judged complete when the known genes appear to account for the biological characteristics or functions of the region.

Acknowledgments

This work is contribution 1651 from the Eleanor Roosevelt Institute and was supported by Grant HD17449 from the National Institute of Child Health and Human Development and Grant VM-122 from the American Cancer Society.

Section III

Patterns of mRNA Expression

[11] cDNA Screening by Array Hybridization

By RADOJE DRMANAC and SNEZANA DRMANAC

Introduction

Unraveling the code of genes and complex genomes in which they reside is a fundamental requirement for continued progress in the areas of molecular biology, biotechnology, molecular genetics, and genetic medicine. To date, the structural and functional analysis of genes has been limited by the speed of the present methods for identification of unknown genes and by the speed of repeated sequencing of genes in many individuals. To catalog, sequence, and map all genes in one organism, there is an increasing need for methods that have the potential for automation and industrial-scale DNA sample processing. The large-scale methods are especially important for identifying functionally integrated gene networks.

One such method is the technique of sequencing by hybridization (SBH). Using the inherent characteristics of DNA to recognize and base pair with complementary sequences, short oligonucleotides of known composition are able to read DNA sequences under complete and match-specific hybridization conditions.[1] This technology can be applied to both genomic DNA and cDNA and has the capability to provide information at increasing levels of detail by the inclusion of more probes; but it is the expressed sequences that are of interest for this chapter.

The use of short oligonucleotide probes (6–8 bases) allows for the detailed analysis of randomly selected cDNA clones in parallel. If the probes are short, a sufficient number of them will hybridize to standard size cDNA clones in various combinations, providing a gene-specific series of positive hybridizations for each clone termed a "hybridization signature" or partial sequence.[2] The use of as few as 100 probes of this length enables the characterization of many more genes owing to this combinatorial nature of positive and negative hybridization signals. The hybridization signatures can then be used to monitor levels of specific transcripts in tissues during normal biological processes and laboratory treatments. The major advantages of this technique are the ability to identify previously unknown genes

[1] R. Drmanac, S. Drmanac, Z. Strezoska, T. Paunesku, I. Labat, M. Zeremski, J. Snoddy, W. K. Funkhouser, B. Koop, L. Hood, and R. Crkvenjakov, *Science* **260,** 1649 (1993).

[2] R. Drmanac, G. Lennon, S. Drmanac, I. Labat, R. Crkvenjakov, and H. Lehrach, *in* "The First International Conference on Electrophoresis, Supercomputing and the Human Genome" (C. R. Cantor and H. A. Lim, eds.), p. 60. World Scientific, Singapore, 1990.

Copyright © 1999 by Academic Press
All rights of reproduction in any form reserved.

0076-6879/99 $30.00

and to define their expression profiles, the sensitivity in resolution of weakly expressed transcripts, and the capability to separate transcripts sharing repetitive sequences. The overall strength of this SBH format is the formation of clone arrays through industrial-scale production and automation of molecular biology techniques, such that all the expressed genes contained within a library can be screened in parallel.

The adaptation of this technology to an industrial setting has already produced 1 million partial cDNA sequences per month and a database of more than 4 million individual cDNA clones in our facility. In this chapter, we discuss the current protocols and the potential application of this technology to the fields of gene discovery and gene function.

cDNA Screening Requirements

The successful use of partial SBH, or any other screening method, for the identification of expressed genes within a given organism is dependent on a number of factors including the number of tissues and clones analyzed and the overall quality of the cDNA libraries used.[3]

To find genes having a low level of expression, and especially to discriminate artifacts by collecting more than 1 corresponding clone for these genes, as many as 100 gene equivalents (10 million clones) may have to be screened overall. Libraries of poorer quality would require screening an even larger number of clones. Previous experiments[4,5] support the expectation[3] that it is mainly statistics, and not the background due to overexpression of several genes, that necessitates the screening of several million clones from ordinary libraries of basic tissues. Thus, we believe that normalization of the library will not significantly speed up gene identification, but it will destroy gene expression information contained in higher clone abundance of more active genes.

To discriminate between housekeeping, tissue-specific and nonfunctional genes requires the analysis of more than 10 million clones collectively from about 30 tissues.[3] To ensure the best possible results from any screening experiments, considerable attention should be paid to the methods of preparation of required libraries. For example, a bias can be introduced

[3] R. Drmanac, S. Drmanac, I. Labat, and N. Stavropoulos, *in* "Identification of Transcribed Sequences" (U. Hochgeschwender and K. Gardiner, eds.), p. 239. Plenum Press, New York, 1994.

[4] S. Drmanac, N. A. Stavropoulos, I. Labat, J. Vonau, B. Hauser, M. B. Soares, and R. Drmanac, *Genomics* **37,** 29 (1996).

[5] A. Milosavljevic, M. Zeremski, Z. Strezoska, D. Grujic, H. Dyanov, S. Batus, D. Salbego, T. Paunesku, M. Bento Soares, and R. Crkvenjakov, *Genome Res.* **6,** 132 (1996).

for short mRNAs by narrow cDNA size selection and for long mRNAs by reduced transformation efficiency.

Much care should be taken with the library preparation to reduce contamination with genomic DNAs and external mRNAs and DNAs, as well as chimeric clones and false primed clones. The partial SBH analysis procedures have been designed to take into consideration the presence of deletions or recombination during clone amplification or polymerase chain reaction (PCR), the presence of alternatively spliced species of the same mRNA, cross-talk between wells, external contamination, and hybridization errors generated during the sequencing processes.[4]

Principles of cDNA Screening by a Generic Set of Short Oligonucleotide Probes

Partial SBH of randomly selected cDNA clones using 7-mer to 8-mer oligonucleotide probes has been proved as a method for the identification of active genes and the simultaneous determination of their transcription levels.[4,5] A small number of these probes can suffice for the identification of distinct genes and gene families. If 1 kb, double-stranded cDNA clones are hybridized with 7-mers two times more frequent in the coding sequence than in the random sequences, 25% of the clones will be positive with each probe, on average. From a set of 100 such probes, a specific subset of 25 will be scored as positive per each clone, providing a clone-specific signature. Additional specificity is provided by scoring hybridization intensity instead of compressing different signal levels to negative or positive response calls.[4] To ensure separation of similar genes, the current protocol uses more than 200 probes that randomly hybridize to various regions of each gene and cover approximately 10% of the clone sequences.

The general schema for the SBH procedure is outlined in Fig. 1. Each step in the procedure has been optimized for expansion to industrial scale. Briefly, clones are grown by normal laboratory procedures on agar plates and picked with a picking tool to quickly generate master cultures in multiwell plates. The clone inserts are then amplified by PCR in plates with a corresponding number of wells. Each PCR clone is spotted on multiple replica membranes using a pin tool that matches the number of wells in the PCR plates. A dot density higher than the well density is achieved by offset blotting from many plates. Probes are hybridized and signals are compared with negative controls (no DNA spotted) and positive controls (previously sequenced clones that are repeatedly spotted). The resulting normalized and properly scaled hybridization intensities of a set of successful probes to a single clone define the "hybridization signature" (Fig. 1, clones 1 through 4).

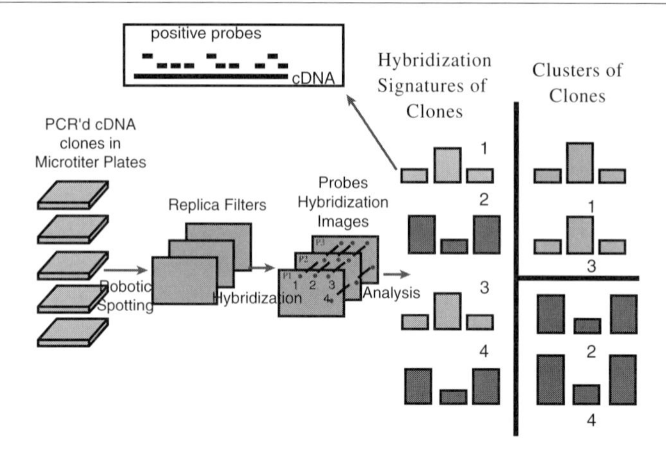

FIG. 1. Sequencing by hybridization for cDNA screening. Schematic representation of the SBH procedure, consisting of multiple industrial-scale steps for full library screening for new and known genes.

The hybridization signatures are used for estimating the sequence similarity between individual clones or the fraction of DNA inserts of one clone that is shared with the other clones. Clones with similar hybridization signatures are grouped into a gene- or a transcript variant-representing cluster (Fig. 1, cluster containing clones 1 and 3). Clusters are used for identification of novel genes, and for estimation of gene expression via calculation of the fraction of analyzed clones that belongs to a given cluster.

The advantage of using short, random, oligonucleotide probes with SBH is that the information is distributed over the entire clone, which allows for an estimate of the overall similarity of clones. This feature generates a smaller number of false-positive and false-negative gene calls as compared with the expressed sequence tag (EST), and provides a foundation to discriminate alternatively spliced, antisense, or other transcript variants of a gene. The use of a series of nonspecific probes identifies all possible transcripts in parallel and provides the capacity to identify all novel genes as opposed to the traditional single-gene identification at a time by a gene-specific probe.

Procedures for Screening Millions of cDNA Clones and Novel Gene Discovery

Large amounts of data can be generated by the parallel treatment of cDNA libraries with three types of dense arrays: an array of wells for the growth and storage of clones, an array of pins for the inoculation and spotting of these clones, and an array of DNA dots for the parallel scoring of positive oligodeoxynucleotide (oligo) hybridizations.[4,6]

Clone Preparation

The plasmid cultures are grown in standard *Escherichia coli* strains (for example, JS5 or DH5α) using LB culture medium and the appropriate antibiotic selection. The cells are transformed by electroporation with an *E. coli* pulser from Bio-Rad (Hercules, CA). A picogram amount of recombinant plasmid is used in one transformation to minimize the possibility of cotransformation. The initial steps of clone transformation and plating are similar to the standard protocols used in most laboratories. A few adaptations of these procedures have increased the scale of each step severalfold.

The transformation mixture is plated in an optimized number and density of the clones to allow large-scale picking of each bacterial colony. To assure cataloging of every gene, even the smallest colonies are picked. The usual practice of picking clones one at a time with an inoculation loop or toothpick has been replaced with a picking device. These colonies are then inoculated into wells of a 384-well plate, greatly reducing time and effort while increasing efficiency. These master plates are grown for 16–24 hr and stored in −70° freezers in freezing medium (63 g of potassium phosphate, 4.4 g of sodium citrate, 0.9 g of magnesium sulfate, 9 g of ammonium sulfate, 18 g of potassium phosphate, and 440 g of glycerol per liter of growth medium).

A trained technician can pick about 1500 colonies per hour with less than 2% of double-clone wells. A sustained, high-quality picking of 1 million colonies per month is achieved in the Hyseq (Sunnyvale, CA) gene discovery program.

Polymerase Chain Reaction of Plasmid Inserts

The extraordinary throughout of 1 million PCRs per month is achieved by performing PCRs in 384-well plates, simultaneous inoculation of 384 reactions with a 384-pin tool, and simultaneous cycling of up to ten 384-well plates in an inexpensive, air-driven temperature cycler equipped with a 5-story rack (BioOven, Beltsville, MD), instead of a single temperature-controlled block.

Master cultures are diluted in water 20- to 50-fold, or a small volume of the cultures is used directly. For the PCRs, the 384 wells are filled with a complete PCR mixture according to standard instructions from Perkin-

[6] S. Drmanac and R. Drmanac, *BioTechniques* **17**(2), 328 (1994).

Elmer (Norwalk, CT) with an optimized concentration of primers and enzyme. Buffer and deoxyribonucleoside triphosphates are obtained from Perkin-Elmer/Applied Biosystems (Foster City, CA). The concentration of primers typically varies from 0.20 to 0.40 μM and the concentration of AmpliTaq (Perkin-Elmer) varies from 2 to 4 units for 100-μl reaction volumes.

The optimal design of primers for a given vector and host *E. coli* strain is critical for the high yield of products and high success rate of reactions. An optimization example is represented by 3 base-extended universal and reverse M13 sequencing primers (5′ GG-GTTTTCCCAGTCACGAC-G 3′ and 5′ CA-CAGGAAACAGCTATGAC-G 3′).

A 1-mm metal pin array is used to inoculate the PCR plates preferentially with a simple robotic station equipped with an automatic tool for flame sterilization. The pins transfer less than 1 μl of the water-diluted cultures into each well of a 384-well plate containing the PCR mixture. Up to 10 plates are incubated in a BioOven. The parameters for 384-well plates are as follows: 90° for 1 min, 45–62° for 1 min depending on the primers, and 70° for 2.5 min. An additional denaturation step of 3 min at 91° is included prior to cycling. Successful PCRs can be performed on a large scale compared with standard tubes and traditional thermocycler machines. This scale allows the routine performance of 50,000 PCRs per day in six BioOvens. Figure 2 represents 96 PCRs (24 samples from 4 different 384-well plates). A majority of wells contain a single product indicating distinct clones. A small percentage contain two distinct bands representing a mixture of clones. The second weak band larger than the first band that is visible in many samples is due to the PCR of two linked inserts and does not indicate a clone mixture. Less than 10% of the wells contain no product, indicating failed reactions. The consistently high amplification rate in the PCRs is aided by the use of longer primers and definition of the optimal concentrations of primers and enzyme. Standard production quality assurance procedures are introduced to minimize unsuccessful runs.

Array Preparation

The current protocol utilizes a robot system (Mega2; Megamation, Lawrenceville, NJ) for the accurate and repeated spotting of 55,296 dots on a 6 × 9 inch membrane (Fig. 3, see color insert). Multiple replica membranes (four to six) are produced in parallel. The technical requirements for the spotting robotic station are as follows: accuracy greater than 100 μm in all three (x, y, and z) directions, sufficient space for laying down plates and membranes, and durability of execution of millions of travels from plates

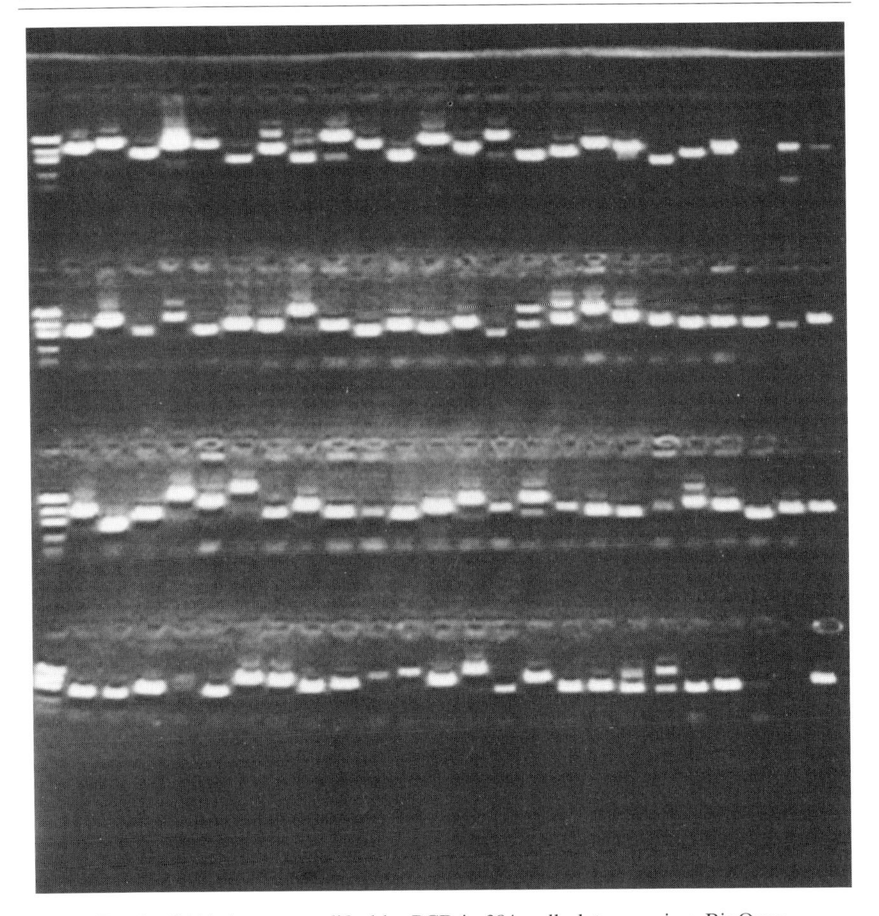

FIG. 2. cDNA inserts amplified by PCR in 384-well plates run in a BioOven.

to membranes and back over the course of 1 year of continuous spotting. It is important to note that the spotting process does not require robots with sensors or visual referencing.

Filter membranes (GeneScreen; Du Pont/NEN, Boston, MA) are previously soaked in denaturation solution (0.5 *M* sodium hydroxide, 1.5 *M* sodium chloride) for spotting PCR products and these membranes are laid down on two 3MM Whatman (Clifton, NJ) papers soaked in the same solution. The wetness of the membranes must be optimized for the production of well-defined dots. The PCR-amplified inserts are deposited onto the membrane by touching the stainless steel pins to each sample and then to the membranes.

The current technique in use is for an 864-pin tool (Fig. 3, inset), which prints DNA samples from sixteen 864-well plates in a 4×4 offset pattern, producing 13,824 spots, 750 μm center to center, in a microtiter plate-size array. Membranes comprise 4 such unit arrays totaling 55,296 dots. Because the current PCR production is performed in 384-well plates, a repipetting step from 384- to 864-well plates is implemented. Efficient depositing of DNA depends on the speed of pulling the pins from the samples and the speed of touching the membranes. The spotting pins are 0.3 mm in diameter, which allows deposit of up to four dots per square millimeter. Further advances such as smaller pin size allows the production of an array of 100,000 dots on the 6×9 inch membrane, which will increase the sample throughput substantially. Current throughput of one Mega2 robot is about 1 million dots per day.

Array Hybridization

Membranes are hybridized with $(N)_{0-2}(B)_{6-10}(N)_{0-2}$ probes; where N represents degenerated positions and B represents specific bases. The currently used set of 300 probes predominantly consists of $N_2B_7N_2$ oligonucleotides. Each probe is end labeled with ^{33}P, which provides three to four times better resolution in dense dot arrays than standard ^{32}P. Probes are kinased in 96-well plates (T4 kinase; Promega, Madison, WI) and used without further purification. Hybridization buffer [0.2 *M* sodium phosphate (pH 7.2), 5% Sarkosyl] and the labeled probe are added to replica filters and hybridization of each probe occurs in a separate box under cooled and shaking conditions (Fig. 4, see color insert). Multiple filters (two to four) with different sets of clones are hybridized in one box. The annealing temperature is at 12° using 0.1–1.0 n*M* probe concentrations. The membranes are then washed in 4× SSC (1× SSC is 0.15 *M* NaCl plus 0.015 *M* sodium citrate) for 15 to 30 min at a temperature range of 0–10°, which is dependent on the length and GC content of the probes. Filters are stripped at 40–60° and can be used more than 50 times.

We routinely hybridize 288 probes per day on 55,000 dot filters, which is only 30% of the hybridization machine capacity of 72 boxes (Fig. 4). This throughput generates about 15 million probe/clone hybridization measurements every day. The same hybridization effort may double the throughput by using arrays of 100,000 dots. The hybridization machine allows automatic selection of hybridization and washing parameters and automatically pipettes and removes hybridization buffer, probes, and washing buffer. The pipetting head is depicted in Fig. 4. Labeled probe plates and filters are bar coded to provide automatic recording of annotation into a database.

After washing, the filters are blotted on paper and in groups of three

FIG. 3. Multiple-pin robotic arm used for offset dotting of clones onto membrane arrays. *Inset:* The 864-pin array head.

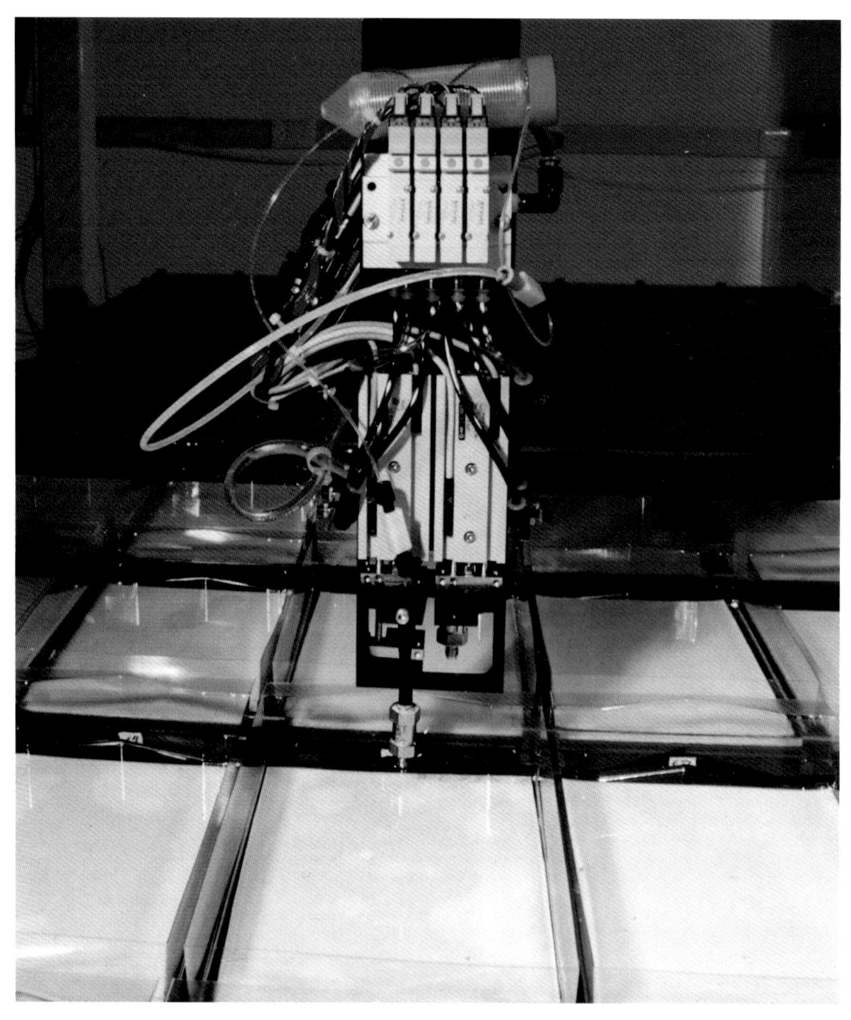

FIG. 4. Multiple-array hybridization machine. Automated large-scale hybridization of many membranes with multiple probes for increased throughput.

FIG. 5. Hybridization image of a 55,000-dot cDNA array.

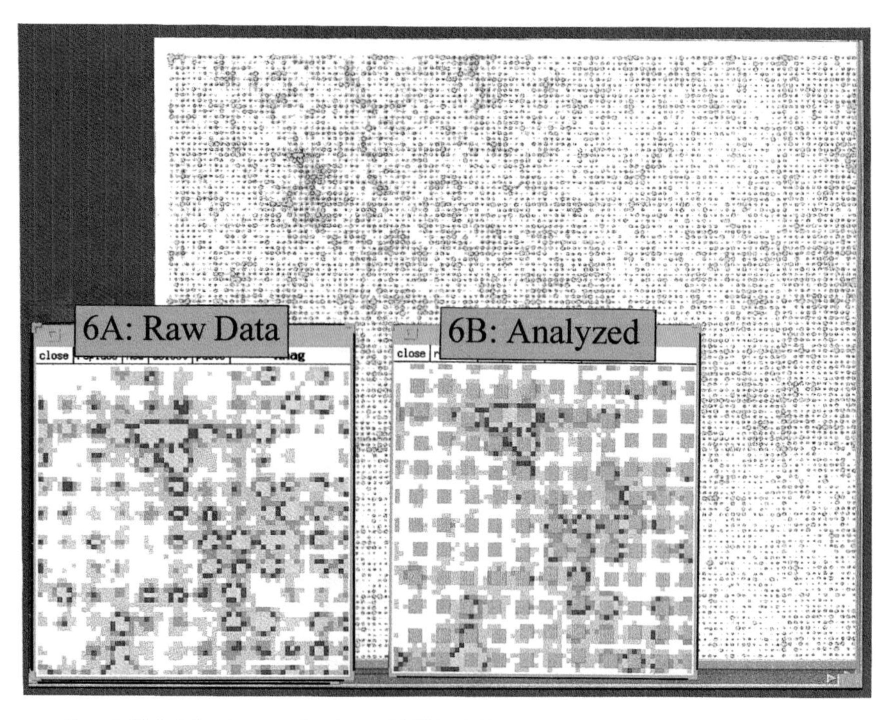

FIG. 6. Hybridization results for a 31,000 cDNA array. *Inset A:* Scanned as raw data. *Inset B:* Computer analyzed for hybridization intensity.

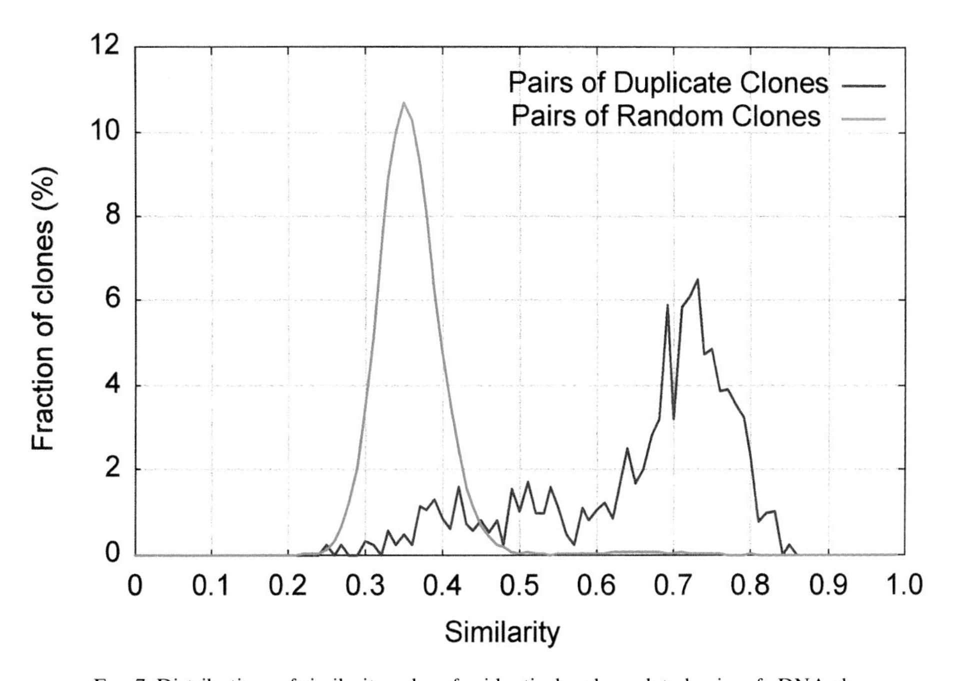

FIG. 7. Distributions of similarity values for identical and unrelated pairs of cDNA clones.

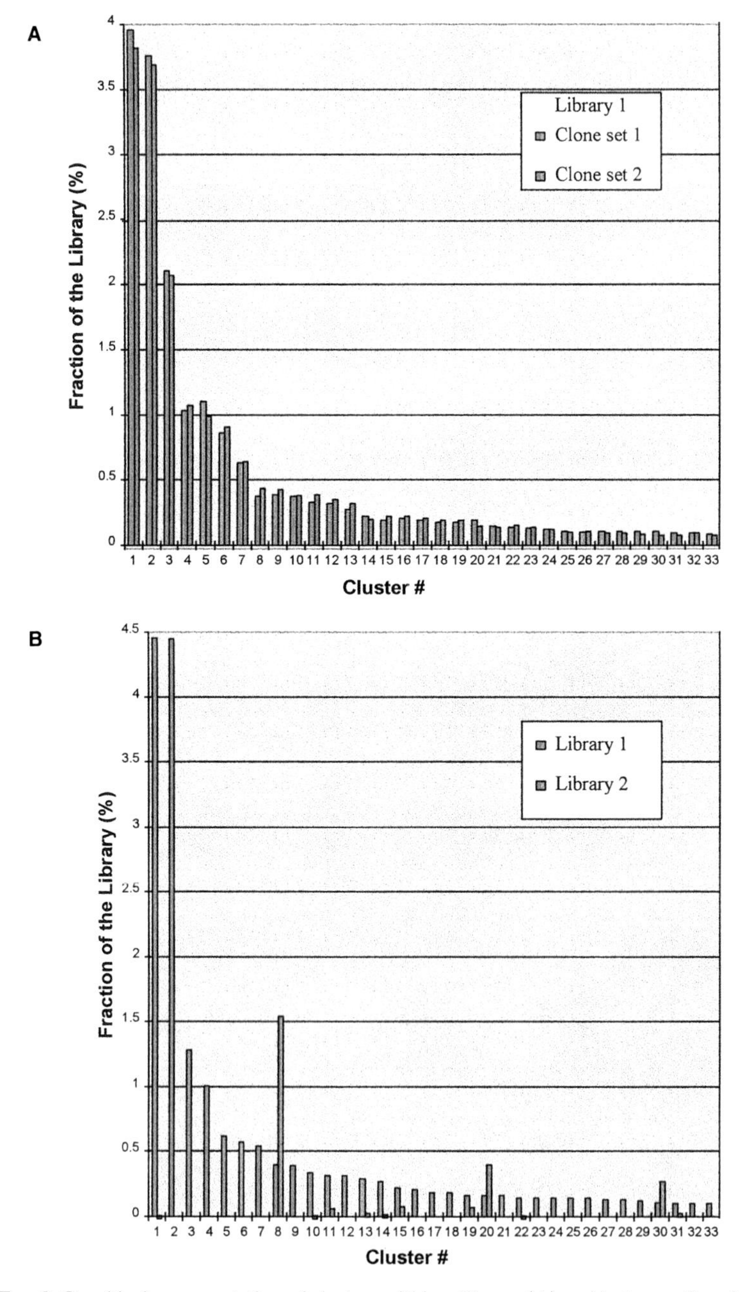

FIG. 8. Graphical representation of clusters within a library (A) and between libraries (B).

exposed for 2–4 hr in PhosphorImager (Molecular Dynamics, Sunnyvale, CA) cassettes. The images (usually 96 per day, total of 288 probes/filter results) are generated by a PhosphorImager scanner (Fig. 5, see color insert), which has a dynamic range of five orders of magnitude. As with most other methods of hybridization and quantification, the presence of a sufficient amount of spotted DNA, appropriate probe concentration, and adequate washing temperature and washing time are critical factors for high-quality data production. Figure 5 depicts a typical, probed array with 55,000 dots in which the color intensity represents the specificity and intensity of hybridization. Hybridization of short and frequently hybridizing probes on high-density arrays of DNA clones allows the screening and sequencing of entire libraries on a small number of filters, quickly and inexpensively.

Data Handling

The image files produced in 1 day occupy about 1 GB of disk memory space. Every dot in these images is represented with about 30 pixels. Because of this enormous memory requirement, the images are immediately processed by an image analysis program and removed from the on-line memory disks.

The precision of acquiring the intensities of hybridization signals from the images in an automatic process is as important as the consistency of the hybridization step itself. The automatic image analysis consists of three critical steps: (1) identification of the position and angle of the first row and column, (2) recognition of the expected centers for all the dots, and (3) determination of the representative hybridization intensity for every dot. The catastrophic error that must be avoided is a row or a column shift, which will lead to the assignment of the wrong intensity to every dot. Our image analysis program consists of many parameters that allow us to determine the correct intensity of the dots, which may be obscured owing to irregularities in the array. The program reports the average pixel value for four central pixels for each dot. The average difference between two readings for the same dot obtained from two exposures of the same filter is about 15%.

To illustrate and visualize a positive hybridization signal, Fig. 6 (see color insert) shows the upper left quadrant of a 31,000 dot membrane array rather than a 55,000 dot array as pictured in Fig. 5. The raw data are represented in five colors: yellow for the strongest signals; pink, green, and blue for the descending intermediate signals; and white for the filter background (Fig. 6, inset A). The analysis of the image is represented in Fig. 6 (inset B). The orange squares indicate 3 $\times$ 3 pixel areas with the highest signal found automatically by the program. Usually, deviations from the straight rows and columns are ±1 pixel, 176 μm in size.

Experimental information (annotation) collected by bar code readers and the hybridization intensities obtained by the image analysis program are stored in a relational database (Sybase; Emeryville, CA). For 1 million clones hybridized with 300 probes a total of 300 million rows is added in the signal storing table and they occupy about 3 GB of hard disk memory. The relational database allows efficient access to any clone/probe result. Currently we run our HyGenomics Database on a Silicon Graphics (San Jose, CA) XL Challenge server with eight processors and 2 GB of random access memory (RAM). One critical feature of data handling is the integration of the database with robotic stations for the automatic retrieval of clones of interest from master plates to perform further analyses of the corresponding genes.

Raw hybridization signals are evaluated for quality and then properly scaled to eliminate run-to-run experimental variations. Data evaluation through statistical functions includes identification of clones that have a short insert or low yield from PCR, replica membranes that have low or uneven amounts of DNA spotted, and hybridization images that have inconsistent signals for repeated control clones. Failed clones are recognized by a small number of strong hybridization signals. The first step of the scaling process consists of transforming absolute hybridization values into relative values that are comparable between repeated hybridizations. Differences in the absolute values are caused by variation of the specific activity of the labeled probes and exposure time. This signal normalization step is accomplished by dividing absolute values by the representative dot value for that image. The second scaling step involves ranking of all normalized hybridization values within each clone. The rank scores of 1 and 100 are assigned to the probes with the smallest and largest relative values, respectively, for the given clone. The rank scores for all other probes that hybridize to that clone are uniformly distributed in between. The rank scaling corrects for DNA concentration differences between clones and reduces experimentally introduced intensity differences of the positive signals. These normalization and scaling processes extract from signatures more accurate and rich clone identity information than is obtainable from the reduction of the signals on the 0/1 or (−/+) values. In transforming raw data in 0/1 values, many definitive errors are introduced and probes with mismatch signals are reduced to the less informative 0 state.

The final products of data evaluation and scaling steps are informative hybridization signatures composed of scores ranging from 1 to 100, obtained with probes successfully hybridized on a selected clone set.

Comparison of Hybridization Signatures

The hybridization signatures, containing scores of more than 200 probes prepared as described above, allow accurate recognition of cDNA clones that share more than 50% of DNA segments as well as clones representing genes that have overall DNA sequence similarity greater than 85%. A few levels of clone relationships can be defined: (i) identical or almost identical, (ii) highly overlapping or similar but distinct, and (iii) minimally overlapped or similar, but recognizable compared with unrelated clones. These relationships are established on the basis of hybridization signature similarity factors calculated for every pair of analyzed clones.

The similarity (SIM) factor for a pair of clones is defined in Eq. (1), where S_i and L_i are the smaller and larger, respectively, for the two rank scores for the ith probe for the given pair of clones.

$$\text{SIM (clone } X, \text{clone } Y) = \left[\sum_{i=1}^{p} (S_i/L_i)^2 \right] \Big/ N, \quad \text{for } L_i \geq T \qquad (1)$$

In Eq. (1), T is the threshold by which probes with low scores in both clones are purposely ignored to eliminate less accurate data and is usually set at 50. N is the number of probes used in the pairwise clone comparison, selected from all probes P, based on the threshold T.

SIM values for randomly generated signatures or unrelated clones have an average value 0.33; infrequently these values can reach 0.50. The low SIM value for most of the pairs of dots from randomly selected positions in the array (Fig. 7, green line; see color insert) demonstrates that this screening procedure can distinguish clones that have little or no similarity, thereby identifying any new clones that represent new genes or new members of existing gene families. The opposite extreme identifies pairs of identical and highly similar or overlapped clones. The average SIM in this case is about 0.75. Figure 7 (blue line) represents the SIM distribution for pairs of dots produced by repeated PCRs on a set of master clones. The data show that the vast majority of the tested pairs retained a high SIM factor, indicating that the entire procedure maintained clone identity. A small proportion of the total number of clone pairs had reduced SIM values (30–55%) owing to failures in PCRs or spotting. The failed clones forming these pairs are recognized and removed from comparisons by the described data evaluation procedures.

Forming Gene-Representing Clone Clusters

After calculating the SIM value for each pair of clones, the best matching clones can be selected for each clone of interest, or all clones that are mutually similar can be recognized. Establishing these groups or clusters of mutually similar clones is a useful process. Clustering allows the researcher to distinguish and count genes and transcript variants and to find the best representative clones for each of them. In addition, the size of a cluster (e.g., the number of clones in that cluster) is proportional to the expression level for the given gene. Because of the multiple benefits and conclusions that could be derived from the clustering results it is extremely important to form gene and transcript variants representing clusters accurately. On the other hand, vast gene and clone diversity as well as many biological and experimental artifacts impose many challenges to this process.

We found that a multistep clustering algorithm is usually necessary to apply to minimize false connections or separations. The first step is used to find the "core" of each of the clusters and then the second step is applied to add clone members that are "outliers." The clones in the core are those with the most prevailing length and splice variants and with no experimental errors. One way to form core clusters is to find every clone that is similar to at least one other clone above the core SIM threshold. This approach is independent of the clone order in the initial clone list. The optimal SIM threshold by which a pair of clones is similar enough to be put into a cluster must be defined so as to minimize the sum of false connections and false disconnections for the clustering of a large number of clones. In the large sets of clones there is a significant chance to find chimeric or cross-contaminated clones that can form bridges between two clusters.

A special procedure is used for clustering sets of clones analyzed across membranes. All clusters are first defined separately for the clone sets on each filter. Every cluster or single clone from the first filter is then compared with every cluster or single clone from the next filter. Two clusters (one from each set) are connected if a majority of clones from one cluster have statistically significant SIM values with majority of clones from the other. The same procedure is used for recognition of additional clusters that represent similar genes or gene variants created by transcription from different promoters or by alternative splicing.

Differential Expression of All Genes

The simultaneous processing of thousands of clones permits a quick screening of comprehensive cDNA libraries from test and reference tissues, providing a differential display for every known and uncharacterized gene active in one of the compared tissues. The analysis of probe hybridization to identify clone signatures and clusters also enables the comparison of transcript abundance for every gene in multiple libraries.

The clone preparation and clone comparison procedures have been optimized to provide a reproducible estimate of mRNA abundance (Fig. 8A, see color insert). The blue bars indicate the size of the 33 largest clone clusters for a represented library. The red bars represent independent analysis of a separate sample from the same library. There is a small and statistically expected fluctuation in the number and size of the clusters between the two sample runs. The graph also demonstrates a 40-fold difference in transcription activity between the first and thirtieth gene ranked by descendent expression. When two samples with the same number of clones are analyzed from two different tissues (Fig. 8B), large differences in the expression of genes are detected. For the majority of the 33 depicted highly expressed genes in the blue library, no single clone is found in the red library. For several genes, an mRNA abundance variation of three- to fivefold is observed. For example, genes represented by cluster 8 in the blue library represent less than 0.5% of the messenger population in that library, but represent 1.5% mRNAs in the red library.

This procedure provides a comprehensive identification and expression ranking of all active genes in the analyzed tissue. In the tissues represented in Fig. 8A, only the 33 genes depicted have mRNA abundance greater than 0.1%. SBH has been shown to be most useful for the analysis of large libraries and multiple libraries from many tissues. The benefit of a larger library size is the improved clustering, more accurate estimate of the expression levels, and identification of clones formed from low-abundance mRNAs. By using this protocol, 500,000 clones each from a test library and a control library can be fully analyzed in 1 month and the protocol has the resolution capacity to identify a single transcript expressed in only a few cells of a body tissue.

Conclusions and Prospects

SBH applied to dense arrays of clones (format 1 SBH) achieves an unprecedented throughput of analyzed samples that can be performed inexpensively. More than 4 million human cDNA clones have been partially sequenced (screened) in our laboratory in less than 1 year. With this capacity, the comprehensive gene surveys in model organisms (e.g., mouse, chimpanzee) or agriculturally important species (e.g., corn, cow) may be completed in a short period of time.

We believe that the availability of a comprehensive set of representative cDNA clones for all human genes or genes of other species will immensely advance genetic understanding of hereditary diseases and phenotypic diversity. For example, representative clones can be spotted in the form of dense arrays for screening by genomic probes or by total mRNA populations expressed under various physiological conditions. The complete gene arrays permit monitoring of all genes involved under any of the tested conditions. The capacity to analyze every gene expressed within a cell or tissue sample could provide valuable information on the gene interactions that are so critical for understanding polygenic disorders. Assembling the complete human "unigene" set requires the screening of tens of millions of cDNA clones and may be completed within 1 year.

Many of the central features of SBH technology can also be used in conjunction with standard gel-based sequencing methods. In SBH, the whole length of each clone is characterized and compared, after which the data may be further complemented with single-pass complete sequencing from the end segments of a small number of selected clones. The obtained ESTs provide an immediate comparison with known genes at the level of both nucleotide and deduced amino acid sequences.

Sequencing by hybridization technology applied on an array of clones provides a progressive approach from expression screening to complete sequencing of complex libraries. The same approach is applicable for mutation detection and complete sequencing of many genes amplified by PCR from thousands of individual samples.[7] A membrane with 50,000 samples represents a batch of 25–50 MB of DNA sequence. The ability to individually access any probe from a premade universal noncomplementary set of 8000 7-mers (currently used in our laboratory) or 32,000 8-mers, allows for the efficient selection of specific probe subsets for following known mutations, screening out normals, or the retesting of questionable sequences. The format 1 SBH technology allows extensive studies of expression modulation and individual allelic variations of genes in donors with documented physiological conditions (individual genomics), the studies that are critical for detailed understanding of gene functions.

Acknowledgments

The authors thank Dr. Patrick Hensler for technical and administrative assistance in the preparation and editing of this manuscript. The authors would also like to thank all group leaders and members of the gene discovery, engineering, and data analysis departments at Hyseq for making this review possible.

[7] S. Drmanac *et al.*, 1999 (in press).

[12] DNA Arrays for Analysis of Gene Expression

By MICHAEL B. EISEN and PATRICK O. BROWN

Introduction

DNA microarrays, microscopic arrays of large sets of DNA sequences immobilized on solid substrates, are valuable tools in areas of research that require the identification or quantitation of many specific DNA sequences in complex nucleic acid samples. DNA microarrays, which come in an ever increasing variety of flavors, have been used in genetic mapping studies, mutational analyses, and in genome-wide monitoring of gene expression,[1–11] and likely will become standard tools in research and clinical applications.

Microarrays are, in principle and practice, extensions of hybridization-based methods that have been used for decades to identify and quantitate nucleic acids in biological samples (e.g., Southern[12] and Northern blots,[13] colony hybridizations, and dot blots[14]). Samples of interest are labeled and allowed to hybridize to the array; after sufficient time for hybridization and following appropriate washing steps, an image of the array is acquired and the representation of individual nucleic acid species in the sample is

[1] J. DeRisi, L. Penland, P. O. Brown, M. L. Bittner, P. S. Meltzer, M. Ray, Y. Chen, Y. A. Su, and J. M. Trent, *Nature Genet.* **14,** 457 (1996).

[2] M. Chee, R. Yang, E. Hubbell, A. Berno, X. C. Huang, D. Stern, J. Winkler, D. J. Lockhart, M. S. Morris, and S. P. Fodor, *Science* **274,** 610 (1996).

[3] J. L. DeRisi, V. R. Iyer, and P. O. Brown, *Science* **278,** 680 (1997).

[4] F. Forozan, R. Karhu, J. Kononen, A. Kallioniemi, and O. P. Kallioniemi, *Trends Genet.* **13,** 405 (1997).

[5] R. A. Heller, M. Schena, A. Chai, D. Shalon, T. Bedilion, J. Gilmore, D. E. Woolley, and R. W. Davis, *Proc. Natl. Acad. Sci. U.S.A.* **94,** 2150 (1997).

[6] D. A. Lashkari, J. L. DeRisi, J. H. McCusker, A. F. Namath, C. Gentile, S. Y. Hwang, P. O. Brown, and R. W. Davis, *Proc. Natl. Acad. Sci. U.S.A.* **94,** 13057 (1997).

[7] L. Wodicka, H. Dong, M. Mittmann, M. H. Ho, and D. J. Lockhart, *Nature Biotechnol.* **15,** 1359 (1997).

[8] M. Schena, D. Shalon, R. Heller, A. Chai, P. O. Brown, and R. W. Davis, *Proc. Natl. Acad. Sci. U.S.A.* **93,** 10614 (1996).

[9] M. Schena, D. Shalon, R. W. Davis, and P. O. Brown, *Science* **270,** 467 (1995).

[10] R. J. Lipshutz, D. Morris, M. Chee, E. Hubbell, M. J. Kozal, N. Shah, N. Shen, R. Yang, and S. P. Fodor, *BioTechniques* **19,** 442 (1995).

[11] R. J. Sapolsky and R. J. Lipshutz, *Genomics* **33,** 445 (1996).

[12] E. M. Southern, *J. Mol. Biol.* **98,** 503 (1975).

[13] J. C. Alwine, D. J. Kemp, and G. R. Stark, *Proc. Natl. Acad. Sci. U.S.A.* **74,** 5350 (1977).

[14] F. C. Kafatos, C. W. Jones, and A. Efstratiatis, *Nucleic Acids Res.* **7,** 1541 (1979).

Copyright © 1999 by Academic Press
All rights of reproduction in any form reserved.

0076-6879/99 $30.00

reflected by the amount of hybridization to complementary DNAs immobilized in known positions on the array.

The idea of using ordered arrays of DNAs to perform parallel hybridization studies is not in itself new; arrays on porous membranes have been in use for years.[15–17] However, many parallel advances have occurred to transform these rather clumsy membranes into much more useful and efficient methods for performing parallel genetic analyses. First, large-scale sequencing projects have produced information and resources that make it possible to assemble collections of DNAs that correspond to all, or a large fraction of, the genes in many organisms from bacteria to humans. Second, technical advances have made it possible to generate arrays with high densities of DNAs, allowing for tens of thousands of genes to be represented in areas smaller than standard glass microscope slides. Finally, advances in fluorescent labeling of nucleic acids and fluorescent detection have made the use of these arrays simpler, safer, and more accurate.

The goal of this chapter is to describe one of the currently used microarray technologies, commonly called "spotting" or "printing" because DNAs, usually larger than oligonucleotides (100 bp and up), are physically spotted on a solid substrate, in contradistinction to the other major current microarray technology, in which short oligonucleotides are synthesized directly on a solid support.[18,19] In standard spotting applications large collections of DNA samples [polymerase chain reaction (PCR) products, plasmids, etc.] are assembled in 96- or 384-well plates. A robot positions a cluster of specially designed tips into adjacent sample wells, draws up approximately 1 μl of DNA, and deposits a small spot of each onto coated glass microscope slides. By printing successive DNA samples in a staggered fashion, thousands of spots can be printed in an area corresponding to a cluster of 4 wells of a 96-well plate or 16 wells of a 384-well plate. Typically 100–200 copies of a given array can be printed simultaneously by successively touching the cluster of printing tips to each slide before washing and drying the tips and reloading with the next set of DNA samples. After a complete set of samples is spotted, the slides are treated to attach the DNA stably to the surface and to minimize nonspecific binding of probes to the slide.

[15] E. M. Southern, G. S. Case, J. K. Elder, M. Johnson, K. U. Mir, L. Wang, and J. C. Williams, *Nucleic Acids Res.* **22,** 1368 (1994).

[16] U. Maskos and E. M. Southern, *Nucleic Acids Res.* **21,** 4663 (1993).

[17] E. M. Southern, U. Maskos, and J. K. Elder, *Genomics* **13,** 1008 (1992).

[18] A. C. Pease, D. Solas, E. J. Sullivan, M. T. Cronin, C. P. Holmes, and S. P. Fodor, *Proc. Natl. Acad. Sci. U.S.A.* **91,** 5022 (1994).

[19] S. P. Fodor, R. P. Rava, X. C. Huang, A. C. Pease, C. P. Holmes, and C. L. Adams, *Nature* (*London*) **364,** 555 (1993).

DNA microarrays have been used for a variety of purposes; essentially any property of a DNA sequence that can be made experimentally to result in differential recovery of that sequence can be assayed for thousands of sequences at once by DNA microarray hybridization. This chapter focuses on the application of DNA microarrays to gene expression studies and discusses general principles of whole genome expression monitoring as well as detailing the specific process of making and using spotted DNA microarrays.

Genome-Wide Expression Monitoring

The ultimate technical goal of whole genome expression monitoring is to be able to determine the absolute representation of every RNA species in any cell or tissue sample of interest. One would like to be able to isolate RNA from a sample, make a labeled representation of the isolated RNA, hybridize this labeled probe to a microarray, and read out the amount of each RNA species in the original sample. However, there is a complex relationship between the amount of input RNA for a given gene and the intensity of the probe signal at a corresponding hybridization target. This relationship depends on a multitude of factors, including the labeling method, hybridization conditions, target features, and the sequence of the gene. Therefore, microarray-based methods are best used to assay the relative representation of RNA species in two or more samples. Fortunately, differences in gene expression between samples, i.e., where and when it is or is not expressed, and how it changes with perturbations, are what matters most about the expression of a gene. Knowing the absolute abundance of the RNA provides only marginal further utility.

Two-Color Hybridizations

To maximize the reliability and precision with which we can quantitate differences in the abundance of each RNA species, we directly compare two samples by labeling them with spectrally distinct fluorescent dyes and mixing the two probes for simultaneous hybridization to one array. The relative representation of a gene in the two samples is assayed by measuring the ratio of the fluorescence intensities of the two dyes at the cognate target element. This ratio is relatively insensitive to the sources of spurious variation discussed above, which may affect the absolute amount of probe that hybridizes to a given target element, but not the relative amount of the two labeled species.

Experimental Questions about Gene Expression to Which DNA Microarrays Can Be Applied

The simplest kind of gene expression survey compares the relative abundance of mRNA corresponding to each arrayed gene, between two different cell or tissue samples. For example, an experiment might compare cells before and after an experimental perturbation, or at successive times during a temporally staged process, or between stages of differentiation, or one might compare the pattern of mRNAs expressed in a mutant cell with that of its wild-type counterpart. However, the flexibility of this experimental approach allows much more precise information about the regulation of each gene to be collected. For example, translation rates can be measured for each gene by comparative hybridization of polysomal RNA and total mRNA to a microarray. The distribution of each polyadenylated RNA between the nucleus and cytoplasm can be assessed by differential hybridization of nuclear and cytoplasmic polyadenlyated RNA samples. A similar approach can allow other kinds of subcellular localization to be surveyed for the entire set of arrayed genes.

Type I and Type II Experiments

Many interesting biological questions can be studied using the simplest form of two-color hybridization experiments, in which two samples are directly compared on a single array (these are now designated Type I experiments; see Ref. 20). However, such direct comparison is impractical for complex questions requiring the comparison of multiple samples. For these, we employ an alternative experimental design using a common reference sample in each of the hybridizations required for an experiment. This maintains the essential internal-control aspect of two-color hybridization, but allows inferred comparisons to be made among large numbers of samples without requiring that every pairwise comparison be performed (such experiments are designated Type II experiments). The most common use of this method has been in time-course experiments (see Ref. 21), in which each time point is compared with an initial time point. Since the amount of a given gene at each time point is measured relative to a fixed reference (the initial time point), the behavior of each gene across the time course can be studied. The reference sample need not be related to the samples being examined. The most important attribute of the reference sample is that it should provide a hybridization signal, and thus a nonzero denomina-

[20] J. DeRisi, L. Penland, P. O. Brown, M. L. Bittner, P. S. Meltzer, M. Ray, Y. Chen, Y. A. Su, and J. M. Trent, *Nature Genet.* **14,** 457 (1996).

[21] J. L. DeRisi, V. R. Iyer, and P. O. Brown, *Science* **278,** 680 (1997).

tor for the hybridization ratio, at each target element on the array. For any large collection of samples to be compared, a convenient approximation of this ideal reference sample is an equal mixture of material from each of the samples. In such a mixture, any gene represented in one of the samples will be represented in the reference.

Hardware

Making and using printed DNA microarrays requires two pieces of hardware: a robot to produce the microarrays (referred to as an *arrayer;* see Fig. 1), and a device to image the hybridized arrays (generally a scanning laser confocal microscope, or *scanner*). The excitement over microarray technology in large research laboratories and the biotech industry has been tempered elsewhere by concerns over the cost and expertise required to perform microarray experiments. To make the necessary hardware accessi-

FIG. 1. The arrayer: Photograph of Brown laboratory arraying robot primarily built by D. Shalon. Movies of the robot printing an array can be viewed at *http://cmgm.stanford.edu/pbrown.*

ble and affordable, we have designed relatively inexpensive hardware that is easy for a molecular biologist without special engineering experience to assemble and operate, and are placing detailed parts lists, designs, and construction and operating instructions on the Web (at *http://genome_www.stanford.edu/hardware*). The total cost for both an arrayer and scanner at the time of writing is about $60,000.

Preparation of Target DNA

Choice of Target

For large-scale gene expression studies, a specific hybridization target is required for every gene to be analyzed. As essentially any double-stranded DNA sample (and probably most single-stranded samples as well) can be printed onto a treated glass surface, the choice of target is largely dictated by the resources available for obtaining representations of the genes to be studied. For organisms whose genomes have been fully sequenced, the most straightforward approach is to amplify every known and predicted open reading frame in the genome (or any subset of this collection of particular interest to the experimenter) using PCR. The ever-decreasing cost of synthetic oligonucleotides (at the time of writing, close to $10 per primer pair), and the availability of 96-well thermocyclers in most departments, make this a viable strategy for even relatively large genomes; our laboratory has used this strategy to print an array containing start-codon to stop-codon PCR products for nearly all of the approximately 6200 open reading frames from the fully sequenced yeast *Saccharomyces cerevisiae* genome.[22] For unsequenced or partially sequenced genomes, or for genomes with large and numerous introns, this strategy is not practical. For many organisms, such as humans and mouse, extensive resources have been directed at determining partial sequences of clones from cDNA libraries. These expressed sequence tags (ESTs) can be used to identify distinct mRNA transcripts,[23] and individual cDNA clones corresponding to each of these transcripts can be used as the source of gene-specific targets in an array. Our laboratory has used this approach to make an array of approximately 10,000 human genes, in which the DNA sequences that were spotted were the PCR-amplified inserts from cDNA clones in the International Molecular Analysis of Genomes and their Expression (I.M.A.G.E.) collection.[24] For unsequenced organisms without a genome sequencing project, or when no sequence is yet available, clones from cDNA

[22] J. L. DeRisi, V. R. Iyer, and P. O. Brown, *Science* **278,** 680 (1997).

[23] M. S. Boguski and G. D. Schuler, *Nature Genet.* **10,** 369 (1995).

[24] G. Lennon, C. Auffray, M. Polymeropoulos, and M. B. Soares, *Genomics* **33,** 151 (1996).

libraries (preferably normalized libraries to minimize redundancy) can be used as targets. Expression data can be collected for each of these clones without any prior information about the sequence, map position, or identity of the gene.

It is likely that some benefits could be found in using DNA fragments smaller than full open reading frames or 1000-base pair (bp) cDNAs to represent each gene. For example, with yeast genes having significant homology to each other, PCR products representing the most divergent parts of these genes could be used in place of the entire coding region to improve discrimination between the two genes. The optimal representation of each gene is a complicated problem currently under investigation.

Polymerase Chain Reaction Amplification

In general, a 100-μl PCR provides sufficient DNA for printing up to 1000 arrays. Most of our reactions are done in this volume in 96-well thermocyclers [Perkin-Elmer (Norwalk, CT) 9600s and 9700s]. The actual parameters of the amplification will depend on the type of template and primers being used. It is important, however, to limit the components of the PCR to template, primers, nucleotides, magnesium, standard buffer, and enzyme. Common additives such as glycerol or gelatin can interfere with the printing process by altering the surface tension of the drop or by competing for surface attachment sites on the slide, and should not be used.

Preparing DNA for Printing

To clean up PCR products and prepare the DNA for printing, we generally isopropanol-precipitate the PCR products and resuspend the DNA at approximately 100 ng/μl in 3× SSC (1× SSC is 0.15 *M* NaCl plus 0.015 *M* sodium citrate). Precipitation can be carried out either in the thermocycler plates used for PCR or in V-bottom 96-well tissue culture plates (Costar, Cambridge, MA). Start by adding, to each well, 10 μl of 3 *M* sodium acetate, pH 5.2, followed by 110 μl of isopropanol (we use isopropanol and not ethanol to reduce the total volume of the precipitation mixture to an amount that will fit in the thermocycler or tissue culture plates). Place the plates at $-20°$ for 1 hr (if time is an issue, this step can be omitted). Precipitate the DNA by centrifugation at 4° [in a Sorvall (Norwalk, CT) RC-3B refrigerated centrifuge with an H6000A rotor fitted with P/N 11267 swinging bucket adapters] at 3500*g* (3500 rpm in this rotor) for 1 hr. Place two or three folded paper towels beneath the plate to prevent cracking. Stacks of up to four V-bottom plates can be spun in one adapter provided that a cushion of paper towels is placed between each plate and that the plates are carefully taped together to prevent slippage. After precipitation, carefully remove the liquid by aspiration or by inverting the plate. Wash the pellets with 100 μl of 70% ethanol (make from 95% ethanol, as 100% ethanol may contain fluorescent impurities) and centrifuge again for 30 min. Aspirate or decant the liquid again, and dry the pellets in a rotary evaporator fitted to accept 96-well plates [we use a Savant (Hicksville, NY) SpeedVac Plus SC210A; if such a device is not available, the pellets can be air dried if care is taken to keep out dust]. Resuspend the pellets in 10–15 μl of 3× SSC (pH 7.0) at 4° for at least 12 hr; care should be taken to seal the plates so that little evaporation occurs, as excessive salt or DNA concentration can interfere with printing. After resuspension, transfer the solution to soft-bottom plates [Fisher (Pittsburgh, PA) MicroTest III flexible assay plates] for printing. Seal the plates and store at 4° or −20° until use. Immediately prior to printing, the plates should be centrifuged briefly to ensure that all material is at the bottom of the well.

Some users have had difficulty obtaining consistent results using this precipitation process. Many additional methods are available for recovering purified DNA from PCRs, and numerous kits for cleaning up PCRs can be purchased. We have not performed a comprehensive survey of these methods for their performance in microarray printing applications. The basic goal of this step is for the final DNA solution to be as free of contamination as possible; salts, detergents, glycerol, gelatin, and particulate matter such as resins from columns can interfere at numerous later stages in the printing process. If these are used, extreme care must be taken to ensure that they are removed from the final DNA solution. Any alternative method for preparing DNA solutions for printing should be tested on a small scale, and taken through the entire printing and hybridization process to ensure that it is compatible. An alternative method that has been used successfully is described in the next section.

Sephacryl Purification

The Sephacryl method is based on using a simple gel-filtration resin that retains salts and other small molecules while excluding PCR products. As the DNA elutes in water, it can simply be dried down, eliminating the need for precipitation. As mentioned above, great care must be taken to prevent the resin from contaminating the printing solutions. To purify 100-μl PCRs, prepare Sephacryl S400 resin [Sigma (St. Louis, MO) S400 HR] by mixing well and diluting 2:1 in double-distilled H_2O (ddH_2O). Aliquot 100 μl of resin into 96-well filter bottom plates [Polyfiltronics (Rockland, MA) UN800-PSC/mepp/D]. Wash four times by adding 400 μl of ddH_2O to plates and aspirating using a vacuum manifold (Polyfiltronics UNIVAC-S). Nest these plates in polypropylene 96-well plates and spin for 5 min at 4° at 2000 rpm [in a Beckman (Fullerton, CA) GS-6 centrifuge with a G.H. 3.8 rotor and Beckman Microplus carriers] to remove all remaining liquid. Add 100 μl of ddH_2O and centrifuge as described above to test recovery. Add 100 μl of PCR product and centrifuge into clean plates. Dry down in a rotary evaporator and resuspend in 3× SSC as described above.

Preparation of Slides

The vast majority of our arrays have been printed on poly-L-lysine-coated glass microscope slides. Many alternate coatings have been tried, including numerous silanes, but none has as yet proved to be as simple and reliable as polylysine.

To prepare polylysine-coated slides, place standard glass microscope slides (Gold Seal MicroSlides, Erie Scientific Corporation, Portsmouth, NH) in metal or glass racks (a wide variety of racks can be obtained from Wheaton, Millville, NJ, or Shandon Lipshaw, Pittsburgh, PA) so that the slides are positioned vertically, and place the rack in an appropriately sized glass chamber. Make sure that slide racks are used that will fit in an available low-speed centrifuge, as they must be spun dry at the end of the coating process, and it is best to avoid having to transfer the slides from one type of rack to another. Although most slides come in boxes labeled "pre-cleaned," it is essential that they be stripped clean prior to coating. Prepare 500 ml of alkaline wash solution by dissolving 50 g of sodium hydroxide in 200 ml of ddH_2O and adding 300 ml of 95% ethanol (again, never use 100% ethanol). Completely submerge the slides in this solution for at least 2 hr. After cleaning, extensively rinse the slides by five cycles of adding clean ddH_2O, rocking for 5 min, and rinsing. Keep the slides submerged in ddH_2O; once the slides are washed, it is essential that they be exposed to air as little as possible, as dust particles will interfere both with the coating and printing processes. Prepare a coating solution containing (for 350 ml) 35 ml of poly-L-lysine [0.1% (w/v) in H_2O; Sigma], 35 ml of sterile-filtered phosphate-buffered saline (PBS), and 280 ml of ddH_2O. Pour this solution into a clean glass chamber, and quickly transfer the slides from the final water wash into the coating solution. Rock the solution gently for 1 hr. Immerse the coated slides in a chamber containing clean ddH_2O and plunge up and down five times. Transfer the slide rack to a low-speed centrifuge (place paper towels below the rack to absorb liquid) and spin the slides dry (5 min, room temperature, 1000 rpm in a Beckman GS-6 centrifuge with a G.H. 3.8 rotor and Beckman Microplus carriers). Wrap the dry slides in aluminum foil to keep out dust. Place the wrapped slide rack in a vacuum drying oven at 45° for 10 min with the vacuum on. Remove the slides from the foil and place in a clean plastic slide rack (wood and cork leave particles on the slides). Store at room temperature with the slide box tightly sealed. The slides are not immediately ready for printing, but must be "cured" for at least a few weeks to allow for the surface to become sufficiently hydrophobic (slides are usually best about 1 month after coating). The hydrophobicity of the coated slide surface is important in maintaining the small size of the printed DNA spots necessary for high-density arraying. A simple test for the readiness of a batch of slides is to take one slide and place 150 μl of H_2O on the surface. Turn the slide 45° to the horizontal and watch as the drop moves down the slide. If it leaves no noticeable trail, the slides are probably ready for use. However, the only reliable way to tell if a batch of slides is ready for printing is to print a test array on a few slides from all available batches (see Printing of DNA Microarrays).

Printing of DNA Microarrays

The arraying robot used in printing DNA microarrays is a variation of a standard "pick-and-place" robot. During standard arraying operations, a large number of microscope slides are placed on and secured to a platter. DNA samples in 3× SSC in 96- or 384-well microtiter plates are placed on a stand. The robot positions a cluster of specialized spring-loaded printing tips into adjacent wells of the DNA source plates, filling the reservoir slot of each tip with approximately 1 μl of DNA solution. The tips are then lightly tapped at identical positions on each slide, leaving a small ($<$0.5-nl) drop of the DNA solution on the polylysine-coated slide. After depositing DNA on every slide the tips are washed and dried and the process is repeated for the next set of DNA samples, with the new spots offset a small distance relative to previous spots to produce a high-density grid. The spacing between spots is determined by the size of the deposited DNA droplet, which is a function of the sharpness and characteristics of the tip and the hydrophobicity of the polylysine surface. We routinely print arrays with spot-center to spot-center spacing of approximately 200 μm and have achieved spacing approaching 100 μm. At 200-μm spacing the entire yeast genome can be printed in a 1.8-cm^2 area, and at 100-μm spacing all of the approximately 75,000 human genes can be printed on a standard 1×3 inch microscope slide. With our current robot, each cycle (wash, dry, load, print on 110 slides) takes about 2 min, and we use a print-tip cluster with 4 tips spaced to fit in adjacent wells of standard 96-well microtiter plates. This results in a printing rate of 120 spots on 110 slides per hour. The cycle time can easily be cut in half with simple improvements in robotics, and we have a similarly sized print-tip cluster of 16 tips spaced for 384-well plates. This configuration results in a printing rate of approximately

1000 spots per 110 slides per hour, or 6 hr for the entire yeast genome. Expanding to 32 tips (and twice the area) will enable a set of all human genes to be spotted on each of 110 slides in about 2 days.

Printing Tips

As the only part of the robot that touches the DNA and the slides, the printing tips are the most critical component of making DNA microarrays. The tips we use (see Fig. 2) operate on the same principle as a quill pen; liquid is drawn up by capillary action and deposited when the tip makes contact with the slide surface. The tips are made by first cutting a thin slot in a cylinder of stainless steel. One end of the cylinder is sharpened to a cone with the slot passing through the point. The two sides of the slot are

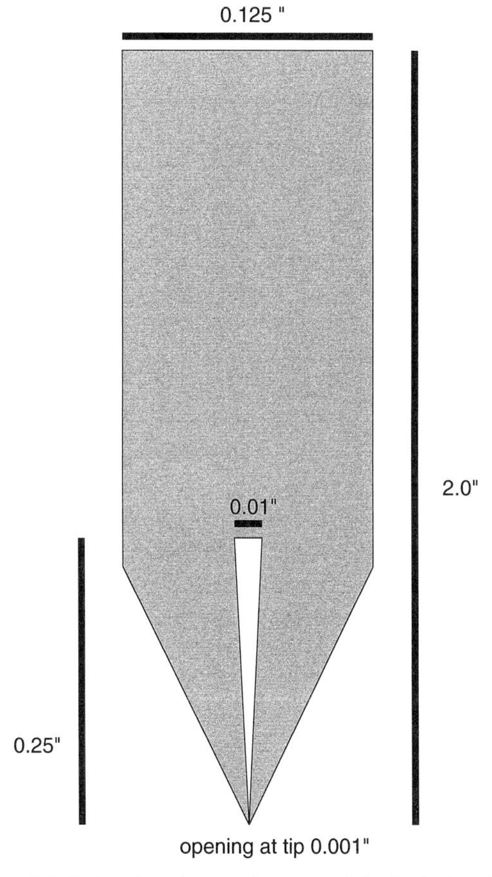

FIG. 2. Printing tip: Diagram (not to scale) of printing tip.

squeezed together until they are almost touching at the point, and the tips are further sharpened by hand. When the tips are immersed in the DNA solution, liquid is drawn up into the reservoir. When the tips are tapped gently onto glass, DNA exits only at the point. While many other factors can influence the size of the spot produced during printing (hydrophobicity of the slides, humidity in the room), in general sharper tips produce smaller spots, so the tips must be kept as sharp as possible. Since we make all of our printing tips in a semimanual manner, they do not behave identically. Furthermore, while good tips can survive hundreds of thousands of taps without noticeable loss of quality, they do eventually need maintenance. Thus, the first important step in printing good microarrays is to maintain high-quality tips. Clean the tips by placing them in a bath sonicator (Branson Ultrasonics Corp., Danbury, CT) with the tips floating, points submerged in clean water. Prime the tips by submerging the point in 3× SSC until it draws up liquid into the reservoir slot. Choose an initial set and place in the print head. Test the tips by doing a "test print." A test print is a print run in which a single slide is placed on the platter, and the tips are submerged in a fresh test solution containing salmon sperm DNA (100 ng/μl in 3× SSC) and tapped a few hundred times on the single slide with the spot-center to spot-center spacing desired for the ultimate printing (the software for the arraying robot should have an option to do such a test print). After the test print, examine the slide. The high salt content of the spotted DNA solution allows for easy visualization of the size and shape of each spot (see Fig. 3). Carefully examine the slide to ensure that (1) each tip is depositing DNA on the slide for at least the same number of taps as there are slides to be printed; (2) the spots are of the desired size and shape and stay roughly the same size and shape across the slide; and (3) all of the tips are producing spots of roughly the same size and shape. When a tip fails to print, the cause is either a failure of the tip to load DNA, or a failure to discharge DNA onto the slide. First, make sure that the tip is touching the glass when printing should occur. Second, make sure that the tip is loading DNA. This can be determined by removing the tip from the holder and slowly submerging it in the salmon sperm test solution described above while holding it up to the light. It should be possible to look through the slot to see if solution is being drawn into the slot. If the tip is not loading, make sure that the DNA solution is not too viscous (if prepared properly and made fresh this solution should be about as viscous as target DNA prepared as described above), and then clean out the slot either by continued sonication or by using a thin (0.01-inch) stainless steel shim (McMaster-Carr, Chicago, IL). If the tip is loading properly and is still not printing, it is likely that the tip is clogged. Remove the tip and examine it under a dissecting scope. The tip can be cleaned again either by sonication

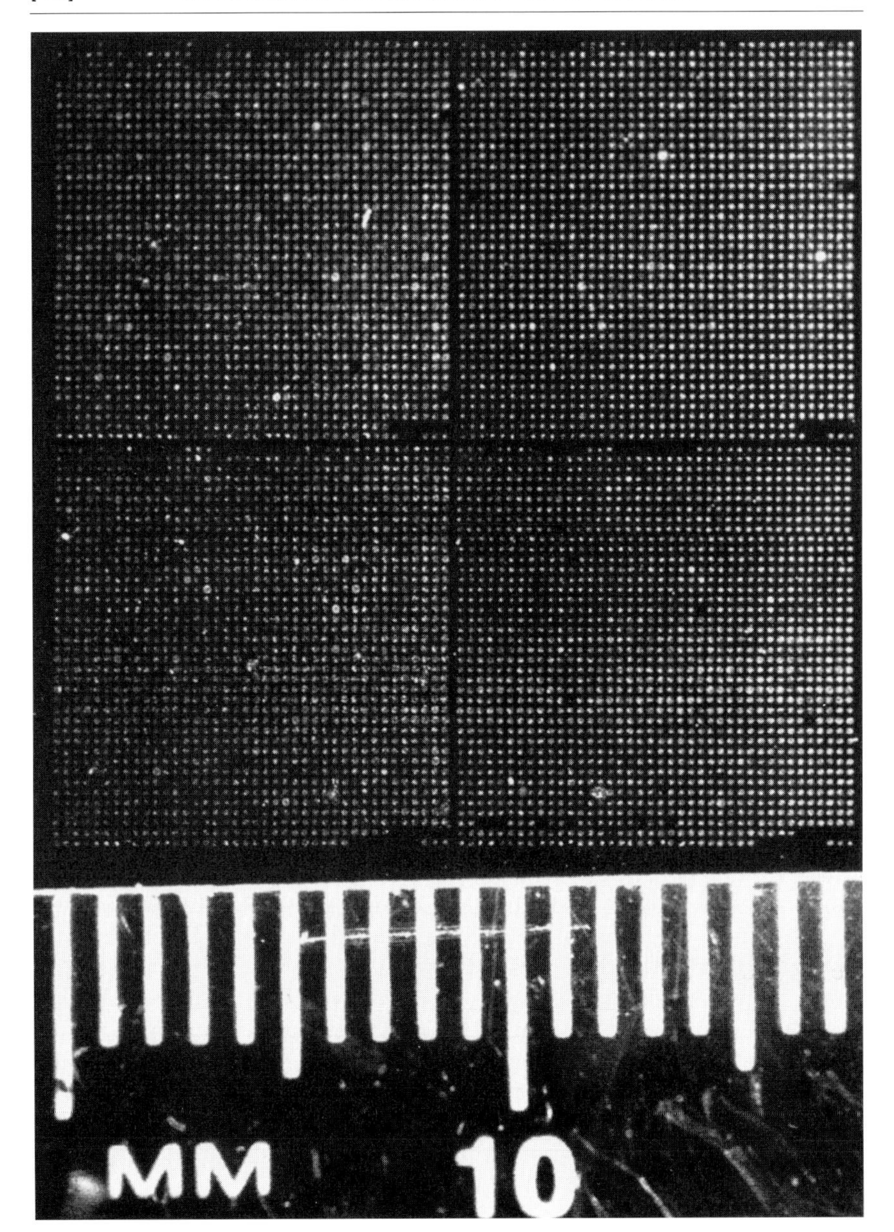

FIG. 3. A microarray: Light micrograph of a DNA microarray with targets representing essentially all open reading frames in the yeast *Saccharomyces cerevisiae.* After printing, the spots are visible owing to their high salt content until they are postprocessed. Each of the four quadrants of the image represents the spots deposited by one tip. [Photograph courtesy of J. DeRisi.]

or by gently pulling a thin piece of metal through the end. Take care not to damage the end of the tip. Make sure that the two ends of the tip are not touching, as this will prevent deposition of DNA.

If the spots are too large, so that adjacent spots run together, the tips may not be sharp enough or they may be hitting the glass too hard. First try to reduce the impact of the tap, and then try sharpening the tips with fine sandpaper or a sharpening stone under a dissecting scope. Large spots can also be caused by insufficiently hydrophobic glass or by surfactant contamination in the DNA solution. While it is aesthetically pleasing for the spots produced by each tip to be the same size, variation in the size and shape of spots is not a serious functional problem so long as the spots remain sufficiently large, and well separated.

If everything looks good in this test print, the process should be repeated at least once, and then the printing should be initiated immediately. Once begun, the printing process involves only feeding plates to the robot and vigilantly monitoring the quality of the printing. As the printing tips behave best when in constant use, printing should be interrupted as infrequently as possible. As the plates are open while they are being printed and will evaporate rapidly (changing the viscosity of the solution and potentially adversely impacting spotting), it is not advisable to design a platter to feed multiple plates to the robot unless it is accompanied by a mechanism either to remove the plate cover immediately before use, or to cool the plates to minimize evaporation.

It is important that the printing process be closely monitored to make sure that spots are still being printed and that they are still an appropriate size and shape. (*Note:* Often too much DNA is deposited on the first few slides; do not stop the print if the remaining slides look acceptable.) The most common problem is that a tip becomes clogged and stops printing. If this occurs, the print run should be interrupted immediately, and the problematic tip should be cleaned as described above. The test printing should be repeated until the tips are working well again. Another problem is caused when a dust particle or fiber becomes stuck in the tip, causing the spots to be irregularly shaped. This too should be cleared up by cleaning the tip. A different type of problem occurs if the tips become dulled during the print, and the spots will become larger, potentially causing adjacent spots to overlap. This can be solved either by replacing the tip or by sharpening it.

After printing is completed, the area containing spots should be delineated in some manner, as the spots will not be visible once the salt has been washed away. We generally do this by marking the boundaries of the area with a diamond-tipped etching device. This pen should also be used to label the slides, as most other forms of labeling (ink pens and labels) will be removed by the solvents used in postprocessing. After the slides are labeled they should be removed from the platter and stored in a clean slide box in a cool, dry place. The slides are stable at this stage for at least 1 month, and need not be postprocessed immediately.

Postprocessing of Slides

Four steps are required to prepare printed arrays for hybridization: rehydration and snap drying, UV cross-linking, blocking, and denaturation.

Rehydration

The spotting process does not, in general, leave DNA evenly distributed throughout the spot. To distribute the DNA more evenly, the spots (which dry rapidly during arraying) are rehydrated and snap dried. This is a delicate process, as insufficient rehydration can produce irregular spot sizes and can reduce overall hybridization intensity, while excessive rehydration can cause spots to fuse. Invert the metal block from a standard benchtop heating unit (VWR Scientific Products, West Chester, PA) to produce a flat metal surface, and set the block to approximately 80°. Fill the reservoir of a plastic slide hydration chamber (Sigma) with hot tap water (approximately 50–60°). Process each slide separately by placing it, array side down, over the reservoir, allowing the vapor to hydrate the elements until they all are glistening (this usually takes between 5 and 15 sec). Immediately place the slide, array side up, on the heat block until the spots have dried (usually about 5 sec). This is a process that requires practice. More consistency can be achieved by carrying out the hydration over a temperature-controlled water bath, although even then the amount of time required to hydrate different arrays, even within one batch, can vary considerably. Some groups prefer to hydrate at lower temperatures for longer periods of time. We have not tried this in our laboratory, but it is likely that the extreme temporal sensitivity of this process is reduced at lower temperatures.

Ultraviolet Cross-Linking

Once the arrays have been rehydrated and snap dried, the DNA is cross-linked to the slide by UV irradiation. While not essential, this step can increase the amount of hybridizable DNA stably attached at each spot, especially when the DNA concentration in the spotting solution is low. UV cross-linking is carried out by placing the slides, array side up, on plastic wrap, in a Stratagene (La Jolla, CA) UV 1800 Stratalinker. Switch from timed mode to total energy mode and expose the slides to 60 mJ of energy (as defined by the Stratalinker).

Blocking

The most critical step in postprocessing is the blocking step, where the remaining free lysine groups are modified to minimize their ability to bind labeled probe DNA. If these groups are not blocked, labeled probe will bind indiscriminately and nonspecifically to the surface and will produce excessively high background. We block by acylation with succinic anhydride (Aldrich, Milwaukee, WI). The charged amino group of the lysine carries out a nucleophilic attack on one of the carboxyl carbons, forming an amide bond and exposing a free carboxylate group on the other end of the chain. In addition to converting the amino groups of the lysines to amides, this new moiety creates a negatively charged surface that further reduces nonspecific binding of DNA.

As the blocking reagent has limited stability in solution, it is important that this process be carried out quickly. A glass chamber, and a slide rack with a handle that fits easily in this chamber, is required. Place the slides in this rack, and place the rack in a fume hood right next to the chamber. The chambers we use hold 350 ml of solution; scale the following reagents to the volume needed to cover the slides completely. Measure out 335 ml of water and pour into a 500-ml glass beaker. Mark the level of the solution with a pen on the outside of the beaker. Remove the water and rinse the beaker with 95% ethanol. Carefully dry the beaker with a paper towel, place a small magnetic stir bar in the beaker, and place on a stir plate in a fume hood. Measure 15 ml of 1 *M* sodium borate, pH 8.0, and place next to the beaker. Weigh out 6 g of succinic anhydride and place in the beaker. Quickly fill the beaker to the 335-ml mark with 1,2-methyl pyrrolidinone (Sigma) and begin stirring at high speed. (The solution should be clear at this point, although it will yellow slightly during the blocking reaction; if it is yellow or orange initially, the solvent has gone bad and should not be used.) Immediately after the succinic anhydride is dissolved, add the sodium borate and stir until the solutions mix (about 5 sec). Pour the solution into the empty slide chamber, immediately immerse the slides in the solution, and plunge the slides up and down five times. Cover the rack (with a glass cover or with aluminum foil) and shake gently for 15 min.

Denaturation

While (or before) the blocking reaction is taking place, prepare a large beaker or glass chamber (large enough so that the slide rack used in the blocking step can easily be submerged) filled with boiling ddH_2O. The volume should be at least twice that of the blocking solution. Turn off the heat shortly before the blocking reaction is complete, and when the water is no longer bubbling quickly transfer the slides to the water, plunge the slides up and down three to five times, and allow the slides to sit for 2 min. Immediately transfer the slides to a chamber containing 95% ethanol (again, never use diluted 100% ethanol as the benzene used to dehydrate this ethanol fluoresces), and again plunge the rack up and down three to five times. Transfer the slides to a centrifuge and spin dry. The slides are now ready for hybridization.

Preparation of Probe

Isolation of RNA

We have had success with RNA prepared by a variety of methods, including hot phenol extraction, guanidinium isothiocyanate stabilization, and others; such methods have been reviewed elsewhere and are not discussed here. We have successfully used Invitrogen (San Diego, CA) Fast-Trak 2.0 mRNA isolation kits for both yeast and human samples. The main criterion is the production of high-quality RNA. For gene expression studies, the best results have been achieved using purified polyadenylated RNA as the template for reverse transcription, although, at least with yeast, total RNA has been used successfully. An experiment requires at least 1 μg (and preferably more) of poly(A) RNA (or 100 μg of total RNA) per sample as template for synthesis of fluorescent cDNA probe. Methods for amplifying smaller amounts of RNA are being examined for their compatibility with the methods discussed here.

Labeling

A variety of methods have been used for preparation of labeled probe from RNA. In most cases we label cDNA directly by incorporating fluorescently labeled nucleotides during oligo (dT)-primed reverse transcription. Where appropriate, random-primed cDNA synthesis can be used for labeling, although it may not be appropriate for all array designs. For example, arrays composed of sequences from the 3′ ends of transcripts are more suited for analysis of oligo (dT)-primed cDNA probes, while arrays whose elements consist of entire cDNA sequences or predicted open reading frames are well suited for probes prepared by either priming method. Samples can also be labeled during second-strand synthesis after first-strand synthesis of unlabeled cDNA. Protocols for all three methods are given below.

Fluorescent Dyes

A large number of fluorescently labeled deoxyribonucleotides are now commercially available. Excellent results have been obtained with Cy3-

and Cy5-dUTP; they are spectrally well separated, can be incorporated with high specific activities with a variety of enzymes, and fluoresce brightly when dry, which simplifies the image acquisition process. Other fluors we have used successfully include R110 (Perkin-Elmer), TAMRA (Perkin-Elmer) and SpectrumOrange (Vysis, Downers Grove, IL).

Oligo (dT) and Random-Primed Labeling

All reagents should be sterilized and treated with diethylpyrocarbonate (DEPC). Start with RNA (1–2 μg of poly(A) RNA or 100–200 μg of total RNA) in 10 μl of H_2O. This solution should include any control RNAs (see section below). Add 2.5 μl (6 μg) of primer [20-mer oligo (dT), anchored oligo (dT), random hexamer or random nonamer: 2.5 μg/μl in H_2O]. Incubate in a 70° water bath for 10 min to denature RNA. Place on ice. Mix 2 μl (400 activity units) of maloney murine leukemia virus (MMLV) reverse transcriptase [from GIBCO-BRL (Gaithersburg, MD) SuperScriptII RNase H-reverse transcriptase kit], 6 μl of 5× first-strand buffer [from SuperScriptII kit; 250 m*M* Tris-HCl (pH 8.3), 375 m*M* KCl, 15 m*M* $MgCl_2$], 3 μl of 0.1 *M* dithiothreitol (DTT)] (also included in the kit), 6 μl of 5× nucleotides (2.5 m*M* dATP, 2.5 m*M* dCTP, 2.5 m*M* dGTP, 1.0 m*M* dTTP), and 3 μl of 1 m*M* Cy3- or Cy5-linked dUTPs (Amersham, Arlington Heights, IL). Add to RNA-primer solution to a final volume of 30 μl. Incubate at 42° for 1 hr. Add 1 μl of SuperScriptII reverse transcriptase and incubate at 42° for an additional hour. Add 1.5 μl of 20 m*M* EDTA to stop the reaction and 15 μl of 0.1 *M* NaOH to degrade the template RNA. Incubate at 70° for 10 min. Immediately add 15 μl of 0.1 *M* HCl to neutralize the sample (the fluors, Cy5 in particular, are sensitive to high pH). Place the samples in a Microcon 30 microconcentrator (Amicon, Danvers, MA) and add 500 μl of TE [10 m*M* Tris (pH 8.0), 1 m*M* EDTA]. Centrifuge for 5 to 10 min in a benchtop microcentrifuge at high speed to a volume of 10 to 20 μl. Discard the flowthrough. Add 500 μl of TE and repeat the centrifugation. After the second centrifugation, the probe retained by the Microcon should be significantly brighter than the flowthrough. This is a strong indicator of a successful labeling reaction. If no color is retained, it is unlikely that the labeling will produce a good hybridization result. Collect the probes by inverting the filter and centrifuging for 1 min. Mix the appropriate Cy3- and Cy5-labeled probes and increase the volume to 500 μl with TE. Centrifuge to a final volume of 8 μl and collect the probe (which should now be purple). Add 2 μl of a blocking solution containing oligo (dA) 20-mers (10 μg/μl), yeast tRNA (10 μg/μl), and, for human probes, human Cot-1 DNA (10 μg/μl; GIBCO-BRL). Add 2.1 μl of 20× SSC (to a final concentration of 3.5×) and 0.4 μl of 10% sodium dodecyl sulfate (SDS; to 0.3%). Be careful that no SDS is present on the outside of the pipette tip when SDS is added to the sample, as extra SDS can adversely affect the hybridization. Denature the sample by placing in a 100° water bath for 1 min. Allow the sample to cool on the benchtop. For human samples the sample should be left at room temperature for 30 min to allow for Cot-1 hybridization to repetitive elements. For yeast, the probe is ready for hybridization as soon as it cools. Twelve microliters is an appropriate volume for arrays of about 2 × 2 cm. A larger final volume should be used for arrays covering a larger area.

Klenow Labeling of cDNA

Prepare a cDNA template using the above procedure, except for the use of a 5× nucleotide solution containing a 2.5 m*M* concentration of each dNTP (i.e., a final concentration of 0.5 m*M* for each dNTP in the labeling reaction) and not including fluorescent nucleotides. Following RNA degradation, wash cDNA three times in Microcon 30 as described above, and recover unlabeled cDNA. Increase the volume of the sample to 20 μl. Add 3 μl of a random hexamer or nonamer (4 μg/μl). Heat to 100° for 1 min. Let cool on the benchtop for 5 min. Add 5 μl of 10× buffer [100 m*M* Tris-HCl (pH 7.4), 50 m*M* $MgCl_2$, 75 m*M* DTT], 5 μl of 10× nucleotides (0.25 m*M* dATP, 0.25 m*M* dCTP, 0.25 m*M* cGTP, 0.09 m*M* dTTP), 1.5 μl of 1 m*M* Cy3- or Cy5-dUTP (Amersham), 2 μl (10 activity units) of exonuclease-free Klenow (U.S. Biochemical, Cleveland, OH), and 13.5 μl of ddH_2O to a final volume of 50 μl. Incubate at 37° for 4 hr. Clean up labeled probe with the same series of Microcon washes described above for direct labeling of RNA, and prepare for hybridization as above.

Hybridization

Centrifuge the probe sample for 1 min in a benchtop microcentrifuge at high speed to pellet any particulate matter. Pipette the probe onto the center of the array, being careful not to dislodge any pelleted material in the tube and to avoid forming bubbles. Place a coverslip (large enough to cover the entire array surface) over the probe, again being careful to avoid the formation of bubbles. Pipette three 5-μl drops of 3× SSC on a separate part of the slide to provide humidity in the hybridization chamber and thus ensure that the probe mixture does not dehydrate during hybridization. Place the slide in a sealed chamber and submerge in a 65° water bath for 4 to 16 hr.

Posthybridization

After hybridization, carefully dry the outside of the slide chamber and remove the slide. Place the slide in a slide rack submerged in wash solution

1 (2× SSC, 0.1% SDS) with the array face of the slide tilted down so that, when the coverslip drops off, it does not scratch the array surface. After the coverslip comes off, plunge the slide up and down in this solution two or three times and transfer to a new slide rack submerged in wash solution 2 (1× SSC). Be careful to minimize the transfer of wash solution 1 to the second chamber, as SDS can interfere with slide imaging. Gently rock the chamber for 2 min, transfer the slide to wash solution 3 (0.2× SSC) for two more minutes, then spin dry the slide. The wash solutions should all be at room temperature. Altering the temperature and/or time of these steps will change the stringency of the hybridization.

Controls

Because of the highly parallel nature of microarray experiments, it is possible, and highly valuable, to include a large number of controls on every array and in every hybridization. Some useful controls are described in the following sections.

Doped RNAs

It is useful to dope in known amounts of control RNAs at various stages of the probe preparation process to monitor and assess various aspects of each experiment. Clearly, doped RNAs should be chosen to have no cross-hybridization with RNA from the organism being studied, but should be similar in their general characteristics [GC content, length, poly(A) tails]. Many yeast genes can be used as controls on human gene expression arrays, as there is limited homology at the nucleotide level between yeast and humans. Hundreds of such genes can be spotted on human arrays with only a nominal increase in the time required for arraying. By cloning these genes into bacterial plasmids with phage RNA polymerase binding sites and artificial poly(A) tails, large amounts of RNA can be made for each gene for incorporation at various experimental stages. Such controls can be used in a number of ways. For example, RNA from a number of such controls can be doped in at various ratios between the Cy3 and Cy5 samples and at varying total concentrations to assay the stringency of the hybridization, the calibration of output ratios to input ratios, and the sensitivity of detection.

RNA Quality Controls

The amount of a given nucleic acid species that will bind to a complementary target element can be sensitive to changes in the size of the complementary region. Thus, the intensity of the hybridization signal for a given cDNA can vary with differences in RNA quality. This can confound microarray hybridization results as, for example, an apparent nonunity ratio can be observed when equal amounts of a given cDNA derived from RNA of differing quality are compared. Judicious choice of target elements can minimize this problem; for oligo(dT)-primed cDNA probe, target elements restricted to the 3′ ends of genes will be minimally sensitive to RNA quality. RNA quality can be easily monitored with appropriate control elements. For example, target elements of DNA fragments of 100–200 base pairs (bp) tiled across an entire gene can be used to assay the relative RNA quality of that gene in the samples being compared [degraded RNA, labeled by oligo(dT)-primed cDNA synthesis, will give a proportionally weaker signal for the more 5′ target elements]. The experimental step that is the source of problems with RNA quality may be identified by making such tilings of genes used as doped controls, and then adding the corresponding RNA at various stages of the RNA and probe preparation process. For example, RNAs can be doped in when cells are harvested, when poly(A) RNA is isolated, when labeled probe is made, and immediately before hybridization.

Blocking Controls

As discussed above, most hybridization reactions require the addition of cold DNA to block hybridization due to non-gene-specific sequence similarities between probe and target, especially those due to repetitive elements. Control target elements corresponding to such potentially confounding sequences can be used to monitor the success of the blocking. For example, on human gene expression arrays, a human Cot-1 DNA spot should have little or no hybridization signal if repetitive elements have been successfully blocked. Other useful negative controls include oligo(dA), oligo(dT), plasmid DNA, and human genomic DNA.

Normalization Controls

As it is difficult to equalize accurately the amount of RNA used to prepare labeled probe for multiple samples, it is useful to have control target elements whose hybridization signal is expected to be the same for both samples used in a hybridization. For example, for yeast gene expression experiments, target elements of total yeast genomic DNA can be used to normalize the signal between the two samples. Human gene expression experiments are more complicated in this regard; as only a small fraction of the human genome is expressed, the hybridization signal to total human genomic DNA should be low.

Slide Imaging and Data Reduction

After hybridization, a fluorescent image (see Fig. 4) of the array must be acquired for both fluorescent dyes used. The device we use is a laser scanning confocal microscope (scanner). The scanner has a laser (or lasers) producing light with a wavelength appropriate for the excitation spectra of the dyes being used (for Cy3 this is green light around 540 nm; for Cy5 red light around 650 nm). The light passes through a standard microscope objective and illuminates a single point on the slide. The emitted light gathered by the objective passes through a series of filters (to remove the excitation beam), a collimating lens, and a pinhole (to minimize noise; this makes it a confocal device) and is quantified in a photomultiplier tube. The slide is rapidly scanned over the laser beam and a raster image of the

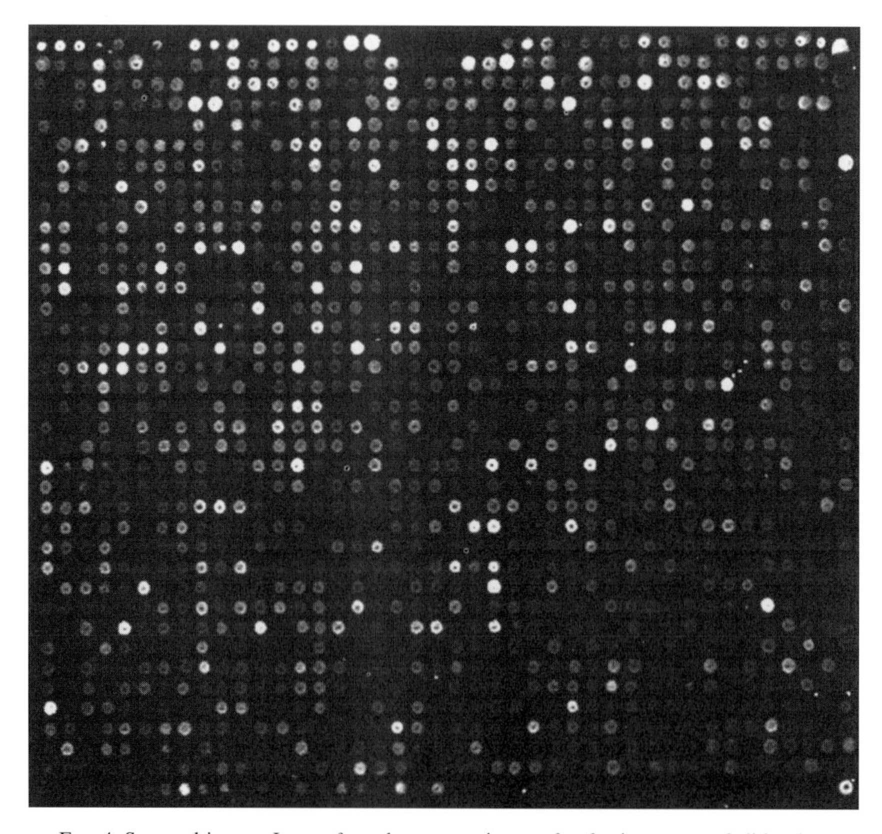

FIG. 4. Scanned image: Image from laser scanning confocal microscope of slide pictured in Fig. 3 (only one quadrant shown). A gray value is assigned to each pixel on the basis of the observed fluorescence intensity (black represents no fluorescence).

array is acquired. It is possible to acquire a similar image using a fixed slide and a pair of galvanometers to position the beam at all possible points on the array surface, or using a wide-field illuminating source and a CCD camera. We have found that the confocal scanning device gives the best results as the signal-to-noise ratios are superior to those achieved with galvanometer- or CCD-based devices.

As discussed above, the experimental goal is to measure, for each spot on the array, the relative amount of fluorescence from each fluor used. There are many ways to accomplish this, the most obvious being to segment the image into areas (boxes) corresponding to each spot and then determine the average fluorescence intensity in each box for each fluor. The relative amount of each element in the two samples is just the ratio of the average fluorescent intensities. This kind of calculation is confounded somewhat by issues involving the compensation for the nonzero fluorescent background observed on most arrays. We have developed software that implements a number of methods for accurately estimating this background, as well as more complex methods for directly determining the relative fluorescent ratios. This software will be made public along with plans for an inexpensive laser scanning confocal microscope as discussed above.

Array Problems

Experiments involving the use of DNA microarrays can encounter technical problems at any step. The opportunity, provided by the microarray format, to include numerous internal controls can greatly facilitate the recognition and correction of many kinds of problems. Indeed, even severely flawed experiments often provide much useful new information. Still, it is wise, whenever possible, in designing an experiment, to plan so that any step of the experiment can be repeated without having to repeat the experiment in its entirety. For example, it is always wise to prepare more mRNA and more arrays than would be necessary if all went well.

Few microarray images are perfect. Many problems that we have encountered are difficult to trace, and do not manifest themselves in detectable ways until the array is scanned. It is advisable to perform multiple levels of quality control before hybridizing valuable experimental samples. Batches of polylysine-coated glass should be tested by printing small arrays that are taken through the entire postprocessing and hybridization process. Only batches of slides that have been used successfully for such small arrays should be used for "real" array printings. Once an array has been printed, postprocessing should be performed on small batches of arrays and the success of the postprocessing should be determined by performing a test hybridization with probe prepared from RNA that is known to produce good results. Finally, all arrays should be scanned prior to hybridization to eliminate slides with obvious blemishes (scratches or anything else that damages the polylysine coating can be detected as differences in the background fluorescence of unhybridized arrays) and to eliminate arrays with improperly high fluorescence prior to hybridization. Polylysine and some other coatings appear to fluoresce weakly at wavelengths used to detect Cy3; for unknown reasons this fluorescence varies from array to array and can occasionally reach intensities that will interfere with the hybridization signal. In addition, printed DNA also fluoresces weakly at these wavelengths and this fluorescence also varies from array to array. If either of these prehybridization sources of fluorescence is at a level near that of the signal on hybridized arrays, these arrays should be discarded. Some of the more common problems (see Fig. 5) are discussed in the following sections.

Spot Shape Problems

One common problem is that some or all of the spots on the array may have "comet tails," that is, the fluorescence of a spot tails off in a direction

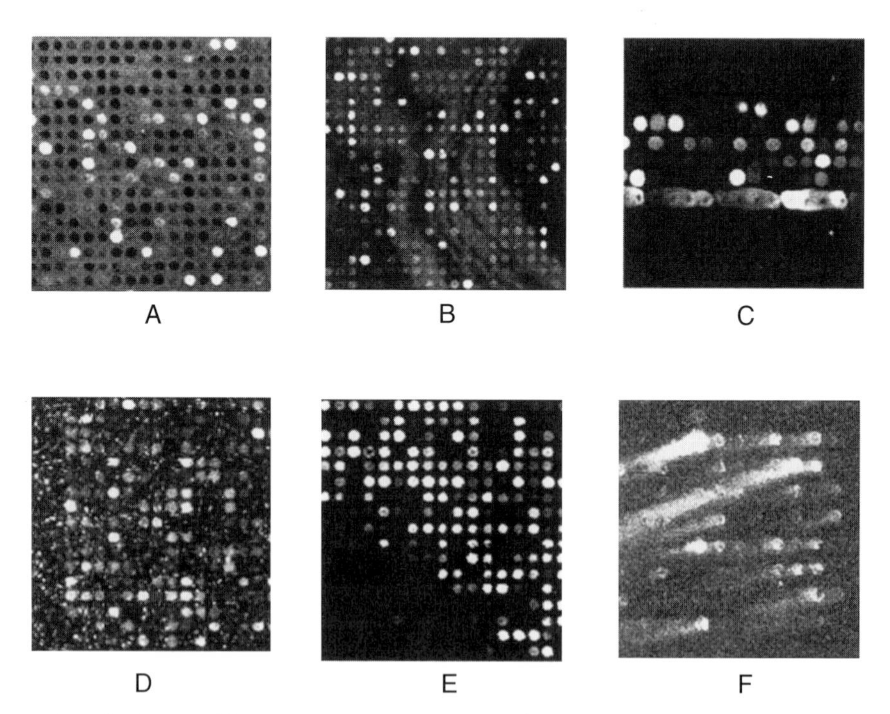

FIG. 5. Array problems: Representative problems: (A) antiprobe; (B) locally high background; (C) spot overlap; (D) precipitate; (E) locally low signal; and (F) comet tails.

that is consistent across the array, and the color of a spot is reflected in the material tailing away from the spot. This phenomenon is likely caused by insufficiently rapid immersion of the slides in the succinic anhydride blocking solution, resulting in DNA binding to the slide as it is being washed off the spot. To the extent that spots (even with comet tails) can be resolved from each other, arrays with comet tails are still usable. Another problem related to general spot appearance consists of "donut holes." When DNA is deposited using tips like those described above, it is not left in a uniform circle; far more material accumulates near the outer edge of the spot, which leaves a relatively empty area in the center. This phenomenon can usually be eliminated by longer rehydration. To a lesser extent than comet tails, donut holes do not make an array unusable and in fact only slightly compromise data quality (by reducing the fluorescent area being analyzed).

High Background

There is a nebulous class of array problems marked by abnormally high fluorescent backgrounds. High backgrounds are observed across the entire hybridized surface as well as in localized regions on the array. Many potential sources of high background can be described, but it is difficult to diagnose the source of a particular type of high background. In attempting to diagnose background problems, it is important to determine if the background is restricted to the area of the slide that was exposed to probe. Unusually high background that covers the whole slide can usually be avoided if arrays are systematically scanned prior to hybridization. High background restricted to the hybridized area indicates that labeled probe is adhering to the slide surface such that it is not being removed during the washing steps. Two possible causes of such an effect are insufficient blocking of the amino groups of the polylysine coating, or precipitation of the labeled probe during hybridization. High background from these sources is usually accompanied by a phenomenon known as "antiprobe," in which the high fluorescent background occurs only in places where no DNA was spotted. This likely reflects the diminished capacity of the polylysine groups in areas containing spotted DNA to bind additional DNA. In some cases, the background is clearly worse toward one side of the array, indicating a relationship to how the slide was positioned in a liquid medium (most likely the blocking solution). Often, high background is observed near the edges of the coverslip, likely indicating that the probe dehydrated during the hybridization. This can be corrected by adding a larger volume of 3× SSC to the hybridization chamber and ensuring that the chambers are well sealed. It is also possible for uniformly high backgrounds to result from poor probe preparation, as small probe fragment and/or unincorporated fluorescent nucleotides can bind to the surface and contribute to background without contributing to the hybridization signal.

Locally Weak Signal

Most of the phenomena that produce high background are also accompanied by a diminution of specific signal, as much of the activity of the probe is being sequestered by the array surface. However, locally decreased signal that is not accompanied by high background is also observed. Our experience suggests that this is due to a local decrease in the thickness of the probe layer in that region. Air bubbles and inhomogeneities in the flatness of the slide and/or coverslip can cause such an effect. Care in choosing and positioning the coverslip can minimize this problem.

Conclusion

DNA microarrays are tools for exploration and discovery on a genomic scale. Biologists have had the ability, for many years, to survey differential expression of a large number of genes under a small number of conditions, or a small number of genes under a larger set of conditions. DNA microarrays make it practical, for the first time, to survey the expression of thousands of genes under thousands of conditions. For organisms whose entire genome is known, (for example, *Saccharomyces cerevisiae*), this technology makes it possible to study the expression of all of the genes at once. For humans, the only limitation to applying the same approach is the number of known gene sequences; efforts to identify and sequence a complete set of human transcripts are likely to approach completion in the next few years. In the past, the limitations of our experimental tools have forced most investigations of gene expression patterns to be designed around specific hypotheses, framed on the basis of the limited knowledge we already have. We can now readily design experiments to explore systematically the vast uncharted areas of biology. The use of DNA microarrays as tools for exploration and discovery relies on their low cost, flexibility, and simplicity. Thus, in continuing to explore the many possible variations on this technology, it is useful to keep in mind that for many applications, where discovery is the goal, simplicity and low cost are critical. In our own efforts to develop this technology, a guiding principle has been to keep the economic, logistical, and psychological barriers to each genomic-scale experiment as low as possible, so that a single laboratory can carry out hundreds to thousands of genome-wide surveys in a year.

Using a single microarrayer and scanner as described here, a single laboratory can, in 1 year, make millions of quantitative measurements of differential expression, or other properties, of specific genes under specific conditions. The opportunity to collect this unprecedented quantity of gene expression information is sufficiently new that appropriate mechanisms and standards for publishing, distributing, and archiving the results lag far behind our ability to produce them. Furthermore, the analytical and computational tools for organizing, visualizing and interpreting this huge volume of information are still in the earliest stage of development. Far from being a deterrent to genome-wide exploration of gene expression and other functions, this new frontier presents a challenging new opportunity for pioneering biologists.

[13] Construction and Analysis of Arrayed cDNA Libraries

By MATTHEW D. CLARK, GEORGIA D. PANOPOULOU, DOLORES J. CAHILL, KONRAD BÜSSOW, and HANS LEHRACH

Introduction

All of life ultimately depends on the selective readout of individual genes from the genome. These appear as primary transcripts, are processed (at least in eukaryotes) to mRNA, and ultimately translated into a protein. Because cDNA libraries represent a sample of the mRNA molecules present in a specific starting material (i.e., a developmental stage, a tissue, or a cell type), their analysis provides important insights into biological processes.

The analysis of cDNA libraries from many different tissues and developmental stages will give us access to most genes long before the genomic sequencing of any vertebrate genome is completed. In fact, such data will be informative long afterward. Despite the experience gained from the large regions of genomic sequence completed, many of the genes in these sequences are not readily identifiable by available algorithms. Incorporating additional data from transcript (cDNA) sequences into the genomic analysis still provides the best way of correctly identifying all coding sequences in a genome. Prediction of the correct splice pattern(s) by computer analysis of genomic sequence has proved to be even more difficult than identifying possible coding sequences. Since cDNA clones represent the final splice product(s), they are often the only possibility for determining the sequence of correctly spliced transcripts, splice variants, and therefore the correct protein products.

Copyright © 1999 by Academic Press
All rights of reproduction in any form reserved.

0076-6879/99 $30.00

In addition to the identification of new genes and of their correct splice products, cDNA libraries can also play an essential role in the analysis of the abundance level of different mRNAs in the starting material, based on the fraction of total cDNA clones derived from a given gene. To be able to identify the genes, and to count their abundance, each cDNA clone must be assigned to its specific gene, typically by some form of sequencing process. A highly efficient partial sequencing approach based on hybridization of short oligonucleotides is described below.

Among cDNA libraries of different types, i.e., "classic" or functional (protein expression), arrayed libraries play an increasingly important role because they effectively immortalize the starting mRNA population. In such an arrayed library each clone is stored independently in separate wells of 96-well or, more efficiently, 384-well microtiter plates of the same size. Thus libraries can be quickly and accurately replicated by inoculating other medium-filled plates using 96- or 384-pinned devices, and thus identical library replicas can be distributed without losing the identity of any clone. This is particularly important because in a cDNA library each clone represents a different original mRNA molecule, and can differ from other clones derived from the same gene in many important aspects, be it different alleles of the same gene, different splice forms, varying clone starts and clone ends, and reading frame from vector-encoded start sites in protein expression library clones. Information on individual clones can be combined only if the clone is the same (or a replica of the same clone), or if all clones have been completely sequenced, and shown to have identical sequences.

Because we can consider replica copies of clones as identical, we can combine data generated from these clones and materials generated from these clones such as high-density membrane grids and polymerase chain reaction (PCR) pools. To allow the systematic generation, collection, and redistribution of such data, our laboratory started in 1989 to distribute membrane grids prepared from a number of different libraries, collecting the data generated from these materials in a separate database: the Reference Library Data Base.[1] This system, now continued in the form of the Resource Center of the German Human Genome Project (RZPD), has been the model for other efforts to systematically generate data on arrayed libraries, e.g., the IMAGE library collection.[2]

To systematically characterize cDNA libraries from many different sources requires the handling of millions of clones. Because it is an impossible and error-prone task to handle millions of samples in a conventional manner, e.g., a set of single tubes labeled by hand, our laboratory has spent

[1] G. Zehetner and H. Lehrach, *Nature* (*London*) **367,** 489 (1994).

[2] G. Lennon, C. Auffray, M. Polymeropoulos, and M. B. Soares, *Genomics* **33,** 151 (1996).

many years automating the technologies necessary for these tasks. The first robot developed, more than 10 years ago at the Imperial Cancer Research Fund in London, generated high-density membrane grids of clones spotted in regular patterns on nylon membranes, allowing simultaneous analysis of thousands of clones by hybridization. Further innovations such as 384-well plates, disposable replicators, high-throughput PCR machines, high-throughput picking robots, a rearraying robot, and sophisticated image and data analysis software, have expanded our capabilities in the type and scale of experiments possible. These machines and materials, many of which are now sold commercially, now form the backbone of the large-scale construction and use of arrayed libraries worldwide. Although this technology now makes the construction and analysis of cDNA libraries of hundreds of thousands of clones feasible, even such large libraries will often not be sufficient to identify clones representing rare transcripts ($\leq 10^{-6}$). The use of arrayed cDNA libraries should therefore be complemented by other techniques if genes represented by rare transcripts are to be identified, including screening of large and random-plated libraries, construction and use of chromosome- and chromosome region-specific libraries by cDNA selection or exon trapping,[3–5] and the cloning of reverse transcription (RT)-PCR products prepared with primers from within genomic sequences predicted to be within exons. Arrayed cDNA libraries of sufficient size (10^5 clones) from a variety of tissues/developmental stages will, however, often be enough to identify and analyze the majority of genes in an organism.

In addition, arrayed cDNA libraries constructed in special vector and host systems will play an increasingly important role in the high-throughput functional analysis of genes. Our laboratory has moved further toward "functional" genomics, especially at the protein level. Our laboratory has developed arrayed *Escherichia coli* protein expression libraries[5a] and is currently integrating phage antibody technology[6–8] to quickly characterize

[3] B. Korn, Z. Sedlacek, A. Manca, P. Kioschis, D. Konecki, H. Lehrach, and A. Poustka, *Hum. Mol. Genet.* **1**(4), 235 (1992).

[4] M. Yaspo, P. Sanseau, D. Nizetic, B. Korn, A. Poustka, and H. Lehrach, *in* "Identification of Transcribed Sequences" (U. Hochgeschwender and K. Gardiner, eds.). Plenum Press, New York, 1994.

[5] M. L. Yaspo, L. Gellen, R. Mott, B. Korn, D. Nizetic, A. M. Poustka, and H. Lehrach, *Hum. Mol. Genet.* **4**(8), 1291 (1995).

[5a] K. Bussow, D. Cahill, W. Nietfeld, D. Bancroft, E. Scherzinger, H. Lehrach, and G. Walter, *Nucleic Acids Res.* **26,** 5007 (1998).

[6] J. McCafferty, A. D. Griffiths, G. Winter, and D. J. Chiswell, *Nature* (*London*) **348**(6301), 552 (1990).

[7] G. Winter, A. D. Griffiths, R. E. Hawkins, and H. R. Hoogenboom, *Annu. Rev. Immunol.* **12,** 433 (1994).

[8] J. L. Harrison, S. C. Williams, G. Winter, and A. Nissim, *Methods Enzymol.* **267,** 83 (1996).

proteins both *in vitro* and *in vivo* (G. Walter and D. Cahill, personal communication, 1998).

Availability of Arrayed Libraries

A large number of arrayed cDNA libraries are already available from both government and commercial sources, with more libraries continually being added to the existing ones. The Resource Center and Primary Database, or RZPD (Berlin, Germany), of the German Human Genome Project is the successor to the first such resource center, the Reference Library Database.[1] It holds the world's largest collection of arrayed cDNA, genomic, and other libraries, many of which are not available elsewhere. This collection covers libraries from humans as well as model and nonmodel organisms. The RZPD has more than 200 arrayed libraries at this time, constituting many millions of clones, and also distributes the IMAGE consortium libraries.[2] All materials are available at cost, or free if part of the German Human Genome Project. A list of libraries can be found on *http://www.rzpd.de.*

Other major distributors of high-density grids include the American Tissue Type Culture Collection (Rockville, MD), Genome Systems (St. Louis, MO), Research Genetics (Huntsville AL), and the UK HGMP Resource Centre (Cambridge, UK). All four have listed their available libraries on the Web (see *http://www.atcc.org, http://www.genomesystems.com, http://ww.resgen.com,* and *http://www.hgmp.mrc.ac.uk,* respectively). Most of these centers focus on arrayed genomic libraries of human or mouse origin; however, most supply IMAGE clones.

Construction of cDNA Libraries

There are a large number of different protocols for the construction of cDNA libraries. The choice of the method used largely depends on the amount of starting material and the uses to which the library will be put. It is beyond the scope of this chapter to document them in detail. However, presented below is an overview of a common protocol, the Superscript plasmid system for cDNA synthesis and plasmid cloning (Life Technologies, Paisley, UK) for the preparation of directionally cloned oligonucleotide (dT)-primed plasmid cDNA libraries used in our laboratory to generate typically more than 1 million clones from starting material of 1 μg of poly(A)$^+$ RNA. Excellent descriptions of how to optimize the steps involved in making such directionally cloned oligonucleotide (dT)-primed plasmid cDNA libraries can be found in Refs. 9 and 10, which also list the quality control steps necessary before further analysis is performed.

Tissue Collection

Tissue should be collected as fresh as possible, and great care should be taken to keep the starting material free from any possible contaminants, such as exogenous tissues and parasites. It is also important to ensure that the samples are physiologically normal. Sterile equipment and materials must be used throughout. Although it is recommended that tissues be prepared just prior to use, snap-freezing using liquid nitrogen and storage at $-80°$ for months does not detectably affect the quality of the material.

RNA Extraction

We find that the acid guanidine–phenol–chloroform (AGPC) extraction procedure detailed by Chomczynski and Sacchi[11] is ideal for a wide range of sample sizes and tissue types, generating excellent-quality total RNA that is largely free of DNA contamination. However, care should be taken not to overload the lysis solution, which leads to much lower yield and quality.

Tissues are ground in a mortar under liquid nitrogen, the resulting powder is added to the AGPC mix and further homogenized using a Dounce (Wheaton, Millville, NJ) homogenizer, and then poly(A)$^+$ purification is performed using either oligonucleotide (dT) magnetic beads (Dynal, Oslo, Norway) or oligonucleotide (dT) cellulose columns.[12] Small amounts of total RNA should then be run on agarose gels to check for size distribution (degradation). A_{280} (protein) and A_{230} (guanidinium) readings should be determined on a UV spectrophotometer to confirm the absence of these common contaminants, as well as A_{260} (nucleic acid).

Reverse Transcription

The Superscript II enzyme (Life Technologies) has been shown to give both a high yield and high proportion of full-length cDNA.[9] Where second-

[9] J. L. Rothstein, D. Johnson, J. Jessee, J. Skowronski, J. DeLoia, D. Solter, and B. Knowles, *Methods Enzymol.* **225,** 587 (1993).

[10] M. D. Adams, A. R. Kerlavage, R. D. Fleischmann, R. A. Fuldner, C. J. Bult, N. H. Lee, E. F. Kirkness, K. G. Weinstock, J. D. Gocayne, O. White, *et al., Nature* (*London*) **377,** 3 (1995).

[11] P. Chomczynski and N. Sacchi, *Anal. Biochem.* **162**(1), 156 (1987).

[12] J. Sambrook, E. F. Fritsch, and T. Maniatis, *in* "Molecular Cloning: A Laboratory Manual" (C. Nolan, ed.), 2nd Ed. Cold Spring Harbor Laboratory Press, Cold Spring Harbor, New York, 1989.

ary structure is a concern, it can be used at temperatures up to 50°. However, we have found that the higher temperatures give a significantly lower yield (data not shown).

To make the cDNA, 1–5 μg of the poly(A)$^+$ RNA is mixed with 1 μg of *Not*I oligonucleotide $(dT)_{15}$ [5′ GACTA GTTCT AGATC GCGAG CGGCC GCCC$(T)_{15}$ 3′], all in diethylpyrocarbonate (DEPC)-treated water. This is heated to 70° for 10 min to denature the mRNA, and quick-chilled on ice. The first-strand buffer and Superscript II reverse transcriptase are added, and synthesis is performed at 37° for 1 hr; 1 μCi of [α-^{32}P]dCTP is added to aid quantification of the product and calculation of the final yield. Size distribution should be checked by running a fraction of the reaction on an alkaline agarose gel, alongside radiolabeled size standards, and exposing it to X-ray film.[12]

Cloning

Second-strand synthesis is performed according to the method of Gubler and Hoffman.[13] The double-stranded cDNA is end polished with T4 DNA polymerase. *Sal*I adaptors are ligated to the blunt-ended cDNA. The cDNA is then cut with *Not*I, and Sepharose column chromatography is performed to isolate cDNA fragments larger than 500 bp. This size fractionation eliminates unwanted excess adaptors and small fragments generated by the *Not*I digestion and allows the largest size fractions to be used, thus optimizing the chance of finding full-length clones.

The fragments now have a *Not*I site at the 3′ end and a *Sal*I site at the 5′ end, and are easily cloned into any plasmid digested with *Not*I and *Sal*I. We use pSport1 as the cloning vector, which is available doubly restricted (Life Technologies), or other specialist vectors that have been modified in-house. The double digestion of the vectors ensure that ligations give a low background, typically 1% or less. We typically transform XL1-Blue strains of *E. coli* (Stratagene, La Jolla, CA), which allow blue-white selection, overexpress the *lac*I gene (to repress LacZ fusion protein expression), give high-quality plasmid preparations,[12] and, importantly, give transformation efficiencies of up to 10^{10} CFU/μg.

Initial test platings for calculating library titer are performed using the blue-white test for recombinants. After additional quality controls to check for average insert size (which should be in the region of 1–2 kb), the transformed cells are plated for arraying under noninducing conditions to minimize the loss of potentially toxic inserts, and thus prevent any change in the representation of the different transcripts in the original tissue.

[13] U. Gubler and B. J. Hoffman, *Gene* **25,** 263 (1983).

Vector Systems

The choice of vector systems needs to be borne in mind during the early stages of the design of the project. Arrayed library technology has been primarily developed for the use of plasmid and not phage (λ) systems. Traditional phage libraries are easy to handle owing to their inherent stability, the high density at which they can be plated for screening, and the high transformation efficiency that can be obtained using commercially available packaging extracts. The same stability and high-titer characteristics become a significant cross-contamination problem in an arrayed library. Moreover, phage libraries are typically plated onto bacterial lawns to grow, and the resulting plaques are more difficult for the picking robot image analysis system to identify than are bacterial colonies.

However, many new λ phage vectors can be excised to form plasmids by infection with M13 helper phage. In the case of high-quality λ phage libraries, especially ones made from rare tissues, it may well be worth mass excising the phage library to convert it into a plasmid library, which can then be easily arrayed. New libraries are preferably made directly in plasmids, because mass excision includes an amplification step that alters the representation of the clones. It is worth noting that it is possible to make a phage library and a plasmid library from the same cDNA preparation, since a λ ZipLox vector (Life Technologies) cut with *Not*I and *Sal*I is also available. The plasmid library can then be arrayed, and the phage library can be used for screening for full-length and rare transcripts, because many millions of clones are more easily screened in this manner.

A wide selection of plasmid vectors is commercially available for cDNA work. The final choice of vector obviously depends on the type of experiments one wishes to do with the library. However, the following is a list of useful properties to consider.

1. Directional cloning ability: Two different and suitable enzyme sites must be present so that they can be used to clone the cDNA in a specific direction. Typically, one would be a rare cutter, used to cleave the cDNA after ligation of adaptors for the second enzyme. The rare cutter site is included at the end of the primer used for reverse transcription. The ability to do directional cloning is of great advantage in subsequent experiments including tag sequencing, protein expression, or generation of sense/antisense RNA transcripts.
2. Phage RNA polymerase sites, e.g., T3, T7, and SP6, to allow sense and antisense transcripts to be made *in vitro*.
3. Sites for common sequencing primers that can be used for PCR and sequencing: If one intends to use automatic sequencers, a vector that allows the use of fluorescently labeled primers (dye primers)

optimized for automatic sequencing use should be chosen. Dye primer use is cost efficient, and gives consistent sequence quality.[10,14] This is especially important in the case of Applied Biosystems (Foster City, CA) machines, where relatively few primers with the necessary mobility data files (correcting for differences in dye mobility) are available. In many cases, the incorporation of phage RNA polymerase sites would solve the primer problem because dye primers are available for T7, T3, and SP6 promoter sequences.

4. Depending on the type of library and its intended use, additional elements may be useful. For example:
   a. Vectors for prokaryotic protein expression libraries may contain *E. coli* or phage promoters, ribosome-binding sites, start codons, and peptide tags for affinity protein purification.
   b. Vectors for eukaryotic protein expression libraries may contain host or viral promoters, start codons, Kozak consensus sequences, polyadenylation sites, peptide tags for affinity protein purification, and selectable markers for eukaryotic cells.

Arraying and Gridding of Library

Arraying of small libraries can be carried out by hand, using, for example, sterile toothpicks. For slightly higher picking requirements, specially constructed picking wheels with 12 spokes can be used, allowing the picking of 12 colonies in a row, followed by inoculation of a row of 12 wells in a single process.[15] For picking of larger libraries, the use of robots is highly recommended, based either on research facilities like the Resource Center, on agreements with other groups with picking robots, or on a commercial basis.

In the following, we describe the protocol used by us, and the RZPD, based on picking robots (Fig. 1) developed by us, and available commercially. Protocols must be appropriately modified if other types of robots are being used.

The primary, unamplified plasmid library is plated at a density of 3000 to 6000 clones per 22 $\times$ 22 cm tray onto LB or 2YT agar[12] containing appropriate antibiotic(s). If ampicillin-resistant plasmids are plated, then

[14] L. D. Hillier, G. Lennon, M. Becker, M. F. Bonaldo, B. Chiapelli, S. Chissoe, N. Dietrich, T. DuBuque, A. Favello, W. Gish, M. Hawkins, M. Hultman, T. Kucaba, M. Lacy, M. Le, N. Le, E. Mardis, B. Moore, M. Morris, J. Parsons, C. Prange, L. Rifkin, T. Rohlfing, K. Schellenberg, M. B. Soares, F. Tang, J. Thierry-Meg, E. Trevaskis, K. Underwood, P. Wohldman, R. Waterson, R. Wilson, and M. Marra, *Genome Res.* **6**(9), 807 (1996).

[15] R. D. Cox, S. Meier-Ewert, M. Ross, Z. Larin, A. P. Monaco, and H. Lehrach, *Methods Enzymol.* **225,** 637 (1993).

FIG. 1. Picker/spotter robot: This robot can be used to pick (array) clones into microtiter plates using the CCD camera mounted on the head to visualize the colonies, and a 96-pin picking gadget (not shown) to pick them. It can also be used to spot colony or PCR membranes using the spotting gadget. Here it is spotting 27,648 PCR products onto 5 membranes to generate high-density PCR membranes.

an equimolar amount of methacillin or carbenicillin should also be included. These derivatives are more stable and will help minimize satellite colony formation during storage. The total concentration of ampicillin and derivatives used is 100 μg/ml for pUC-derived cDNA vectors such as pSport1. Plates are then grown at 37° for 16 hr, or until the colonies have grown to a diameter of between 0.5 and 1 mm. The maximum size of the colonies is dictated by the colony density, because larger colonies will overlap, and consequently be ignored by the picking robot system.

Images of the plating trays are captured using a charge-coupled device (CCD) camera, and an image analysis package calculates the positions of the clones, ignoring touching or overlapping clones. The clones are picked by a 96-pin picking device, with each pin picking an individual clone. The 96 clones, 1 on each pin, are then transferred to a 96- or 384- well microtiter plate. The picking device (or gadget) is sterilized in an ethanol bath before the next 96 clones are picked. Operator intervention is needed only every few hours to change plating trays, to remove inoculated microtiter plates, and to add fresh plates. Picking rates of more than 3000 clones per hour are routinely achieved (see Refs. 15–20 for more detail).

Colony Membranes. Although manual techniques have been described to interleave clones from four 384-well plates onto a membrane,[21] generation of higher density membranes invariably involves the use of robots because of the much higher accuracy and, correspondingly, increased density of the automated procedure. After the first robot developed by our laboratory in 1987 to generate such high-density membrane grids, our laboratory has continued to participate in the development of robots of increasing accuracy, throughput, and complexity. In addition, a number of other general-purpose laboratory robots have been adapted to this use,[22] and many other ancillary machines have also been developed by us.

When a library has been picked, and several copies made, high-density membranes are then spotted by a robot using a 384-pinned spotting gadget. Up to 15 identical membranes can be made per run, with 3 such robot runs achievable per day. During spotting the robot transfers small amounts (~0.1 μl) of culture medium onto Hybond N+ nylon membranes (Amersham International, Little Chalfont, Bucks, UK), which are on top of blotting paper prewetted in medium. At the end of each run the membranes are placed on agar containing the appropriate antibiotics, and grown for 16–20 hr at 37°. Densities of 57,600 spots per 22 × 22 cm membrane are easily achievable, whereas at higher densities the colonies tend to merge. Typically 57,600 spots per membrane are used, with each clone spotted in 2 adjacent positions. This has the double advantage of easing the scoring process and limiting the number of false positives due to "hot spots" often caused by dust on membranes.

Colony membranes are processed by the method of Nizetic *et al.*,[23] using a modified alkaline lysis protocol, followed by protease digestion, drying, and UV fixation. Such colony membranes contain 10 to 50 ng of DNA per spot, and can be hybridized about 20 times.

Protein Membranes. It is worth noting that the arraying technology used

[16] S. Meier-Ewert, E. Maier, A. Ahmadi, J. Curtis, and H. Lehrach, *Nature* (*London*) **351,** 375 (1993).

[17] E. Maier, S. Meier-Ewert, A. Ahmadi, J. Curtis, and H. Lehrach, *J. Biotechnol.* **35,** 191–203, 1994.

[18] S. Meier-Ewert, J. Rothe, R. Mott, and H. Lehrach, *in* "Identification of Transcribed Sequences" (U. Hochgeschwender and K. Gardiner, eds.). Plenum Press, New York, 1994.

[19] H. Lehrach, D. Bancroft, and E. Maier, *Interdisc. Sci. Rev.* **22**(1), 37 (1997).

[20] E. Maier, S. Meier-Ewert, D. Bancroft, and H. Lehrach, *Drug Discovery Today* **2,** 315 (1997).

[21] B. S. Wong, P. J. de Jong, and M. A. Batzer, *Anal. Biochem.* **216**(1), 237 (1994).

[22] A. S. Olsen, J. Combs, E. Garcia, J. Elliott, C. Amemiya, P. de Jong, and G. Threadgill, *Biotechniques* **14**(1), 116 (1993).

[23] D. Nizetic, R. Drmanac, and H. Lehrach, *Nucleic Acids Res.* **19,** 182 (1991).

for cDNA libraries has been adapted to handle prokaryotic (*E. coli*) protein expression libraries. Protein expression libraries are addressed in a separate section below.

High-Density Polymerase Chain Reaction Membranes. For PCR amplification cDNA clones that have been picked and grown in 384-well microtiter plates are amplified by transferring a small amount of culture, using a 384-pin polypropylene gadget, into polypropylene plates prefilled with PCR mix. The plates are then heat sealed with a bilaminar nylon/polypropylene film, using a commercial plate-sealing device (Genetix, Christchurch, Dorset, UK). This seal is easily removed after the PCR is finished.

The plates are cycled between preheated water baths in a basket moved by a robot arm, which is controlled by a computer. Using our current robot (Fig. 2) up to 135 microtiter plates, or 51,840 reactions, can be cycled at a time, with 2 such runs possible per day.[16–20] Heat transfer with these microtiter plates is relatively slow compared with the ultra-thin PCR tubes used in benchtop machines, and this becomes a limiting step. We have optimized the reaction to be a two-step process: a denaturation is followed by a combined annealing and extension step. For this, longer primers are used so that the annealing and extension steps occur at the same temperature.

Fig. 2. Water bath robot: This robot is used to cycle PCR plates between three water baths set at different temperatures. The basket holds up to 135 plates, each with 384 wells, for a total of more than 50,000 PCRs.

The reaction mix contains 50 m*M* KCl, 50 m*M* Tris (pH 8.4), 1.5 m*M* $MgCl_2$, a 200 μM concentration of each dNTP, with 0.5 unit of *Taq* DNA polymerase and 5 pmol of each primer per 40-μl reaction. For the pSport1 vector we use M13FSP (a 32-mer: 5′ GCTATTACGCCAGCTGGC GAAAGGGGGATGTG 3′) at the 3′ end, and M13RSP (a 32-mer: 5′ CCCCAGGCTTTACACTTTATGCTTCCGGCTCG 3′) at the 5′ end. The plates are cycled between baths at 96° (3 min) and 72° (5 min), for 30 such cycles. Success rates of up to 90%, with yields up to 100 ng/μl, are routinely achieved, and >7-kb inserts have been amplified in this way. The addition of a proofreading DNA polymerase (such as *Pfu*) increases both the success rate and the yield, especially of the longer products.

When spotting purified DNA or PCR products, densities up to 230,400 spots per membrane have been achieved, but reliability and, more importantly, the difficulty in scoring membranes of such high density, limits their use. Thus PCR products are spotted onto nylon membranes, prewetted in denaturant, similar to the procedure described above. Each PCR product is spotted in duplicate in a 5 × 5 block spotting pattern (57,600 spots per membrane). For membranes to be used for complex hybridizations, or oligo fingerprinting (see below), concentrated genomic salmon sperm DNA is spotted at the center of each 5 × 5 block to act as a guide spot, or beacon, in the later image analysis process (see below). To increase the amount of DNA in each spot, we use 10 rounds of spotting for each PCR product, transferring a total of 10 to 100 ng of DNA per spot. After spotting, the membranes are neutralized in 1 *M* Tris-HCl (pH 7.4) and 1.5 *M* NaCl, dried, and the DNA fixed by UV irradiation. Such membranes can be used up to 30 times.

Analysis of Arrayed cDNA Libraries

Typically, the main question posed in the analysis of clone libraries, and especially of cDNA libraries, is the identification of clones carrying a specific sequence. This includes the identification of cDNA clones corresponding to, or relating to, a previously isolated DNA fragment, containing specific sequence motifs, or clones with a sequence contained in a complex sequence mixture (e.g., cDNA made from mRNA from a specific tissue or stage of development). In principle, these questions can be answered by a number of different approaches. If the sequence of all components has already been determined then sequence matching may be carried out by computer algorithms. Similarly, the use of PCR screening techniques again identifies a specific clone, based on its unique sequence. In contrast, the use of high-density membrane grids or DNA chips from arrayed libraries offers the possibility to carry out such sequence matching on a large number of samples in parallel.

The most common application of gridded libraries is to use a single-copy probe to screen for a particular gene of interest. Such data, and the subsequent further analysis of individual clones, can be entered into a database as previously described.[1] With many laboratories using the same library, the database can integrate all available data, generating a more complete picture than any one laboratory could achieve.

While arrayed libraries are undoubtedly useful for single-copy probe hybridization experiments, the full potential of the high-density gridded format is realized only in larger scale experiments, aiming at exhaustive characterization of the starting material tissue. The most efficient approach for large-scale analysis is, again, hybridization based.[24–30] Although it is possible to work systematically through the library, plate by plate, generating expressed sequence tags (ESTs) for each clone, as has been done with several arrayed libraries,[10,14] the sequencing of thousands of clones is time consuming and costly. Two techniques developed in our laboratory to analyze arrayed libraries, oligonucleotide fingerprinting and the use of complex probes, have proved enormously successful in characterizing libraries, and therefore, tissues, both quickly and inexpensively.[16–20,24,25,30–33]

The techniques are both based on the principle of using a large number of probes that independently hybridize to a large proportion of the clones. In the case of oligonucleotide fingerprinting, these sequences are short, computer-designed oligonucleotides, whereas complex probes are derived from an entire mRNA population of a cell or tissue type, and as such consist of tens of thousands of transcripts. Hybridization experiments using

[24] A. Poustka, T. Pohl, D. P. Barlow, G. Zehetner, A. Craig, F. Michiels, E. Ehrich, A.-M. Frischauf, and H. Lehrach, *Cold Spring Harbor Symp. Quant. Biol.* **LI,** 131 (1986).

[25] H. Lehrach, R. Drmanac, J. Hoheisel, Z. Larin, G. Lennon, A. P. Monaco, D. Nizetic, G. Zehetner, and A. Poustka, *in* "Genome Analysis, Vol. 1: Genetic and Physical Mapping," pp. 39–81. Cold Spring Harbor Laboratory Press, Cold Spring Harbor, New York, 1990.

[26] G. G. Lennon and H. Lehrach, *Trends Genet.* **7**(10), 314 (1991).

[27] J. D. Hoheisel, G. Lennon, G. Zehetner, and H. Lehrach, *J. Mol. Biol.* **220,** 903 (1991).

[28] J. D. Hoheisel, E. Meier, S. Meier-Ewert, and H. Lehrach, *Ann. Biol. Clin.* **50,** 827 (1993).

[29] J. D. Hoheisel, M. Ross, G. Zehetner, and H. Lehrach, *J. Biotechnol.* **35,** 121 (1994).

[30] J. D. Hoheisel, *Trends Genet.* **10**(3), 79 (1994).

[31] T. M. Gress, J. D. Hoheisel, G. Lennon, G. Zehetner, and H. Lehrach, *Mamm. Genome* **3,** 609 (1992).

[32] T. M. Gress, F. Müller-Pillasch, G. Adler, G. Zehetner, and H. Lehrach, *Eur. J. Cancer* **30A**(9), 1391 (1994).

[33] T. M. Gress, F. Muller-Pillasch, M. Geng, F. Zimmerhackl, G. Zehetner, H. Friess, M. Buchler, G. Adler, and H. Lehrach, *Oncogene* **13,** 1819 (1996).

such probes give a high hit rate, and can be used to fingerprint cDNA libraries, partitioning them into groups having similar or identical sequence.

Small-Scale Analysis: Single-Copy Probes

When colony membranes are used for the first time, multiple changes of hybridization buffer may be necessary to remove agar and colony debris from the membranes; this reduces background. With PCR product membranes, one round of prehybridization is sufficient. Two general principles of hybridization experiments are to use stepwise stringency washes and exposures (allowing weak signals to be seen and some false positives to be excluded) and to make sure that the membranes do not dry out at any point during the experiments, as this may make the probe difficult to remove. We typically radiolabel 10–50 ng of template with ^{32}P, and aim for 0.1–10^6 counts per nanogram, following the method of Feinberg and Vogelstein.[34] The nucleotides and small fragments can be removed simply by probe precipitation with ammonium acetate and ethanol precipitation, using tRNA or another such coprecipitating carrier molecule for small amounts of nucleic acid. After precipitation the incorporated radioactivity can be measured with a scintillation counter.

High-Stringency Radioactive Hybridizations. Membranes are prehybridized at 65° for a minimum of 30 min in modified Church buffer[35] containing 0.25 *M* Na_2HPO_4 (pH 7.2), 7% sodium dodecyl sulfate (SDS), and 1 m*M* EDTA. If required, the probe can be competed against total or Cot-1 DNA of the probe or target organism, using a 0.5–1.5-mg/ml concentration of DNA in 0.12 *M* Na_2HPO_4 (pH 7.2).[36] The membranes are hybridized in Church buffer at 65° for at least 16 hr, using a probe concentration of 1 million cpm/ml. Membranes are then washed twice in either 2× SSC (1× SSC is 0.15 *M* NaCl plus 0.015 *M* sodium citrate) with 0.1% SDS (lower stringency), or in 40 m*M* Na_2HPO_4 (pH 7.2) with 0.1% SDS (higher stringency) at room temperature, followed by two washes in the same buffer at 65° for 20 min.

The membranes can be exposed to X-ray film for 1–3 days at −70°, or for a few hours on a phosphor screen. We recommend the use of a PhosphorImager (Molecular Dynamics, Sunnyvale, CA), which gives a linear response over four to five orders of magnitude[37] and the ability to alter the gray scale viewed on a computer screen, thus avoiding many of the

[34] A. P. Feinberg and B. Vogelstein, *Anal. Biochem.* **132,** 6 (1983).

[35] G. M. Church and W. Gilbert, *Proc. Natl. Acad. Sci. U.S.A.* **81,** 1991 (1984).

[36] S. Baxendale, G. P. Bates, M. E. MacDonald, J. F. Gusella, and H. Lehrach, *Nucleic Acids Res.* **19**(23), 6651 (1991).

[37] R. F. Johnston, S. C. Pickett, and D. L. Barker, *Electrophoresis* **11,** 355 (1990).

problems typically associated with film exposures, such as preflashing and over- or underexposure.

Low-Stringency Radioactive Hybridization. Low-stringency hybridization is typically used to identify clones corresponding to genes previously identified in other organisms. Because hybridization of arrayed libraries usually gives at least some variation in signal owing to uneven background, careful control of the required stringency of hybridization, and critical interpretation of the data, are required. Inclusion of a Southern blot containing digested genomic DNA from different species (a zoo blot), ideally including the species from which the arrayed library has been prepared, is highly recommended in order to compare the signal expected for the real homolog with the signal on the membrane. It should, however, be kept in mind that the comparison of hybridization results obtained on zoo blots and cDNA membranes can be difficult owing to the possible distribution of exons of the cDNA on a number of different genomic fragments.

The hybridization mix commonly used in our laboratory contains accelerating and blocking agents designed to increase the signal-to-noise ratio when low homology probes are used, such as homologous genes from divergent species. The mix consists of 30% deionized formamide, 4× SSC, 1 m*M* EDTA, 50 m*M* Na_2HPO_4 (pH 7.2), 1% SDS, 8% dextran sulfate, and sheared salmon sperm DNA (50 μg/ml) in 10× Denhardt's solution.[12] This same mix can also be used for high-stringency hybridizations if the formamide concentration is increased to 50%.

Prehybridization is performed at 42° for at least 4 hr, preferably longer. The same mix is used for the hybridization step, with random-primed probes, and at a concentration of about 1 million cpm/ml. Hybridization is performed at 42° for at least 16 hr. Using the same buffer, the membranes are then washed in 2× SSC and 0.1% SDS, twice for 20 min at room temperature, and twice for 20 min at 55° or 65°. After exposure, a more stringent wash should be done using 0.1× or 0.5× SSC at 55° or 65°.[12]

Nonradioactive Hybridization. For nonradioactive detection systems, our laboratory uses biotin- or digoxigenin (DIG)-labeled probes. DIG labeling gives a better signal-to-noise ratio on colony membranes, presumably because of the background of endogenous biotin and biotinylated proteins in the *E. coli* debris. For this reason, our laboratory has optimized the use of DIG labeling,[38] and by using this detection system with plastic-backed (laminated) membranes[39] has developed a high-throughout hybridization system in which one technician can monitor more than 1 million probe–

[38] E. Maier, H. R. Crollius, and H. Lehrach, *Nucleic Acids Res.* **22,** 3423 (1994).

[39] D. R. Bancroft, J. K. O'Brien, A. Guerasimova, and H. Lehrach, *Nucleic Acids Res.* **25**(20), 4160 (1997).

clone interactions per day. However, biotin-labeled probes are equally effective when using PCR or purified DNA membranes.

Random-primed or PCR DIG-dUTP-labeled probes are hybridized to the membranes at a concentration of 20 ng/ml in 0.25 *M* Na_2HPO_4 (pH 7.2), 7% SDS, and 1 m*M* EDTA, at 65° for 16 hr. The membranes are then washed in 40 m*M* Na_2HPO_4 (pH 7.2) and 0.1% SDS, first at room temperature for 20 min, and then at 65° for another 20 min. Blocking of the membranes is performed by incubating them in 5% (w/v) nonfat milk powder in 1× phosphate-buffered saline (PBS) at room temperature for 45 min, before addition of the anti-DIG alkaline phosphatase conjugate (0.15 units/ml) in the same blocking solution, which is incubated at room temperature for 1 hr. The membranes are then washed twice in PBS[12] for 30 min at room temperature, and twice in 0.1 *M* Tris (pH 9.5) and 1 m*M* $MgCl_2$ for 10 min at room temperature. Color development is performed in the same buffer, and a one-fifth volume (i.e., about 2 ml for a 22 × 22 cm membrane) of 5 m*M* Attophos (JBL Scientific, San Luis Obispo, CA) in 2.4 *M* diethylamine (pH 9.2) is sprayed on evenly. The membranes should be incubated at 37° for 1–4 hr, until color develops. Attophos is a commercial alkaline phosphatase substrate, 2′-(2-benzothiazolyl)-6′-hydroxybenzothiazole phosphate, which is extremely fluorescent on dephosphorylation, and also has a high Stokes shift,[40,41] allowing attomole sensitivity in hybridization experiments.[38,40] Data capture is achieved by illuminating with long-wave UV light and a high-resolution CCD camera, which enables the image analysis of DNA arrays at densities of up to 120,000 clones on a 22 × 22 cm nylon membrane.

Stripping Membranes. Membranes can be reused at least 10 times, providing the probe(s) from the previous experiment has been removed/stripped off. This can be done (in order of increasing harshness) with 50% deionized formamide, 40 m*M* Na_2HPO_4 (pH 7.2), and 0.1% SDS at 65–75°, with 1 m*M* EDTA and 0.1% SDS at 65–75°, or with 200 m*M* NaOH and 0.1% SDS at room temperature or at 65°. The membranes should last longer if the less harsh protocols are used; membranes are frequently used more than 20 times in our laboratory. Membranes with larger insert clones, i.e., genomic P1-derived artificial chromosome (PAC) libraries, are even more stable, presumably because of the larger number of DNA–nylon interactions per molecule.

[40] R. J. Cano, M. J. Torres, R. E. Klem, and J. C. Palomares, *Biotechniques* **12**(2), 264 (1992).

[41] J. L. Cherry, H. Young, L. J. Di Sera, F. M. Ferguson, A. W. Kimball, D. M. Dunn, R. F. Gesteland, and R. B. Weiss, *Genomics* **20**(1), 68 (1994).

Large-Scale Library Characterization

Image Analysis of High-Density Grids. (See Fig. 3.) A critical step in the use of high-density grids, especially in large-scale analysis, is the interpretation or scoring of the results. For small-scale users the scoring is done by the *x* and *y* coordinates of positives on the membranes. These are submitted to the distributing center (i.e., RZPD), and a simple computer algorithm calculates the plate well position using the spotting order file for that membrane. An alternative is to use an overlaying transparency with a printed grid.

For the large-scale user, the scoring of many membranes, or even the scoring of one membrane containing many positives, is a time-consuming process. To expedite hybridization-based experiments, our laboratory[16–20,31,42] and others[43] have developed a sophisticated set of automated scoring packages. The high-density grids are divided into blocks, each of which contains multiple duplicated spots, with each pair in a different orientation to allow easy identification of the different clones. At the center of every block and, therefore, at regular intervals over the membrane, are guide spots, which serve as beacons for the grid finding program (see Fig. 3). Such an automated system is essential for any large-scale analysis.

We use two image analysis packages, one completely automatic and another semiautomatic. Both share the same core simulated annealing algorithm, which finds the best fit for a grid pattern onto a hybridization image, homing in on the guide spot signals in the center of every block. The automatic system sends values into other data analysis programs, whereas in the semiautomatic system signals are defined as positive or negative by a human operator.

The semiautomated system is used when there are few positives, and these are qualitative results. It automatically aligns a grid onto the membrane image, and calculates the plate well coordinates for each clone highlighted by a click with the mouse. The human mind is excellent at pattern recognition of this type, and it is difficult to define a positive for use in a conventional algorithm when one realizes problems such as the degree to which signal-to-noise ratios can vary across single membrane hybridizations. Our laboratory continues to train neural networks for this job; in the meantime unskilled personnel can score membranes with 10 positives from a total of 27,648 clones in 2–3 min.

The fully automated system is used for quantitative analysis, and uses the same simulated annealing algorithm. First, the package finds the blocks

[42] K. Hartelius, Ph.D. thesis. Technical University of Denmark, Lyngby, Denmark, 1996.

[43] S. Granjeaud, C. Nguyen, D. Rocha, R. Luton, and B. R. Jordan, *Genet. Anal.* **12,** 151 (1996).

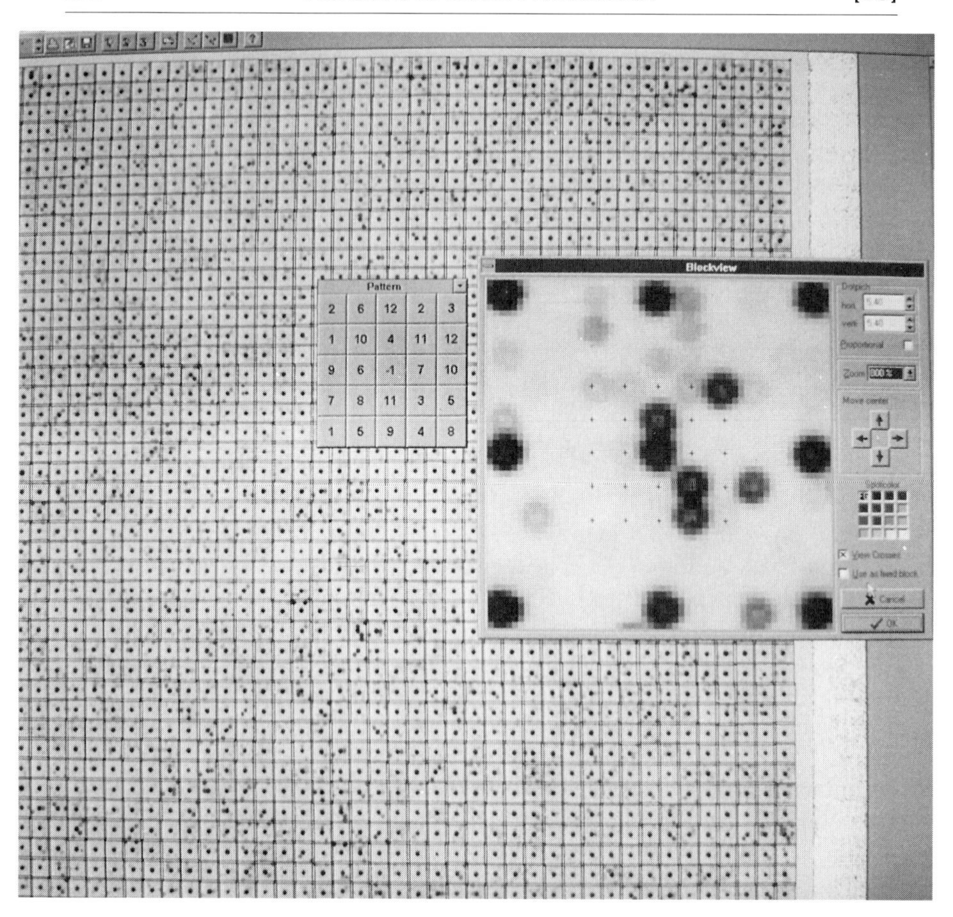

FIG. 3. Image analysis of high-density membrane: A section of a high-density PCR membrane of a zebrafish gastrula cDNA library is shown here hybridized with an 8-mer oligonucleotide. A regular pattern of guide spots (concentrated salmon sperm DNA) across the membrane is visible, as are positive clones, the automatically aligned grid, the duplicate spotting pattern, and a magnified section.

by aligning the grid, and then finds individual spots within each block, writing their intensities to another file along with values for general and local background. The correlation factor between the duplicate spots is used as a quality assessment for the results of each clone.

Oligonucleotide Fingerprinting. (See Fig. 4.) Oligonucleotide fingerprinting is the one technique that can take full advantage of the vast amount of information stored in an arrayed cDNA library in a quick and efficient manner. The ideal oligonucleotide probe will hybridize to 50% of all clones.

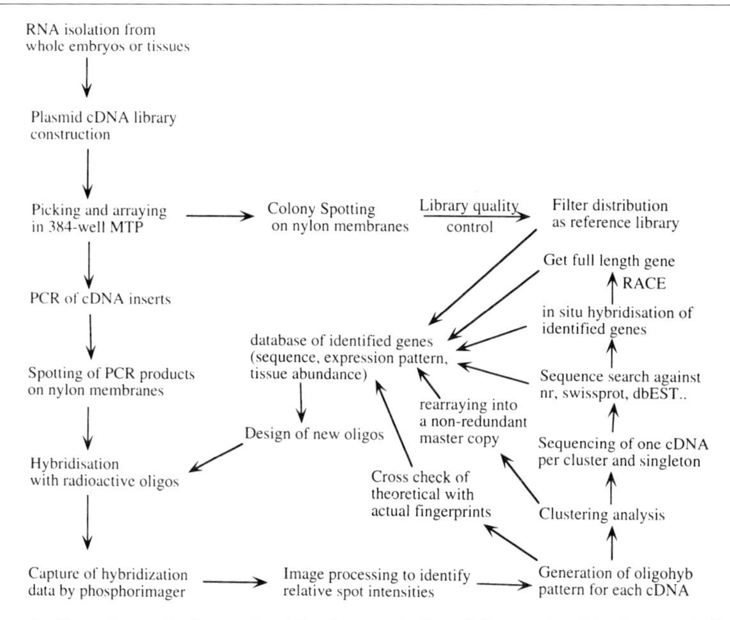

FIG. 4. Overview of oligonucleotide fingerprinting: Oligonucleotide fingerprinting is a central part of our laboratory's approach to identifying and analyzing genes as cDNAs.

A set of 20 such probes could partition a library into 2^{20} (1,048,576) unique fingerprints, a number at least 1 magnitude greater than the vertebrate gene set, which is estimated to contain up to 100,000 genes.[44] In practice, the use of oligonucleotides as short as 6 bases have been shown to hybridize to 10 to 20% of large cDNA libraries in a reliable and sequence-specific manner.[16–20,45–49] The unique fingerprints are then used to identify the overlapping DNA segments (cDNA clones from the same gene), and can be matched to GenBank sequences using BLAST-type algorithms. Both Milosavljevic *et al.*[47] and Drmanac *et al.*[45] have fingerprinted EST sequenced

[44] C. Fields, *et al., Nature Genet.* **7,** 345 (1994).

[45] S. Drmanac, N. A. Stavropoulos, I. Labat, J. Vonau, B. Hauser, M. B. Soares, and R. Drmanac, *Genomics* **37**(1), 29 (1996).

[46] A. Milosavljevic, Z. Strezoska, M. Zeremski, D. Grujic, T. Paunesku, and R. Crkvenjakov, *Genomics* **27,** 83 (1995).

[47] A. Milosavljevic, M. Zeremski, Z. Strezoska, D. Grujic, H. Dyanov, S. Batus, D. Salbego, T. Paunesku, M. B. Soares, and R. Crkvenjakov, *Genome Res.* **6,** 132 (1996).

[48] R. Drmanac, Z. Strezoska, I. Labat, S. Drmanac, and R. Crkvenjakov, *Cell Biol.* **9**(7), 527 (1990).

[49] R. Drmanac, S. Drmanac, I. Labat, and N. Stavropolous, *in* "Identification of Transcribed Sequences" (U. Hochgeschwender and K. Gardiner, eds.), pp. 239–251. Plenum Press, New York, 1994.

IMAGE libraries and have achieved clustering comparable to that achieved by the sequencing approach, yet at a fraction of the cost. Both Milosavljevic *et al.*[47] and our laboratory[49a] have also correctly matched many of the largest clusters to GenBank sequences, using the sequence fingerprint information. With oligonucleotide fingerprinting, one person can easily hybridize 10 oligonucleotides to 100,000 cDNAs (4 high-density grids) in a single day, generating 1 million bits of information, or 800,000 bases (10% hit rate with an 8-mer probe). To obtain the same data using automated gel-based technology, one would have to generate 2000 to 3000 ESTs per day, a number achievable only by large sequencing laboratories with many more personnel, and at a much higher cost. The ability of oligonucleotide fingerprinting to process at least 10 times as many clones for the same cost gives a higher probability of identifying low-level transcripts, with the potential of identifying all transcripts in libraries and, therefore, the organism. Importantly, sequence data are generated over the entire length of the clone, and not only over a few hundred base pairs, which may not contain informative sequence, as is the case with ESTs. Oligonucleotides for use are selected from a list of all possible sequences, using an algorithm designed to partition complete coding sequences from GenBank, based on their theoretical fingerprints. Impractical oligonucleotides can be removed from the starting set by considering G/C and A/T content, palindromic sequences, and other problematic sequences. The selected oligonucleotides should be sequence specific, and effectively partition coding sequences into different groups.

Below, we describe the oligonucleotide fingerprinting procedure as used in our laboratory; others have reported the use of similar methods, with the majority of differences between groups lying largely in the computer algorithms used to analyze the data.

MEMBRANE GENERATION. The purity and concentration of the DNA are critical concerns. Contaminating bacterial or vector sequences can also hybridize oligonucleotides; even 1% bacterial genome contamination masks other signals because of its large number of oligonucleotide sites. In addition, signal intensities are dependent on DNA amount and probe concentration; the more DNA the stronger the signal, thus more stringent hybridization and washing conditions can be used, and more accurate results can be obtained. The most efficient way to generate pure and concentrated insert DNA is by PCR, and so high-density PCR membranes as previously described are used.

MEMBRANE HYBRIDIZATION. To avoid problems with spot resolution, the oligonucleotides are labeled with ^{33}P at the 5′ end, rather than with

[49a] S. Meier-Ewert, J. Lange, H. Gerst, R. Herwig, A. Schmitt, J. Freund, T. Elge, R. Mott, B. Herrmann, and H. Lehrach, *Nucleic Acids Res.* **26,** 2216 (1998).

the less expensive ^{32}P. ^{33}P is a low energy β particle emitter and reduces signal spread, thus allowing the use of high spotting densities. Thirty picomoles of each oligonucleotide is labeled with T4 polynucleotide kinase in a 30-μl reaction containing 50 μCi of [γ-^{33}P]dATP.[12] The reaction mixture is incubated at 37° for 45 min and terminated by adding EDTA to a final concentration of 50 m*M*. Labeling efficiency is assessed for every oligonucleotide by polyethyleneimine chromatography, in which 0.5 μl of the reaction is spotted onto a PEI Polygram (Macherey-Nagel, Düren, Germany), and run for 10–20 min with an acid running buffer (i.e., 0.75 *M* KH_2PO_4, pH 3.5), and the amount of acid-insoluble (incorporated) nucleotides, which stay at the loading end, is compared with free-running nucleotides at the buffer front to give a simple measure of the efficiency.

The PCR membranes are prehybridized in 1× SSARC buffer (4× SSC, 7.2% sodium lauryl sarcosinate, and 1 m*M* EDTA) for at least 20 min at 4°. Hybridizations are carried out in a manner similar to those described previously,[48] using a 4 n*M* concentration of an oligodeoxynucleotide (oligo) and hybridizing at 4° for 3–16 hr. Membranes are subsequently washed in a large volume of 1× SSARC buffer for 20–40 min at 4°, dried briefly on blotting papers, wrapped in Saran Wrap,[18] and exposed on phosphor storage screens for 3–16 hr. Bound oligonucleotides are removed by washing membranes in 0.1× SSARC at 65° for 10–30 min. Membranes can be used for at least 20 hybridization-stripping cycles without significant loss of signal strength.

Data Capture. The phosphor storage screens are scanned at 176-μm resolution on a Molecular Dynamics PhosphorImager. The resulting 16-bit TIF format image files, 5 MB each, are transferred to a local DEC Alpha UNIX workstation, where they are image analyzed as described above. Absolute intensity values for the membranes are stored and passed into other analysis programs.

Data Analysis. The quality of every individual hybridization result is assessed by using 1920 control clones, which are spotted on every membrane with the cDNA clones; these clones constitute about 2 Mbp of known DNA sequence. The hybridization ratio, given by comparing the full match and mismatch values from these known sequences, is calculated for every experiment. Any hybridization giving a low ratio, i.e., with a low or nonspecific signal, is discarded and the hybridization repeated.

The intensity values of each cDNA clone are normalized for experimental variables before fingerprint analysis. If such normalizations were not carried out, incorrect data resulting from, e.g., weakly hybridizing clone sequences with a large amount of DNA counting as positive signals relative to perfectly hybridizing clones with less DNA, would enter into the later analysis. Dividing all hybridization values by the values obtained with a common oligonucleotide contained within the vector sequence has not proved to be the best system for normalization, because it places a huge weighting on the result and scoring of a small set of hybridizations, and fails to normalize for other variables. An alternative is to use a ranking procedure, such as the double-ranking method.[46] Double ranking consists of two signal-ranking steps:

Step 1: In the first ranking, signals of all clones with the same oligonucleotide in a single membrane hybridization experiment are compared and ranked. The strongest clone is assigned a value of 1, the weakest 0, and all other clones at regular intervals in between, i.e., at intervals of $1/N$ (where N equals the number of clones). This is repeated for every membrane hybridization experiment. Such a system means that clones are only ever compared with other clones in the same experiment, thus probe-specific activities, washings, and exposure times are always identical.

Step 2: In the second ranking, all of the values for every oligonucleotide for a clone are ranked, thus the strongest oligonucleotide is assigned a value of 1, and the weakest 0. In this normalization, because only oligonucleotide results on the same clone are compared, the amount of DNA is constant.

Normalized data are passed onto the next stage. Pairwise comparisons between every clone identify statistically similar fingerprint patterns using BLAST-type algorithms. Clones grouped together owing to similarities in their fingerprint are termed "clustered" (Fig. 5), whereas clones that have no statistically similar fingerprint neighbors are called "singletons." Consensus fingerprints of whole clusters, and those of singletons, can be matched to theoretical fingerprints of the GenBank sequences and putative identities assigned.

Fingerprinted libraries from different tissues, or even different developmental stages, can be compared to study changes in transcript levels.[49a] Importantly, the analysis can be repeated, and multiple *in silico* subtractions, normalizations, etc., carried out with different parameters. The results can be checked electronically using lists of previously identified clones, and the best method of analysis can be chosen. This can be repeated each time another set of data from a new library is generated. Using the fingerprint information, one can also pick clones most unlike others studied, in an effort to identify novel genes, and concentrate valuable resources.

Complex Probes. Surprisingly little has changed since the first use of complex probes to identifying differentially expressed transcripts.[50,51] The main advances include the ability to sequentially hybridize identical arrayed

[50] T. P. St. John and R. W. Davis, *Cell* **16,** 443 (1979).

[51] J. G. Williams and M. M. Lloyd, *J. Mol. Biol.* **129,** 19 (1979).

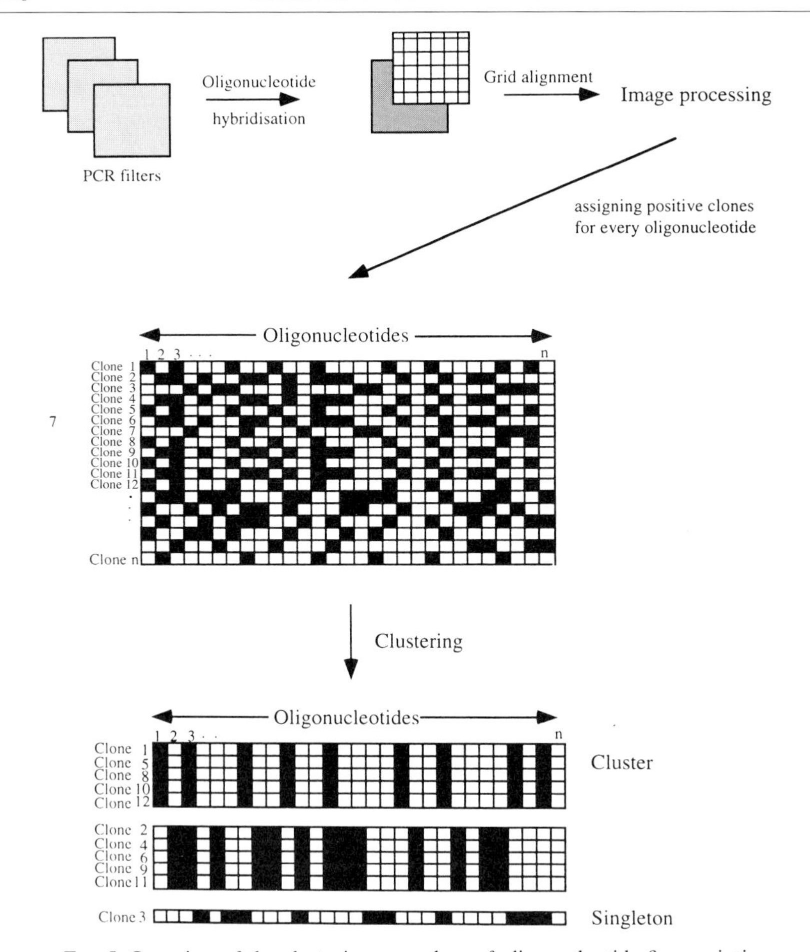

FIG. 5. Overview of the clustering procedure of oligonucleotide fingerprinting.

target DNAs with many different probes (allowing systematic characterization of probe and target transcripts) and the availability of huge numbers of EST sequences and sequenced cDNA clones.

While experiments with complex probes can yield useful expression data they can be difficult to interpret correctly, if neither the probe nor library sequences are known. Repeats, long poly(A) tails, or closely related sequences can all give false positives, and are difficult to block completely. False negatives can occur from lack of target DNA, as well as low probe activity or sensitivity. Because it is easier to know the composition of the target rather than the probe, such probes are best used on a set of well-characterized gene sequences. This allows the comparison of different cell and tissue types by which transcripts and at what level each transcript is expressed. The list of sequences to be arrayed for complex hybridizations can be derived from different libraries previously characterized by either EST or oligonucleotide hybridization sequencing approaches. The availability of so many transcripts, as clones and sequences in databases, has led to the rapid development of a new field: expression profiling.

Thus a prerequisite for such an experiment is a known set of genes. With the huge numbers of ESTs deposited in public databases, several attempts[10,52] have been made to assemble these ESTs by clustering them together, in a manner similar to oligonucleotide fingerprint clustering. Because the vast majority of clones from which these ESTs are derived are available from the IMAGE libraries, representative clones from each cluster can be rearrayed into new plates, either by hand or robot. Some programming and hardware modifications to our picker–spotter robot allows it to be used for rearraying. Great care must be taken in choosing clones, however, because previous work[53] has shown that resequencing previously tag-sequenced clones identifies large numbers (~6% in this case) that were chimeric or did not match the original sequence. It is not yet known if such figures are continued across the IMAGE libraries, or if different libraries have better, or worse, error rates. There are also the problems of nonspecific hybridization due to repeats, low-complexity sequences such as long poly(A) tails, and shared motifs from related but distinct genes. These problems can be reduced by careful selection of the sequences to be used, either by resequencing of clones or by using unique sequence regions of each gene.

Once the clones are chosen, they can be gridded onto nylon membranes[31–33,53,54] or glass slides.[55–62] For nylon membranes, clones can either

[52] M. S. Boguski and G. D. Schuler, *Nature Genet.* **10**(4), 369 (1995).

[53] G. Pietu, O. Alibert, V. Guichard, B. Lamy, F. Bois, E. Leroy, R. Mariage-Sampson, R. Houlgatte, P. Soularue, and C. Auffray, *Genome Res.* **6**(6), 492 (1996).

[54] C. Nguyen, D. Rocha, S. Granjeaud, M. Baldit, K. Bernard, P. Naquet, and B. R. Jordan, *Genomics* **29**(1), 207 (1995).

[55] M. Schena, D. Shalon, R. W. Davis, and P. O. Brown, *Science* **270**(5235), 467 (1995).

[56] M. Schena, D. Shalon, R. Heller, A. Chai, P. O. Brown, and R. W. Davis, *Proc. Natl. Acad. Sci. U.S.A.* **93**(20), 10614 (1996).

[57] J. DeRisi, L. Penland, P. O. Brown, M. L. Bittner, P. S. Meltzer, M. Ray, Y. Chen, Y. A. Su, and J. M. Trent, *Nature Genet.* **14**(4), 457 (1996).

[58] J. L. DeRisi, V. R. Iyer, and P. O. Brown, *Science* **278**(5338), 680 (1997).

[59] D. A. Lashkari, J. L. DeRisi, J. H. McCusker, A. F. Namath, C. Gentile, S. Y. Hwang, P. O. Brown, and R. W. Davis, *Proc. Natl. Acad. Sci. U.S.A.* **94**(24), 13057 (1997).

[60] D. Shalon, S. J. Smith, and P. O. Brown, *Genome Res.* **6**(7), 639 (1996).

[61] R. A. Heller, M. Schena, A. Chai, D. Shalon, T. Bedilion, J. Gilmore, D. E. Woolley, and R. W. Davis, *Proc. Natl. Acad. Sci. U.S.A.* **94**(6), 2150 (1997).

[62] M. Schena, *Bioessays* **18**(5), 427 (1996).

be grown as colonies or PCR amplified and spotted, as previously described. Our laboratory uses PCR product membranes, as they give less background than colony membranes, and are therefore more sensitive. Glass slide microarray generation (applying purified PCR products to specially treated glass slides) is a much newer field, one that is rapidly changing as new techniques are adopted. However, at present, while higher densities are achieved than by spotting on nylon membranes, the total numbers of clones are low, and thus there is much interest in developing gridding technologies newer than pin-transfer, such as ink-jet, which should allow higher densities.[63,64] Such glass slides are designed for monitoring with direct fluorescent probes, using a confocal laser scanning microscope.[55–62]

An alternative to using cDNAs is the generation of oligonucleotide chips containing sets of 11- to 30-mer oligonucleotides designed by computer to give unique oligonucleotide fingerprints for every gene. These detect fluorescently labeled complex cDNA or cRNA.[65] This is essentially oligonucleotide fingerprinting with the cDNAs as the probes, rather than the oligonucleotides. Such chips are currently expensive to make and use, because they must contain many thousands of oligonucleotides assembled by special patented photochemistry, and require specialized confocal laser scanning microscopes and software for scanning. These chips will be free of incorrect clone assignment problems, and when carefully designed, relatively free of nonspecific hybridization. However, they are still susceptible to some chimeric ESTs and bad clustering. At present they represent the highest density achievable, because the oligonucleotides are synthesized directly on the glass surface, using a tightly focused laser beam to initiate each nucleotide addition. This avoids the engineering difficulties of nano- and picoliter drop transfer.

A complex probe can be generated by a number of means, either by direct labeling during first-strand cDNA synthesis with random primers[50–54] or oligonucleotide(dT).[55–62] Random-primed labeling of first-strand cDNA[31–33] and the generation of cRNA by T7 RNA polymerase from double-stranded cDNA with a T7 promoter at the end of the first-strand oligonucleotide(dT) primer[65] have also been used. The cRNA approach allows the use of smaller amounts of starting material, which are then amplified without using PCR. Random-primed first-strand synthesis seems to be the best for whole cDNAs spotted on nylon, because it maximizes signal and minimizes the poly(A) tail effect.[53,54] When unique sequence areas are chosen, such as with oligonucleotide chips, and small amounts of

[63] M. J. O'Donnell-Maloney and D. P. Little, *Genet. Anal. Biomol. Eng.* **13,** 151 (1996).

[64] A. M. Castellino, *Genome Res.* **7**(10), 943 (1997).

[65] D. J. Lockhart, H. Dong, M. C. Byrne, M. Follettie, M. Gallo, M. Chee, M. Mittmann, C. Wang, M. Kobayashi, H. Horton, and E. Brown, *Nature Biotechnol.* **14,** 1675 (1996).

tissue are available, the use of cRNA would seem to be advisable, because RNA:DNA hybrids are more stable than DNA:DNA. However, any amplification step is potentially problematic, particularly when one seeks to be quantitative.

Hybridization conditions are dependent on the matrix to which the target DNA is bound. Nylon membrane hybridizations use the dextran-containing mix, as described earlier, but with 50% formamide and supplemented with poly(A) [or poly(T)] and salmon/herring sperm DNA to block nonspecific signals. For such hybridizations we typically label 1 μg of poly(A)$^+$ mRNA with 50–100 μCi of ^{33}P; ^{33}P is used for the same reason as in oligonucleotide fingerprinting—high spatial resolution. Glass slide hybridizations avoid the use of accelerating agents because of the small volumes used and, therefore, the much higher probe concentration. Hybridization mixes containing 2× to 6× SSC and 0.2% SDS supplemented with poly(A) have been used and probes are labeled with fluorescent dyes such as fluorescein or Cy5.[55–62,65]

The results of such hybridizations can be collected using a PhosphorImager (radioactive labeled probe) or confocal microscope (fluorescent labeled probe) and the images analyzed using software such as that described above. Values for each transcript are then compared between different probes and can be normalized by a number of methods, such as by dividing all signals by the mean signal value. In general the sensitivity with all approaches is reported to be in the region of 1/100,000, i.e., transcript at the level of 1 molecule per 100,000 messenger RNA molecules can be detected.[31–33,53–62,65] This seems sensitive enough for all but the rarest of transcripts, because this is a level of a few molecules (about three) per cell.

Readers are directed to the excellent papers of Pietu *et al.*[53] and Nguyen *et al.*,[54] which contain many experimentally determined parameters for complex hybridizations on nylon membrane-based arrayed libraries and that also discuss the analysis of such data.

Arrayed Protein Libraries

With the determination of increasing numbers of biologically and medically interesting genomes and gene sets, attention is now shifting toward determining the function of the novel genes revealed by these sequences. For comprehensive genome-wide functional analysis at the levels of transcript and protein, novel techniques are being developed to overcome the problem of identifying the cognate gene from minute amounts of protein. Arrayed protein expression cDNA libraries provide an excellent opportunity to integrate such proteome work with transcript studies. They allow the application of high-throughput technologies, both DNA based (as described above) and protein based, using phage antibody screening technologies.[6–8] To address this problem and to adapt the technologies currently available to proteome analysis, our laboratory have been successful in developing high-density protein arrays directly on membrane surfaces, using arrayed cDNA expression libraries.[5a]

cDNA Library Construction

There are a number of expression vectors available for protein expression cDNA library construction. We have successfully used a modified pQE protein expression vector (Qiagen, Chatsworth, CA) that features a hexahistidine tag, a T5 promoter, and two *lac* operator sequences, allowing repression during the growth phase by overexpression of the Lac repressor protein, and simple induction of protein expression with isopropyl-β-D-thiogalactopyranoside (IPTG). The same strategy for making the library was followed as previously described. An oligo(dT)-primed, directionally cloned cDNA library was prepared from human fetal brain tissues and, after transformation of *E. coli* cells, colonies were picked by a robot into 384-well microtiter plates.

High-Density Protein Membranes

Protein membranes are prepared like the colony membranes described above, except that the protease digestion step is omitted and polyvinylidene difluoride (PVDF) (Hybond-P; Amersham) membranes are used (pretreated with a brief wash in absolute ethanol and distilled water). Clones are spotted in duplicate at a density of 27,648 clones per membrane, as before. After overnight colony growth, protein membranes are transferred to agar plates containing 1 m*M* IPTG to induce protein expression. After 3 hr at 37°, the membranes are placed on blotting paper soaked in 0.5 *M* NaOH, 1.5 *M* NaCl for 10 min; then twice for 5 min on 1 *M* Tris-HCl (pH 7.5), 1.5 *M* NaCl; then on 2× SSC for 15 min; and finally air dried and stored at room temperature. Protease inhibitors such as phenylmethylsulfonyl fluoride (PMSF) are unnecessary.

Antibody Screening of High-Density Membranes

(See Fig. 6.) Before screening, protein membranes are soaked in 100% ethanol for 1 min and then washed in distilled water for 1 min. Bacterial debris is wiped off with paper towels in TBST-T buffer [20 m*M* Tris-HCl (pH 7.5), 0.5 *M* NaCl, 0.05% Tween 20, 0.5% Triton X-100]. All subsequent steps are carried out in a large volume, at room temperature, with shaking, unless otherwise stated.

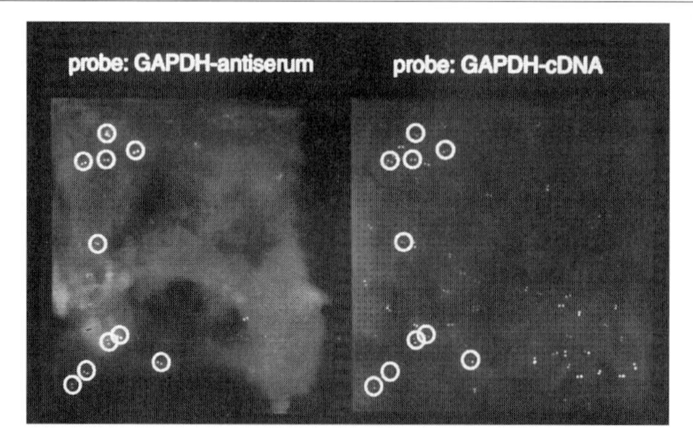

FIG. 6. *Right:* Nonradioactive hybridization of human GAPDH gene to the high-density membrane (colony filter on a nylon membrane) of a human fetal brain library. *Left:* Screen of an identical high-density membrane (protein membrane) using anti-GAPDH antibody. A subset of the clones positive in the hybridization are in the correct reading frame and thus the protein is expressed (circled). DNA sequence analysis and SDS–PAGE of protein extracts showed these positives to be true GAPDH clones, to be expressing the protein, and in many cases full-length.[5a]

Membranes are blocked for 1 hr in blocking buffer [3% nonfat dry milk powder in TBS—150 m*M* NaCl, 10 m*M* Tris-HCl (pH 7.5)], washed twice in TBST-T, and once in TBS buffer. Membranes are incubated for 2 hr with the primary antibody (against the protein of interest, in this case from mouse) diluted in blocking buffer according to the manufacturer instructions. After washing twice in TBST-T and once in TBS, the membranes are incubated with secondary antibody (anti-mouse conjugated to alkaline phosphatase) diluted in blocking buffer for 1 hr. After washing three times in TBST-T buffer and once each in TBS and in alkaline phosphatase (AP) buffer [1 m*M* $MgCl_2$, 0.1 *M* Tris (pH 9.5)], the membranes are then sprayed with 0.5 m*M* Attophos (JBL Scientific) in AP buffer and the resulting fluorescent enzyme product visualized with a CCD camera system and a UV lamp. Computer images of the membranes are analyzed using the previously described image analysis program for semiautomated scoring.

Summary

For any attempt to understand the biology of an organism the incorporation of a cDNA-based approach is unavoidable, because it is a major approach to studying gene function. The complete sequence of the genome alone is not sufficient to understand any organism; its gene regulation, expression, splice variation, posttranslational modifications, and protein–protein interactions all need to be addressed. Because the majority of vertebrate genes have probably been identified as ESTs[44,52] the next stage of the Human Genome Project is attributing functional information to these sequences. In most cases hybridization-based approaches on arrayed pieces of DNA represent the most efficient way to study the expression level and splicing of a gene in a given tissue. Similar technology, now being applied at the protein level using protein expression libraries, high-density protein membranes, and antibody screening, should allow studies of protein localization and modifications.

Coupled to these approaches is the use of technologies, which although lacking the highly parallel nature of hybridization, can potentially characterize large numbers of samples individually and with high accuracy. Automated gel-based DNA sequencing is an example of such a technique; protein sequencing and mass fingerprinting are further examples. In the case of mass spectroscopic analysis, the speed and sensitivity are vastly superior to that of gel-based approaches; however, the preparation of samples is more tedious. Our laboratory is developing a system to characterize DNA samples by mass spectrometry, allowing more rapid genotyping than is currently possible using gel-based technologies (■. Gut, ■. Berlin and H. Lehrach, personal communication, 1998). Such technology would make information on gene polymorphisms widely accessible.

Data generated using all of these techniques at the DNA and protein level will be connected by both protein expression libraries and database comparisons; finally, two hybrid library screens will identify many of the protein–protein interactions, linking genes together.[66–68] In this way we will start to understand the interplay between genes on a global scale, both at the level of molecular interaction and the biological processes they regulate.

Acknowledgments

We thank S. Meier-Ewert, E. Maier, and G. Walter in particular, and everyone in the laboratory, past and present, who has worked hard to the develop the technologies described here. We also thank D. Bancroft, C. Gauss, L. Schalkwyk, and M. V. Wiles for critical reading of the manuscript.

[66] P. James, J. Halladay, and E. A. Craig, *Genetics* **144**(4), 1425 (1996).
[67] P. L. Bartel, J. A. Roecklein, D. SenGupta, and S. Fields, *Nature Genet.* **12**(1), 72 (1996).
[68] M. Fromont-Racine, J. C. Rain, and P. Legrain, *Nature Genet.* **16**(3), 277 (1997).

[14] Principles of Differential Display

By KATHERINE J. MARTIN and ARTHUR B. PARDEE

Introduction

Differential display (DD) and related techniques, including RNA arbitrarily primed polymerase chain reaction (RAP–PCR) and display of 3'-end restriction fragments, are powerful methods that allow the comparison of similar cells or tissue types and the identification and isolation of differentially expressed genes.[1–3] These methods produce visually appealing results that allow side-by-side comparisons of gene expression levels. Gene fragments from the differentially expressed genes can be excised from the DD gel, identified, and used to prepare gene tags (probes) for the study of gene expression levels using hybridization-based assays and for full-length gene cloning by library screening methods.

The introduction of DD and related techniques has contributed to the recent shift in focus from DNA genetics to expression genetics, especially in the field of cancer research.[4] Whereas mutations in relatively few oncogenes and tumor suppressor genes likely provide the genetic basis for tumor formation, expression levels of a large number of genes are altered in tumors. Results show that expression levels of about 1000 genes are altered in breast cancer[4] and similar numbers of genes are aberrantly expressed in colon cancer.[5] For the most part, these genes are not themselves mutated, but rather their expression is incorrectly controlled, possibly many steps downstream from the initiating oncogenic mutation.[4] Information on the identities and functions of these differentially expressed genes provides new insights into the molecular mechanisms that generate the tumor phenotype and new opportunities for therapeutic intervention.

DD is distinguished from related methods by a low stringency, competitive PCR step that uses primer pairs to target the 3' ends of messenger RNAs. More than 1000 studies using DD have been reported to date. This enormous focus on the technique has led to its evolution into a versatile

[1] P. Liang, and A. B. Pardee, *Science* **257,** 967 (1992).

[2] J. Welsh, K. Chada, S. Dalal, R. Cheng, D. Ralph, and M. McClelland, *Nucleic Acids Res.* **20,** 4965 (1992).

[3] Y. Prashar, and S. M. Weissman, *Proc. Natl. Acad. Sci. U.S.A.* **93,** 659 (1996).

[4] R. Sager, *Proc. Natl. Acad. Sci. U.S.A.* **94,** 952 (1997).

[5] L. Zhang, W. Zhou, V. E. Velculescu, S. E. Kern, R. H. Hruban, S. R. Hamilton, B. Vogelstein, and K. W. Kinzler, *Science* **276,** 1268 (1997).

Copyright © 1999 by Academic Press
All rights of reproduction in any form reserved.

0076-6879/99 $30.00

and streamlined method, supported by a large academic information base and commercial suppliers of reagents in kit form.

This chapter describes the principles of DD, including specific methods for working with the primers described by Zhao *et al.*,[6] which we have used extensively.[7–9] It is divided into three sections: (1) production of a DD gel, (2) strategies for isolating and identifying DD band cDNAs, and (3) information on performing high-throughput DD.

Principles of Method

The basic steps of DD are diagrammed in Fig. 1. First, total cellular RNA is isolated from two or more comparison cell types. The RNA is then reverse transcribed from an "anchor primer" containing a poly(dT) region, which targets the 3′ poly(A) region of eukaryotic messenger RNAs. PCR is performed using the anchor primer in combination with an "arbitrary primer." PCRs are run initially at low stringency to generate mismatching at a controlled frequency such that approximately 100 different mRNAs are amplified, each producing (in principle) products of different sizes, which are subsequently resolved by electrophoretic methods. Improvements on the original method of Liang and Pardee[1] have shown that the use of longer primers and a two-step PCR protocol reduces false positives and facilitates cloning of the PCR products.[6,10] The amplified cDNAs from different cell types are displayed in adjacent lanes of a sequencing gel. Differentially expressed genes appear as bands present in the track generated from one but not the other cell type(s). These bands can be excised, the genes identified by DNA sequencing and database matching, and gene tags made. The gene tags allow confirmation of differential expression using a hybridization-based assay, such as Northern blotting and screening of panels of cell types or tissues for expression of the gene, and can be used to clone the full-length message.

One of the major criticisms of the DD technique has been the high number of false positives obtained in some studies.[11] In contrast to studies reporting false-positive rates of 50–90% and higher, the methods described here have generated false-positive rates of approximately 5%.[8] Attention in several arcas is important in avoiding false positives, including (1) use

[6] S. Zhao, S. L. Ooi, and A. B. Pardee, *BioTechniques* **18,** 842 (1995).
[7] K. J. Martin, C.-P. Kwan, and R. Sager, *Methods Mol. Biol.* **7,** 77 (1997).
[8] K. J. Martin, C.-P. Kwan, A. B. Pardee, and R. Sager, *BioTechniques* **24,** 1018 (1998).
[9] K. J. Martin, C.-P. Kwan, K. Nagasaki, X. Zhang, M. J. O'Hare, C. M. Kaelin, R. Burgeson, A. B. Pardee, and R. Sager, *Mol. Med.* **4,** 602 (1998).
[10] M. H. Linskens, J. Feng, W. H. Andrews, B. E. Enlow, S. M. Saati, L. A. Tonkin, W. D. Funk, and B. Villeponteau, *Nucleic Acids Res.* **23,** 3244 (1995).
[11] C. Debouck, *Current Opin. Biotech.* **6,** 597 (1995).

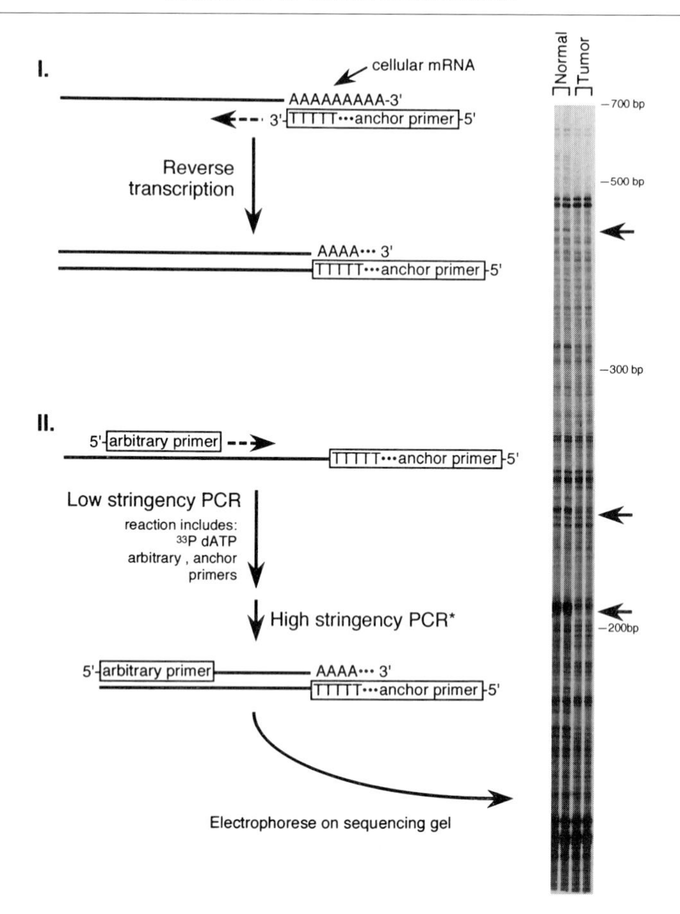

FIG. 1. Diagram of the DD method. (I) Total cellular RNA is reverse transcribed from the anchor primer, whose poly(dT) stretch targets the 3′ poly(A) region of mRNAs. (II) A low-stringency PCR is then performed with the anchor primer and an arbitrary primer, in the presence of isotopically labeled dATP. In some protocols, high-stringency PCR is subsequently performed. Reaction products are then electrophoresed on a sequencing gel. Gel shows duplicate reactions performed with 76N normal breast epithelial cells (Normal) and the metastatic breast carcinoma cell line MDA-MB-435 (Tumor). Arrows indicate some of the products expressed preferentially in the normal cells.

of high-quality RNA, (2) reduction and control of intersample variability in the DD-PCR reaction, (3) isolation of a homogeneous probe for the band of interest, and (4) use of a reliable assay to confirm differential expression. Each of these issues is discussed in the appropriate sections.

Producing the Differential Display Gel

Appropriate Cells and Tissues for Differential Display Comparisons

DD is most successful when the cells to be compared are homogeneous populations and are relatively closely matched. Homogeneous cell types include cultured cell lines and sorted primary cultures. Closely matched cell types include diseased versus normal cells and cells treated versus untreated with various agents. Tissue specimens are often successfully used for DD[12]; however the presence of multiple cell types has the potential to obscure differential signals contributed by an individual constituent cell type and to produce elevated rates of false positives. Our approach to identifying differentially expressed genes in breast tumors has been to obtain tags for many genes differentially expressed between a highly malignant breast tumor cell line and normal breast epithelial cells and then to use these tags to screen gene expression levels in panels of patient-isolated breast tumor specimens.[8,9] We discuss below clinical tissue sources that yield high-quality RNA suitable for DD, although subject to shortcomings due to cell type heterogeneity, as well as information on performing DD with low numbers of cells (see the next section).

A particularly successful design for a DD experiment compares more than one sample of each of the "normal" and "differential" cells and selects bands that are consistently differentially expressed. For example, one can compare two different normal cell lines with two different tumor cell lines, or analyze several time points following drug addition. This approach controls for factors such as expression differences that are not relevant to the system being studied, differences in the quality of the extracted RNA, and differences in the reverse transcription (RT) or PCR conditions.

Methods for RNA Isolation

RNA quality is one of the most critical factors in DD. RNA that is partially degraded, contaminated with RNases or particulates, or inaccurately quantitated can produce DD tracks that are not reproducible between duplicate reactions, result in high rates of false positives, or produce DD patterns that terminate or fade at the higher molecular weights. High-quality RNA is routinely obtained using the guanidine isothiocyanate

[12] For example: F.-L. Wang, Y. Wang, W. K. Wong, Y. Lui, F. J. Addivinola, P. Liang, L. B. Chen, P. W. Kantoff, and A. B. Pardee, *Cancer Res.* **56,** 3634 (1996); R. K. Blanchard and R. J. Cousins, *Proc. Natl. Acad. Sci. U.S.A.* **93,** 6863 (1996).

(GUT)/Cs cushion centrifugation method of Chirquin *et al.*[13,14] adapted as follows:

1. To a minced tissue specimen (~100 mg), washed cell pellet (~10^6 cells), or cell monolayer (100-mm dish), add 3 ml of GUT–2-ME [4 *M* guanidine isothiocyanate, 25 m*M* sodium citrate (pH 7), passed through a 0.2-μm pore size nylon filter; immediately prior to use add 7 μl of 2-mercaptoethanol per milliliter]. Homogenize tissue specimens, using three or four strokes of a Teflon-pestle Dounce homogenizer (Wheaton, Millville, NJ), triturate the cell pellets, or vigorously scrape the cell monolayers.
2. The lysates are layered on top of 1.5 ml of CsCl solution [5.7 *M* CsCl, 25 m*M* sodium acetate, 0.1% diethylpyrocarbonate (DEPC), stirred for 1 hr, and then autoclaved for 45 min] in a 13 × 51 mm centrifuge tube. The tubes are balanced with extra GUT–2-ME, then centrifuged at 40,000 rpm at 20° for 18–24 hr.
3. The supernatant is decanted and the RNA pellet air dried in the inverted tube for 1 hr. The RNA is dissolved in 400 μl of SETS buffer [10 m*M* Tris-HCl (pH 7.4), 5 m*M* EDTA, 1% sodium dodecyl sulfate (SDS) in DEPC-treated distilled H_2O (dH_2O)], extracted with 400 μl of phenol–chloroform (1 : 1, v/v), then centrifuged at 12,000 rpm for 2 min at room temperature.
4. The upper phase is precipitated by adding 40 μl of 3 *M* DEPC-treated sodium acetate, then 960 μl of 100% ethanol, and centrifuged at 12,000 rpm for 5 min at room temperature. Pellets are washed with cold (−20°) 70% ethanol and centrifuged again.
5. RNA pellets are dried under vacuum for 3–5 min, then allowed to dissolve in DEPC-treated dH_2O for at least 30 min at room temperature.
6. One microliter of the RNA is diluted in 500 μl of dH_2O and quantitated spectrophotometrically using the equation: 1 OD_{260} = 40 μg/ml. RNA integrity is determined by running 1–2 μg on a small agarose–formaldehyde gel.[15] The 18S (~2-kb) and 28S (~4-kb) rRNAs should be sharp and clearly visible against a background of cellular mRNAs.

RNA that is acceptable for DD can also be produced by the more rapid methods that use commercial products such as Trisolv, RNAzol, or Ultraspec (all Biotecx Laboratories, Houston, TX). Although these meth-

[13] J. M. Chirqwin, A. E. Przybyla, R. J. MacDonald, and W. J. Rutter, *Biochemistry* **18,** 5294 (1979).

[14] R. E. Kingston, *in* "Current Protocols in Molecular Biology" (F. M. Ausubel, R. Brent, R. E. Kingston, D. D. Moore, J. G. Seidman, J. A. Smith, and K. Struhl, eds.), pp. 4, 2, 3. John Wiley & Sons, New York, 1995.

[15] T. Brown and K. Mackey, *in* "Current Protocols in Molecular Biology" (F. M. Ausubel, R. Brent, R. E. Kingston, D. D. Moore, J. G. Seidman, J. A. Smith, and K. Struhl, eds.), pp. 4, 9, 2. John Wiley & Sons, New York, 1995.

ods may lead to variable RNA quality, especially with tissue samples, they are often more amenable than the cesium method to samples with low numbers of cells or when working with many samples. Methods to control for problems due to differences in RNA quality are described below (see Reverse Transcription Reactions).

It is often important for DD studies to maintain clinical relevance. To maintain RNA integrity, clinical tissue samples should be processed or frozen in liquid N_2 immediately after dissection of the specimen. In practice, we have found that gross surgical samples (e.g., breast lumpectomy tissue) can remain intact at room temperature for as long as 45 min. Once the tissue is dissected, the tissue pieces (e.g., tumor or adjacent normal tissue) must be immediately processed to extract RNA or flash frozen and stored at $-80°$. To extract RNA from frozen tissue, mince the tissue into 1-mm^3 pieces while still frozen, then homogenize in GUT solution as described above. A breast tumor piece of 100 mg will provide approximately 30–50 μg of total cellular RNA, which is sufficient for a DD study or for several reverse Northern assays. We have found that adjacent normal tissue of the breast yields nearly an order of magnitude less RNA per unit of weight relative to the tumor specimens. These measures likely provide approximate guidelines for other tissue types and actual RNA yields may vary depending on the source.

Archived tissue samples are also suitable for DD and hybridization-based assays, depending on the method of preservation. While RNA from paraffin frozen sections is degraded to about 200 bases in length, OCT (Miles, Elkhart, IL) frozen samples will generally yield intact RNA that is suitable for DD and reverse Northern analysis.[16,17] OCT cryosections should be sliced to remove excess mounting material, minced into pieces of approximately 1 mm^3, then homogenized in 3 ml of GUT solution using three or four strokes of a Teflon-pestle Dounce. Continue to process the sample as outlined above. We prefer the GUT/Cs cushion centrifugation method for RNA extraction from tissue specimens, because the GUT solution effectively inactivates RNases, unlike the commercial reagents for rapid RNA isolation. Because of the likelihood of cell type heterogeneity in clinical samples, it may be useful to perform DD using cells cultured under the same conditions of growth and to subsequently screen patient-isolated specimens for expression levels of the selected genes.

It is important to remove residual DNA from RNA preparations by performing a DNase I reaction followed by phenol–chloroform extraction

[16] J. A. E. Irving, G. Cain, A. Parr, M. Howard, B. Angus, and A. R. Cattan, *J. Clin. Pathol.* **49,** 258 (1996).

[17] H. Ford and A. B. Pardee, personal communication (1998).

and ethanol precipitation.[18] The RNA is then quantitated, diluted to 0.1 μg/ml, aliquoted, and stored at $-80°$. Repeated cycles of thawing and freezing should be avoided. Basic laboratory practices for working with RNA should be followed including wearing gloves, keeping RNA samples on ice and minimizing their time out of the freezer, and autoclaving all labware; in addition, treat all solutions with 0.1% DEPC and autoclave prior to use.

When only a small number of cells is available, amplification of cellular RNA is required prior to DD analysis. For each compared cell type, a typical DD study using 30–60 primer pairs requires 1–5 μg of total cellular RNA, and hence approximately 10^6 cells. Methods for performing DD with 10^3 cells have been reported.[19] The method involves first synthesizing an amplified cDNA pool from a small RNA sample. A poly(dT) anchor primer with a 3′ degenerate anchor base is used for first-strand synthesis and a primer with four 3′ degenerate bases is used for the second-strand synthesis. The cDNA is then amplified using the same primer set without 3′ degenerate bases. A kit (Clontech, Palo Alto, CA) for preparing full-length libraries from small amounts of RNA is also available that is useful for this purpose.

Primer Selection

The primers originally used for DD by Liang and Pardee[1] were 9- or 10-mer arbitrary primers and 14-mer anchor primers (Table I). These primers were predicted to be short enough to hybridize to many sites on mRNAs, while in practice they were found to be long enough to prime the PCR efficiently. Criteria important in the design of the arbitrary primers included maintaining a GC content of approximately 50% and a G or C at the immediate 3′ position. These primers have been used by many researchers, applied to diverse biological systems, and have led to the identification of a number of novel differentially expressed genes.[20,21] Methods for their use have been described[22] and primers are available in kits from GeneHunter Corporation (Nashville, TN).

Results of studies have found that lengthening the DD primers in combination with using a two-step PCR protocol (see the next section) improves the reproducibility of banding patterns, increases the sizes of DD band

[18] P. Liang, L. Averboukh, and A. B. Pardee, *Methods Mol. Genet.* **5,** 3 (1994).

[19] S. Zhao, G. Molnar, J. Zhang, L. Zheng, L. Averboukh, and A. B. Pardee, *BioTechniques* **24,** 824 (1999).

[20] Example: U. Utans, P. Liang, L. R. Wyner, M. J. Karnovsky, and M. E. Russell, *Proc. Natl. Acad. Sci. U.S.A.* **91,** 6463 (1994).

[21] Review: F. J. Livesey, and S. P. Hunt, *Trends Neurosci.* **19,** 84 (1996).

[22] P. Liang, D. Bauer, L. Averboukh, P. Warthoe, M. Rohrwild, H. Muller, M. Strauss, and A. B. Pardee, *Methods Enzymol.* **254,** 304 (1995).

TABLE I
EXAMPLES OF PRIMER SETS FOR USE IN DIFFERENTIAL DISPLAY

Anchor	Arbitrary
	Original DD Primers[a]
$T_{11,12}MN$[b]	Ltk3: 5′-CTTGATTGCC-3′
	Ldd1: 5′-CTGATCCATG-3′
	Ldd2: 5′-CTGCTCTCA-3′
	OPA-1: 5′-GCAGGCCCTTC-3′
	OPA-2: 5′-GTGCCGAGCTG-3′
	OPA-3: 5′-GAGTCAGCCAC-3′
	OPA-4: 5′-AATCGGGCTG-3′
	OPA-5: 5′-AGGGGTCTTG-3′
	OPA-6: 5′-GGTCCCTGAC-3′
	OPA-7: 5′-GGAAACGGGTG-3′
	OPA-8: 5′-GGTGACGTAGG-3′
	OPA-9: 5′-GGGTAACGCC-3′
	OPA-10: 5′-GTGATCGCAG-3′
	Extended DD Primers[c]
$TGCCGAAGCT_{11}M$[d]	LH-A1: 5′-TGCCGAAGCTTACCAGTC-3′
	LH-A2: 5′-TGCCGAAGCTTGATCGCT-3′
	LH-A3: 5′-TGCCGAAGCTTGAGCCTG-3′
	LH-A4: 5′-TGCCGAAGCTTCGAGATC-3′
	LH-A5: 5′-TGCCGAAGCTTCGATGCA-3′
	LH-A6: 5′-TGCCGAAGCTTGCAGCGA-3′
	LH-A7: 5′-TGCCGAAGCTTGCGATCA-3′
	LH-A8: 5′-TGCCGAAGCTTGACTGGA-3′
	LH-A9: 5′-TGCCGAAGCTTGATCCAG-3′
	LH-A10: 5′-TGCCGAAGCTTATGCGAC-3′
	LH-A11: 5′-TGCCGAAGCTTAGCAGGT-3′
	LH-A12: 5′-TGCCGAAGCTTGCCACCT-3′
	LH-AP3: 5′-TGCCGAAGCTTTGGTCAG-3′
	E1-OPA-1: 5′-CGTGAATTCGCAGGCCCTTC-3′
	E1-OPA-2: 5′-CGTGAATTCGTGCCGAGCTG-3′
	E1-OPA-3: 5′-CGTGAATTCGAGTCAGCCAC-3′
	H3-OPA-4: 5′-TGCCGAAGCTTATTCGGGCTG-3′
	H3-OPA-5: 5′-TGCCGAAGCTTAGGGGTCTTG-3′
	H3-OPA-6: 5′-TGCCGAAGCTTGGTCCCTGAC-3′
	E1-OPA-7: 5′-CGTGAATCGGAAACGGGTG-3′
	E1-OPA-8: 5′-CGTGAATTCGGTGACGTAGG-3′
	H3-OPA-9: 5′-TGCCGAAGCTTGGGTAACGCC-3′
	H3-OPA-10: 5′-TGCCGAAGCTTGTGATCGCAG-3′
	E2-AP2: 5′-CGTGAATTCGGACCGCTTGT-3′
	E2-AP3: 5′-CGTGAATTCGAGGTGACCGT-3′

[a] Original DD primers[1] are available commercially (GeneHunter).
[b] M = G, A, or C and N = G, A, T, or C.
[c] Extended primer sets include *Eco*RI or *Hin*dIII restriction sites (underlined) (from Ref. 6).
[d] M = G, A, or C.

cDNAs, and can allow for the inclusion of restriction sites to facilitate cloning.[6,23] The importance of these issues is discussed in the appropriate sections below (see the next section, Direct Sequencing, and Cloning). Further, improving the balance of GC content and hence melting temperatures (T_ms) between the anchor and arbitrary primers results in a higher percentage of bands being generated by both the anchor and arbitrary primers, as opposed to having the arbitrary primer at both ends.[24] This improves the distribution of fragments and is important for direct sequencing (see Direct Sequencing). The sequences of the LH-A and OPA series of extended DD primers described by Zhao *et al.*[6] are shown in Table I.

The extended LH-A and OPA series of primers shown in Table I include both one-base and two-base anchored primer sets. The use of one-base anchored primers, which have a single nucleotide (G, A, or C) following the poly(dT) stretch, requires fewer anchor primers for full coverage of mRNAs.[25] On the other hand, two-base anchored primers generally produce fewer DD bands per track, hence improving band resolution, and appear to slip less, thus reducing multiplet bands. Selection between one-base and two-base anchored primers may be a matter of personal preference; we have used the one-base anchors.

Still further variations in primer design have been reported. For example, third-generation primers increase primer sizes to 25 bases or more, incorporate T7 or SP6 universal sequencing primer sites, or incorporate degenerate oligonucleotides. Such primers are tailored for specific uses or provide incremental improvements on the original and second-generation primers. Sets of long primers are available from commercial sources, including Genomyx Corporation (Foster City, CA).

Reverse Transcription Reactions

To produce the first-strand cDNA, a series of reverse transcription (RT) reactions are performed, using all of the combinations of anchor primers and RNAs. In setting up these reactions, we use a procedure that minimizes pipetting and hence variability between reactions. The specific conditions given have been adapted from Zhao *et al.*[6] and optimized for use with the 21-mer LH-A series of anchor primers designed and described by these authors (Table I).

[23] M. H. Linskens, J. Feng, W. H. Andrews, B. E. Enlow, S. M. Saati, L. A. Tonkin, W. D. Funk, and B. Villeponteau, *Nucleic Acids Res.* **23,** 3244 (1995).

[24] D. Graf, A. G. Fisher, and M. Merkenschlager, *Nucleic Acids Res.* **25,** 2239 (1997).

[25] P. Liang, W. Zhu, X. Zhang, Z. Guo, R. P. O'Connell, L. Averboukh, F. Wang, and A. B. Pardee, *Nucleic Acids Res.* **22,** 5763 (1994).

1. A stock mixture of sufficient volume for the number of reactions to be run is prepared consisting of reaction buffer, dNTPs, and dithiothreitol (DTT), all in DEPC-treated dH_2O. For each reaction to be run, the stock should contain 18.5 μl of DEPC-dH_2O, 10 μl of 5× RT buffer (as supplied with reverse transcriptase), 4 μl of a mix containing a 250 μM concentration of each dNTP, and 5 μl of 0.1 M DTT. The actual volumes to use in a given experiment are calculated by multiplying the given volumes by the number of reactions to be run and adding an additional ~20% for pipetting.
2. This stock mixture is then aliquoted into a separate tube for each anchor primer being used, 37.5 μl per reaction to be run using that primer, plus an additional ~10% for pipetting. Add 5 μl of 10 μM anchor primer in DEPC-dH_2O per reaction to be run (plus the percentage of extra volume added in the previous step) to each tube.
3. Each mixture is then aliquoted into a separate tube for each RNA being transcribed, 42.5 μl per reaction. Add 5 μl of RNA (0.1 μg/μl) per tube.
4. To run the reaction, heat to 65° for 5 min for primer annealing, reduce the temperature to 37°, and incubate for 10 min, then add 2.5 μl of reverse transcriptase (Superscript II; Promega, Madison, WI) to each reaction and incubate for a further 60 min at 37°.
5. Hydrolyze the RNA and inactivate the RT by heating at 95° for 5 min, then place the tubes on ice or store at −20°. The final reaction volume is 50 μl.

We recommend keeping all reagents used in the RT reaction, especially the 2 μM anchor primer solutions, separate from those used in the subsequent PCR step, where RNase contamination is not an issue.

We have found that RT reactions are highly reproducible. However, in some situations, it may be useful to verify reproducibility by running duplicate reactions. Prepare twice the volume needed of reaction mix (including the primer and RNA), using the procedure above, then aliquot final mixtures into two tubes and perform the incubations. The resulting duplicate cDNAs are subsequently analyzed by DD as described below. Because pipetting errors are controlled through the use of a core mix, problems are more likely to arise from differences in RNA quality. It is often useful to perform duplicate (or more) RNA extractions and to compare final DD banding patterns following parallel RT-PCR reactions. The same is accomplished if the DD experiment is designed to compare more than one sample of each of the "normal" and "differential" cells or conditions. For example, two different normal cell lines can be compared with two different tumor cell lines, or several time points can be analyzed following the addition of a drug. Inaccurate quantitation of RNA can be controlled for by comparing the DD patterns generated from different concentrations of RNA and selecting only those bands whose appearance is concentration independent.[26]

Differential Display Polymerase Chain Reaction Step

The DD-PCR step is sensitive to small differences in reaction conditions between samples. Intersample variability here can produce differences in DD banding patterns that do not accurately represent differences in first-strand cDNA abundance, i.e., false positives. This issue can be dealt with at several levels. First, using longer DD primers in combination with a two-step PCR protocol has been shown to increase the reproducibility of DD banding patterns.[6,27] These studies have substituted primers approximately 20 bases in length for the original DD primers[1] of 10–14 bases. The use of longer primers enables the use of a two-step PCR protocol, which reduces the low stringency cycles to an initial 1 to 4 cycles and follows these with 25 to 35 high-stringency cycles. Next, we recommend a pipetting protocol for preparing the DD reactions that minimizes the pipetting of individual reagents into individual tubes (protocol described below). And third, all PCR should be performed in duplicate, electrophoresed in adjacent lanes, and only those bands that are reproducibly differentially displayed should be selected for analysis.

To perform the DD-PCR, we have used a core mix procedure similar to that described for the RT reaction, which minimizes pipetting and hence variability between reactions. The general scheme is diagrammed in Fig. 2. The conditions given have been adapted from Zhao *et al.*[6] and optimized for use with their LH-A and OPA series of primers (Table I). We routinely prepare duplicate PCR for each cDNA and primer pair to be tested. Hence, when determining the total number of reactions being run, the number of cDNAs times primer pairs should be multiplied by two.

1. A stock mixture of sufficient volume for the number of reactions to be run is prepared consisting of PCR buffer, dNTPs, [α-^{33}P]dATP, and *Taq* polymerase. For each reaction to be run, the stock contains 9.95 μl of dH_2O, 2 μl of 10$\times$ PCR buffer (as supplied with the polymerase), 1.6 μl of a mix containing a 25 μM concentration of each dNTP, 0.25 μl of [α-^{33}P]dATP, and 0.2 μl of AmpliTaq (Perkin-Elmer, Norwalk, CT). Actual volumes for a given experiment are calculated by multiplying the given volumes by the number of reactions to be run and adding approximately 50% for pipetting.

[26] J. Welsh, and M. McClelland, *Nucleic Acids Res.* **18,** 7213 (1990).

[27] M. H. Linskens, J. Feng, W. H. Andrews, B. E. Enlow, S. M. Saati, L. A. Tonkin, W. D. Funk, and B. Villeponteau, *Nucleic Acids Res.* **23,** 3244 (1995).

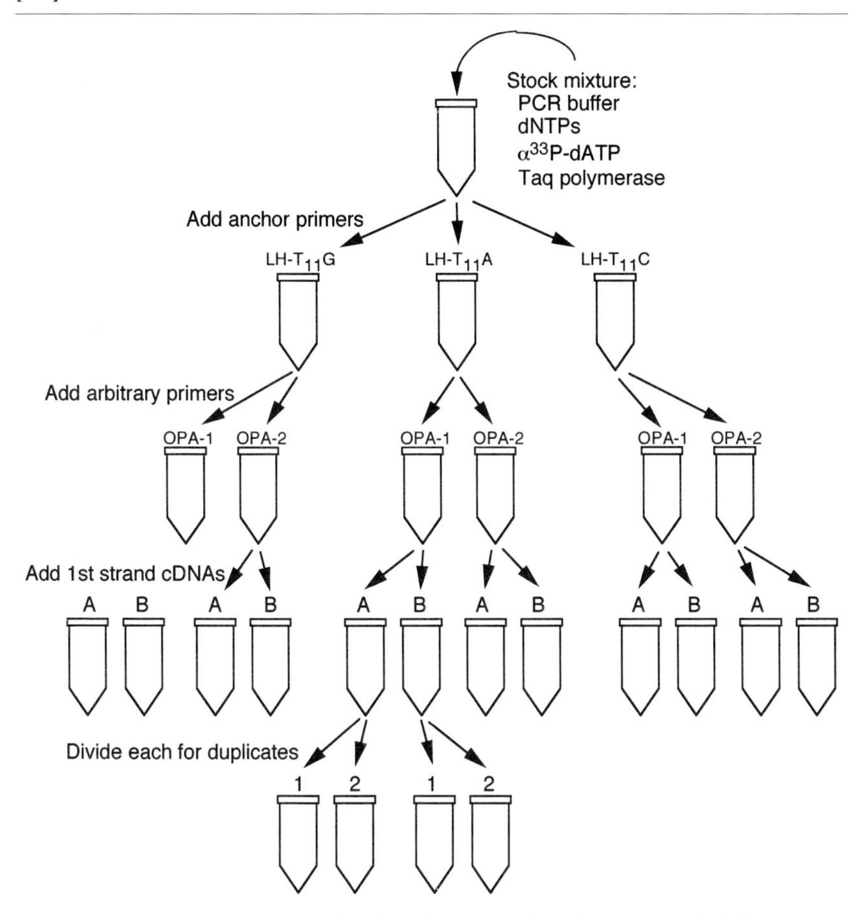

FIG. 2. Diagram of sequential aliquoting protocol used to set up DD-PCRs.

2. This mixture is then aliquoted into a separate tube for each anchor primer being used, 14 μl per reaction to be run with that primer, plus ~40% for pipetting. To each tube, add 2 μl of 2 μM anchor primer per reaction to be run (plus the percentage of extra volume used in the previous step).

3. Aliquot each mixture into a separate tube for each arbitrary primer being used, 16 μl per reaction to be run with that primer, plus ~30% for pipetting. To each tube, add 2 μl of 2 μM arbitrary primer per reaction to be run (plus the percentage of extra volume used in the previous step).

4. Aliquot each mixture into a separate tube for each first-strand cDNA being used, 18 μl per reaction to be run with that cDNA, plus ~20% for pipetting. To each tube, add 2 μl of RT reaction products per reaction to be run (plus the percentage of extra volume used in the previous step).

5. To make the duplicate reactions, aliquot each mixture to two tubes, 20 μl per tube. Add 2 drops (or 100 μl) of paraffin oil per tube and perform the PCR.

We have used the two-step PCR protocol described by Zhao *et al.*[6] Perform 1 low-stringency cycle of 94°, 1 min; 40°, 4 min; 72°, 1 min, followed by 35 high-stringency cycles of 94°, 1 min; 60°, 4 min; 72°, 1 min. (For the LH-A arbitrary primers, which are 3 or 4 bases shorter than the OPA arbitrary primers, the annealing temperature for the high-stringency cycles should be reduced to 58° to avoid obtaining fragments primed by the anchor at both ends.) The extension reaction is continued for another 5 min at 72° and samples are then prepared for electrophoresis.

Electrophoresis

A variety of polyacrylamide gel types and electrophoresis apparatuses have been successfully used for DD, including denaturing and nondenaturing gels, acrylamide concentrations ranging from 4.5 to 6%, simple gel boxes, and programmable, temperature-controlled units. We use denaturing 6% acrylamide gels and the Genomyx programmable electrophoresis unit,[28] with plates approximately 20% longer than standard sequencing gels. Wells should be formed with a square-toothed comb with adequate separation between lanes. Allow the gels to polymerize for 1 hr, and then rinse the wells with water immediately following comb removal, and again just prior to loading the samples with electrophoresis buffer. To prepare PCR samples for electrophoresis, mix 3.5-μl samples with 2 μl of standard sequencing dye,[29] incubate at 92° for 3 min to denature, chill on ice, centrifuge the samples, and then load them onto gels. Electrophoresis should continue until bands less than approximately 150 kb have run off the bottom, i.e., the xylene cyanol dye band will be at or just off the gel. Dry the gels to the glass (as per the manufacturer instructions, if using the Genomyx unit), or as we prefer, transfer the gels to filter paper and dry under vacuum. Expose overnight to autoradiography film.

DNA size markers can be prepared from a commercially available 100-bp ladder (e.g., Promega) by performing a kinase reaction in the presence of [γ-^{32}P]ATP. Add 20 μl of standard sequencing dye to each 35 μl

[28] L. Averboukh, S. A. Douglas, S. Zhao, K. Lowe, J. Maher, and A. B. Pardee, *BioTechniques* **20,** 918 (1996).

[29] C. D. Earl, P. Heinrich, and B. T. Nixon, *in* "Current Protocols in Molecular Biology" (F. M. Ausubel, R. Brent, R. E. Kingston, D. D. Moore, J. G. Seidman, J. A. Smith, and K. Struhl), pp. 7, 4, 16. John Wiley & Sons, New York, 1995.

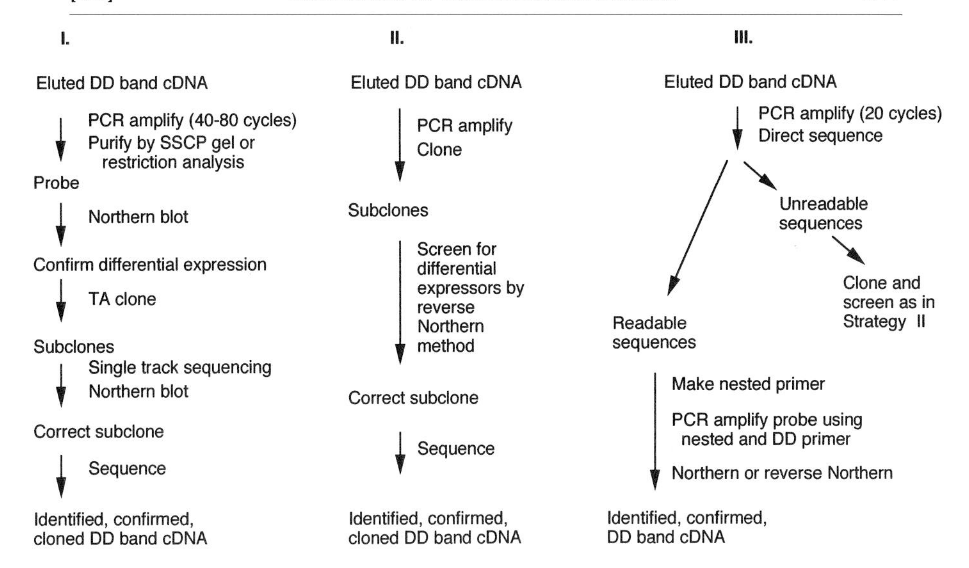

FIG. 3. Three different strategies used to identify and confirm differential expression of DD band cDNAs. Strategy I represents the original method described by Liang and Pardee[1]; strategy II is described by Zhang *et al.* (1996); strategy III is described by Martin *et al.*[7,8]

of completed reaction, incubate at 92° for 3 min to denature, chill on ice, centrifuge the sample, then load 4 μl onto a gel and electrophorese adjacent to DD reactions.

Analysis of Differential Bands

Strategies

Several different strategies for analyzing DD bands have been reported and three of these are diagrammed (Fig. 3). The original strategy used by Liang and Pardee[1] (Fig. 3, strategy I) involves the following.

1. PCR amplification of excised DD band cDNA using the DD primers and PCR conditions: This PCR step may require multiple rounds of 40-cycle PCR for amplification of sufficient DNA. Amplified cDNA can be purified at this point to remove contaminating nondifferentially expressed sequences using a single-strand conformational polymorphism (SSCP) gel or restriction analysis.[30,31]

[30] S. Zhao, S. L. Ooi, F.-C. Yang, and A. B. Pardee, *BioTechniques* **20,** 400 (1996).

[31] F. Mathieu-Daude, R. Cheng, J. Welsh, and M. McClelland, *Nucleic Acids Res.* **24,** 1504 (1996).

2. Northern blotting is then performed with a probe generated from the PCR-amplified DNA. The results of this assay are used to confirm differential expression.

3. cDNAs from bands giving positive results in the Northern assay are cloned using the TA cloning method (InVitrogen, San Diego, CA). Clones can be sorted to identify those representing different sequences by performing single-track sequencing reactions.[32]

4. Clones are then sequenced and used to prepare Northern probes to confirm cloning of the correct cDNA.

More recent approaches have been developed that streamline the process and reduce the rates at which false positives are obtained. These newer strategies incorporate more recently introduced techniques, including direct sequencing of DD bands and/or reverse Northern assays, and revise the order in which steps are performed.

A strategy making use of the reverse Northern assay (also Southern assay or hybridization arrays) was first reported by Zhang *et al.*[33] (Fig. 3, strategy II). Briefly, the reverse Northern assay involves binding gene tags to a nylon membrane and hybridizing these with ^{32}P-labeled cDNA prepared from total cellular RNA (see Verification of Differential Expression). In the strategy of Zhang *et al.*, the reverse Northern assay is used to screen subclones for differential expressors. It is performed on colony transfers prepared following bacterial transformation with the ligated cDNA mixture. This approach assures selection of the correct subclone, if the band indeed represents a differentially expressed gene, even if it is contaminated with nondifferentially expressed sequences.

The strategy we have developed and used[7,8,34] is summarized here and described in detail in later sections:

1. Excised DD band cDNA is PCR amplified using the DD primers, stringent PCR conditions, and as few cycles of PCR as possible (generally 20–25 cycles).

2. Amplified DD band cDNA is then directly cycle sequenced using one of the primers used for DD. These can be long or short primers as described below.

3. A nested primer is designed and synthesized on the basis of the sequence of the database cDNA. This nested primer is positioned to produce a DNA fragment at least 100 bp in length when used for PCR in

[32] S. Zhao, S. L. Ooi, F.-C. Yang, and A. B. Pardee, *BioTechniques* **20,** 400 (1996).
[33] H. Zhang, R. Zhang, and P. Liang, *Nucleic Acids Res.* **24,** 95 (1996).
[34] D. K. Biswas, L. Averboukh, S. Sheng, K. J. Martin, J. Lamb, D. S. Ewaniuk, T. F. Jawde, F. Wang, and A. B. Pardee, *Mol. Med.* **4,** 454 (1998).

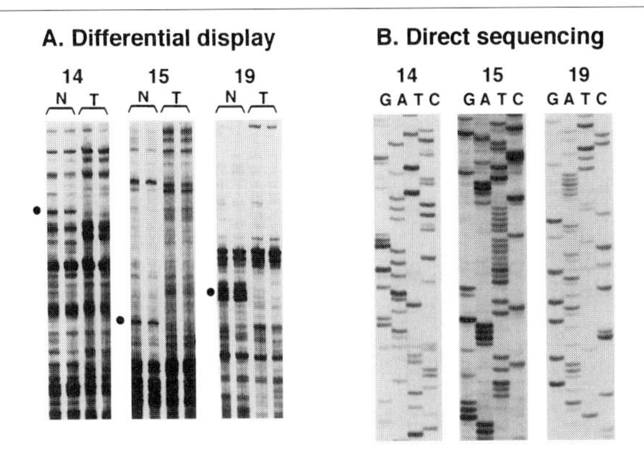

FIG. 4. Examples of DD and direct sequencing results. (A) DD bands 14, 15, and 19, indicated by dots, were excised from DD tracks shown, PCR-amplified for 20 cycles, and directly sequenced. (B) Direct sequencing tracks of DD bands 14, 15, and 19. Sequences identified bands as α_1-antichymotrypsin (14), transcription factor C/EBP δ (15), and a new gene with an EST match (19).

combination with one of the DD primers. This step produces a homogeneous gene tag from the potentially heterogeneous DD band cDNA.

4. Differential expression is confirmed using the gene tag in reverse Northern hybridization assays or as a probe for Northern blotting. The gene tags can be used to gain information on expression patterns in a variety of tissues.

5. Cloning is required only for DD bands that cannot be directly sequenced (approximately 10%, in our experience)[7,8] or when screening libraries for full-length transcripts of selected genes.

The advantages of this strategy are that it has produced negligible rates of false positives by using only homogeneous probes for hybridization assays and that it avoids the lengthy procedure of cloning of DD bands until bands are prioritized for full-length cloning.

Direct Sequencing

Direct sequencing of DD band cDNA was first described by Wang and Feuerstein[35] and has since been used in many DD studies. A cycle sequencing protocol is used to allow the detection of small amounts of DNA. Direct sequencing can be successful when primers ranging from short 10-mers[36,37] to long 21-mers are used as DD and sequencing primers (Fig. 4). Long

[35] X. Wang, and G. Z. Feuerstein, *BioTechniques* **18,** 448 (1995).

[36] M. Buess, C. Moroni, and H. H. Hirsch, *Nucleic Acids Res.* **25,** 2233 (1997).

[37] D. K. Biswas, L. Averboukh, S. Sheng, K. J. Martin, J. Lamb, D. S. Ewaniuk, T. F. Jawde, F. Wang, and A. B. Pardee, *Mol. Med.* **4,** 454 (1998).

primers (>25-mers) may not produce good sequencing tracks if their T_ms exceed the optimum range for PCR (55–80°)[38] and hence sequencing will require the use of a shorter primer that anneals to a region of the long primer. The most important factors in obtaining good sequencing tracks are to adjust the annealing temperature of the cycle sequencing reaction to approximately 5° higher than the T_m of the primer in use, preferably by testing a range of temperatures and prior to sequencing, to thoroughly remove the primers and nucleotides remaining after PCR amplification of the template DNA.

Direct sequencing is performed as follows.

1. The autoradiogram is lined up with the DD gel using either radioactive ink markings or the pattern created by the edge of the gel.
2. Cut out the band by excising the minimum amount of gel and verify band removal by reexposing the gel to film. If the gel is dried onto glass, first wet the gel with a small amount (5 μl) of water. *Note:* Excessive hydration may cause tearing.
3. Extract DNA by boiling for 10 min in TE or by soaking,[39] and then ethanol precipitate using 1 μl of glycogen (1 mg/ml in dH_2O) as a carrier.
4. PCR amplify the eluted DD band cDNA using approximately 20 cycles of PCR.[7] A comparison of PCR cycle numbers can be performed to determine the optimum number. The annealing temperature of the PCR is, in general, the same as that used for the high-stringency step of DD. A comparison of annealing temperatures may be useful for producing optimum quality sequencing tracks from the particular primers being used. We have used annealing temperatures of 60° for the 21-mer LH-A anchor primers and 45° for the original 10-mer arbitrary primers from GeneHunter (Fig. 4).
5. Perform a purification step to remove the PCR primers: (1) isopropanol precipitation,[7,40] (2) spin column (e.g., QIAquick PCR purification kit; Qiagen, Chatsworth, CA), or (3) agarose gel electrophoresis and extraction.[41] We have found the spin columns to be the preferable method.
6. Cycle sequence using a commercially available kit (e.g., VentSequenase; New England BioLabs, Beverly MA). We resuspend the isopropanol-precipitated DNA in 10 μl and use 2 μl to generate the four sequencing tracks.

[38] M. A. Innis and D. H. Gelfand, *in* "PCR Protocols: A Guide to Methods and Applications" (M. A. Innis, D. H. Gelfand, J. J. Sninsky, and T. J. White, eds.), p. 9. Academic Press, San Diego, California, 1990.

[39] X. Wang, and G. Z. Feuerstein, *Methods Mol. Biol.* **7,** 69 (1997).

[40] M. A. D. Brow, *in* "PCR Protocols: A Guide to Methods and Applications" (M. A. Innis, D. H. Gelfand, J. J. Sninsky, and T. J. White, eds.), p. 189. Academic Press, San Diego, California, 1990.

[41] X. Wang, and G. Z. Feuerstein, *Methods Mol. Biol.* **7,** 69 (1997).

Direct sequencing is subject to problems that can produce either of two results: faint sequence tracks or tracks with unreadable sequence information due to banding in all lanes. Faint tracks are often due to loss of the DNA template at some point, especially during the isopropanol precipitation if supernatants are not removed carefully, or to the selection of a faint DD band. Excessive banding may be due to excessive PCR amplification or to the use of an annealing temperature too low for the primer being used. Unreadable direct sequence will also be produced if DD fragments were generated by a single DD primer at both ends. For example, fragments generated by the anchor primer at both ends will produce two superimposed sequences when the anchor primer is used for direct sequencing and no sequence information when the arbitrary primer is used. To avoid this problem, we recommend performing two control DD-PCR reactions, one reaction using only a DD anchor primer and another using only an arbitrary primer. These reactions should be electrophoresed adjacent to the dual-primer PCR reaction for ready comparison of banding patterns. If necessary, adjust PCR conditions or T_ms of primers[42] to avoid the generation of single-primer fragments in the dual-primer reaction.

Direct sequencing will generally provide several hundred bases of DNA sequence, even though some bases scattered throughout the ladder may be obscured by PCR artifacts. Theoretically, 12–15 bases of sequence are required to unambiguously identify an individual gene. In practice, we have required that a match include 95% identical bases over a stretch of at least 30 bases. This allows for errors that may occur in reading or transcribing the sequences. Databases of nonredundant (Nr) entries from GenBank, EMBL, and DBJ can be queried through the server at NCBI (*http://www.ncbi*) using the BLAST algorithm.[43] A database of expressed sequence tags (DBEST) can also be queried through this server. In our experience, more than 90% of sequences found by DD currently match at least one entry in the NCBI EST database. Hence this database provides a useful tool for confirming the accuracy of direct sequences. In addition, this database overcomes what has been described as a shortcoming of DD,[44] in that it targets the 3′ untranslated region (UTR), which is a region whose sequence is not always reported in databases of known genes. Shorter DD tags that do not extend into the coding regions can often be identified by matching with long ESTs that overlap both the DD tag and the reported coding region of a known gene.

[42] D. Graf, A. G. Fisher, and M. Merkenschlager, *Nucleic Acids Res.* **25,** 2239 (1997).

[43] S. F. Altschul, W. Gish, W. Miller, E. W. Myers, and D. J. Lipman, *J. Mol. Biol.* **215,** 403 (1990).

[44] C. Debouck, *Current Opin. Biotech.* **6,** 597 (1995).

Verification of Differential Expression

DD is subject to the generation of artifacts and hence each selected differential band must be considered as only a candidate for a truly differentially expressed gene and verified as such by a complementary assay. The most often used methods are the hybridization-based Northern and reverse Northern assays. Performing these assays requires the synthesis of a "probe" or "gene tag" from the DD band cDNA.

DD bands in practice often include multiple DNA sequences of nearly identical lengths and hence it is important to purify bands prior to performing hybridization-based assays. The use of unpurified PCR-amplified DD bands as probes for Northern blots can lead to the inaccurate classification of DD bands as false positives and has likely contributed to the high false-positive rates reported in some DD studies. A number of methods for purifying DD cDNAs have been reported including SSCP gels,[45,46] cloning the cDNA and then screening the subclones for differentially expressed messages,[47] and PCR amplification of the DD band with a nested primer.[7,8] We prefer and have used extensively the nested primer approach, in which a new primer that hybridizes to a single component of the DD band cDNA is designed on the basis of direct sequence information and then used in combination with one of the DD primers to PCR amplify a homogeneous gene tag (Fig. 5). This strategy has produced negligible rates of occurrence of false positives[8] and is described below.

To PCR amplify a homogeneous gene tag (probe) by the nested primer approach, a nested primer with a T_m similar to that of the primers used for DD is designed on the basis of the sequence of the database match. We have synthesized 20-mers with a GC content of approximately 50% that contain a G or C at their 3′ terminus. The primers can be positioned near either end of the DD cDNA. In practice, we have positioned nested primer sites near the arbitrary DD primer because this region tends to be less A/T rich and allows the nested primer to be used in combination with one of three anchor primers (Fig. 5). Sites should be selected such that resulting tags (probes) are at least 100 bp in length and preferably >200 bp. PCR amplify a small amount of ethanol-precipitated DD band cDNA using both the nested primer and the appropriate DD primer. For 20-mers, we use an annealing temperature of 60°. Confirm the amplification of a fragment of the correct size by agarose gel electrophoresis and purify if necessary.

[45] S. Zhao, S. L. Ooi, F.-C. Yang, and A. B. Pardee, *BioTechniques* **20,** 400 (1996).
[46] F. Mathieu-Daude, R. Cheng, J. Welsh, and M. McClelland, *Nucleic Acids Res.* **24,** 1504 (1996).
[47] H. Zhang, R. Zhang, and P. Liang, *Nucleic Acids Res.* **24,** 95 (1996).

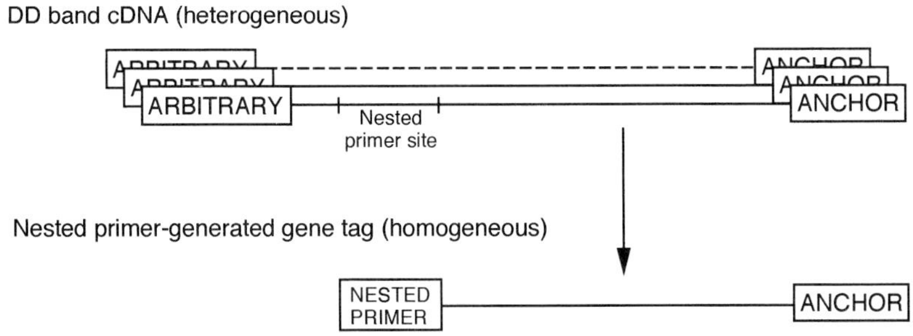

DD band cDNA (heterogeneous)

ARBITRARY

Nested primer site

ANCHOR

Nested primer-generated gene tag (homogeneous)

NESTED PRIMER

ANCHOR

FIG. 5. Diagram showing construction of a nested primer gene tag. The nested primer site is selected just inside the arbitrary primer site and is based on the sequence of the database match to the direct sequence. Nested primer gene tags are generated by PCR from the DD band cDNA, using the nested primer and the DD anchor primer.

For Northern analysis, homogeneous DD cDNA can be labeled with ^{32}P by the random-priming method[48] and used as a probe following standard Northern transfer and hybridization methods.[49]

Reverse Northern dot blotting allows one to simultaneously measure the expression levels of many genes at the expense of information on message length. Briefly, the reverse Northern assay involves binding gene tags to a nylon membrane and hybridizing these with ^{32}P-labeled cDNA prepared from total cellular RNA. To perform this assay:

1. Purify gene tags to remove primers and nucleotides, then apply to a nylon membrane (e.g., ZetaProbe; Bio-Rad, Hercules, CA) using a slot- or dot-blotting manifold (e.g., Bio-Rad BioDot 96-well manifold) according to the manufacturer instructions. Each gene tag should be applied in at least two different dilutions. We have applied tags either to (1) four spots as a 3-fold serial-dilution series using ~100 ng of DNA in the first spot or (2) two spots using ~100 ng of DNA in the first spot and 3 ng of DNA in the second (Fig. 6). Sets of identically prepared membranes are made in parallel: one membrane for each RNA preparation to be assayed in a single experiment.
2. Replicate membranes are hybridized in parallel to different preparations of ^{32}P-labeled cDNA, made by reverse transcribing total cellular RNA

[48] S. Tabor, and K. Struhl, *in* "Current Protocols in Molecular Biology" (F. M. Ausubel, R. Brent, R. E. Kingston, D. D. Moore, J. G. Seidman, J. A. Smith, and K. Struhl, eds.), pp. 3, 5, 8. John Wiley & Sons, New York, 1995.

[49] T. Brown and K. Mackey, *in* "Current Protocols in Molecular Biology" (F. M. Ausubel, R. Brent, R. E. Kingston, D. D. Moore, J. G. Seidman, J. A. Smith, and K. Struhl, eds.), pp. 4, 9, 1. John Wiley & Sons, New York, 1995.

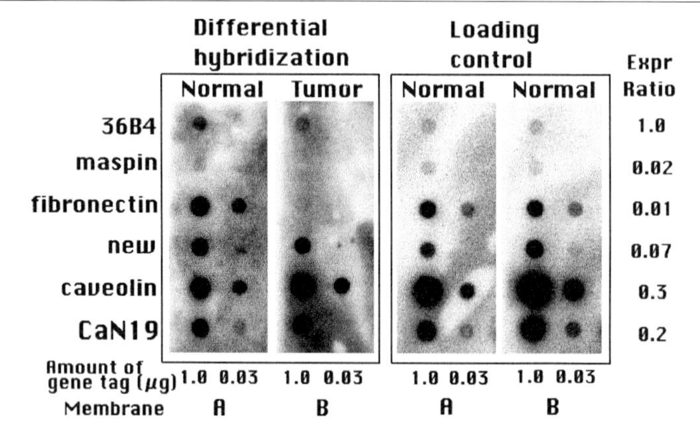

FIG. 6. Example of reverse Northern dot-blot results. Membranes were loaded with two spots, 1.0 and 0.03 μg of each nested primer-generated gene tag as indicated below the panels. Identities of gene tags are indicated at left. The differential hybridization panel shows two replicate membranes, one hybridized with ^{32}P-cDNA generated from normal breast lumenal epithelial cells (Normal), the other from the metastatic breast tumor cell line, ZR75-1 (Tumor). The loading control panel shows the same two membranes as in the differential hybridization panel, following stripping and rehybridization of both membranes with normal breast lumenal epithelial cells (Normal). Expression ratios (listed at right) were obtained by imaging and represent the ratio of normal/tumor signal intensities normalized to the 36B4 signal and corrected for loading differences.

from different cell types. Membranes are hybridized in bags or roller bottles using standard Northern procedures.[50]

3. Evaluation of the results by phosphor-imaging, rather than by standard autoradiography and densitometry methods, is preferable to accommodate a wide range of signal intensities. Background-corrected intensities are calculated for individual spots using PhosphorImager software (Molecular Dynamics, Sunnyvale, CA). With the aid of a spreadsheet (e.g., Excel), values are normalized to the intensity of a nondifferentially expressed control gene, a tag for which must be included on each membrane. For example, the ribosomal protein gene 36B4 is equally expressed in tumor and normal breast epithelial cells.[51] Relative differences are calculated for dots with equal amounts of tag loaded and final differential values are obtained from spots producing the lowest detectable signals. In practice, we have defined a detectable signal as one that is at least threefold above the background.

4. Replicate spotting of any given gene tag onto the different mem-

[50] T. Brown and K. Mackey, *in* "Current Protocols in Molecular Biology" (F. M. Ausubel, R. Brent, R. E. Kingston, D. D. Moore, J. G. Seidman, J. A. Smith, and K. Struhl, eds.), pp. 3, 5, 8. John Wiley & Sons, New York, 1995.

[51] P. Masiakowski, R. Breathnach, J. Bloch, K. Gannon, A. Krust, and P. Chambon, *Nucleic Acids Res.* **10,** 7895 (1982).

branes is of central importance to the reverse Northern assay. In practice, however, it may be difficult to accurately reproduce spots such that identical amounts of denatured gene tag are bound to the membrane and available for hybridization. Hence, it is important to control for differences in the amounts of any given gene tag on the replicate membranes. (We note that the different gene tags need not be present at levels equal to each other when measuring relative levels of gene expression between different cell types.) To control for differences in replicate spotting, each set of membranes should be stripped and rehybridized in parallel with labeled cDNA prepared from one source. For example, after hybridizing a set of membranes with cDNA generated from normal versus tumor RNA, strip the membranes and hybridize them all with cDNA generated from normal RNA. If each given tag was loaded equally, signals for this tag from the replicate membranes should be equal. Final differential values are adjusted appropriately to correct for unequally loaded tags. In practice, we discount values if their loading differs by more than 10-fold.

5. All experiments should be performed at least twice, noting that different tags will produce different degrees of reproducibility. We have found that the membranes can be stripped and reused repeatedly (approximately five to eight times, as can Northern membranes) until signal intensities become too low and providing sets of membranes are handled identically. Likewise, labeled sets of cDNAs can also be reused in parallel until signals are too low (approximately 3 weeks for ^{32}P).

Other assays that have been used to verify differential expression of DD bands include the RNase protection assay and quantitative PCR, which may be more sensitive than the Northern-based assays. For an additional level of sensitivity, amplified antisense RNA (aRNA) can be synthesized with T7 polymerase by transcription from a T7 promoter, isotopically labeled, and used to probe blots.[52]

Cloning

In most DD studies the amplified DD bands are routinely cloned prior to any further analysis. PCR-amplified DD bands can be directly cloned into a TA-cloning vector, such as pCR 2.1 (InVitrogen), or, if restriction sites are present on the DD primers, amplified DD-band cDNA can be restriction digested and cloned into a convenient vector such as pBluescript (Promega). However, our approach, based on nested primers, has been to clone only the approximately 10% of DD bands that cannot be directly sequenced.[7,8] Because cloning is one of the most time-consuming steps of DD-band analysis, the nested primer approach streamlines band analysis.

[52] S. A. Mackler and J. E. Eberwine, *Proc. Natl. Acad. Sci. U.S.A.* **91,** 385 (1994).

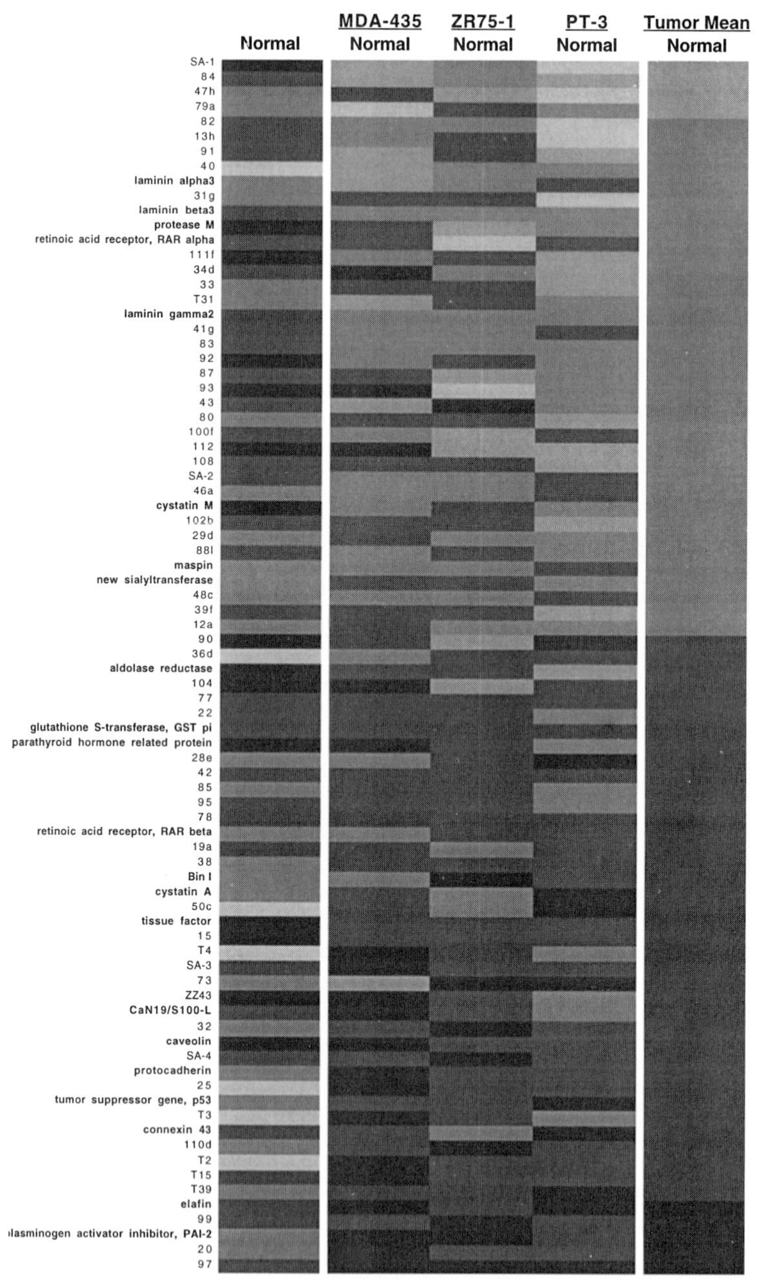

Normal
MDA-435 Normal
ZR75-1 Normal
PT-3 Normal
Tumor Mean Normal
SA-1
84
47h
79a
82
13h
91
40
laminin alpha3
31g
laminin beta3
protease M
retinoic acid receptor, RAR alpha
111f
34d
33
T31
laminin gamma2
41g
83
92
87
93
43
80
100f
112
108
SA-2
46a
cystatin M
102b
29d
88l
maspin
new sialyltransferase
48c
39f
12a
90
36d
aldolase reductase
104
77
22
glutathione S-transferase, GST pi
parathyroid hormone related protein
28e
42
85
95
78
retinoic acid receptor, RAR beta
19a
38
Bin I
cystatin A
50c
tissue factor
15
T4
SA-3
73
ZZ43
CaN19/S100-L
32
caveolin
SA-4
protocadherin
25
tumor suppressor gene, p53
T3
connexin 43
110d
T2
T15
T39
elafin
99
plasminogen activator inhibitor, PAI-2
20
97
Normal expression key:
10-300
1.0-9.0
0.1-0.9
0.01-0.09
0.001-0.009
Tumor/normal expression key:
0.5-2.0
0.125-0.5
0.032-0.125
0.008-0.032
0.002-0.008
0.0005-0.002

High-Throughput Differential Display

The DD procedure, including performing the RT-PCR reactions and running the DD gels, is adaptable to liquid handlers and robotics, which can increase throughput and may reduce intersample variability.[53] DD-band analysis is also adaptable to robotics, including miniprep machines for cloning and automated DNA sequencers (Applied Biosystems, Foster City, CA) for direct sequencing.[54] A bottleneck to the development of high-throughput mechanized DD currently is the process of band excision. However, it is the process of prioritizing DD bands for future study that may have the greatest need for mechanization.

Band Prioritization Using Reverse Northern Hybridization Arrays

A consequence of the power of DD for many laboratories, even in the absence of robotics, is the accumulation of an enormous number of tags representing differentially expressed genes, which must then be prioritized for further study. This situation has precipitated the development of high-throughput, hybridization-based reverse Northern dot-blot assays. A similar assay currently under development is the glass microchip hybridization array assay. Both assays are based on the same experimental principles but differ in the technologies involved. Reverse Northern dot blots typically allow one to assay up to 96 genes per membrane while microarrays will allow the analysis of thousands of genes on a single glass microchip.[55]

Hybridization assays accomplish two different goals. First, they allow the prioritization of tags into those that are more or less interesting for future study on the basis of the prevalence of differential expression across panels of cell lines or tissues. Products of the selected genes may be suitable as candidate drug development targets or as genes central to a particular molecular process under study. Second, hybridization assays produce a "fingerprint" or "profile" composed of the expression level of each of the DD-selected genes across a panel of cell lines or tissue samples (Fig. 7). In cancer research,

[53] S. W. Jones, D. Cai, O. S. Weislow, and B. Esmalaeli-Azad, *BioTechniques* **22,** 536 (1997).
[54] M. Buess, C. Moroni, and H. H. Hirsch, *Nucleic Acids Res.* **25,** 2233 (1997).
[55] Reviewed in *Nature Genet.* **14,** 267 (1996). [Editorial]

FIG. 7. Gene expression profiles of normal breast and breast tumor cells. Gene identities are indicated at left. Normal profile represents the signal intensity obtained by reverse Northern analysis, relative to that of control gene 36B4, and hence is an estimate of the level of gene expression in normal breast lumenal epithelial cells. Three tumor profiles are shown: MDA-MB-435, ZR75-1, and PT-3; each is expressed relative to gene expression levels in the normal breast cells. Only downregulated genes are included. MDA-MB-435 and ZR75-1 are metastatic breast tumor cell lines, and PT-3 was derived from a surgical breast tumor specimen. Tumor mean values represent geometric means of the relative gene expression levels of the three tumor cell types.

these profiles will likely provide sensitive marker systems for categorizing patient tumors for diagnostic, prognostic, and therapeutic information.

Multicolored or gray-scale displays provide a convenient means for displaying expression profiles (Fig. 7) and can be generated using applications developed with the Microsoft Visual Basic programming environment for Excel. Relative expression data entered into an Excel spreadsheet and sorted appropriately are first matched to categories, then all values within a given range are assigned a common identifying value. Once categorized, the different values are each assigned a customized color format and the numerical data are cleared.

Acknowledgments

The authors thank P. Liang, S. Xhao, B. Koh, X. Zhang, and L. Borodyansky for their comments on the manuscript.

[15] READS: A Method for Display of 3′-End Fragments of Restriction Enzyme-Digested cDNAs for Analysis of Differential Gene Expression

By YATINDRA PRASHAR and SHERMAN M. WEISSMAN

Introduction

A gel-based method for analysis of differential gene expression has been described[1] and has since been given the acronym READS[2] or *r*estriction *e*ndonucleolytic *a*nalysis of *d*ifferentially expressed *s*equences. This method uses stringent polymerase chain reaction (PCR) conditions to amplify only the extreme 3′-end fragment of a restriction endonuclease-digested cDNA molecule without employing methods for physical separation of the fragments before or after PCR amplification. For different cDNAs, restriction digestion produces 3′-end fragments of different sizes. Therefore, a pool of cDNAs such as the one prepared from total messenger RNA of a cell can be systematically resolved into gel patterns of 3′-end restriction fragments by using a set of restriction endonucleases. Because a given restriction enzyme can target only a subset of cDNA molecules in a given population,

[1] Y. Prashar and S. M. Weissman, *Proc. Natl. Acad. Sci. U.S.A.* **93,** 659 (1996).

[2] READS is the trademark of Gene Logic, Inc. (Gaithersburg, MD) and this technology is covered by U.S. patent No. 5712126.

Copyright © 1999 by Academic Press
All rights of reproduction in any form reserved.

0076-6879/99 $30.00

different restriction enzymes generate characteristic gel patterns, each utilizing a different set of cDNA molecules within a given pool. A comprehensive analysis of differentially expressed mRNAs between two physiological states can be done by comparing the gel patterns of their 3′-end cDNA restriction fragments. No prior knowledge of the transcript sequences is needed so the method does not depend on genomic sequence, expressed sequence tag (EST), or cDNA data.

Conventionally, subtractive hybridization and differential screening[3–6] of cDNA libraries have been employed to study differential gene expression. Liang and Pardee[7] have described a gel-based method for the display of cDNA fragments that are differentially expressed. In this method, first-strand cDNA synthesized by using a two-base anchored oligo(dT) primer is PCR amplified by using 10-base-long oligonucleotides of arbitrarily chosen sequence as 5′ primers[7,8] and the same anchored oligo(dT) as the 3′ primer. Subsequently, several variations and improvements of this technique have emerged.[8–10] In addition to the use of arbitrary primers for differential gene expression analysis, other methods that employ alternative strategies were also described.[11–14] Some of these alternative approaches involve the use of class II S restriction enzymes,[11] or the use of conserved motif sequence primer and restriction enzymes to study the family of genes that are differentially expressed,[12] and PCR amplification of 3′-end restriction fragments of biotinylated cDNA that are isolated with avidin beads.[13] Alternatively, other methods for expression analysis include sequencing of multiple cloned cDNA fragments [serial analysis of gene expression (SAGE) technique[14]] and hybridization to arrays of oligonucleotides or cDNAs.

The general scheme for READS is depicted in Fig. 1. A dinucleotide-anchored oligo(dT) primer[8] that has an additional sequence at its 5′-end referred to as a "heel" is used for first-strand cDNA synthesis from total RNA using reverse transcriptase. This is followed by second-strand cDNA

[3] S. M. Hedrick, D. I. Cohen, E. A. Nielsen, and M. M. Davis, *Nature (London)* **308,** 149 (1984).

[4] T. Koyama, L. R. Hall, W. G. Haser, S. Tonegawa, and H. Saito, *Proc. Natl. Acad. Sci. U.S.A.* **84,** 1609 (1987).

[5] O. D. von Stein, W. G. Thies, and M. Hoffmann, *Nucleic Acids Res.* **25,** 2598 (1997).

[6] J. Arnau and K. I. Sorensen, *Gene* **188,** 229 (1997).

[7] P. Liang and A. B. Pardee, *Science* **257,** 967 (1992).

[8] P. Liang and A. B. Pardee, *Curr. Opin. Immunol.* **7,** 274 (1995).

[9] P. Liang, W. Zhu, X. Zhang, Z. Guo, R. P. O'Connel, L. Averboukh, F. Wang, and A. B. Pardee, *Nucleic Acids Res.* **22,** 5763 (1994).

[10] S. Zhao, S. L. Ooi, and A. B. Pardee, *BioTechniques* **18,** 842 (1995).

[11] K. Kato, *Nucleic Acids Res.* **23,** 3685 (1995).

[12] A. Fischer, H. Saedler, and G. Theissen, *Proc. Natl. Acad. Sci. U.S.A.* **92,** 5331 (1995).

[13] N. B. Ivanova and A. V. Belyavsky, *Nucleic Acids Res.* **23,** 2954 (1995).

[14] V. E. Velculescu, L. Zhang, B. Vogelstein, and K. W. Kinzler, *Science* **270,** 484 (1995).

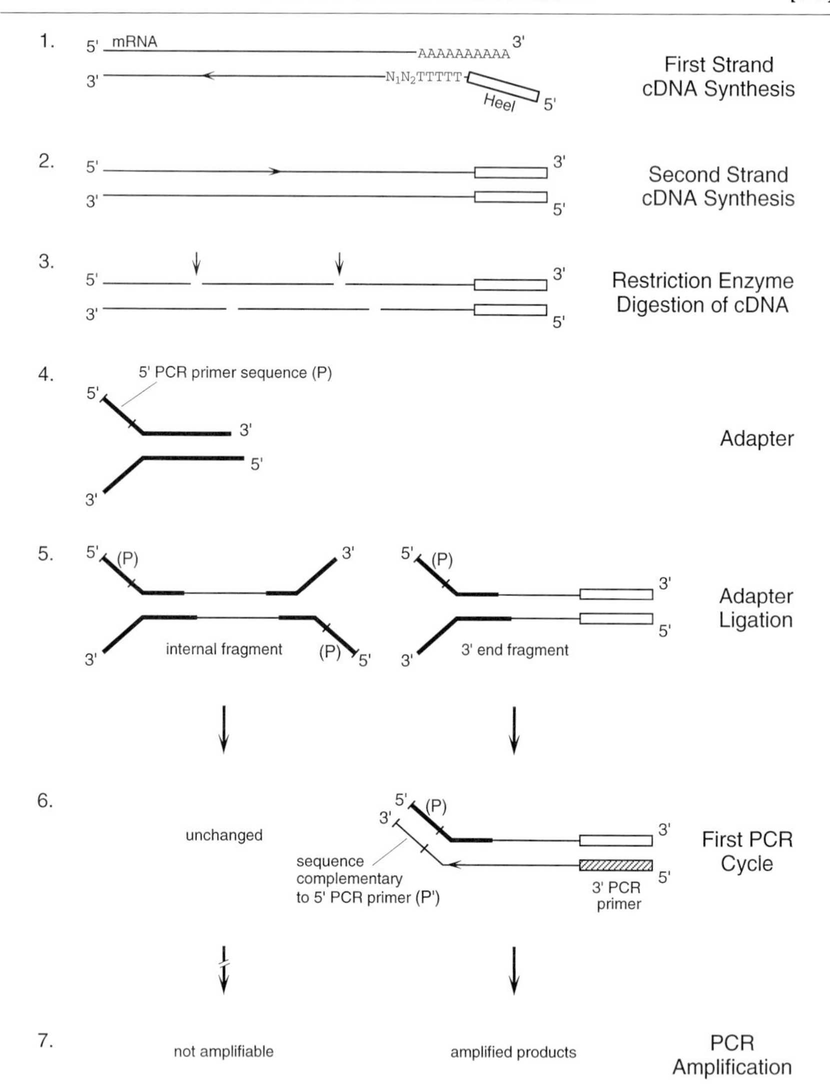

FIG. 1. Schematic of READS. [Adapted from Y. Prashar and S. M. Weissman, *Proc. Natl. Acad. Sci. U.S.A.* **93,** 659 (1996). Copyright 1996, National Academy of Sciences, USA.]

synthesis by the Gubler–Hoffman method.[15] Therefore, all of the cDNA molecules thus synthesized have a common 3′-end heel sequence that plays a critical role in the amplification of only the 3′-end fragments in subsequent steps. This cDNA is digested with a six base-cutting restriction endonuclease

[15] U. Gubler and B. J. Hoffman, *Gene* **25,** 263 (1983).

and ligated to a Y-shaped adapter.[1] This adapter consists of two DNA strands and has three distinct regions. On its 3′-end, the adapter has an overhang corresponding to that produced by a restriction enzyme; in the middle portion the sequence is complementary on opposite strands; and on the 5′-end, it has a stretch of noncomplementary sequences on opposite strands that give rise to the Y shape of the adapter. The adapter ligation step is followed by PCR amplification. The 5′ PCR primer has the same sequence as that of the noncomplementary region at the 5′-end of the top strand of the adapter. The 3′ PCR primer has the same sequence as that of the heel of the anchored oligo(dT) primer used during the first-strand cDNA synthesis.

During PCR, only the extreme 3′-end restriction fragments of the cDNAs that have the Y-shaped adapter ligated on the 5′-end and have a heel sequence on the 3′-end are amplified. In contrast, there is no amplification of the internal cDNA fragments that have been ligated with the adapter to one or both ends. The 3′ PCR primer does not anneal with these internal cDNA fragments because they lack a heel while the 5′ primer (for which the complementary sequence does not exist) cannot anneal to either one of the Y-shaped adapters ligated on both ends of the internal fragments. However, during the first PCR cycle, the 3′ primer anneals to its complementary heel sequence on the top strand of the 3′-end fragment and initiates DNA synthesis, which extends into the Y region of the adapter (Fig. 1, step 6), synthesizing a new sequence that is complementary to the top strand of the Y adapter (the same sequence as that of 5′ PCR primer). Consequently, the 5′ PCR primer can now anneal to this newly synthesized sequence. Therefore, in the subsequent PCR cycles, both of the primers continue to amplify only the 3′-end fragments. The 5′ PCR primer is radiolabeled or fluoresceinated in order to detect the PCR-amplified bands on the display gel.

A typical cell expresses about 10,000 to 15,000 mRNAs. A six base-cutting restriction endonuclease cuts approximately 8% of the cDNAs at positions between 50 and 500 bases from the poly(A) tract[1]; therefore, about 10–12 six base-cutting restriction enzymes will be needed to achieve an average single hit on every cDNA molecule when each enzyme is used with all 12 different dinucleotide-anchored oligo(dT) primers.

One of the important features of READS is the reproducibility of gel patterns in repeat experiments. Because the method relies on specific amplification of only the 3′-end restriction fragments under stringent PCR conditions, reproducibility of gel patterns can be expected because the set of cDNA molecules synthesized from a particular total RNA by a given anchored primer will always be the same and will, therefore, always generate the same pattern of 3′-end restriction fragments. Owing to the reproducibility of the gel patterns, the differences observed between the intensities of bands under two conditions being compared turn out to be a true representation of a differential level of mRNA expression that exists between that particular condition under comparison as seen in our earlier and subsequent studies.[1,16] Furthermore, little redundant representation of a given cDNA is observed because only one 3′-end fragment is amplified from each cDNA molecule. Another advantage of READS is that the size of a known cDNA fragment and, hence, its position on the display gel can be predicted.[1,16] Because a set of 12 anchored oligo(dT) primers can utilize all of the poly(A) mRNAs for cDNA synthesis and about 12–24 six base-cutting restriction enzymes can bring about their complete digestion, a comprehensive analysis of differential gene expression can be carried out for a given cell type in a fairly rapid and highly reproducible manner. Finally, any hidden differences that arise in adjacent lanes of a display gel owing to comigration of bands of different sequences but similar size can be resolved by additional restriction digestion of the adapter-ligated cDNA fragments prior to PCR amplification.[1]

As a result of availability of parameters, such as the fragment length, the sequence of the two bases before the poly(A) tail of mRNA, and the restriction enzyme used, READS presents an opportunity to carry out a direct search in the databases for the fragments seen on the display gel. In other words, identity of the cDNA fragments whose full-length sequence already exists in databases can be established without sequencing them by ascertaining the different sizes expected with different restriction enzymes. In the following sections, a detailed description of the method used to perform READS analysis on control and activated[1] Jurkat T cells (Fig. 2) is described.

Materials and Methods

RNA Isolation

Jurkat cells are synchronized for their cell cycle state by serum starvation and activated with phytohemagglutinin and phorbol ester as described before.[1] Total RNA is isolated from control and activated Jurkat T cells using Trizol reagent (GIBCO-BRL, Gaithersburg, MD). Poly(A)$^+$ purified RNA is not used as it could become contaminated with oligo(dT) during its purification, which could potentially interfere in the subsequent steps. Also, poly(A)$^+$ RNA purification could also introduce a bias toward quantitative recovery of some mRNAs, especially the less abundant species. The

[16] Y. V. B. K. Subrahmanyam, Y. Prashar, N. Hoe, C. Whitney, J. D. Goguen, P. E. Newburger, and S. M. Weissman, in preparation (1999).

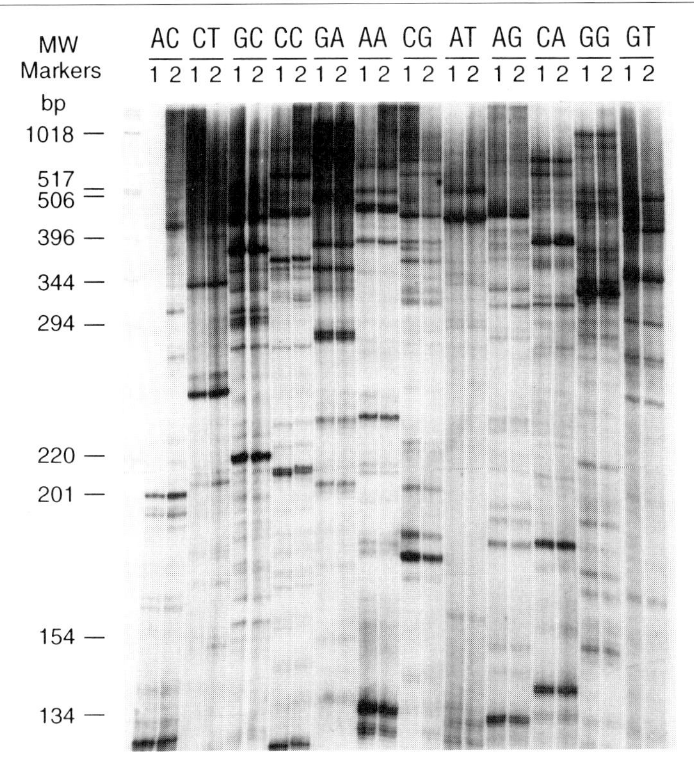

FIG. 2. READS analysis of control and activated Jurkat T cell line. Restriction endonuclease *Nco*I was used for the display of control (lanes 1) and activated (lanes 2) Jurkat cDNAs. Sequence of the dinucleotide (N_1N_2) anchor of the oligo(dT) primer is shown above each set of the control and activated sample. Jurkat cells (10^6/ml) were serum starved for 24 hr, followed by a 4-hr activation with 1 μM ionomycin and phorbol 12-myristate 13-acetate (PMA, 10 ng/ml). RNA was isolated from these cells by conventional methods. Using 10 μg of total RNA for each reaction, 12 cDNAs for each control and activated Jurkat were synthesized with each 1 of the 12 N_1N_2 anchored oligo(dT) primers. The cDNA synthesis and their subsequent processing for READS analysis were performed according to the procedures described in this chapter.

ratio between the absorbance at 260 and 280 nm for the RNA samples is 1.8 to 2.0. The intactness of RNA is determined by running the total RNA on a formaldehyde gel. The ratio between 28S and 18S ribosomal RNA should be about 2:1. Semiquantitative reverse transcriptase (RT)-PCR[17] is performed on 1 μg of total RNA from control and activated Jurkat cells and an equal level of β-actin expression is confirmed. A similar analysis

[17] B. Li, P. Sehajpal, A. Khanna, H. Vlassara, A. Cerami, K. H. Stenzel, and M. Suthanthiran, *J. Exp. Med.* **174,** 1259 (1991).

confirms a high level of interleukin 2 (IL-2) expression in the activated but not in the control Jurkat cells.

Oligonucleotides

All oligonucleotides are purified either by gel electrophoresis or high-performance liquid chromatography (HPLC). It has been found that the contaminating short failure sequences in the unpurified oligonucleotides interfere with the annealing of the adapter and can contribute to smearing in the display gel. These impurities also affect the final yield during cDNA synthesis. The following oligonucleotide sequences are used in this protocol. Adapters: A1, TAGCGTCCGGCGCAGCGACGGCCAG; A2, GATCCTGGCCGTCGGCTGTCTGTCGGCGC. Anchored oligo(dT) primers with heel: RP 8.3, TGAAGCCGAGACGTCGGTCG$(T)_{18}$AC; primers RP 8.4, 8.5, and 8.6 have the same heel sequence as RP 8.3 and have CT, GC, and CC, respectively, as their respective anchor nucleotides. Anchor primer RP 9.2, CAGGGTAGACGACGCTACGC$(T)_{18}$GA; primers RP 9.3, 9.4, and 9.5 share the same heel sequence as RP 9.2 and have AA, CG, and AT, respectively, as their respective anchor nucleotides. Anchor primer RP 10.2, TGGTGGATGGCGTTCCAGGG$(T)_{18}$AG; primers RP 10.3, 10.4, and 10.5 share the same heel sequence as RP 10.2 and have CA, GG, and GT, respectively, as their anchor primers; 5′ PCR primer: A1.1, TAGCGTCCGGCGCAGCGAC.

The adapter and heel sequences have been chosen such that they do not match any known sequence in the human genome. For this purpose, these sequences have been chosen from a viral genome and analyzed against all known sequences in databases. As an additional test, the radiolabeled PCR primers do not amplify any fragment from human genomic DNA in a PCR. Also, the anchor nucleotides are assigned to various heel sequences (without any preference) in order to avoid any possible bias of a heel sequence in its ability to prime the first-strand synthesis.

Synthesis of cDNA

In each reaction for first-strand cDNA synthesis, 10 μg of total RNA is used along with 1 of the 12 two base-anchored oligo(dT) primers. The amount of double-stranded cDNA finally synthesized from this RNA quantity has been found to be sufficient for analysis with as many as 32 restriction enzymes. Therefore, 120 μg of total RNA is required for 12 cDNA synthesis reactions with each 1 of the 12 two base-anchored oligo(dT) primers. The reaction mixture for first-strand synthesis includes 10 μg of total RNA, 2 pmol (i.e., 1 μl of 2 μM solution) of the two base-anchored oligo(dT) primer with a heel sequence, 4 μl of 5× first-strand synthesis buffer [250 mM

Tris-HCl (pH 8.3), 375 mM KCl, 15 mM $MgCl_2$], 2 μl of 0.1 M dithiothreitol (DTT), 1 μl of 10 mM dNTPs, and water in a final volume of 18 μl. This mixture is overlayered with a drop of mineral oil and incubated at 65° for 7 min followed by 50° for another 7 min on a thermocycler or a heating block. At this stage, 2 μl of reverse transcriptase SuperScript (200 units/μl; GIBCO-BRL) is added by removing each tube individually and quickly replacing it back in the heating block so that the temperature of the tube contents does not drop significantly. After the enzyme is added, each tube is again removed briefly to mix its contents with a pipette tip and quickly replaced on the heating block to allow the reaction to proceed at 50° for 1 hr. The reverse transcriptase SuperScript (GIBCO-BRL) is stable at 50° unlike some other reverse transcriptases, and is therefore useful for first-strand synthesis for READS, as this temperature allows stringent annealing of the two base-anchored oligo(dT) primers. At this temperature, these oligo(dT) primers are less likely to anneal to the stretches of poly(A)s found in the internal sequence of the mRNAs. Evaporation of samples is an important issue to consider during first-strand cDNA synthesis, especially when handling 24 reaction tubes that arise as a result of making cDNAs for 2 test RNA samples using all 12 two base-anchored oligo(dT) primers. To avoid this, each reaction tube is prepared individually and overlayered with mineral oil before moving on to the next reaction tube.

Second-strand cDNA synthesis is performed according to the Gubler–Hoffman method[15] described in the GIBCO-BRL cDNA synthesis kit. The reaction mixture for second-strand cDNA synthesis consists of 30 μl of 5× second-strand synthesis buffer [100 mM Tris-HCl (pH 6.9), 450 mM KCl, 23 mM $MgCl_2$, 0.75 mM β-NAD^+, 50 mM $(NH_4)_2SO_4$ (cDNA synthesis kit manual, GIBCO-BRL)], 93 μl of water, 3 μl of 10 mM dNTPs, 18 μl of the first-strand synthesis reaction, 1 μl of RNase H (2 units/μl; GIBCO-BRL), 4 μl of *Escherichia coli* DNA polymerase I (10 units/μl; GIBCO-BRL), and 1 μl of *E. coli* DNA ligase (10 units/μl; GIBCO-BRL). The reaction is assembled on ice and then incubated at 16° for 2 hr. An alternative to removing 18 μl of the first-strand reaction mixture is to add the second-strand synthesis reaction mixture directly in the tubes after carrying out the first-strand synthesis reaction. The mineral oil carried over from the first-strand reaction would be removed during phenol–chloroform extraction of the double-stranded cDNA. Following second-strand cDNA synthesis, the reaction mixture is extracted with an equal volume of Tris-HCl (pH 8.3)-saturated phenol and all of the aqueous phase is recovered and extracted with an equal volume of water-saturated chloroform. To avoid chloroform contamination, only about 125 to 130 μl of the aqueous phase is recovered. To this aqueous phase, one-half volume of 7.5 M ammonium acetate is added followed by the addition of 2.5 μl of glycogen (20 mg/ml). To this, 2.5 times the volume of −20° chilled ethanol is then added. The tube contents are mixed by vortexing and centrifuged immediately at 14,000 rpm in a microcentrifuge for 30 min. To the cDNA precipitate, about 100 μl of 70% ethanol is added and the tube is centrifuged at 14,000 rpm for 5 min without prior vortexing so that the cDNA pellet is not disturbed. Following the wash with 70% ethanol, the cDNA pellet is vacuum dried and dissolved in 170 μl of water for subsequent analysis. It is possible to quantitate cDNA yield by incorporating ^{32}P-labeled α-dCTP in the synthesis reaction (as described in the GIBCO-BRL cDNA synthesis manual). However, this step is tedious and inconsistent. Instead, we have found that equal dilution of the cDNAs synthesized from the same amount of RNA results in reliable quantitative display patterns.

Restriction Digestion of cDNAs

Only six base-cutting restriction enzymes are used since four-base cutters produce far more 3′-end fragments than six-base cutters from a given pool of cDNAs. A large number of 3′-end fragments results in crowding on the display gel and produces smeary patterns. The digestion reaction consists of 5 μl of the cDNA, 1 μl of the 10× buffer, and 1.5 units of restriction enzyme (dilution of enzymes is carried out only in the 1× reaction buffer) in a final volume of 10 μl and is incubated for 1 hr, usually at 37°. After each reaction, it is helpful to briefly centrifuge the reaction tube to recover the condensed contents of reaction. For enzymes that require higher temperatures for their activity, the reaction mixture is overlayered with mineral oil to avoid evaporation during incubation.

Adapter and Its Assembly

The adapter consists of a top strand and a bottom strand (for example, A1 and A2, respectively). Except in the region of restriction site overhang, the adapters for different restriction enzymes have the same sequence. As a result, a common 5′ PCR primer (A1.1) can be used for different restriction enzymes. For restriction enzymes producing 5′ overhangs, the bottom strand of the adapter carries the overhang sequence; and for the restriction enzymes producing 3′ overhangs, the top adapter strand has the enzyme overhang. To ensure that both strands of the adapter are ligated to the incoming cDNA fragment, the bottom strand of the adapter is phosphorylated at the 5′-end. One microgram of the bottom strand is mixed with 5 μl of 10 m*M* ATP, 1 μl of 10× T4 polynucleotide kinase buffer, and 4 units of T4 polynucleotide kinase (New England BioLabs, Beverly, MA) in a total volume of 10 μl at 37° for 30 min. After phosphorylation, the enzyme is inactivated by heating at 65° for 20 min. Following denaturation, the reaction tube is centrifuged to collect the condensed contents of the tube and 1 μg of the top adapter strand is added followed by the addition of 2 μl of the 10× annealing buffer [1 *M* NaCl, 100 m*M* Tris-HCl (pH 8.0), and 10 m*M* EDTA (pH 8.0)]. The final volume of the reaction mixture is brought to 20 μl with water. This tube is then incubated at 65° for 10 min, followed by slow cooling at room temperature for 30 min, resulting in formation of the Y-shaped adapter at a concentration of 100 ng/μl.

Adapter Ligation

For the ligation reaction, it is essential that 5 μl of the restriction-digested cDNA be first mixed thoroughly with 100 ng of the adapter (which contains approximately a 40-fold excess of adapter as compared with the cDNA). This is followed by the addition of a master mix that provides 1× ligation buffer [50 m*M* Tris-HCl (pH 7.5), 10 m*M* $MgCl_2$, 10 m*M* DTT, 1 m*M* ATP, and 25 μg of bovine serum albumin], additional 1 m*M* ATP, and 4 units of T4 DNA ligase (NEB; diluted in 1× reaction buffer) in a final reaction volume of 15 μl. This reaction mixture is incubated at 16° for 16 hr, followed by inactivation of T4 DNA ligase by heating at 65° for 10 min. For PCR amplification, this mixture is diluted with water to a final volume of 100 μl, of which 5 μl is taken for PCR.

Polymerase Chain Reaction Amplification

The quantity of template is crucial for PCR amplification, as less than the critical amount of template gives rise to inconsistent gel patterns. However, the optimal quantity of template always gives rise to consistent display patterns in repeat PCRs. Display patterns remain consistent when a higher than optimal amount of template is taken for PCR, except that the patterns begin to appear smeary at high template concentration.

The 5′ primer for PCR is A1.1 and the 3′ primer is either the two base-anchored oligo(dT) with heel or the heel sequence alone. The advantage of using longer sequences such as A1 and the entire two base-anchored oligo(dT) with heel is that stringent PCR conditions can be used such as two-temperature PCR (94° for 30 sec and 68° for 2 min 30 sec).

Because the 5′ PCR primer is common, irrespective of the restriction enzyme or the anchor primer used, it is radiolabeled (or fluorescently labeled) for the purpose of detection of PCR-amplified bands on the display gel. For radiolabeling 24 pmol (1 μl of a 24 μ*M* solution) of the 5′ primer is mixed with 15 μl of [γ-^{32}P]ATP (3000 Ci/mmol; Amersham, Arlington Heights, IL), 1 μl of 10× polynucleotide kinase buffer, and 5 units of T4 polynucleotide kinase in a total volume of 20 μl and incubated at 37° for 30 min. After inactivating the polynucleotide kinase at 65° for 20 min, the radiolabeled primer is diluted with 2 μM unlabeled primer solution to a final volume of 80 μl. The PCR mixture (20 μl) contains 5 μl of the diluted ligation reaction mixture as template, 2 μl of the 10× PCR buffer [100 mM Tris-HCl (pH 8.3) and 500 mM KCl], 2 μl of 15 mM $MgCl_2$ (1.5 mM magnesium concentration has been found to be optimal), 200 μM dNTPs, 200 nM each of the 5′ (2 μl of the radiolabeled primer solution) and 3′ PCR primers, and 1 unit of AmpliTaq (Perkin-Elmer, Norwalk, CT). Primers and dNTPs are added after preheating the reaction mixture, containing the rest of the components overlayered with mineral oil, to 85°. This "hot start" PCR avoids artifactual amplification arising out of arbitrary annealing of the PCR primers at lower temperatures during transition from room temperature to 94° in the first PCR cycle. Another way to achieve a hot start is by replacing regular AmpliTaq with AmpliTaq Gold (Perkin-Elmer; a variant of AmpliTaq that must be heated at 96° for at least 18 min in order to become active) in the reaction mixture. It can be added along with all the other components of the PCR mixture. The regular amplification steps are preceded by one incubation at 96° for 18 min. The PCR consists of 28 to 30 cycles of 94° for 30 sec, 56° for 2 min, and 72° for 30 sec. This number of PCR cycles, used for Jurkat cells, has been found to be optimal because a lesser number of cycles results in a weak signal on the display gel and a higher number of PCR cycles gives rise to smeary gel patterns. However, for cDNAs prepared from other sources of RNA, the optimal number of PCR cycles may vary. As expected, the 5′ primer A1.1 alone does not amplify any fragments from the adapter-ligated cDNA, confirming that internal cDNA restriction fragments that have adapters ligated to both the ends are not amplified. Ideally, PCR primers should not amplify any fragments in the absence of DNA template in a PCR. If, however, any fragments do amplify, steps should be taken to contain DNA contamination in the system. To avoid this contamination in the PCR, care should be taken to use fresh solutions, maintain a clean working environment, and clean the pipettes periodically with 10% bleach solution. After PCR amplification, 2.5 μl of the reaction mixture is mixed with an equal volume of gel loading buffer (100% formamide, bromphenol blue, and xylene cyanol), heat denatured, and loaded on a 6% denaturing polyacrylamide–8 M urea gel (sequencing gel). The gel is dried and exposed to X-ray film for autoradiography. Normally, the gel patterns are clear and reproducible, not only in successive PCR but also in overall repeat experiments. Faint and inconsistent display patterns observed in repeat experiments indicate less than an optimal quantity of template in the PCR. This inconsistency of display patterns can be overcome by testing serial dilutions of template for the PCR and choosing the dilution that gives consistent patterns in repeat PCRs. A low quantity of template in the PCR may be due either to excessive dilution of the postligation mixture or to a low yield in the cDNA synthesis. The latter could be due either to poor quality of the reagents or to impure oligo(dT) primers, in which contaminating short failure sequences compete with the original primer for mRNA. Also, faint display patterns with few bands arise if the restriction enzyme in use is not a frequent cutter of a particular cDNA, e.g., using Jurkat cDNA six base-cutting restriction enzymes *Mlu*I and *Sal*I always produces weak display patterns with few bands whereas the same quantity of Jurkat cDNA produces clear and reproducible patterns with *Bgl*II or *Hin*dIII. Dark and smeary display patterns result when either too high a quantity of template is used in the PCR or when more than an optimum number of PCR cycles is performed. The optimum template quantity for PCR may differ for different cell types or tissues because 10 μg of total RNA from different sources may carry varying amounts of poly(A)$^+$ RNA and yield varying amounts of cDNAs. The protocol described above works successfully with Jurkat cells, human osteoblasts, human and mouse fibroblasts, human neutrophils, and rat brain (our unpublished data, 1998).

Uncovering Hidden Differences

When cDNA fragments of the same size but different sequence comigrate in adjacent lanes, they could be easily mistaken for a cDNA with an equal level of expression under both conditions and mask a true difference. Such hidden differences can be resolved by redigesting the adapter-ligated cDNA with another restriction enzyme before PCR amplification. If a site for this secondary restriction enzyme in use is present in one of the two comigrating bands, the band will be digested and will not be amplified. The other comigrating fragment would be amplified, resolving a difference that was hidden because of comigration of the two bands.[1] This secondary digestion could be performed with single or multiple restriction enzymes. Because adapter-ligated cDNA becomes further diluted during secondary restriction digestion and the digestion reaction mixture typically contains 10 mM magnesium, care is taken to maintain the concentrations of cDNA template and magnesium in the PCR at the same level as in the regular PCR. To achieve that, 8.25 μl of the adapter-ligated and diluted cDNA template is digested with the secondary restriction enzyme(s) in a final reaction volume of 12.5 μl. After the digestion, 7.5 μl of the reaction mixture is taken and the PCR is set up in a final volume of 50 μl. These adjustments are necessary to maintain a final volume of the cDNA template in the PCR equivalent to 5 μl of the original untreated template and to maintain the final magnesium concentration at 1.5 mM. It is essential that the 10× PCR buffer used here not contain magnesium, which is contributed by the restriction enzyme buffer. The control PCR is also processed in the same manner, except that the restriction enzyme is substituted with water in the digestion reaction mixture. The need for secondary digestion should not arise when a large number of restriction enzymes (25–30) is used for the overall analysis. This is because a hidden difference arising with one restriction enzyme has more chances (depending on the number of restriction enzymes being used) of being resolved by another restriction enzyme.

Recovery of Fragments from Display Gel

For the purpose of band recovery, the dried display gel is marked asymmetrically with a colored marker pen that carries radioactive ink (such a marker pen can be made by injecting any ^{35}S radiolabel in the marker pens commonly used in laboratories). These marks are covered with a tape to avoid radioactive contamination of the X-ray film or the film cassette. The impressions created on film by these marks are aligned with the original markings on the dried gel and the relevant bands are excised directly through the film, using a sharp razor blade. The cut portion of the gel is boiled in 100 μl of water for 10–15 min and the solution is then precipitated with a one-tenth volume of 3 *M* sodium acetate (pH 5.0) and 2.5 vol of ethanol in the presence of 3 μl of glycogen (20 mg/ml). After chilling the sample at $-70°$ for 30 min (or overnight at $-20°$), it is centrifuged at 14,000 rpm in a microcentrifuge, and the pellet is washed in 70% ethanol, vacuum dried, and dissolved in 10 μl of water. Five microliters of this sample is reamplified by PCR with the set of primers used in the original amplification reaction. It is important to establish the identity of the reamplified band because contamination of the PCR with other DNA is quite common at this stage. Identity of the excised band is established by radiolabeling the reamplified band with polynucleotide kinase and rerunning it on the sequencing gel alongside the original display sample. Under these conditions, the purified excised band should comigrate with the band of interest in the original display sample. A clean product can be directly sequenced by using either one of the PCR primers. At times, when the excised band is contaminated with other DNA fragments, it cannot be sequenced directly. Therefore, it becomes necessary to clone the amplified products in a suitable plasmid vector. Moreover, sequencing of the cloned product allows an accurate determination of the fragment size, which can then be corroborated with the fragment size from the display gel besides confirming the restriction site at the 5′-end and the N_1N_2 at the 3′-end. After cloning, several white colonies are picked, and colony PCR is performed on individual white colonies using the same set of PCR primers. The size of the amplified insert is compared with the original band of interest as described earlier. A colony that carries the desired band is grown and a plasmid miniprep is carried out to sequence the plasmid containing the right insert. For cloning of fragments, a PCR script kit (Stratagene, La Jolla, CA) is used. The differential expression of 3′-end fragments recovered from display gel is validated by Northern blot or semiquantitative RT-PCR. Northern blot analysis could become a limitation if a sufficient quantity of RNA is not available to test a large number of 3′-end fragments identified for differential expression. Under these circumstances, semiquantitative RT-PCR[17] serves as an alternative for validation of these differences. For RT-PCR, first-strand cDNA is prepared from 1 μg of total RNA by using oligo(dT) or random hexamer primer. It is important that the extent of first-strand cDNA template dilution in the RT-PCR and the number of PCR cycles be the same as in the display PCR to ensure that RT-PCR is performed in the linear phase of amplification. Owing to this dilution of template, the amplified products may not be visible on the regular agarose gel stained with ethidium bromide. Therefore, either more sensitive dyes, such as Syber Green (Molecular Probes, Eugene, OR), or radiolabeling of the PCR primers is required to detect the amplified products. When using radiolabeled primers, a denaturing acrylamide gel offers better resolution and quantitation than do agarose gels. Primers for RT-PCR are derived from sequences of the recovered 3′-end fragment under test. As an internal control, β-actin PCR primers[17] are used.

Comments

The display patterns from a given RNA sample were consistently reproducible not only in successive PCR but also in overall repeat experiments. Moreover, this reproducibility of display patterns serves as a preliminary validation of the differential expression pattern observed for a given 3′-end restriction fragment of a cDNA. The absence of any DNA precipitation between various steps (such as restriction digestion, adapter ligation, and PCR) and performing the PCR in the linear phase of amplification helps maintain a better quantitative relationship between the samples under comparison. The absence of any physical separation steps such as isolation of biotinylated cDNA fragments also helps maintain the quantitative relationship.

Because only the 3′-end fragment is amplified from a restriction-digested cDNA, each cDNA band on the display gel is represented by one band or at most a few bands. However, for a gene that has alternate spliced forms, each individual spliced form will be represented as a separate 3′-end fragment on the display gel. Similar expression patterns will be observed for each spliced form if the pattern of their regulation is the same. However, a differential regulation of any one of the alternate spliced forms that might be different from that of others will be detectable.

It has been shown that when a mixture of cDNA fragments is PCR amplified, such as in the differential display techniques, the reaction kinetics favor rehybridization of high-abundance cDNA fragments in the later cycles of PCR instead of primers annealing to these fragments.[18] Consequently, amplification of more abundant fragments declines as the PCR progresses. The less abundant cDNA fragments are able to utilize the PCR primer owing to reduced competition for these primers by the high-abundance cDNA fragments. During analysis of differential gene expression by the READS protocol, such reaction kinetics during PCR should allow amplification of low-abundance cDNA fragments.

Owing to the ability of this technique to generate signature patterns for individual mRNAs in a cell, it is possible to monitor subtle changes such as in a time course study of a physiological phenomenon with high accuracy.

[18] F. Mathieu-Daude, J. Welsh, T. Vogt, and M. McClelland, *Nucleic Acids Res.* **24,** 2080 (1996).

[16] A Modified Method for the Display of 3′-End Restriction Fragments of cDNAs: Molecular Profiling of Gene Expression in Neutrophils

By Y. V. B. K. SUBRAHMANYAM, NAMADEV BASKARAN, PETER E. NEWBURGER, and SHERMAN M. WEISSMAN

Introduction

Polymorphonuclear leukocytes (PMNs) are the most abundant cell type among circulating white cells. These cells are of considerable biomedical importance as PMNs provide the earliest cellular responses to many bacterial infections and these cells are also involved in the pathology of several clinical processes.[1,2] Although the rate of macromolecular synthesis per cell may be low, these cells can synthesize large amounts of various biologically active molecules.[3] There are several advantages in working with neutrophils as a model system to study gene expression. PMNs can be readily and

[1] J. A. Smith, *J. Leukocyte Biol.* **56,** 672 (1994).
[2] T. M. Ainsworth, E. B. Lynam, and L. A. Sklar, *in* "Cellular and Molecular Pathogenesis" (A. E. Sirica, ed.), p. 37. Lippincott-Raven, Philadelphia, 1996.
[3] M. A. Castella, *Immunol. Today* **16,** 21 (1995).

Copyright © 1999 by Academic Press
All rights of reproduction in any form reserved.

0076-6879/99 $30.00

rapidly isolated from peripheral blood and, hence, minimal disturbance occurs to the physiologic state during the purification. Circulating neutrophils are postmitotic cells in which variation in stages of the cell cycle does not contribute to the heterogeneity of the cell population. Finally, a number of physiologic stimulants for this cell type are known[4,5] and these cells can also be exposed under controlled conditions to various microorganisms that they encounter *in vivo*.

To study gene expression in PMNs we chose to evaluate the changes in expression of most of the mRNAs in response to various stimuli, and initially to analyze the mechanisms responsible to these changes by inhibiting RNA synthesis.[6,7] For this purpose it was desirable to use a method that could demonstrate levels of unknown RNA species as well as those already represented in the databases, and that could conveniently be applied to multiple RNA samples. We have developed such a technique by modification of the previously described 3′-end cDNA restriction fragment display analysis method.[8]

With some refinements this modified method of differential display has proved to be easy and useful in uncovering changes in mRNA as well as in detecting mRNAs currently undocumented in the public databases. Changes of about twofold in the relative levels of individual amplification products could commonly be detected and mRNAs whose relative abundance ranged over three to four orders of magnitude could be measured. Sequences from the cDNA bands could be readily obtained in most cases, and were useful not only in identifying unique sequence mRNAs, but also in distinguishing RNA species containing common repetitive sequences. In the following sections we present the detailed protocol that we use for most of our studies.

Materials and Methods

Oligonucleotides. The oligonucleotides (gel purified) used are listed in Table I.

Fly Adapter. Fly adapters are Y-shaped adapters prepared by annealing

[4] J. H. Liu and J. Y. Djeu, *in* "Human Cytokines: Their Role in Disease and Therapy" (B. B. Aggarwal and R. K. Puri., eds.), p. 71. Blackwell Science, Cambridge, Massachusetts, 1995.

[5] F. Colotta, F. Re, N. Polentarutti, S. Sozzani, and A. Mantovani, *Blood* **80,** 2012 (1992).

[6] I. Tamm and P. B. Sehgal, *Adv. Virus Res.* **22,** 187 (1978).

[7] R. Zandomeni, B. Mittleman, D. Bunick, S. Ackerman, and R. Weinman, *Proc. Natl. Acad. Sci. U.S.A.* **79,** 3167 (1982).

[8] Y. Prashar and S. M. Weissman, *Proc. Natl. Acad. Sci. U.S.A.* **93,** 659 (1996).

the FA-1 and FA-2 adapter oligonucleotides (see Table I). The annealed fly adapter has a four-nucleotide overhang that is complementary to that generated by the six nucleotide-recognizing restriction enzyme, which makes a staggered cut at the site. By changing the four nucleotides on FA-2 we can generate overhangs for any restriction enzyme of interest. This fly adapter also has a region where both FA-1 and FA-2 are annealed together, and a region where these oligonucleotides are not complementary. As a result they do not anneal to each other in this region, giving a Y shape to the adapter (see Fig. 1).

Reagents

NEUTROPHIL ISOLATION. Dextran-70 is purchased from McGaw (San Diego, CA) and Ficoll-Paque from Pharmacia Biotech (Piscataway, NJ). All reagents are lipopolysaccharide (LPS) free by *Limulus* amebocyte assay.

TABLE I
OLIGODEOXYNUCLEOTIDES USED IN THIS STUDY

Oligo name	Sequence
	Heel sequence — Oligo(dT)$_{18}$ — N_1N_2
RP 8.0 AA	5′-TGA AGC CGA GAC GTC GGT CG TTT TTT TTT TTT TTT TTT AA-3′
RP 8.0 AC	5′-TGA AGC CGA GAC GTC GGT CG TTT TTT TTT TTT TTT TTT AC-3′
RP 8.0 AG	5′-TGA AGC CGA GAC GTC GGT CG TTT TTT TTT TTT TTT TTT AG-3′
RP 8.0 AT	5′-TGA AGC CGA GAC GTC GGT CG TTT TTT TTT TTT TTT TTT AT-3′
RP 5.0 CA	5′-CTC TCA AGG ATC TTA CCG CT TTT TTT TTT TTT TTT TTT CA-3′
RP 5.0 CC	5′-CTC TCA AGG ATC TTA CCG CT TTT TTT TTT TTT TTT TTT CC-3′
RP 5.0 CG	5′-CTC TCA AGG ATC TTA CCG CT TTT TTT TTT TTT TTT TTT CG-3′
RP 5.0 CT	5′-CTC TCA AGG ATC TTA CCG CT TTT TTT TTT TTT TTT TTT CT-3′
RP 6.0 GA	5′-TAA TAC CGC GCC ACA TAG CA TTT TTT TTT TTT TTT TTT GA-3′
RP 6.0 GC	5′-TAA TAC CGC GCC ACA TAG CA TTT TTT TTT TTT TTT TTT GC-3′
RP 6.0 GG	5′-TAA TAC CGC GCC ACA TAG CA TTT TTT TTT TTT TTT TTT GG-3′
RP 6.0 GT	5′-TAA TAC CGC GCC ACA TAG CA TTT TTT TTT TTT TTT TTT GT-3′
RP 8.0 heel	5′-TGA AGC CGA GAC GTC GGT CG-3′
RP 5.0 heel	5′-CTC TCA AGG ATC TTA CCG CT-3′
RP 6.0 heel	5′-TAA TAC CGC GCC ACA TAG CA-3′
FA-1 (fly adapter 1)	5′-TAG CGT CCG GCG CAG CGA CGG CCA G-3′
FA-2 (fly adapter 2)	5′-GAT CCT GGC CGT CGC TGT CTG TCG GCG C-3′
T7-Sal-Oligo(dT)$_{18}$V[a]	5′-ACG TAA TAC GAC TCA CTA TAG GGC GAA TTG GGT CGA C TTT TTT TTT TTT TTT TTT V-3′

[a] V = A, C, or G.

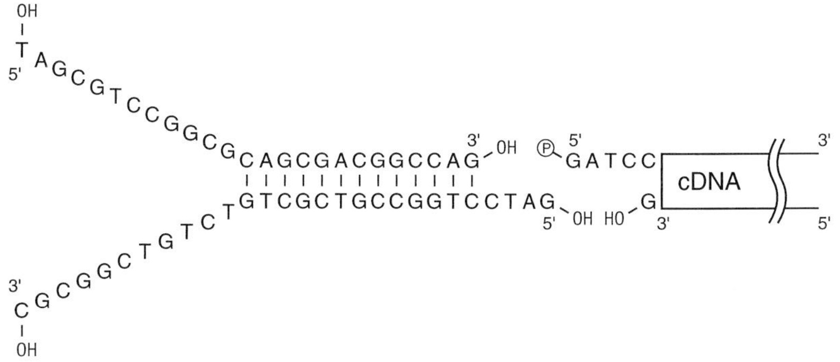

FIG. 1. Diagrammatic representation of the annealed adapter with 5′-GATC overhang and the 3′-end restriction fragment of the cDNA with the compatible overhang (5′-GATC, generated by *Bgl*II, *Bcl*I, or *Bam*HI enzyme).

cDNA SYNTHESIS. Superscript II reverse transcriptase (200 U/μl) along with 5× first-strand buffer [250 m*M* Tris-HCl (pH 8.3), 375 m*M* KCl, 15 m*M* $MgCl_2$] and 0.1 *M* dithiothreitol (DTT) are from GIBCO-BRL (Gaithersburg, MD). RNase inhibitor (RNasin, 40 U/μl) is purchased from Promega (Madison, WI). Deoxyribonucleoside triphosphates (dNTPs) at a concentration of 100 m*M* are obtained from New England BioLabs (Beverly, MA). The 5× second-strand buffer [100 m*M* Tris-HCl (pH 6.9), 23 m*M* $MgCl_2$, 450 m*M* KCl, 0.75 m*M* β-NAD^+, and 50 m*M* $(NH_4)_2SO_4$] can either be made or purchased from GIBCO-BRL. DNA polymerase I (10 U/μl), *Escherichia coli* DNA ligase (10 U/μl), and ribonuclease H (3 U/μl) are purchased from GIBCO-BRL.

RESTRICTION ENZYME DIGESTION AND LIGATION. Restriction enzymes and T4 DNA ligase (400 U/μl) are from New England BioLabs.

END LABELING OF PCR PRIMER FA-1. [γ-^{32}P]Adenosine triphosphate (ATP, 3000 Ci/mmol) is purchased from Amersham (Arlington Heights, IL). T4 polynucleotide kinase (T4PNK, 10 U/μl) is purchased either from New England BioLabs or from GIBCO-BRL. Quick Spin (Sephadex-G25) columns are purchased from Boehringer Mannheim (Indianapolis, IN).

POLYMERASE CHAIN REACTION. AmpliTaq Gold (5 U/μl) with GeneAmp 10× PCR buffer [100 m*M* Tris-HCl (pH 8.3), 500 m*M* KCl, 15 m*M* $MgCl_2$, 0.01% (w/v) gelatin] from Perkin-Elmer (Norwalk, CT) is used for polymerase chain reaction (PCR) amplification of the adapter (see Table I)-ligated cDNA. For amplification of the bands recovered from the gel, either AmpliTaq Gold or Platinum *Taq* polymerase (5 U/μl) from GIBCO-BRL can be used. KlenTaq 1 DNA polymerase (25 U/μl) is from Ab Peptides (St. Louis, MO) and native *Pfu* DNA polymerase (2.5 U/μl) is from Stratagene (La Jolla, CA). Nujol mineral oil is purchased from Perkin-Elmer.

GEL ANALYSIS AND AUTORADIOGRAPHY. The sequencing Gel-Mix 6 is from GIBCO-BRL. SeaKem GTG agarose is from FMC (Philadelphia, PA). Sigmacote is from Sigma (St. Louis, MO). Ammonium persulfate and *N*,*N*,*N*′,*N*′-tetramethylethelenediamine (TEMED) are purchased from Bio-Rad Laboratories (Hercules, CA). Bio Max MR X-ray film is purchased from Eastman Kodak (Rochester, NY).

OTHER REAGENTS. Glycogen (20 mg/ml) and shrimp alkaline phosphatase (1 U/μl) are from Boehringer Mannheim. Exonuclease I (10 U/μl) is from United States Biochemical (Cleveland, OH). The pCR-Script cloning system from Stratagene and pGEM-T Easy cloning system from Promega are used for cloning the PCR products. Gel extraction kits and PCR purification kits are from Qiagen (Chatsworth, CA). Diethyl pyrocarbonate (DEPC), complement factor C7-deficient human serum, and 5,6-dichloro-1-β-D-ribofuranosylbenzimidazole (DRB) are purchased from Sigma. Betaine monohydrate is from Fluka Biochemica (Buchs, Switzerland).

Isolation of Neutrophils

Several methods are available in this series[9–11] for isolation and handling of neutrophils. We adapted a method[12] for isolation of neutrophils under lipopolysaccharide (LPS)-free conditions. All the reagents, serum, buffers, and media are free of LPS (<0.01 ng/ml.).

1. Draw about 450 ml of blood from healthy volunteers into 60-ml syringes containing 6 ml of acid citrate dextrose [ACD: 2.45 g of dextrose (hydrous), 2.2 g of sodium citrate (hydrous), and 0.730 g of anhydrous citric acid in a final volume of 100 ml] and 12 ml of 6% (w/v) Dextran-70 solution prepared in normal saline.
2. Mix the contents of the syringe by gentle inversion. Tape the syringes in an upright position at room temperature and let the red blood cells (RBCs) settle for 90 min.
3. Collect the leukocyte-rich plasma into 50-ml Falcon tubes (Becton Dickinson Labware, Lincoln Park, NJ) by pushing down each of the syringes and drawing off the supernatant above the settled RBCs through tubing attached by Luer-Lok to the upright syringes.

[9] R. F. Rest and D. P. Speert, *Methods Enzymol.* **236,** 91 (1994).
[10] R. F. Rest, *Methods Enzymol.* **236,** 119 (1994).
[11] J. R. Kalmer, *Methods Enzymol.* **236,** 108 (1994).
[12] B. M. Babior and H. J. Cohen, *in* "Leukocyte Function" (M. J. Cline, ed.), p. 1. Churchill-Livingstone, New York, 1981.

4. Centrifuge the leukocyte-rich plasma at 300 *g* in a refrigerated centrifuge for 8 min at 4°.

5. Resuspend each cell pellet in 5 ml of Hanks' balanced salt solution (HBSS without Ca^{2+} and Mg^{2+}). Layer the cell suspension over a cushion of Ficoll-Paque (4 ml) and centrifuge at 600 *g* for 30 min at 4°.

6. Each pellet contains neutrophils and a few contaminating RBCs. Carefully aspirate the supernatant and the interface down to each pellet. Wash the pellets twice by resuspending in HBSS and centrifuging at 300 *g* for 8 min at 4°.

7. If necessary, the contaminating RBCs may be removed by gentle, brief hypotonic lysis. After readjusting to normal osmolarity, the PMNs are separated by centrifugation at 300 *g* for 8 min at 4°. Resuspend the neutrophils in HBSS, and calculate the cell density and yield by counting in a hemocytometer.

Yield Specifications. We usually recover about 1–2 × 10^9 cells from about 450 ml of blood. More than 95–97% of these cells are neutrophils. A variable number (3–5%) of eosinophils are present in the preparation, but no monocytes are detectable. On the basis of trypan blue staining, more than 95% of the cells are found to be viable. Once the cells are isolated, it is important to use them immediately.

Incubation of Neutrophils with 5,6-Dichloro-1-β-D-ribofuranosylbenzimidazole and Escherichia coli

Aliquot neutrophils (about 2.6 × 10^8 cells each) in 100 ml of RPMI 1640 medium with 10% heat-inactivated fetal bovine serum (low endotoxin) into four different Corning (Corning, NY) sterile disposable polycarbonate 250-ml Erlenmeyer flasks with screw caps. Treat the cells in each flask as follows.

Flask 1: Use the cells in this flask as untreated control

Flask 2: Incubate the cells with *E. coli* R594supO. [Preopsonize the bacteria with complement factor C7-deficient human serum (20%, v/v).[11,13–16] The ratio of bacteria to neutrophils is 20:1]

Flask 3: Incubate the cells in this flask with 50 μM 5,6-dichloro-1-β-D-ribofuranosylbenzimidazole (DRB)[6,7]

Flask 4: Incubate the cells in this flask with opsonized *E. coli* R594supO and 50 μM DRB

[13] D. Rennell and A. R. Potete, *Genetics* **123,** 431 (1989).

[14] S. L. Newman and L. K. Mikus, *J. Exp. Med.* **161,** 1414 (1985).

[15] D. L. Gordon, J. Rice, J. J. Finlay-Jones, P. J. McDonald, and M. K. Hostetter, *J. Infect. Dis.* **157,** 697 (1988).

[16] L. M. Madsen, M. Inada, and J. Weiss, *Infect. Immun.* **64,** 2425 (1996).

Incubate all of the flasks at 37° with gentle rotation for 2 hr. At the end of the incubation centrifuge the cells at 450 *g* for 10 min and wash once with HBSS. Process the neutrophils for RNA isolation as described below.

Isolation of Total RNA from Neutrophils

We isolated the RNA from peripheral blood neutrophils using the guanidine-HCl method described by Strohman *et al.*[17] with some modifications.

1. Resuspend the cell pellet in an ice-cold solution containing 19 parts 6 *M* guanidine hydrochloride and 1 part 2 *M* potassium acetate (pH 5.0) and vortex vigorously to lyse the cells. The volume to be added depends on the number and type of cells. For neutrophils add 4 ml of the reagent for every 1–5 $\times$ 10^8 cells.
2. Sonicate the lysate (four 15-sec bursts on ice) and centrifuge at 3000 rpm for 15 min at 4°. Collect the supernatant into 1.5-ml Eppendorf tubes (several if required), add glycogen (1 μl of a 20-mg/ml solution) as a carrier, and precipitate by adding 0.5 vol of ice-cold ethanol, and leave the tubes overnight at −20°.
3. Collect the precipitate by centrifugation for 30 min at top speed in an Eppendorf centrifuge and dissolve the pellet in 500 μl of a solution containing 19 parts 6 *M* guanidine hydrochloride, 1 part 2 *M* potassium acetate (pH 5.0), and 1 part 0.5 *M* EDTA (pH 8.0). Bring into solution by forcing the pellets through a 20-gauge $\times$ 1.5 inch needle on a 3-cm^3 syringe at least 10 times, combining all pellets of each group into one tube. Precipitate once again with 0.5 vol of ethanol at −20° overnight, collect and dissolve the precipitate as described above, and repeat the precipitation step once again.
4. Wash the final pellet twice with 75% (v/v) ethanol, redissolve each pellet in 300 μl of DEPC-treated water with a needle as described above, and add 75 μl of 5$\times$ TNES buffer [5$\times$ TNES is 100 m*M* Tris-HCl (pH 8.0), 500 m*M* sodium chloride, 50 m*M* EDTA (pH 8.0), and 0.1% sodium dodecyl sulfate].
5. Extract the RNA sample once with 0.5 ml of phenol (with hydroxyquinoline, 1 mg/ml)–chloroform–isoamyl alcohol (25 : 25 : 0.5, v/v/v). Vortex the mixture vigorously and centrifuge to separate the phases. Collect the upper aqueous phase into a new microcentrifuge tube. Reextract the phenol–chloroform tube with another 200 μl of 1$\times$ TNES buffer. Extract the pooled aqueous phase once with chloroform–isoamyl alcohol (no phe-

[17] R. C. Strohman, P. S. Moss, J. Micou-Eastwood, D. Spector, A. Przybyla, and B. Paterson, *Cell* **10,** 265 (1977).

nol), add 0.2 vol of 10 *M* ammonium acetate and 1 μl of glycogen to the aqueous phase, and precipitate with 2 vol of chilled ethanol at $-20°$ overnight.

6. Centrifuge the RNA precipitate, wash with 70% ethanol, and dissolve the pellet in a small volume of water with 5 m*M* dithiothreitol (DTT) and RNasin (100 U/ml). Store the RNA at $-70°$.

Yield Specifications. The average yield of RNA is approximately 10–20 μg per 10^8 neutrophils. An aliquot of the RNA is analyzed by electrophoresis on a formaldehyde–agarose gel[18] to ensure its quality and the integrity of the 28S and 18S ribosomal RNA bands by assessing their relative intensities (at a ratio of 2:1).

cDNA Synthesis

First-Strand Synthesis. We use 6 μg of total RNA for every cDNA synthesis reaction.[19] This amount is enough to obtain a display of 10 enzyme digests.

1. Take 6 μg of total RNA in a 10-μl volume of water in a 0.5-ml microcentrifuge tube (nonstick; USA Scientific Plastics, Ocala, FL) and add 1 μl (200 ng) of T7-Sal-Oligo$(dT)_{18}$V (see Table I for sequence).

2. Mix the contents on ice, heat the sample to 65° for 5 min, and chill it on ice for 5 min. Repeat this denaturation and annealing once again and keep the tubes on ice.

RNA in water	10 μl (6 μg)
T7-Sal-Oligo$(dT)_{18}$V	1 μl (200 ng)

3. Set up the first-strand cDNA synthesis reaction as follows.

First-strand buffer (5×)	4 μl
DTT, 0.1 *M*	2 μl
dNTPs, 10 m*M*	1 μl
RNase inhibitor (40 U/μl)	1 μl

Note: For multiple RNA samples make a reaction cocktail with the preceding four components and dispense 8 μl per tube to maintain consistency.

4. Warm the contents to 45°, initiate the cDNA synthesis by adding 1 μl (200 units) of Superscript II reverse transcriptase, and continue the incubation (at this stage the final reaction volume will be 20 μl) at 45° for 1 hr. Do this step in an air incubator instead of a water bath to avoid evaporation.

[18] J. Sambrook, E. F. Fritsch, and T. Maniatis (eds.), "Molecular Cloning: A Laboratory Manual." Cold Spring Harbor Laboratory Press, Cold Spring Harbor, New York, 1989.

[19] GIBCO-BRL, "SuperScript Choice System for cDNA Synthesis: Instruction Manual." GIBCO-BRL, Gaithersburg, Maryland.

Second-Strand Synthesis. At the end of the first-strand synthesis chill the tubes on ice and centrifuge briefly to collect all of the contents at the bottom of the tube. Set up the second-strand reaction[19,20] on ice in the same tube as follows:

First-strand reaction	20 μl
Water	91 μl
Second-strand buffer (5×)	30 μl
dNTPs, 10 m*M*	3 μl
E. coli DNA polymerase (10 U/μl)	4 μl
E. coli DNA ligase (10 U/μl)	1 μl
RNase H (3 U/μl)	1 μl
	150 μl

and incubate the tubes at 16° for 2 hr.

For multiple samples a cocktail can be made on ice in a 1.5-ml microcentrifuge tube (nonstick). The enzymes are added to the cocktail and mixed gently just prior to use. Aliquot 130 μl of the cocktail to each tube containing first-strand reaction sample (20 μl) to give a final volume of 150 μl. Mix gently and incubate at 16° for 2 hr.

At the end of incubation:

1. Stop the reaction by adding 10 μl of 0.5 *M* EDTA, pH 8.0.
2. Extract once with phenol–chloroform (1 : 1, v/v) and once with chloroform. Presaturate the chloroform with nuclease-free water.
3. Precipitate the cDNA by adding 0.5 vol of 7.5 *M* ammonium acetate and 2.5 vol of ethanol (−20°). At this stage the sample can be left overnight at −20°. *Note:* Prior to precipitation of the cDNA, 1 μl (20 μg) of glycogen may be added as a carrier.
4. Collect the cDNA precipitate by centrifugation in an Eppendorf centrifuge at top speed for 15 min.
5. Without disturbing the pellet carefully remove the ethanol. Wash the pellet with 70% ethanol and centrifuge again for 15 min.
6. Remove the supernatant and dry the pellet at room temperature.
7. Dissolve the cDNA in 20 μl of water or TE buffer [10 m*M* Tris-HCl (pH 8.0), 1 m*M* EDTA].

Restriction Enzyme Digestion

The procedure described below is for digesting the cDNA with *Bgl*II enzyme; however we can use any restriction enzyme with its corresponding 10× buffer to set up the digestion in the same way.

[20] U. Gubler and B. J. Hoffman, *Gene* **25,** 263 (1983).

1. Prepare the reaction cocktail as follows on ice.

Water	33 μl
Enzyme (*Bgl*II) buffer (10×)	5 μl; mix well and add
Restriction enzyme (*Bgl*II)	2 μl (20 units)
	40 μl

Note: 10× *Bgl*II buffer (buffer 3, New England BioLabs) consists of the following: 500 m*M* Tris-HCl (pH 7.9), 100 m*M* $MgCl_2$, 1 *M* NaCl, and 10 m*M* dithiothreitol.

2. Set up the digestion of the cDNA by adding 8 μl of the reaction cocktail to 2 μl of cDNA and incubate the sample at 37° for 2 hr. At the end of incubation heat inactivate the enzyme at 65° for 20 min and keep the tubes on ice.

Ligation of Digested cDNA with Fly Adapter

Annealing of Fly Adapter Oligonucleotides FA-1 and FA-2

1. Take 5 μg of adapter FA-1 and adapter FA-2 (see Table I) into a 1.5-ml microcentrifuge tube (nonstick).

2. Add 10 μl of 10× annealing buffer [100 m*M* Tris-HCl (pH 8.0), 10 m*M* EDTA (pH 8.0), 1 *M* sodium chloride] and adjust the volume with water to 100 μl.

3. Heat the sample in a boiling water bath for 5 min. Turn off the burner and allow the bath to reach room temperature. Briefly centrifuge the contents to the bottom of the tube and store the annealed fly adapter at −20°. At this stage the concentration of the fly adapter is 100 ng/μl.

Ligation Procedure

Take 2 μl of restriction enzyme (*Bgl*II)-digested cDNA in a 0.5-ml microcentrifuge tube on ice and add 1 μl of fly adapter with suitable overhang (5'-GATC overhang for *Bgl*II). Mix gently and add 2 μl of ligase cocktail (see below). Mix the contents and incubate at 16° overnight.

Ligase cocktail is prepared as follows:

1. Take 88 μl of nuclease-free water in a 0.5-ml microcentrifuge tube on ice, add 10 μl of 10× T4 DNA ligase buffer [500 m*M* Tris-HCl (pH 7.8), 100 m*M* $MgCl_2$, 100 m*M* dithiothreitol, 10 m*M* ATP, bovine serum albumin (250 μg/ml)] and mix well. Add 2 μl of T4 DNA ligase (400 U/μl) and mix gently. This is referred to as diluted ligase.

2. In another tube on ice take 20 μl of water and 10 μl of 10× T4 DNA ligase buffer, mix well, and add 10 μl of diluted ligase (final volume at this stage is 40 μl). This is referred to as ligase cocktail.

Polymerase Chain Reaction Amplification

For every cDNA sample 12 PCRs are to be set up using N_1N_2-anchored oligodeoxynucleotide (oligo) primer as the 3′ primer and ^{32}P-labeled FA-1 as the 5′ primer.

^{32}P Labeling of Primer FA-1. Take 2 μl (1 μg) of FA-1 primer (see Table I) in a 0.5-ml microcentrifuge tube and set up the reaction as follows.

FA-1 primer	2 μl (1 μg)
T4 polynucleotide kinase buffer (10×)	2 μl
[γ-^{32}P]ATP (3000 Ci/mmol)	15 μl (150 μCi)
T4 polynucleotide kinase (10 U/μl)	1 μl
	20 μl

Incubate at 37° for 1 hr. *Note:* 10× T4 polynucleotide kinase buffer (New England BioLabs) consists of 700 m*M* Tris-HCl (pH 7.6), 100 m*M* $MgCl_2$, and 50 m*M* dithiothreitol.

1. At the end of the incubation add 20 μl water to the tube and inactivate the kinase enzyme by heating at 65° for 15 min. Chill the tube on ice and centrifuge briefly.
2. Purify the ^{32}P-labeled primer FA-1 from unincorporated [γ-^{32}P]ATP by using a Quick Spin (Sephadex G-25, fine) column.
3. Dilute 40 μl of the labeled oligonucleotide by adding 1 μg of unlabeled FA-1 primer and adjust the final volume to 80 μl. At this point the concentration of this primer is 25 ng/μl and is ready for use in a PCR.

Polymerase Chain Reaction. Set up the PCRs (20-μl final volume) in 0.5-ml thin-wall PCR tubes as follows. For every cDNA we need to have 12 tubes as there are 12 anchored primers.

1. Add 2 μl of a 2 μ*M* stock solution of 3′-anchored primer [N_1N_2-anchored oligo$(dT)_{18}$ with heel; see Table I] to a PCR tube and mix with 18 μl of the reaction cocktail (see below). Overlay the tube with mineral oil.
2. The reaction cocktail will have the following components per tube:

GeneAmp 10× PCR buffer (with $MgCl_2$)	2 μl
dNTPs, 2 m*M*	2 μl
^{32}P-Labeled primer FA-1	2 μl
Water	9 μl

At this stage mix everything gently, then add

AmpliTaq Gold (5 U/μl)	0.7 units/reaction
Fly adapter-ligated cDNA template (diluted 1:20 in water)	2 μl
	18 μl (volume adjusted with water)

Note: It is always better to make a cocktail mixture for the total number of tubes to maintain consistency and to avoid tube-to-tube variation. While making the cocktail mixture add all of the required components except the enzyme and mix gently. Add the enzyme last, mix gently again, and dispense into individual tubes.

3. PCR conditions:

Temperature (°C)	Time	No. of cycles
94	12 min	1
94	30 sec	
55	2 min	5
72	1 min	
94	30 sec	
60	2 min	25
72	1 min	

At the end of the PCR the samples can be stored at −20°, prior to gel analysis.

Purification of Polymerase Chain Reaction Products. The PCR product (2 μl) is analyzed on a 6% polyacrylamide–7 *M* urea sequencing gel. Although we use the same amount of the template and ^{32}P-labeled primer FA-1 for every PCR with each anchored primer, there is some variation among these anchored primers in terms of the amount of radioactivity incorporated. To evaluate the differences associated with the expression of individual cDNAs under different experimental conditions, it is important to load an equal amount of radioactivity for each sample. This can be achieved by purifying each of the PCR-amplified cDNAs by using a PCR purification kit from Qiagen, following the instructions provided by the supplier. Purification of the PCR products also reduces the background. Elute the sample into a volume of 30 μl and it is free of unincorporated primer. At this stage use 1 μl of the sample to determine the counts in a scintillation counter. Load equal amounts of the radioactivity (in terms of Cerenkov counts) on the gel.

Gel Analysis of Polymerase Chain Reaction Products

PCR-amplified samples are analyzed on a 6% polyacrylamide–7 *M* urea sequencing gel. Along with the samples it is important to load ^{32}P-labeled molecular weight markers [1-kb or 1-kb Plus ladder (GIBCO-BRL) or pBR322/*Msp*I digest (New England BioLabs) for ^{32}P-labeling] or sequencing ladders in one of the lanes, to size the PCR products on the gel.

^{32}P-Labeling of Molecular Weight Marker DNA. Label the marker DNA by using an exchange reaction. In a 0.5-ml microcentrifuge tube add 12 μl of water, 5 μl of 5× exchange reaction buffer [250 m*M* imidazole-HCl buffer (pH 6.4), 60 m*M* $MgCl_2$, 5 m*M* 2-mercaptoethanol, 350 μM adenosine diphosphate], 2 μl (2 μg) of 1-kb Plus marker DNA (GIBCO-BRL) and 5 μl (50 μCi) of [γ-^{32}P]ATP (3000 Ci/mmol). Initiate the reaction by adding 1 μl (10 units) of T4 polynucleotide kinase enzyme (GIBCO-BRL) and incubate at 37° for 30 min. Heat inactivate the kinase enzyme at 65° for 15 min, and remove the unincorporated [γ-^{32}P]ATP using a Quick Spin (Sephadex G-25 fine) column.

To 1 μl of the ^{32}P-labeled marker add 5 μl of stop solution and 4 μl of water. Denature the DNA at 95° for 3 min and chill it on ice. Load 2 μl of this sample as a marker on the gel.

Preparation of Sequencing Gel and Electrophoresis

1. Siliconize the smaller of the two sequencing gel plates (GIBCO-BRL, S2001 sequencing system) with Sigmacote and rinse both the plates with ethanol. Allow the plates to dry for a few minutes.
2. Assemble the glass plates using 0.4-mm spacers and a gel-casting boot as described in the instruction manual.
3. Cast the gel using Gel-Mix 6 (GIBCO-BRL) acrylamide gel solution. The comb used here has 8-mm-wide teeth and 2-mm spaces between the teeth (29 wells; this can be custom made by GIBCO-BRL or can be purchased from Eastman Kodak. The comb can be cut according to the size of the gel unit using scissors). Alternatively, a 36-well comb with a well width of 6.8 mm is used.
4. Assemble the gel in the electrophoresis apparatus with a cooling base, load 2 μl of stop solution each in two wells, and do a prerun for 1 hr using 1× Tris–borate–EDTA (1× TBE) buffer at 2000-V constant voltage.
5. Prepare the samples for loading by mixing 2 μl of the PCR product with 2 μl of stop solution (see below). Heat denature the samples at 95° for 3 min and quick-chill them on ice. This can best be done by taking the samples into 0.5-ml PCR tubes and heating the tubes in a PCR machine.

   Stop Solution: Formamide (95%, v/v), 20 m*M* EDTA (pH 8.0), bromophenol blue (0.05%, w/v), xylene cyanol FF (0.05%, w/v).

   Tris–borate–EDTA buffer (1× TBE)[18]: 90 m*M* Tris–borate (pH 8.3), 2 m*M* EDTA
6. Disconnect the power supply to the gel unit and wash the wells, using a syringe to remove the urea diffused from the wells. Load samples using sequencing gel sample loading tips. Continue the electrophoresis at 2000-V constant voltage.

7. Let the run continue till the xylene cyanol FF dye reaches the bottom of the gel or has just run out of the gel.

8. At the end of the run, remove the gel. Separate the glass plates and transfer the gel to a precut sheet of filter paper (3MM; Whatman, Clifton, NJ). Cover the gel with extra-wide Saran Wrap (this may be purchased from any wholesale warehouse club). It is important to have another layer of a precut sheet of filter paper (3MM; Whatman) below to protect the gel from picking up any contaminants from the gel drier. Dry the gel in a gel drier.

Autoradiography

Once the gel is dried and cooled to room temperature, carefully remove the gel from the drier. Trim the edges of the filter paper sheet and seal the edges of the gel along with the Saran Wrap, using a tape. This helps in keeping the wrap intact on the gel surface. Cut a few pieces of Radtape (Diversified Biotech, Boston, MA), write any details about the gel and/or arrow marks on the tape, and stick them at the corners of the gel (this will help orient the X-ray film on the gel for recovering the cDNA bands of interest for further analysis). Expose the gel to Bio Max MR X-ray film (Kodak) at room temperature overnight. The dried gels are stored carefully so that we can go back to the gels to recover the cDNA from the bands at a later date. It is also important to make sure that these gels do not shrink or accumulate any contaminants and/or dirt during storage that could potentially cause problems in subsequent PCR amplifications. We use Zip-Seal polyethylene bags (13 × 18 inches, PGC Scientific, Frederick, MD) for this purpose.

Recovery of cDNA Bands from Dried Gels for Sequencing and Analysis

Specific bands of interest from the display gel, as determined on the autoradiogram, can be excised from the gel and reamplified for further analysis.[8]

1. With a fine-point marker, make arrow marks to the bands of interest.

2. Mount the gel on a soft wooden board covered with two sheets of filter paper (3MM; Whatman) below as padding. Place the autoradiogram on top of the gel and align the marks on the X-ray film with the corresponding marks on the dried gel. Hold the gel and the X-ray film in place by using several thumbtacks.

3. Using a sharp razor blade, cut the specific cDNA band directly through the X-ray film and collect the gel piece into a 1.5-ml microcentrifuge

tube. Several bands can be cut from the gel in the same way, using a separate razor blade for every band.

4. Add 120 μl of TE to every tube and let the tubes stand at room temperature for 10 min. Incubate the tubes in a boiling water bath for 10 min. Cool the tubes to room temperature and centrifuge for 10 min in an Eppendorf centrifuge at top speed to remove any particulate gel material.

5. Carefully collect the eluate from the tube (at this stage there will be about 100 μl, as filter paper absorbs some TE) into a fresh tube. Precipitate the cDNA from the eluate by adding 1 μl (20 μg) of glycogen as carrier, 0.1 vol of 3 *M* sodium acetate (pH 6.0), and 2.5 vol of ethanol. Keep the tubes at $-20°$.

6. Centrifuge the samples for 15 min, carefully remove the ethanol, and layer 200 μl of 70% ethanol on the top of each pellet without disturbing the pellets. Centrifuge again, remove the ethanol, and dry the pellets at room temperature.

7. Dissolve the pellets in 10 μl of water and centrifuge at top speed for 10 min to remove any insoluble material. Use 2 μl of this supernatant for PCR amplification.

Purification of Gel-Eluted cDNA

The gel-eluted cDNA band may not be pure and may have some trace amounts of other cDNAs as contaminants that may be coamplified during PCR. These contaminants are removed by another round of PCR amplification with the ^{32}P-labeled FA-1 and the anchored primer. Instead of the anchored primer its heel primer can also be used. Separate the products on a sequencing gel.

Polymerase Chain Reaction

1. Take 2 μl of gel-eluted cDNA and 2 μl of 2 μM stock solution of the corresponding 3′-anchored (or heel) primer in a 0.5-ml thin-walled PCR tube. Add 16 μl of the reaction cocktail (see below) and overlay with mineral oil.

The reaction cocktail is prepared as follows: for every tube add

GeneAmp 10× PCR buffer	2 μl
dNTPs, 2 m*M*	2 μl
^{32}P-Labeled primer FA-1	2 μl
AmpliTaq Gold (5 U/μl)	0.7 units/reaction
Water	10 μl
	16 μl (adjusted with water)

Always add the enzyme last after adding the required amount of water and gentle mixing.

2. The PCR conditions are as follows:

Temperature (°C)	Time	No. of cycles
94	12 min	1
94	30 sec	30
60	2 min	
72	1 min	

Alternatively, a two-temperature PCR can also be used with the following conditions*:

Temperature (°C)	Time	No. of cycles
94	12 min	1
94	30 sec	30
68	2 min, 30 sec	

*Both conditions work well. If the heel primers are used, then it is better to anneal at 60° followed by extension at 72° as indicated above.

3. Take 2 μl of the PCR sample, add 2 μl of the stop solution, and analyze on a sequencing gel exactly as described above.

4. Dry the gel on a Bio-Rad gel dryer and process it for autoradiography. Expose the gel to Bio Max MR X-ray film for 1 hr at room temperature.

5. At this stage we can see intense bands and several faint contaminating bands. The specific cDNA band can be identified by its position relative to the molecular size markers and in comparison with the original autoradiogram.

6. Excise the gel band and elute the cDNA from the gel piece as described above. Dissolve the cDNA in 10 μl of water.

Polymerase Chain Reaction Amplification of cDNA with Unlabeled Primers for Gel Analysis and Sequencing. At this stage we need to amplify a sufficient amount of cDNA either for direct sequencing or for cloning of the PCR product.

1. Take 5 μl of the gel-eluted cDNA from the preceding step and set up a 50-μl PCR as follows.

GeneAmp 10× PCR buffer	5 μl
dNTPs, 2 m*M*	5 μl
Primer FA-1 (25 ng/μl)	5 μl
3′-Anchored primer *or* heel primer (2 μ*M* stock)	5 μl
AmpliTaq Gold (5 U/μl)	1.5 units/reaction
Water	to 50 μl

Always add the enzyme last after adding the required amount of water and gentle mixing.

2. Overlay the sample with mineral oil and start the PCR.

Polymerase Chain Reaction. The PCR conditions are exactly as described under Purification of Gel-Eluted cDNA, above.

At the end of the PCR, the samples can be processed in any of the following ways for sequencing as described.

Sometimes a given cDNA band recovered from the display gel may not be amplified, probably because of its size or GC content. Under such conditions these cDNA bands can be amplified by a modified method developed in our laboratory.[21]

Take 5 μl of gel-eluted cDNA and 5 μl of 2 μM stock solution of specific N_1N_2-anchored 3′-primer (or heel primer) in a 0.5-ml thin-walled PCR tube. Add 40 μl of the reaction cocktail (see below) without dNTPs, overlay with mineral oil, and perform a hot start PCR as described below, by adding 5 μl of 2 mM dNTPs for a 50-μl reaction.

The reaction cocktail contains the following components per tube:

LA-PCR buffer (10×)	5 μl
Betaine, 5 M	10 μl
Bovine serum albumin (7.5 mg/ml)	1 μl
Primer FA-1 (25 ng/μl)	5 μl
Water	18.8 μl
Enzyme LA-16	0.2 μl
	40 μl

LA-PCR buffer (10×): 200 mM Tris-HCl (pH 9.0), 160 mM ammonium sulfate, and 25 mM $MgCl_2$

Betaine (5 M): Dissolve 13.5 g of betaine monohydrate (F.W.135.16) in autoclaved water in a final volume of 20 ml and filter the solution, using a disposable filtration unit

LA-16 enzyme[21]: Mix Klentaq 1 DNA polymerase (25 U/μl) and native *Pfu* DNA polymerase (2.5 U/μl) gently on ice, at a ratio of 15:1 (v/v)

PCR conditions:

Temperature (°C)	Time	No. of cycles
94	1 min	1
80	10 sec	

When the tube temperature reaches 80°, pause the machine at this

[21] N. Baskaran, R. P. Kandpal, A. K. Bhargava, M. W. Glynn, A. Bale, and S. M. Weissman, *Genome Res.* **6,** 638 (1996).

temperature, add 5 μl of 2 m*M* dNTP stock solution to each tube, and continue the PCR.

Temperature (°C)	Time	No. of cycles
94	30 sec	35–40
60	2 min	
68	1 min	

Alternatively, a two-step PCR can also be done as described under Purification of Gel-Eluted cDNA, Polymerase Chain Reaction, above.

This modified PCR step can be used with the ^{32}P-labeled primer FA-1 as a first step after recovering the bands from the display gel and/or at the level of amplification with cold primers for the agarose gel electrophoresis and sequencing.

Processing of Polymerase Chain Reaction-Amplified Samples for Sequencing

Direct Sequencing of Polymerase Chain Reaction Products. Analyze 10 μl of the 50-μl PCR product on a 2% agarose gel. This is to ensure that there are no coamplified contaminating bands and also to make sure that the PCR product is of the right size. If the product is pure enough and of the expected size, the sample can be processed for sequencing by following method I.[21a]

METHOD I

1. Place 0.5-ml microcentrifuge tubes (nonstick) on ice and add shrimp alkaline phosphatase (1 U/μl) and exonuclease I (10 U/μl) at a ratio of 10:1 (v/v).
2. Place 5 μl of the PCR-amplified product in the 0.5-ml tubes. Add 2 μl of the preceding enzyme mixture and incubate the tubes at 37° for 15 min. Following incubation, heat inactivate the enzymes at 80° for 15 min. Dilute the samples to approximately 10 ng of DNA for every 100 bp, in a final volume of 15 μl with water and subject the samples to automated sequencing using the 5′-adapter oligonucleotide (FA-1) as the primer.

If the sample presents some contaminating bands along with the correctly sized band, then the specific DNA band can be excised from the gel and sequenced by following method II.

METHOD II

1. Analyze the PCR products by electrophoresis on a 2% agarose gel. Stain the gels with ethidium bromide to visualize the DNA.

[21a] E. Werle, C. Schneider, M. Renner, M. Volker, and W. Fiehn, *Nucleic Acids Res.* **22,** 4354 (1994).

2. Excise the specific band and extract the DNA from the agarose gel by using a Qiagen gel extraction kit.

3. The eluted DNA can either be used directly for sequencing with the 5′-adapter oligonucleotide (FA-1) as the primer, or can be used for cloning into a plasmid vector.

Cloning of Gel-Purified DNA into Plasmid Vector. For cloning the PCR products into plasmid vectors, we have used two different cloning systems and both work well.

pCR-Script Cloning Kit from Stratagene. This method requires that the ends of the PCR product be polished using *Pfu* DNA polymerase. The agarose gel-purified DNA can be polished according to the supplier protocol. We generally polish the DNA immediately following the PCR by adding the *Pfu* DNA polymerase to the reaction and continuing the incubation at 72° for 30 min.

Following the PCR, take 25 μl of the sample into a fresh PCR tube. Add 1 μl (0.5 units/μl) of *Pfu* DNA polymerase, overlay with mineral oil, and continue the incubation at 72° for 30 min. At the end, analyze the PCR product on a 2% agarose gel and recover the DNA from the specific band as described above, using a Qiagen gel extraction kit. The eluted DNA is then ethanol precipitated in the presence of 0.1 vol of 3 *M* sodium acetate. Recover the precipitate by centrifugation at top speed in an Eppendorf centrifuge, wash the pellet once with 70% ethanol, and air dry the pellet by leaving the tube at room temperature. Dissolve the DNA in 10 μl of water. Use about 2 to 3 μl of the DNA sample and follow all of the other steps for cloning exactly as suggested by the supplier.

pGEM-T Easy (Promega) Vector System. There is no need to polish the ends of the PCR products. Analyze the PCR product directly on a 2% agarose gel and excise the specific DNA band from the gel. Extract the DNA from agarose, using a Qiagen gel extraction kit, and precipitate with ethanol. Collect the precipitate by centrifugation, wash once with 70% ethanol, and air dry the pellet. Dissolve the pellet in 10 μl of water and use it for cloning, following the supplier protocol.

Data Documentation

The autoradiograms can be scanned for analysis by using a UMAX-Mirage D-16L X-ray film scanner (Fremont, CA) connected to a personal computer. Quantitative information on the relative band intensities from the dried sequencing gel can be obtained by phosphor-image analysis using a Molecular Dynamics (Sunnyvale, CA) PhosphorImager with Imagequant software.

Technical Comments

The quality of RNA is important for cDNA synthesis. Several methods and kits are available to isolate good-quality, intact RNA with ease and reproducibility. The choice of RNA isolation method depends on the source of the starting material. We use Trizol reagent (GIBCO-BRL) for isolation of RNA from several cell lines. However, for neutrophil RNA we have used the method described by Strohman *et al.*[17] with some modifications (see Materials and Methods). The RNA preparations sometimes may contain trace amounts of genomic DNA. This may lead to spurious bands on the display gels owing to the priming from the poly(A) stretches. It is desirable to make sure that the RNA preparation is free of this genomic DNA. One way to do this is by incubating the RNA preparation with RNase-free deoxyribonuclease I followed by phenol extraction and ethanol precipitation. It is important to ensure the quality of the total RNA by analyzing an aliquot of the preparation on a formaldehyde–agarose gel[18] and quantitating the 28S and 18S ribosomal RNA bands.

In the original procedure described by Prashar and Weissman,[8] the RNA was primed for first-strand cDNA synthesis, using an anchored oligo(dT) primer, which has a heel sequence. The anchor used was a dinucleotide (N_1N_2),[22,23] resulting in 12 different possible primers. This requires 12 individual cDNA preparations from every RNA sample by priming with each of the 12 primers. Therefore a substantial amount of RNA is required. Sometimes there could be limitations in obtaining sufficient amounts of RNA from every experimental condition or from the clinical sample we would want to analyze. We were able to overcome this limitation by introducing the modifications described in Materials and Methods. The RNA is primed with oligo$(dT)_{18}$V (where V is A or C or G), so that we are using a mixture of A/C/G anchors at the time of making the cDNA. Therefore we need to make only one cDNA preparation for every RNA sample. The first-strand synthesis reaction temperature was lowered to 45° as we are using only a single-nucleotide anchor. The cDNA made by this approach is digested with a restriction enzyme, ligated with the fly adapter as described, and subjected to PCR amplification. At the PCR stage we use each of the N_1N_2-anchored primers to set up 12 individual PCRs. We used 6 μg of total RNA to make cDNA although it is possible to use much smaller amounts of RNA. The cDNA made (from 6 μg of RNA) is sufficient for generating the display patterns for about 10 restriction enzymes using all of the anchored primers. Apart from being able to use small amounts of

[22] P. Liang and A. B. Pardee, *Science* **257,** 967 (1992).
[23] P. Liang and A. B. Pardee, *Curr. Opin. Immunol.* **7,** 274 (1995).

RNA for the analysis, this approach conveniently avoids the need to make a large number of cDNA preparations and associated tube-to-tube variations that can occur during pre- and post-cDNA synthesis steps.

It is possible to use small amounts of RNA for cDNA synthesis and subsequent display analysis (see [3] in this volume[24]). First-strand synthesis can also be done at higher temperatures with Superscript II reverse transcriptase, by using thermostabilizing compounds such as D-(+)-trehalose[25] in order to increase the specificity of the interaction with the anchored oligo(dT). This may also help in avoiding nonspecific priming at the internal poly(A) stretches.

The FA-2 adapter was phosphorylated in the original procedure[8] to ensure that both FA-1 and FA-2 are ligated to the 5′-overhang of the 3′-end restriction fragments of the cDNA. This step is not required and in our procedure, at the end of ligation, only the FA-1 becomes ligated to the cDNA (Fig. 1). This works equally well. By altering the overhang bases of the FA-2 we could design the fly adapters to be suitable for other restriction enzymes of interest. Some of the enzymes, such as *Pst*I, result in a 3′-overhang and in that case the FA-1 carries the corresponding four-base overhang instead of FA-2. In either case we could use ^{32}P-labeled FA-1 for the PCR amplification. There are some restriction enzymes that are not heat inactivated subsequent to digestion, therefore it is desirable to design the overhang on FA-2 such that once the fly adapter is ligated to the cDNA it does not regenerate the enzyme site.

During PCR amplification of the fly adapter-ligated cDNA, the first five cycles of low annealing temperature help the anchored N_1N_2 primers to anneal to the template efficiently. Following this we increased the annealing temperature to help maintain the specificity. We prefer to use the T7-Sal-Oligo(dT)$_{18}$V primer for cDNA synthesis. This is because, if the anchored primers with heel sequence are used, there will be no annealing of the rest of the primer to the template, which might enhance the mispriming. We used AmpliTaq Gold for all our display analyses. This enzyme resulted in clean patterns as compared with the regular AmpliTaq enzyme. We have also used Platinum *Taq* polymerase from GIBCO-BRL. We found that while this enzyme requires lesser amounts of template it gives a better amplification of the high molecular weight cDNA bands. Hence this enzyme is also a good choice for display analysis with some optimization.

[24] M. Liu, Y. V. B. K. Subrahmanyam, and N. Baskaran, *Methods Enzymol.* **303,** [3], 1999 (this volume).

[25] P. Carninci, Y. Nishiyama, A. Westover, M. Itoh, S. Nagaoka, N. Sasaki, Y. Okazaki, M. Muramatsu, and Y. Hayashizaki, *Proc. Natl. Acad. Sci. U.S.A.* **95,** 520 (1998).

Either of the enzymes helps in setting up the hot start PCR with ease and convenience.

It is important to optimize the amount of cDNA template used for PCR amplification. This is because too much template results in smeary patterns, whereas too little template results in faint and/or missing bands. Generally, when we start with about 6 μg of total RNA (we consistently attain about 30% efficiency of cDNA synthesis), 2 μl of 1:20 diluted template works well for a 20-μl PCR. It is desirable to try different dilutions and select the optimal dilution and quantity of the template for the PCR.

We observed that under identical conditions of PCR, there is some variation in the incorporation of radioactivity with respect to the anchored primers used. While the PCR conditions described above work well for all of the anchored primers, it is possible to employ higher annealing temperature or reduce the number of cycles with some of the anchored primers (as is the case with GG, GC, CG, and CC). The annealing temperature could be lowered or the number of cycles could be increased for GT-, AT-, CT-, and GA-anchored primers.

Direct sequencing of the purified PCR product results in good-quality sequence. However, a given PCR product may occasionally be a mixture of more than one species of cDNA of the same size, and hence may yield an unreadable sequence. Under such conditions the cDNA products need to be cloned. While this approach results in good-quality sequence, it is important to sequence the DNA from a significant number of colonies. This ensures that a given sequence is represented in a statistically significant number of colonies in order to assign its specific position on the display gel. If the sequence obtained corresponds to the one represented in the database, then the sequence can be analyzed for the N_1N_2 and the restriction enzyme site that were used in generating the display gel. We can predict and confirm the exact position on the gel; this is quite helpful. In general we prefer to do sequencing of the PCR products. However, if a given product results in an unreadable sequence, we clone the product for sequencing.

Sometimes a specific cDNA band recovered from the display gel does not become amplified in the subsequent PCR steps. This could be due to a variety of reasons including the size and GC content. As the original cDNA template is a complex mixture, PCR amplification of such a mixture behaves differently as compared with that of an isolated cDNA band. Occasionally we do experience problems in amplifying larger size bands. Sometimes a given band does not become amplified even though the size is small. This could possibly be due to the GC content of the cDNA band. Under these conditions the modified PCR method using betaine[21] works quite well.

Results and Discussion

As seen in Fig. 2, in response to a biologically relevant stimulus such as interaction with *E. coli,* neutrophils exhibit distinct changes in the pattern of gene expression. A part of the gel presented in Fig. 2 is enlarged to show these changes clearly (Fig. 3). The mechanistic aspects of these changes were addressed by using an inhibitor of transcription, such as DRB. This approach provides a tool to distinguish between new transcription versus mRNA stabilization. It should be possible to survey the entire cell for all possible changes in the message levels which are increased or decreased owing to stabilization/destabilization, and which of them are due to new transcription. We are presently analyzing neutrophils in this manner, and are investigating the biological relevance of these changes.[26]

Overall the cDNA displays obtained by the present protocol have been readily reproducible, even when the same sample of the cDNA is analyzed at different times. Differences of 1.5-fold in the relative intensities of the bands can be detected visually for the darker bands. Estimates of the relative amounts of material in different bands by use of PhosphorImager quantitation indicate that the amount of radioactive cDNA in perceptible bands can vary by nearly four orders of magnitude. The most abundant mRNAs observed in the present study are unlikely to represent more than 1–2% of the total cellular mRNA. These factors suggest that the approach is able to detect bands corresponding to mRNAs present at less than 1 part per 100,000 of total mRNA, corresponding to approximately 1 mRNA molecule per cell for a typical cell.

The measurement of absolute amount of individual mRNAs by this approach is certainly inaccurate, owing to the contribution of a number of factors. "Normalization" effects can occur with the most abundant cDNAs, in which self-annealing competes with primer annealing to cDNA strands, decreasing the amplification probability per cycle of PCR.[27] Longer cDNAs are amplified by PCR less efficiently than shorter cDNAs. Amplification of cDNAs of similar length may vary because of base composition or other undefined factors. These considerations are of lesser importance when examining shifts in the ratio of cDNA bands as a function of the changing physiological status of the cell.

The present method is relatively efficient in terms of sequencing effort. Nevertheless the same mRNA can give rise to several bands that increase the total amount of sequencing needed to understand a display pattern.

[26] Y. V. B. K. Subrahmanyam, Y. Prashar, N. P. Hoe, C. Whitney, J. D. Goguen, P. E. Newburger, and S. M. Weissman, in preparation (1999).

[27] F. Mathieu-Daude, J. Welsh, T. Vogt, and M. McClelland, *Nucleic Acids Res.* **24,** 2080 (1996).

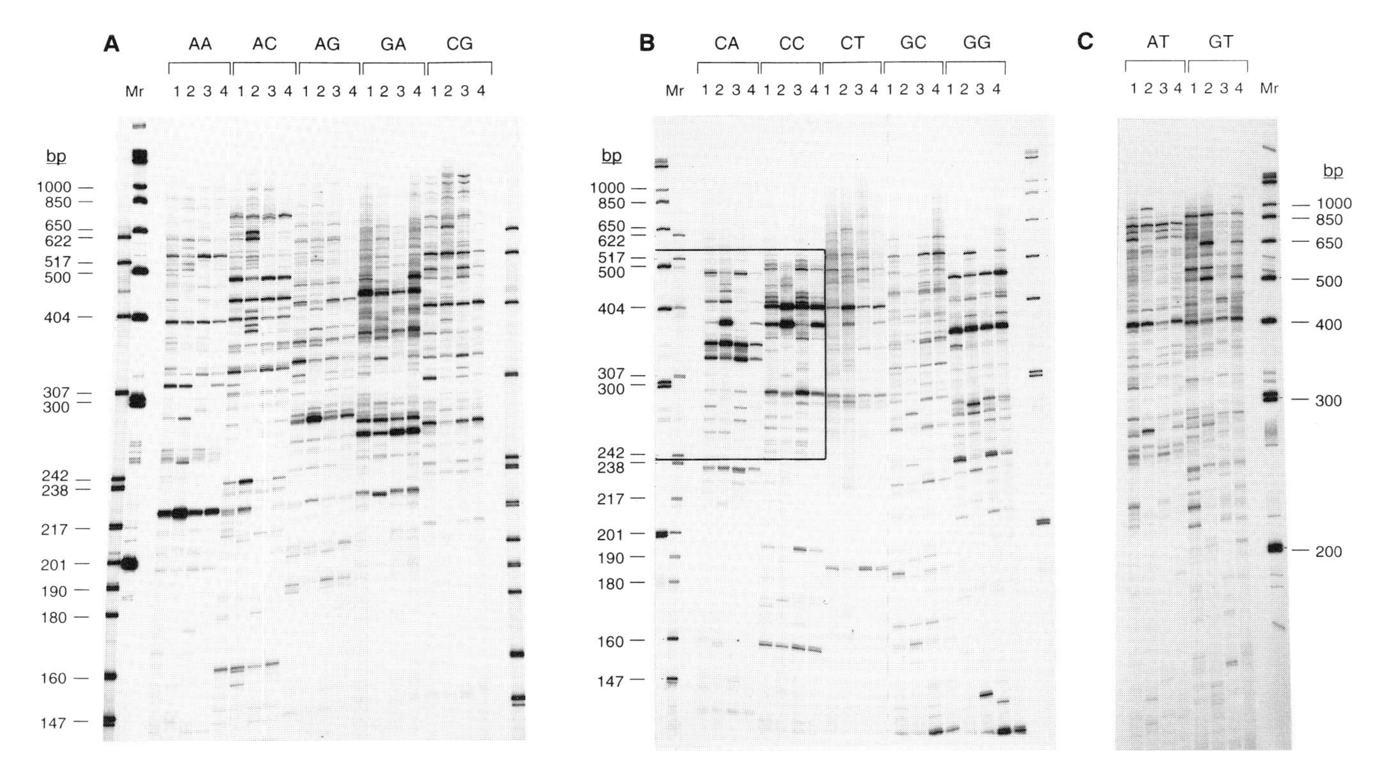

FIG. 2. Display analysis of differential mRNA expression in neutrophils. Total RNA isolated from neutrophils was primed with oligo$(dT)_{18}V$ to synthesize cDNA. This cDNA was digested with *Bgl*II, ligated to fly-adapter, and subjected to differential display using the anchored primers as described in text. (A–C) Three different gel runs under identical conditions covering all the anchored primers. The PCR products were purified using a PCR purification kit and an equal amount of radioactivity was loaded in each lane. Lane 1, Untreated control neutrophils; lane 2, neutrophils incubated with opsonized *E. coli;* lane 3, neutrophils incubated with DRB; lane 4, neutrophils incubated with opsonized *E. coli* and DRB. Specific N_1N_2 anchors are indicated at the top of the lanes. Two different molecular weight markers (1-kb Plus and pBR322 digested with *Msp*I) were used as size standards.

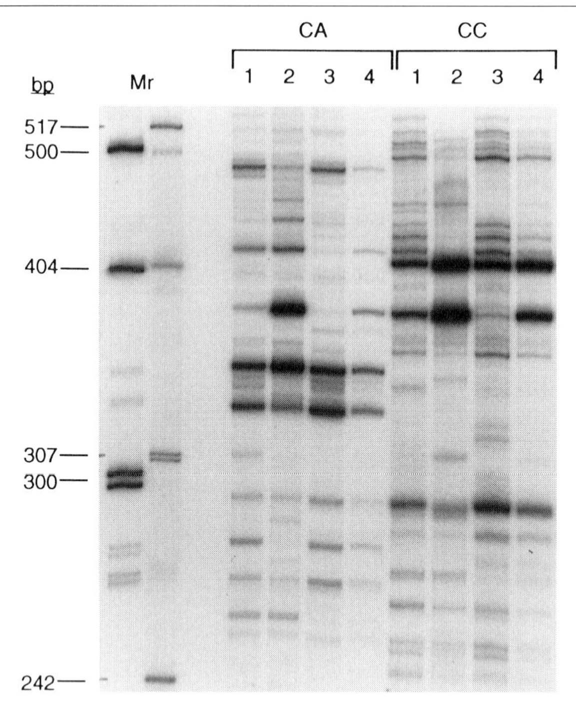

FIG. 3. Magnified region of the display gel shown as inset in Fig. 2B. Lanes 1–4 are as described in Fig. 2.

Redundant bands can occur if several enzymes cut the same cDNA, although in principle this could be anticipated once the first sequence was obtained. Redundant bands could also arise either because of the use of alternate polyadenylation signals or because of inaccuracy at the site of addition of poly(A) to the mRNA. We have seen multiple cases attributable to these causes. In addition, slippage can occur during reverse transcriptase priming, producing poly(A) that is either copied by an oligo(dT) primer with a mismatch at one of the two 3′-terminal bases, or because of looping out of an internal mismatch base. These phenomena are seen particularly with the most abundant mRNA species, and presumably are responsible for a significant part of the background in display gels.

In any display method that depends on oligo(dT) priming of cDNA from total cellular RNA rather than cytoplasm RNA, there is a considerable risk that some of the products will be generated from intranuclear RNA species, including incompletely spliced products and even intergenic transcripts. In particular, priming might occur at poly(A) tracts that are often associated with interspersed repetitive sequences such as Alu elements.[28,29] The finding of a matching sequence in the EST databases does not give absolute assurance that the sequence is not derived from one of the above artifacts, owing to the common features of the cDNA preparation methods.

Repetitive sequences represent a particular problem in interpretation of display gel bands. Undoubtedly a certain fraction of these sequences do represent 3′ ends of the legitimate mRNAs. However, a substantial fraction of the transcript fragments that contain only repetitive sequences lack a definitive polyadenylation signal appropriately placed relative to the apparent polyadenylation site. Sometimes the repetitive sequence is identical over hundreds of bases with an EST sequence in the database, making it probable, although not certain, that the cDNA clone is available and could be used to obtain unique sequences for the mRNA. In some cases this can be a difficult call. For example, we have found several occurrences of a band sequence that differs only by one or two nucleotides from the sequence of an Alu element apparently in an intron of the database NF-κB mRNA precursor. This could be an incompletely spliced or even a prematurely terminated form of NF-κB mRNA. The repetitive sequence bands that do not exactly match database entries represent a particular problem. These sequences cannot be used to probe cDNA libraries because of their repetitiveness and their evaluation would require either relatively intense work or more genomic sequence.

Even without considering repetitive sequence bands, about 15% of the bands we obtained did not match any database entries. Some of these lacked classic polyadenylation signals. This does not mean that these sequences do not correspond to mRNAs, but does signal the need for caution. However, some of these sequences had adequate or perfect polyadenylation signal sequences at appropriate positions. This suggests that, even for a cell such as neutrophil, which may be present at some level in many tissues, the databases are not entirely complete. The cDNA display methods will remain useful in proportion to the frequency of such uncatalogued transcripts, even if array methods for oligonucleotide hybridization emerge that are able to conveniently detect all known ESTs from a species.

Acknowledgments

The authors gratefully acknowledge Dr. Jon Goguen and Dr. Nancy Hoe for their help and suggestions on experiments involving the use of bacteria. We also thank Constance Whitney and Padmakumari Yerramilli for technical help and Dr. A. Raghunathan for critical reading of the manuscript.

[28] C. W. Schmid, *Prog. Nucleic Acid Res. Mol. Biol.* **53,** 283 (1996).

[29] A. M. Weiner, P. L. Deininger, and A. Efstratiadis, *Annu. Rev. Biochem.* **55,** 631 (1986).

[17] Fluorescent Differential Display: A Fast and Reliable Method for Message Display Polymerase Chain Reaction

By TAKASHI ITO and YOSHIYUKI SAKAKI

Introduction

Identification of genes expressed in a spatially and/or temporarily specific manner has been among the major interests of many molecular biologists. While various methods can be used for the purpose, they can be classified into three categories based on their principles.[1] The first is the most traditional hybridization-based approach, including differential screening of cDNA libraries and the subtractive hybridization-based methods. The second category is based on a novel concept of sampled cDNA sequencing, which is realized by the development of automated fluorescence DNA sequencers ensuring high-throughput DNA sequencing. This is a straightforward approach, in which the frequency of each cDNA species in a 3′-directed library is counted by random tag sequencing as a faithful measure of the abundance of the corresponding mRNAs.[2] For further acceleration of such sampled cDNA sequencing, more complex methods termed SAGE (serial analysis of gene expression) have been developed by using the concatemerization of short tags generated with the class IIS restriction enzymes.[3]

The third category includes the so-called message display technologies. It was first introduced as "differential display" (DD) by Liang and Pardee.[4] Its basic concept is to sample a defined subset of transcripts as 3′-end cDNA fragments by means of polymerase chain reaction (PCR), and display the selected cDNAs as a gel fingerprint.[4] In brief, the first-strand cDNA is synthesized using a 3′-anchored oligo(dT) primer to reduce the complexity of the cDNA and to fix the priming site to the beginning of the poly(A) tail. The second-strand synthesis and subsequent amplification of 3′-end portions of further selected cDNAs are accomplished by PCR with the anchor and arbitrary primers. The amplified products are then resolved on a high-resolution polyacrylamide gel to obtain a fingerprint pattern. Then, simple comparison among the fingerprints from different samples allows

[1] T. Ito and Y. Sakaki, *Essays Biochem.* **31,** 11 (1996).

[2] K. Okubo, N. Hori, R. Matoba, T. Niiyama, A. Fukushima, Y. Kojima, and K. Matsubara, *Nature Genet.* **2,** 173 (1992).

[3] V. E. Velculescu, L. Zhang, B. Vogelstein, and K. W. Kinzler, *Science* **270,** 484 (1995).

[4] P. Liang and A. B. Pardee, *Science* **257,** 967 (1992).

Copyright © 1999 by Academic Press
All rights of reproduction in any form reserved.

0076-6879/99 $30.00

one to readily identify the differentially displayed bands that are assumed to be derived from the differentially expressed transcripts. Since its first introduction in 1992, many modifications and variants have been reported, and DD and its relatives have been extensively used in various fields of modern biology in which the identification of differentially expressed genes is of particular interest and importance.[5] The beauty of these message display PCR techniques is that they can compare a number of samples in parallel to simultaneously reveal transcripts of various behaviors, including those of low abundance, from a limited amount of starting RNAs.

On the other hand, DD has some drawbacks. The use of arbitrary primers in DD inevitably increases the number of primer combinations to be tested for statistical coverage of complex transcript populations. For instance, a simple calculation based on the assumption of random sampling of 100 independent cDNAs by each DD reaction tells us that one must test ~450 reactions for 95% coverage of 15,000 transcripts assumed to be expressed in a typical mammalian cell.[6] To run this scale of analysis using multiple specimens, one must increase the speed of each analysis, in particular the time-consuming and tedious steps involved in postrun gel processing as well as those for signal detection. The safety problem inherent to radioactive DD would not be negligible in such cases.[7] Also, the original DD protocol was reported to be plagued with a high incidence of false signals.[8]

To overcome these drawbacks, we established an improved DD protocol using novel primer design and reaction conditions to ensure highly reproducible and reliable DD reactions.[6,9–11] Furthermore, these protocols were optimized so as to be compatible with various fluorescent detection systems. They thus allow one to perform fluorescent DD (FDD), which has higher throughput and much improved operational safety than conventional radioactive DD and nonradioactive, but rather time-consuming, DD. We developed two protocols, termed protocols S and L. The protocol S uses short arbitrary 10-mer and modified anchor primer $GT_{15}N$ (N = A, C, or G), which has much improved priming efficiency compared with the original anchor primers ($T_{12}MN$, where M = A, C, or G; N = A, C, G, or T) and hence improved the reproducibility of the reactions. In protocol L, longer anchor primers ($CCCGGATCCT_{15}N$) and usually oligonucleotides of ~20-

[5] M. McClelland, F. Mathieu-Daudé, and J. Welsh, *Trends Genet.* **11,** 242 (1995).

[6] T. Ito and Y. Sakaki, *Methods Mol. Genet.* **8,** 229 (1996).

[7] S. M. Trentmann, E. van der Knaap, and H. Kende, *Science* **267,** 1186 (1995).

[8] C. Debouck, *Curr. Opin. Biotechnol.* **6,** 597 (1995).

[9] T. Ito, K. Kito, N. Adati, Y. Mitsui, H. Hagiwara, and Y. Sakaki, *FEBS Lett.* **351,** 231 (1994).

[10] T. Ito and Y. Sakaki, *in* "Fingerprinting Methods Based on Arbitrarily Primed PCR" (M. R. Micheli and R. Bova, eds.), p. 305. Springer-Verlag, New York, 1997.

[11] T. Ito and Y. Sakaki, *Methods Mol. Biol.* **85,** 37 (1997).

mers (originally synthesized for another purpose) are recruited as the arbitrary primer. In this protocol, the initial PCR cycle is performed at low stringency to encourage second-strand cDNA synthesis on multiple targets even with the use of long upstream primers, followed by stringent PCR in the subsequent cycles to ensure efficient and reproducible amplification.

For fluorescence detection, we used fluorescently labeled anchor primers in the reaction, and detect the signals by DNA sequencer or fluorescence image analyzer. The use of a DNA sequencer allows one to perform FDD analysis in a semiautomated manner. The use of half-length gel plates allows the performance of 4 sample loadings a day, thereby enabling the recording of more than ~10,000 cDNA bands with minimum operator interaction. While the sequencer-based FDD system is essentially for use by a single user, the image analyzer-based system is, albeit off-line, suitable for multiple users, because a conventional slab electrophoretic apparatus can be used for gel running, thus occupying the machine only for 5–10 min for the scanning.

Described below are our protocols for FDD and the postdisplay steps,[6,9–11] careful use of which has brought successful identification of differentially expressed or polymorphic messages from various biological sources in our laboratory.[12–15]

Basic Fluorescent Differential Display

RNA Isolation

Isolate total RNAs from the samples to be compared. We routinely use TRIzol reagent (BRL, Gaithersburg, MD) according to the instructions from the supplier. The RNA samples should be treated with RNase-free DNase (Promega, Madison, WI) to remove contaminating genomic DNAs. Examine the quantity and quality of each RNA sample by spectrophotometry as well as by agarose gel electrophoresis. We routinely adjust the concentration of working RNA solution to 2.5 $\mu g/\mu l$ in DEPC (diethyl pyrocarbonate)-treated water and store it at $-80°$ until use.

[12] N. Adati, T. Ito, C. Koga, K. Kito, Y. Sakaki, and K. Shiokawa, *Biochim. Biophys. Acta* **1262,** 43 (1995).

[13] K. Kito, T. Ito, and Y. Sakaki, *Gene* **184,** 73 (1997).

[14] C. Furihata, M. Oka, M. Yamamoto, T. Ito, M. Ichinose, K. Miki, M. Tatematsu, Y. Sakaki, and K. Reske, *Cancer Res.* **57,** 1416 (1997).

[15] Y. Hagiwara, M. Hirai, K. Nishiyama, I. Kanazawa, T. Ueda, Y. Sakaki, and T. Ito, *Proc. Natl. Acad. Sci. U.S.A.* **94,** 9249 (1997).

Reverse Transcription

1. Preheat a thermal cycler or a programmable heat block to 70°.
2. Mix the following reagents in 0.5-ml tubes:

RNase-free water	8.0 μl
Anchor primer (50 μ*M*)	1.0 μl
Total RNA (2.5 μg/μl)	1.0 μl

Use FITC (fluorescein isothiocyanate)–$GT_{15}N$ and FITC–$CCCGGATCCT_{15}N$ for protocol S and L, respectively. Place the tubes in the thermocycler heated to 70° for 5–10 min.

3. During the heat denaturation, assemble enough 2× reverse transcription (RT) solution by mixing the following components from the BRL preamplification kit:

DEPC-treated water	2.0 (× *N*) μl
PCR buffer, 10×	2.0 (× *N*) μl
$MgCl_2$, 25 m*M*	2.0 (× *N*) μl
Dithiothreitol (DTT), 0.1 *M*	2.0 (× *N*) μl
dNTP, 10 m*M*	1.0 (× *N*) μl
SuperScript II (200 U)	1.0 (× *N*) μl

The mixture should be mixed thoroughly by gentle pipetting and placed on ice.

4. Quench the heated tubes containing the RNA and anchor primer by soaking in an ice–water bath. Centrifuge the tubes briefly. Add 10 μl of 2× RT solution to each tube and mix gently but thoroughly.
5. Place the tubes in the thermal cycler cooled to 25°. Start the following thermal program for reverse transcription:

25° for 10 min
42° for 50 min
70° for 15 min

6. Following the reverse transcription, centrifuge the tubes briefly. Add 80 μl of TE [10 m*M* Tris-HCl (pH 8.0), 1 m*M* EDTA] to each tube. Mix thoroughly and store the cDNA solution at −20° until use.

Polymerase Chain Reaction

1. Prepare PCR mix I from the following components:

Distilled H_2O (dH_2O)	13.0 (× *N*) μl
10× GeneTaq universal buffer	2.0 (× *N*) μl
dNTP, 2.5 m*M*	1.6 (× *N*) μl
Anchor primer (50 μ*M*)	0.2 (× *N*) μl
cDNA solution	2.0 (× *N*) μl

Note that in determining the value of N, the loss incurred during repeated pipetting in the following steps must be taken into account.

2. Dispense 1.0 μl of arbitrary primer solution (10 μM) into each well of a 96-well thermal plate (HI-TEMP 96; Techne, Cambridge, UK), or into each PCR tube. As arbitrary primers, 10-mers and (typically) ~20-mers are used in protocols S and L, respectively.

3. Make PCR mix II by mixing the following components:

PCR mix I	18.8 ($\times N$) μl
GeneTaq DNA polymerase (Nippon Gene, Toyama, Japan)	0.1 ($\times N$) μl
Taq DNA polymerase (BRL) or AmpliTaq DNA polymerase (Perkin-Elmer, Norwalk, CT)	0.1 ($\times N$) μl

Be sure to mix the components thoroughly by gently inverting the tube.

4. Dispense 19.0 μl of PCR mix II to each well containing 1.0 μl of arbitrary primer solution, and overlay a drop of mineral oil.

5. Place the 96-well plate in the thermal cycler (Techne PHC-3), and subject it to the following thermal cycling.

For protocol S, which uses FITC–$GT_{15}N$ (N = A, C, or G) and arbitrary 10-mers as anchor and upstream primers, respectively:

94° × 3 min + 40° × 5 min + 72° × 5 min	1 cycle
95° × 15 sec + 40° × 2 min + 72° × 1 min	~25 cycles
72° × 5 min	1 cycle

For protocol L, which uses FITC–$CCCGGATCCT_{15}N$ (N = A, C, or G) and (18- to 24-mers) as anchor and upstream primers, respectively:

94° × 3 min + 37° × 5 min + 72° × 5 min	1 cycle
95° × 15 sec + 55° × 1 min + 72° × 1 min	~25 cycles
72° × 5 min	1 cycle

Gel Electrophoresis

1. Prepare a 6% polyacrylamide–8 M urea gel (200 × 330 × 0.35 mm) in 1 × TBE (Tris–borate–EDTA) buffer. Prerun the gel for ~1 hr at 1000 V.

2. Mix 4 μl of PCR products with 4 μl of formamide including appropriate dyes, and heat at 90° for 3 min followed by rapid chilling in ice–water.

3. Load 6 μl of the sample into each well formed with a shark-tooth comb. Run the gel at 1500 V for ~1 hr until bromphenol blue migrates to the bottom of the gel.

4. Remove the gel slab from the electrophoretic chamber. Clean the glass plates, and scan the gel slab (without removing the glass plates) with a FluorImager 595 (Molecular Dynamics, Sunnyvale, CA) using the high-

sensitivity mode. This first scan will provide the fingerprint image, focusing mainly on low molecular weight species.

5. Place the gel slab back in the electrophoretic chamber, and run the gel again for an additional hour until xylene cyanol FF reaches to the buttom. Scan the gel again to visualize high molecular weight bands. A typical FDD pattern recorded by Fluorlmager is shown in Fig. 1.

Cloning of Bands of Interest

1. Examine the reproducibility of the band of interest using different batches of RNA samples. (If it is impossible to prepare new RNA batches, repeat the experiment from the reverse transcription step to confirm the reproducibility.)

2. Run the gel of appropriate concentration that would maximize the separation of the target band. Remove the upper glass plate, and scan the gel as described above.

3. Print the gel image by its actual size, and place the gel onto the printed image precisely.

4. Carefully excise the band of interest, using a razor blade. Scan the gel again to see how precisely the band was excised.

5. Place one-half of the excised gel piece into a PCR tube. Although boiling or other steps are usually unnecessary, cracking the gel piece may improve reamplification of long fragments. The other half should be stored at $-20°$.

6. Add 100 μl of PCR reaction mix II (see above) with lower primer concentrations (0.25 μM each), and subject to reamplification PCR using the following thermal program.

Protocol S:

$94° \times 3$ min	1 cycle
$94° \times 30$ sec $+ 40° \times 2$ min $+ 72° \times 1$ min	20 cycles
$72° \times 5$ min	1 cycle

For protocol L:

$94° \times 3$ min	1 cycle
$94° \times 30$ sec $+ 55° \times 1$ min $+ 72° \times 1$ min	20 cycles
$72° \times 5$ min	1 cycle

7. Examine a portion of the reamplified product by polyacrylamide gel electrophoresis. If some extra bands are obvious, use the other half of the gel piece for reamplification with reduced thermal cycles.

8. Following ethanol precipitation, clone the fragment into an appropriate cloning vector. Because most of the reamplified products bear additional dA at their 3′ termini, we routinely use pT7 T-vector (Novagen, Madison, WI).

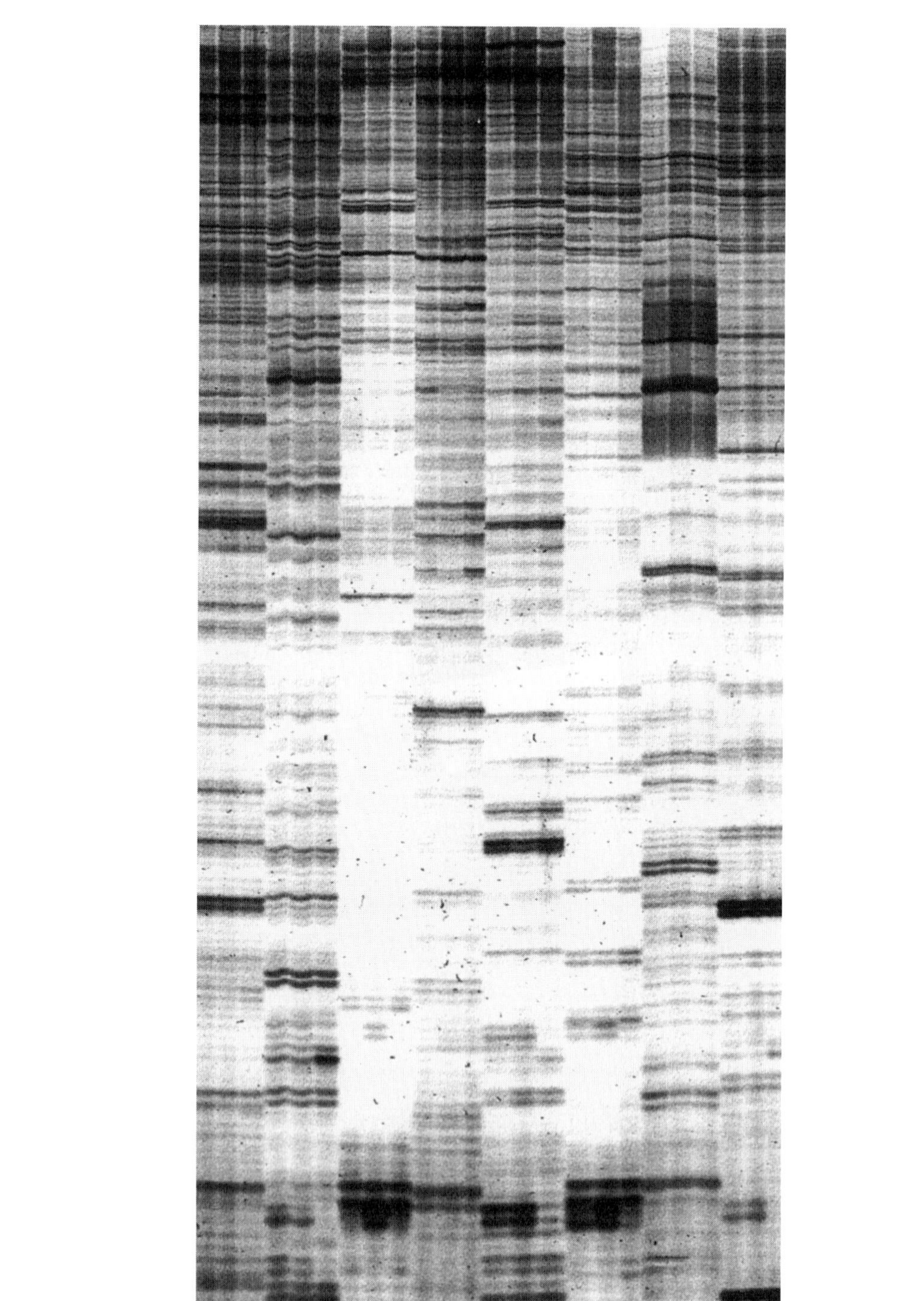

Selection of Correct Clones

1. Suspend each colony in 40 μl of L broth. Use 1.0 μl of suspension for PCR (20 μl), using conditions identical to those for reamplification PCR described above.
2. Run 0.1–1.0 μl of colony PCR product in parallel with the original FDD reaction, using the sequencing gel ("comigration test"). Select clones bearing inserts that comigrate precisely with the target band.
3. Isolate plasmid DNAs from the selected clones, and determine their nucleotide sequences.
4. From the obtained sequence data of each candidate clone, search for appropriate restriction enzyme sites that can discriminate one candidate from the others. Using these enzymes, digest the amplified insert of each candidate clone and the original FDD product, and run the digested products with an undigested control in parallel ("restriction test"). Select the clone bearing an insert that shows a digestion pattern identical to that of the target band in the original FDD products, and use it for the confirmation of expression pattern by Northern blot hybridization, RNase protection, or quantitative RT-PCR.

Comments

RNA Isolation

Any standard RNA isolation methods can be used for DD. It is critical to use RNA samples with similar qualities for the comparison. If some samples show weaker band intensity or shorter amplification length, we recommend omitting them from further comparison and preparing the same samples again.

Another important point concerns the elimination of contaminating genomic DNA. At the beginning of a series of experiments, we always include control experiments using mock-reverse transcribed cDNAs and several different primer combinations to confirm that virtually no amplification is observed in the absence of reverse transcription.

FIG. 1. An example of a typical FDD pattern. From left to right in each triad, total RNAs from yeasts bearing a wild type, a null, and a gain-of-function allele for the *PDR1* gene (generous gift from A. Goffeau) were compared using FDD protocol S and the FluorImager (Molecular Dynamics). Different *Mbo*I primers are used in individual triads. Several differentially displayed bands were observed.

Reverse Transcription

Anchor primers for protocols S and L are FITC–$GT_{15}N$ and FITC–$CCCGGATCCT_{15}N$, respectively. They must be prepared with special care to avoid contamination by RNase, and dissolved in DEPC-treated water.

For the reverse transcription, we routinely use SuperScript II reverse transcriptase (BRL). However, we have found that other mouse mammary leukemia virus (MMLV) reverse transcriptases give comparable results when used as recommended by the manufacturer.

We prefer to use the thermal cycler for reverse transcription, because then we do not have to prepare many heat blocks or water baths of various temperatures.

Polymerase Chain Reaction

The PCR conditions in our protocol are much more standard than those in the original protocol, which used a much higher primer concentration to compensate for the poor priming efficiency of anchor primers and a low dNTP concentration to maintain specific activity of internally labeled PCR fragments. In our protocol, the modified anchor primers have much improved priming efficiency and can be used at much reduced concentration. Also, the terminal labeling of the anchor primers allows one to use dNTPs at more standard concentrations. We assume that the use of these more standard PCR conditions contributes to the improved reproducibility as well as the reliability of our DD protocol.

For message display PCR, we prefer to use modified *Taq* DNA polymerases with a large N-terminal deletion, because they generally generate much stronger signals than the conventional full-length enzyme, in particular for DNA fragments smaller than 500 bp, the range for DD fragments. We thus use GeneTaq DNA polymerase (Nippon Gene). Another choice would be the AmpliTaq Stoffel fragment (Perkin-Elmer). We also found that the combined use of such a modified enzyme with the conventional full-length enzyme is sometimes quite useful for the enhancement of signals in the higher molecular weight range.

Although arbitrary primers have been used in DD-PCR, we are currently shifting to the use of rationally designed 10-mers to minimize the fraction of transcripts escaping detection by FDD. Based on the observation that more than 90% of mammalian mRNAs seem to carry at least one *Mbo*I site (GATC) within 1 kb of their polyadenylation sites,[16] we prepared

[16] N. B. Ivanova, and A. V. Belyavsky, *Nucleic Acids Res.* **23,** 2954 (1995).

the following 32 primers, termed *Mbo*I primers, that should theoretically cover all possible GATC sites as discussed previously.[11]

GGN XYZ GAT C
N = mixture of A, C, G, and T
X = R (purine) or Y (pyrimidine)
Y, Z = one each of A, C, G, or T

Although these primers have the identical 3′-end sequence GATC, each primer gives a distinct FDD pattern under our experimental conditions. Our data on mismatches between the arbitrary primers and the corresponding transcript showed the lesser importance of positions 1 and 2. Because we observed G–T mismatches even in the 3′ portions, the primers may cover not only all of the GATC sites but other sequences, such as AATC or GACC. The other advantage of these primers is that they enable the rational use of the secondary primer set comprising the following 64 primers for the nested PCR, which is quite effective in revealing transcripts of low abundance or those hidden by thick bands.

GNX YZG ATC W
N, X = mixture of A, C, G, and T
Y, Z, W = one each of A, C, G, or T

Gel Electrophoresis

The two-step scanning procedure on a fluorescence image analyzer described here enables the scanning of a greater size range of cDNA bands than is possible by conventional autoradiographic detection, which involves stopping electrophoresis and drying the gel at some fixed time point for autoradiographic signal detection; typically smaller fragments have already migrated out of the gel whereas the larger fragments are still close to the top of the gel and thus poorly resolved. It should be emphasized that this is another advantage of FDD. The same is the case with FDD using DNA sequencers, in which all of the bands that migrate in front of the detector located at the bottom of the gel will be recorded automatically. The detailed protocol for FDD on DNA sequencers has been described previously.[6,9]

For much simpler FDD analysis, we use a nondenaturing gel system using a discontinuous buffer system [6–7% native polyacrylamide gel in 375 m*M* Tris-HCl (pH 8.9) run in 25 m*M* Tris–192 m*M* glycine] and poststaining with SYBR Green I (1:10,000 dilution) (FMC BioProducts, Rockland, ME).[10,12,14] For this system, 5 μl of PCR products is mixed with 5 μl of 125 m*M* Tris-HCl (pH 6.8)–10% glycerol and applied to the gel without denaturation. Since SYBR Green I can be excited by conventional UV (254 nm), this system dye-staining method would be a simple alternative for FDD, which requires no special equipment but still has considerable resolution. Note that the discontinuous system described here gives much sharper bands than does the conventional continuous TBE buffer system.[17]

We also confirmed that other fluorescent dyes, such as Texas Red and Rhodamine-X, can be used for FDD in our protocol using another fluorescent image analyzer, FMBIO-II (Hitachi Software, South San Francisco, CA).

Cloning of Bands of Interest

Molecular cloning of the target band is the other critical step for successful DD experiments. Although the excised bands are occasionally pure enough to be subjected to direct cycle sequencing using the anchor primer as the sequencing primer, they are often contaminated with neighboring or comigrating bands, which should be eliminated from the true target by molecular cloning or by another mode of electrophoretic separation such as SSCP (single-strand conformation polymorphism).[18] Prior to proceeding to the next step, such as Northern blotting, it is necessary to confirm that the target band, but not the contaminating bands, is cloned. For this purpose, we first screen candidate clones by precise sizing, followed by DNA sequencing and restriction mapping. In addition to our protocol described here, various other approaches would be plausible for the same purpose. For instance, differential screening of candidate clones with cDNA probes generated from the RNA samples to be compared (i.e., reverse Northern hybridization) or DD-PCR products.[19] It is also possible to perform Southern hybridization to a DD gel, using each candidate clone as a probe.[20]

Conclusion

The FDD protocols described above have been used successfully in our laboratory. Even with these improved protocols, one cannot be too careful in performing DD experiments. It is necessary to confirm differentially displayed bands using independent RNA samples as well as the selection of correct clones. We believe, on the basis of our experience, that careful

[17] T. Ito, H. Hohjoh, and Y. Sakaki, *Electrophoresis* **14,** 278 (1993).

[18] F. Mathieu-Daudé, R. Cheng, J. Welsh, and M. McClelland, *Nucleic Acids Res.* **24,** 1504 (1996).

[19] H. Zhang, R. Zhang, and P. Liang, *Nucleic Acids Res.* **24,** 2454 (1996).

[20] K. K. Wong and M. McClelland, *Proc. Natl. Acad. Sci. U.S.A.* **91,** 639 (1994).

use of our FDD protocols and postdisplay procedure is the key to successful DD cloning.

Acknowledgments

We thank Fumihito Miura for providing the data in Fig. 1. This work was supported by grants-in-aid from the Ministry of Education, Science, Sports and Culture, Japan, and "Research for the Future" Program by The Japan Society for the Promotion of Science.

[18] Identification of Differentially Expressed Genes Using RNA Fingerprinting by Arbitrarily Primed Polymerase Chain Reaction

By FRANÇOISE MATHIEU-DAUDÉ, THOMAS TRENKLE, JOHN WELSH, BARBARA JUNG, THOMAS VOGT, and MICHAEL MCCLELLAND

I. Introduction

Genomic fingerprinting by arbitrarily primed polymerase chain reaction (AP-PCR)[1,2] is a sensitive and efficient method for generating a large number of molecular markers. Based on the selective amplification of DNA sequences flanked, by chance, by sequences matching an arbitrarily chosen primer, AP-PCR reveals sequence polymorphisms between different template DNAs. The ability to detect differences between fingerprints of closely related organisms made this approach a useful tool for studying genetic diversity,[3,4] population biology,[5,6] epidemiology,[7] and genetic mapping.[8]

[1] J. Welsh and M. McClelland, *Nucleic Acids Res.* **18,** 7213 (1990).

[2] J. G. K. Williams, A. R. Kubelik, K. J. Livak, J. A. Rafalski, and S. V. Tingey, *Nucleic Acids Res.* **18,** 6531 (1990).

[3] D. Kersulyte, J. P. Woods, E. J. Keath, W. E. Goldman, and D. E. Berg, *J. Bacteriol.* **174,** 7075 (1992).

[4] D. Ralph, M. McClelland, J. Welsh, G. Baranton, and P. Perolat, *J. Bacteriol.* **175,** 973 (1993).

[5] J. Welsh, C. Pretzman, D. Postic, I. Saint Girons, G. Baranton, and M. McClelland, *Int. J. Syst. Bacteriol.* **42,** 370 (1992).

[6] F. Mathieu-Daudé, J. Stevens, J. Welsh, M. Tibayrenc, and M. McClelland, *Mol. Biochem. Parasitol.* **72,** 89 (1995).

[7] A. van Belkum, W. van Leeuwen, J. Kluytmans, and H. Verbrugh, *Infect. Control Hosp. Epidemiol.* **16,** 658 (1995).

[8] S. M. Al Janabi, R. J. Honeycutt, M. McClelland, and B. Sobral, *Genetics* **134,** 1249 (1993).

Copyright © 1999 by Academic Press
All rights of reproduction in any form reserved.

0076-6879/99 $30.00

AP-PCR has also been used to detect somatic genetic alterations in cancer.[9,10]

The more recent application of AP-PCR fingerprinting to RNA, differential display,[11] or RNA-arbitrarily primed PCR (RAP-PCR)[12] has resulted in an interesting tool for the detection of differential gene expression. RNA fingerprinting provides a complex molecular phenotype reflecting changes in the abundances of a sample of transcripts. This approach allows detection of differences between RNA populations in a wide variety of situations, such as different tissues or developmental stages or cells subjected to different treatments.[13–16] RNA fingerprinting has been used in cancer research to detect differences between normal and carcinoma cells in human ovarian and colon cancer,[17,18] or alteration of transcriptional control by ultraviolet light in human melanocytes.[19]

Several methods have been described for the detection of differential gene expression, including subtractive hybridization, differential screening, and those based on cDNA library sequencing. Serial analysis of gene expression (SAGE)[20] is an interesting but laborious variation of this latest approach. Among the differential screening methods, array-based approaches have evolved from spotting multiple clones or cDNAs onto nylon membranes[21,22] to spotting on modified glass microscope slides.[23] A sensitive and potentially powerful variation based on hybridization to small, high-

[9] M. A. Peinado, S. Malkhosyan, A. Velazquez, and M. Perucho, *Proc. Natl. Acad. Sci. U.S.A.* **89,** 10065 (1992).

[10] Y. Ionov, M. A. Peinado, S. Malkhosyan, D. Shibata, and M. Perucho, *Nature (London)* **363,** 558 (1993).

[11] P. Liang and A. Pardee, *Science* **257,** 967 (1992).

[12] J. Welsh, K. Chada, S. S. Dalal, D. Ralph, R. Cheng, and M. McClelland, *Nucleic Acids Res.* **20,** 4965 (1992).

[13] D. Ralph, M. McClelland, and J. Welsh, *Proc. Natl. Acad. Sci. U.S.A.* **90,** 10710 (1993).

[14] K. K. Wong and M. McClelland, *Proc. Natl. Acad. Sci. U.S.A.* **91,** 639 (1994).

[15] M. McClelland, F. Mathieu-Daudé, and J. Welsh, *Trends Genet.* **11,** 242 (1995).

[16] F. Mathieu-Daudé, J. Welsh, C. Davis, and M. McClelland, *Mol. Biochem. Parasitol.* **92,** 15 (1998).

[17] K. K. Wong, S. C. H. Mok, J. Welsh, M. McClelland, S. W. Tsao, and R. S. Berkowitz, *Int. J. Oncology* **3,** 13 (1993).

[18] M. Ricote, P. Geller, and M. Perucho, *Mutation Res.* **374,** 153 (1997).

[19] T. Vogt, J. Welsh, W. Stolz, F. Kullmann, B. Jung, M. Landthaler, and M. McClelland, *Cancer Res.* **57,** 3554 (1997).

[20] V. E. Velculescu, L. Zhang, B. Vogelstein, and K. W. Kinzler, *Science* **270,** 484 (1995).

[21] T. M. Gress, J. D. Hoheisel, G. G. Lennon, G. Zehetner, and H. Lehrach, *Mamm. Genome* **3,** 609 (1992).

[22] N. Zhao, H. Hashida, N. Takahashi, Y. Misumi, and Y. Sakaki, *Gene* **156,** 207 (1995).

[23] M. Schena, D. Shalon, R. W. Davis, and P. O. Brown, *Science* **270,** 467 (1995).

density arrays containing tens of thousands of synthetic oligonucleotides has been described.[24] Nevertheless, RAP-PCR and differential display have the desirable features of simplicity of use, low cost, high sensitivity, and the ability to compare many different RNA samples in parallel. Consequently, RNA fingerprinting is often the tool of choice to investigate differential gene expression and coordinate regulation. Moreover, several new technologies, such as automated sequencing or the use of fluorescent-tagged primers, are adaptable to RNA fingerprinting and we can expect, in the near future, more technological developments to facilitate further the use of this approach.

It is important to consider the limitations of the method. One limitation is the bias toward sampling more abundant transcripts. Also, using RNA fingerprinting to search for a particular differentially expressed gene would be in vain. Rather, the method is more appropriate when many differentially expressed genes are anticipated or when there is no a priori argument allowing one to predict the extent of involvement of differential gene regulation. The difficulty of sampling rarer transcripts or transcripts from particular genes might be overcome by various adaptations of the method toward priming at particular motifs or sequences statistically overrepresented among a given family of genes.[25–29]

Differential display uses an anchored oligo(dT) for reverse transcription, while RAP-PCR uses an arbitrary primer. Most of the points discussed in the present chapter apply to both approaches (see Section V,A,1). However, in this chapter, we focus on the use of RAP-PCR fingerprinting, the method that we use in our laboratory. We describe the method and provide our current protocols for the different steps from initial RNA preparation through sequence analysis and confirmation of differential expression of the transcripts identified.

II. Principles of RNA Arbitrarily Primed Polymerase Chain Reaction

The RAP-PCR protocol was designed by extension of AP-PCR[1] by adding an initial step of arbitrarily primed reverse transcription (RT) for

[24] D. J. Lockhart, H. Dong, M. C. Byrne, M. T. Follettie, M. V. Gallo, M. S. Chee, M. Mittmann, C. Wang, M. Kobayashi, H. Horton, and E. L. Brown, *Nature Biotech.* **14,** 1675 (1996).

[25] B. Stone and W. Wharton, *Nucleic Acids Res.* **22,** 2612 (1994).

[26] T. Yoshikawa, G. Q. Xing, and S. D. Wadleigh, *Biochim. Biophys. Acta* **1264,** 63 (1995).

[27] C. E. Lopez-Nieto and S. K. Nigam, *Nature Biotech.* **14,** 857 (1996).

[28] R. Fislage, M. Berceanu, Y. Humboldt, M. Wendt, and H. Oberender, *Nucleic Acids Res.* **25,** 1830 (1997).

[29] G. Pesole, S. Liuni, G. Grillo, P. Belichard, T. Trenkle, J. Welsh, and M. McClelland, *BioTechniques* **25,** 112 (1998).

first-strand cDNA synthesis. This synthesis, achieved by reverse transcriptase, is initiated from sites in the RNA that best match an arbitrary primer. Second-strand synthesis is achieved by a thermostable polymerase and the same or a second arbitrary primer, at sites where it finds the best match (see Section V,A,2). These two steps result in products flanked by the arbitrary primer (or primers), that serve as templates for PCR amplification. The PCR products are then displayed on a polyacrylamide sequencing-type gel, resulting in a fingerprint specific for both the particular template RNA and the particular primers used. Comparisons of RAP-PCR fingerprints reveal differences in transcript abundances for differentially regulated genes.

Figure 1 shows examples of RAP-PCR fingerprints obtained with arbitrary 10- and 11-mer primers.

III. Equipment and Reagents

- Rotor–stator homogenizer, or Dounce (Wheaton, Millville, NJ) homogenizers for RNA preparation from tissues
- QIAshredder columns (Qiagen, Chatsworth, CA) for homogenization of tissues or concentrated cell lysates
- RNeasy total RNA purification kit (Qiagen)
- DNase I, RNase free (Boehringer Mannheim Biochemicals, Indianapolis, IN)
- RNase inhibitor (Boehringer Mannheim Biochemicals)
- TE buffer: 10 m*M* Tris-HCl, 1 m*M* EDTA (pH 7.5)
- 10× RT buffer: 500 m*M* Tris (pH 8.3), 500 m*M* KCl, 40 m*M* $MgCl_2$
- Dithiothreitol stock solution (100 m*M*)
- Murine leukemia virus (MuLV) reverse transcriptase (Promega, Madison, WI)
- dNTP mix (20 m*M*): Each dNTP 5 m*M*
- Arbitrary primers (25 and 100 μM), 10- to 20-mers [Operon Technologies (Alameda, CA) or Genosys Biotechnologies (The Woodlands, TX)]
- PCR buffer (10×): 100 m*M* Tris-HCl, 100 m*M* KCl (pH 8.3). For use with AmpliTaq DNA polymerase Stoffel fragment; for AmpliTaq DNA polymerase use 100 m*M* Tris-HCl, 500 m*M* KCl (pH 8.3)
- $MgCl_2$ stock solution (100 m*M*)
- *Taq* polymerase: AmpliTaq DNA polymerase and AmpliTaq DNA polymerase Stoffel fragment (Perkin-Elmer-Cetus, Norwalk, CT)
- [α-^{32}P]dCTP (3000 Ci/mmol)
- Thermal cycler (e.g., GeneAmp PCR system 9600; Perkin-Elmer-Cetus)

Multichannel micropipette
TBE buffer (10×): 0.89 *M* Tris–borate, 20 m*M* EDTA (pH 8.3)
Polyacrylamide (4–5%)–urea (43%) gel prepared in 1× TBE buffer
HydroLink MDE gel solution (J.T. Baker, Phillipsburg, NJ)
Ammonium persulfate (10%, w/v) and *N,N,N′,N′*-tetramethylethylenediamine (TEMED) (Sigma, St. Louis, MD)
Sequencing gel electrophoresis apparatus with plates and combs
Denaturing loading buffer: 96% formamide, 0.1% bromophenol blue, 0.1% xylene cyanol FF, 10 m*M* EDTA (or 10 m*M* NaOH for MDE gel)
Power supply and gel dryer with vacuum pump
X-ray films [Eastman Kodak (Rochester, NY) X-Omat or BioMax] and cassettes

IV. Methods and Protocols

A. RNA Preparation

Using a homogenizer or a Dounce, tissues are blended in lysis buffer (RNeasy; Qiagen). If using a Dounce, samples can be homogenized by passing them through a Qiashredder column (Qiagen). Cultured cells are recovered by scraping the petri dishes in the presence of lysis buffer and can be homogenized through Qiashredder columns if the cell concentration is high. Total RNA can be purified by different methods. Guanidinium thiocyanate–cesium chloride centrifugation[30] gives good yield, but it is time consuming and only a few samples can be processed at the same time. We routinely use the RNeasy total RNA purification kit (Qiagen). On average, 7×10^6 cells (confluent growth on a 100-mm diameter petri dish) yield 35 μg of total RNA. RNA is treated with DNase [0.08 U/ml in 20 m*M* Tris, 10 m*M* $MgCl_2$ buffer in the presence of RNase inhibitor (0.32 U/μl)] for 30 min at 37°, and cleaned using the RNeasy kit. This step is important because small amounts of genomic DNA can easily dominate the fingerprints. The quantity is measured by spectrophotometry, and RNA samples are adjusted to approximately 200 ng/μl in TE buffer. They are checked for quality and concentration by agarose gel electrophoresis, and stored at −80°. Three serial dilutions of each RNA sample are then prepared for reverse transcription. RNA samples are checked for DNA contaminants by including a reverse transcriptase-free control in initial RAP-PCR experiments (see Section V,B and protocol below).

[30] J. M. Chirgwin, A. E. Przybyla, R. J. McDonald, and W. J. Rutter, *Biochemistry* **18,** 5294 (1979).

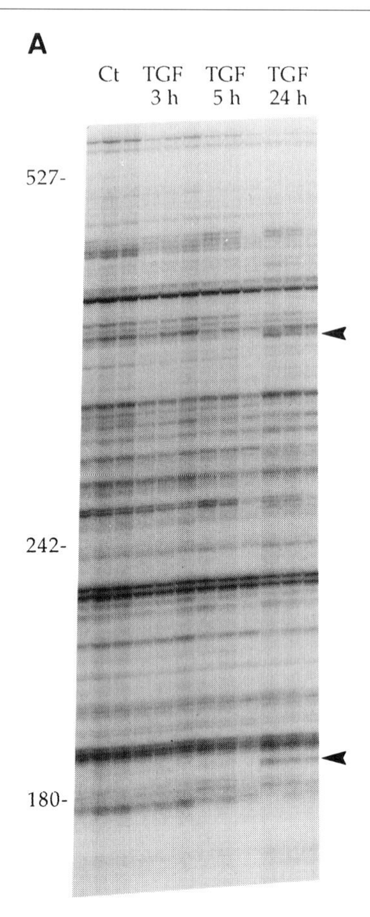

FIG. 1. Autoradiograms of RAP-PCR fingerprints involving two different 10- and 11-mer primer combinations, showing cell responses to different treatments. Arrows indicate differentially amplified products subsequently isolated and sequenced. Molecular sizes (in bp) are shown on the left. (A) TF1 cells showing a late response to TGF-β. Ct, Control; TGF, transforming growth factor β treatment (1 ng/μl of medium) for 3, 5, and 24 hr. Primers are NUC1 (5′-ACGAAGAAGAG-3′) and US9 (5′-GTGACAGACA-3′). The reaction was performed with three different concentrations of the same RNA template, using (from left) 500, 250, and 125 ng of RNA in the RT reaction. (B) PC12 cells showing a secondary response to nerve growth factor (NGF + CX). Ct, Control; CX, cycloheximide; NGF, nerve growth factor; EGF, epidermal growth factor. Primers are TYRP1 (5′-GTGGCGTTGAT-3′) and OPN25 (5′-GGGGCACCAG-3′). The reaction was performed with two different concentrations of the same RNA template, using 500 ng (on the left) and 250 ng.

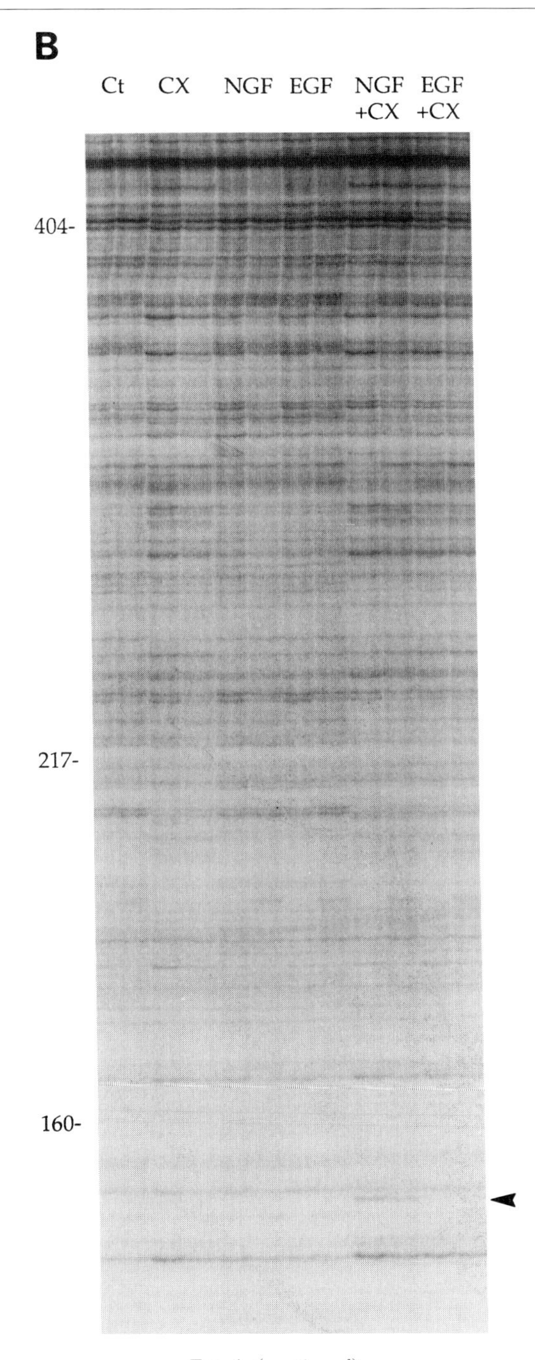

FIG. 1. (*continued*)

B. Choice of Primers and Enzymes

The oligonucleotides used in the RAP-PCR experiments are arbitrarily selected. The actual sequence is not important, besides the usual consideration of choosing oligonucleotides that do not contain palindromes or 3′ sequences that will promote primer dimers. The advantage of using short primers is that they can be inexpensively purchased or synthesized. Also, the use of 10- or 11-mer primers in combination with the Stoffel fragment of *Taq* polymerase rather than the full-length enzyme results in fingerprints that have a greater number of fragments. Variations on the original protocol have been devised involving the choice of at least one "nonarbitrary" primer. These adaptations attempt to prime a conserved region in mRNAs or to sample particular sequences that are conserved or statistically present among a number of genes (see Section V,C).

C. RNA Fingerprinting

1. Reverse transcription is performed on total RNA using three concentrations per sample (500, 250, and 125 ng per reaction) and a 10- or 11-mer oligonucleotide primer of arbitrary sequence. RNA (5 μl) is mixed with 5 μl of RT mixture for a 10-μl final reaction containing 50 m*M* Tris (pH 8.3), 50 m*M* KCl, 4 m*M* $MgCl_2$, 10 m*M* dithiothreitol (DTT), a 0.2 m*M* concentration of each dNTP, 2 μ*M* primer, and 16 U of MuLV reverse transcriptase.
2. The reaction is performed at 37° for 1 hr (after a 5-min ramp); the enzyme is inactivated by heating the samples at 94° for 3 min, and then the newly synthesized cDNA is diluted fourfold in water.
3. PCR is performed using the same or another oligonucleotide primer of arbitrary sequence. Diluted cDNAs (10 μl) are mixed with the same volume of PCR mixture for a 20-μl final reaction containing 10 m*M* Tris (pH 8.3), 10 m*M* KCl, 4 m*M* $MgCl_2$, a 0.2 m*M* concentration of each dNTP, 4 μ*M* oligonucleotide primer, 1 μCi of [α-^{32}P]dCTP, and 3 U of AmpliTaq polymerase Stoffel fragment (Perkin-Elmer-Cetus).
4. Thermocycling is performed using 30 cycles of 94° for 1 min, 35° for 1 min, and 72° for 2 min, followed by a final 72° extension for 7 min.
5. An aliquot of the amplification products (4 to 5 μl) is mixed with 15 μl of formamide dye solution, denatured at 68° for 15 min (or at 94° for 3 min), and chilled on ice. A 2.2-μl volume is loaded onto a polyacrylamide–urea gel, prepared in 1× TBE buffer. The PCR products resulting from the three different concentrations of the same RNA template are loaded side by side on the gel.
6. Electrophoresis is performed at 1500 V or at a constant power of 50–60 W until the xylene cyanol FF tracking dye reaches the bottom of

the gel (approximately 4 hr). The gel is dried under vacuum and placed on an Eastman Kodak BioMax X-ray film for 16 to 48 hr. It is necessary to tape several luminescent labels on the dried gel to facilitate the alignment of the autoradiogram with the gel when excising interesting bands out of the gel.

D. Isolation and Purification of Differentially Amplified Products

Once fingerprints have been generated and display interesting differences in abundances of some products, the next steps are the isolation and characterization of the PCR fragments representing differentially amplified products. Initially, the bands were excised from the gel, reamplified using the primers originally employed for the fingerprinting, cloned, and sequenced. However, the problem with this approach is that the reamplification of the tiny portion of the gel usually generates multiple products of almost the same size, as a result of comigration of products in the original fingerprint. Thus in the next steps, reamplified material contains a mixture of desirable and undesirable products. After cloning of this mixture, it is necessary to sequence many clones, and it is still difficult to know a priori which one of the sequences corresponds to the gene of interest. Assuming that the band of interest vastly predominates after PCR, the statistically predominant sequence should be the targeted sequence. However, we and others found out that during PCR, the "Cot effect" tends to reduce the amplification rate of the most abundant products compared with the less abundant ones, resulting in a partial normalization (see Section V,D).[31,32] One way to address this problem is to reduce the number of PCR cycles (20 cycles) to minimize the Cot effect. A considerable amount of wasted effort can also be avoided by the use of a native acrylamide gel to separate DNAs in the reamplified mixture based on single-strand conformation polymorphisms (SSCP). We have described this approach,[33] in which SSCP gels are used to purify the cDNA product of interest away from other products of different sequence but of similar size, after reamplification. This gel allows the classification of the PCR products into (1) those that can be directly cloned because they are relatively pure, (2) those that will be excised from the SSCP gel and reamplified, and (3) those that consist of a too-complex mixture and can be rejected for further study. This latter approach is the method of choice to isolate and characterize products from old fingerprints that were achieved for later use. However, for new RAP-

[31] F. Mathieu-Daudé, J. Welsh, T. Vogt, and M. McClelland, *Nucleic Acids Res.* **24,** 2080 (1996).

[32] M. T. Suzuki and S. Giovannoni, *Appl. Environ. Microbiol.* **62,** 625 (1996).

[33] F. Mathieu-Daudé, R. Cheng, J. Welsh, and M. McClelland, *Nucleic Acids Res.* **24,** 1504 (1996).

PCR fingerprinting experiments, for which gels are still radioactive, problems associated with reamplification of the initial PCR product can be avoided by the direct purification of the product of interest from a preparative denaturing polyacrylamide gel, followed by SSCP, prior to reamplification. This approach is faster, but must be done while the products are still radioactive.

Here we describe this latter protocol.

1. The RAP-PCRs containing the bands of interest are reloaded onto a preparative polyacrylamide–urea gel in multiple adjacent lanes and electrophoresed as in Section IV,C. This gel, which can be up to 1 mm thick, allows the resolution of a large quantity of the PCR product of interest. This extra gel is not necessary if the original gel was made with a wide well comb and the band to be excised has a sufficient mass.

2. The autoradiogram and the gel are aligned and the band of interest is excised from the gel. This gel is reexposed to X-ray film to check for the accuracy of the excision.

3. The band is eluted from the tiny piece of gel in 100 μl of TE at 65° for 2 to 3 hr and at room temperature for a few more hours.

4. After ethanol precipitation of the eluate, the radioactive pellet is resuspended in 4 μl of the loading dye containing 10 m*M* NaOH, denaturated at 94° for 3 min, cooled on ice, and loaded onto a HydroLink MDE gel prepared in 0.6× TBE buffer, according to the manufacturer instructions. Electrophoresis is performed overnight (for approximately 16 hr) at 8 W.

5. The dried gel, with luminescent labels, is autoradiographed using an intensifying screen. Each PCR product is expected to produce two SSCP bands, one for each of the two DNA strands. Frequently, only one band is observed, either because the two strands have the same mobility, or because only one strand was excised from the polyacrylamide denaturing gel. Occasionally, the two bands have different intensities, or more than two bands are observed. In these cases, the most intense band is always from the product of interest.

6. The product resolved on the SSCP gel is excised and eluted as described above in 50 μl of TE. Five microliters of the eluted solution is reamplified using the same primers as in the original fingerprint. The template is mixed with 15 μl of PCR mixture for a 20-μl final reaction containing 10 m*M* Tris (pH 8.3), 50 m*M* KCl, 3 m*M* $MgCl_2$, a 0.2 m*M* concentration of each dNTP, a 1 μM concentration of each primer, and 2 U of AmpliTaq polymerase.

7. Thermocycling is performed using 25 cycles of 94° for 30 sec, 37° for 30 sec, and 72° for 1 min, followed by a final 72° extension for 5 min.

8. PCR products can be run on low melting point agarose gel (NuSieve

GTG; FMC BioProducts, Rockland, ME), cleaned with the QIAquick gel extraction kit (Qiagen), and used for direct sequencing or cloning followed by sequencing.

E. Cloning, Sequencing and Southern Blots

Fragments can be cloned directly in pCR 2.1 vector (original TA cloning kit; Invitrogen, San Diego, CA). Plasmids or purified PCR products are sequenced using the ABI Prism dye terminator cycle sequencing kit and the ABI 373 automatic sequencer (Perkin-Elmer-Applied Biosystems, Foster City, CA). For direct sequencing of the PCR products using 10-mer primers, we modified the cycling conditions by lowering the annealing temperature to 35° and adding a ramp to reach the elongation temperature. The cloning of the correct fragment can be confirmed by performing a Southern blot against the original RAP fingerprint gel. Capillary transfer of the polyacrylamide gels onto nylon membranes (Hybond N+; Amersham, Buckinghamshire, England) is performed under standard conditions. Clones that serve as probes are labeled and hybridized to the membranes using the nonradioactive ECL direct nucleic acid-labeling and detection system (Amersham) according to the manufacturer instructions.

F. Confirmation of Differential Expression

Confirmation of differential expression is an obligatory step following identification of differentially amplified transcripts from RAP-PCR fingerprints. Conventional methods for determining the relative abundances of RNAs between samples include Northern blotting using standard protocols[34] or RT-PCR using an internal control.[35,36] However, these methods require either a large amount of template RNA, or a useful control. To sidestep these limitations, we have developed a simple way to determine relative abundances of specific RNA transcripts between samples that is particularly useful for cells or microorganisms for which low quantities of RNA are available, or for which an invariant internal standard is not known. In this method, two specific primers are used in RT-PCR under low-stringency conditions, similar to those used to generate RAP-PCR fingerprints. In addition to the product of interest, a pattern of arbitrary products is generated. These extra PCR products are largely invariant and

[34] J. Sambrook, E. F. Fritsch, and T. Maniatis, "Molecular Cloning: A Laboratory Manual." Cold Spring Harbor Laboratory Press, Cold Spring Harbor, New York, 1989.

[35] G. Gilliland, S. Perrin, K. Blanchard, and H. F. Bunn, *Proc. Natl. Acad. Sci. U.S.A.* **87,** 2725 (1990).

[36] M. Bouaboula, P. Legoux, B. Pessegue, B. Delpech, X. Dumont, M. Piechaczyk, P. Casellas, and D. Shire, *J. Biol. Chem.* **267,** 21830 (1992).

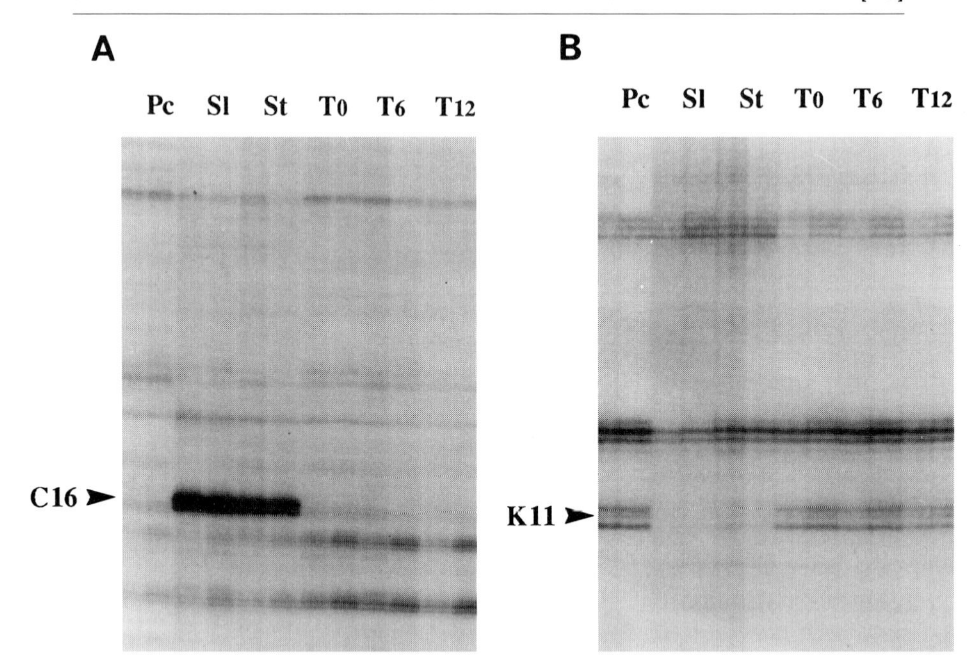

FIG. 2. Confirmation of differential gene expression by low-stringency RT-PCR. The RNA samples are from the three different stages of the life cycle of *Trypanosoma brucei* (Protozoa): the insect procyclic form (Pc), the mammalian bloodstream slender (Sl) and stumpy (St) forms, and the procyclic form, subjected to heat shock for 6 and 12 hr (T6 and T12, T0 being the control). RT was performed with a mixture of anchored dT primers (see Protocols). PCR was performed under low-stringency conditions (45°) using specific primers of the target transcripts, and 20 amplification cycles. Some arbitrary amplified products behave as internal controls for RNA quantities and RT efficiency. (A) Transcript C16 is upregulated in the two bloodstream forms of the parasite (Sl and St), but no response to the temperature elevation is observed. (B) Transcript K11 is upregulated in the insect procyclic form (Pc) and downregulated in the bloodstream forms (Sl and St). The abundance of the transcript is not affected by the elevation of the temperature. [Reprinted from Mathieu-Daudé *et al.,* Differentially expressed genes in the *Trypanosoma brucei* life cycle identified by RNA fingerprinting, *in* "Molecular and Biochemical Parasitology," with kind permission from Elsevier Science, NL, Sara Burgerhartstraat 25, 1055 KV Amsterdam, The Netherlands.[6]].

behave as internal controls for RNA quality and quantity, and for reverse transcription efficiency. Figure 2 shows examples of low-stringency RT-PCR experiments confirming the differential expression of two transcripts that were identified on RAP-PCR fingerprints.

Here we describe an RT-PCR protocol using low-stringency conditions that has proved to be effective and relatively simple. We also provide an RT-PCR protocol using high-stringency conditions and an internal control (e.g., actin or "housekeeping" gene mRNA), as an alternative method. In both cases, the number of PCR cycles must be low (14 to 24, according to the abundance of the product) to preserve the differences in starting template mRNA abundances. Indeed, because of the Cot effect, rehybridization of abundant products during the PCR inhibits their amplification, and the difference in product abundances diminishes as the number of PCR cycles increases (see Section V,D).

1. Low-Stringency Reverse Transcriptase-Polymerase Chain Reaction

1. The reverse transcription is performed under the same conditions as in the RAP-PCR protocol, using either a mixture of three different anchored oligo(dT) primers [$(T)_{15}$-G, A, or C], or a mixture of random 9-mer primers (Stratagene, La Jolla, CA).

2. The PCR is performed using two primers (19- to 24-mers) that are specific for the sequences characterized by sequencing reactions. They are chosen to generate PCR products of 100 to 250 base pairs (bp) and have melting temperatures of at least 60°.

3. The PCR conditions are the same as in the RAP-PCR fingerprint protocol, except that each primer is used at 1.5 μM concentration, and 0.5 μCi of [α-^{32}P]dCTP is used in the 10-μl final volume.

4. The following low-stringency thermal profile is used: 94° for 30 sec, 35° for 30 sec, and 72° for 1 min, for 20 cycles. Alternatively, if the transcript is highly abundant, the reaction can be carried out in two sets of tubes and cycled for 15 and 20 cycles.

5. PCR products are run, under the same conditions as in the RAP-PCR fingerprint protocol, on a 6% polyacrylamide–43% urea gel. A longer exposure of the gel (40 to 72 hr) is required.

2. High-Stringency Reverse Transcriptase-Polymerase Chain Reaction. For this protocol, two pairs of primers are selected. One pair of primers is specific for the transcript of interest (see low-stringency RT-PCR protocol); the other pair is specific for the mRNA used as internal standard. If the two mRNAs are of similar abundances, the PCR can be carried out using the two primer pairs in the same tube. If the standard is much more abundant than the target transcript (as is often the case), diluted cDNA is divided into two sets of tubes and the PCR is performed simultaneously, but in two different tubes, one for each primer pair. The PCR should be sampled at different time points during the exponential phase. We recommend using two or three different numbers of cycles.

1. Reverse transcription is performed under the same conditions as in the RAP-PCR fingerprint protocol, but using an oligo(dT) including two ribonucleotides, 5′-$(T)_{10}$rUrUTTT [which block elongation through the oligo(dT) sequence during subsequent PCR amplification].

2. cDNA is diluted eightfold in water, allowing multiple second step reactions.

3. The PCR is performed in triplicate for each pair of primers (three sets of tubes per primer pair, or three sets of tubes for both pairs together). Five microliters of diluted cDNA is mixed with the same volume of PCR mixture for a 10-μl final reaction containing 10 mM Tris (pH 8.3), 50 mM KCl, 1.75 mM $MgCl_2$, a 0.15 mM concentration of each dNTP, a 0.5 μM concentration of each primer, 0.5 μCi of [α-^{32}P]dCTP, and 0.5 U of AmpliTaq polymerase (Perkin-Elmer-Cetus).

4. Thermocycling conditions are 94° for 30 sec, 65° to 70° (according to the primer pair) for 30 sec, and 72° for 50 sec. Samples are taken out of the PCR machine at different cycles, from 14 to 24 cycles (3 different times).

5. Amplification products are resolved on acrylamide gels, using the same conditions as described above.

3. Quantitation of Polymerase Chain Reaction Products for Semiquantitative Analysis. PCR products can be excised from the polyacrylamide gel and disintegrations of ^{32}P counted in 4 ml of Ecolume scintillation fluid (Du Pont-NEN, Boston, MA). We directly quantify by scanning the polyacrylamide gel using an Ambis instrument and software (Ambis Core Software, version 4.0; Ambis, San Diego, CA) or by using a PhosphorImager (Molecular Dynamics, Sunnyvale, CA). Ratios of the radioactive counts of the product to the radioactive counts of an internal control are calculated and compared. For the low-stringency RT-PCR experiment, the standard is a piece of fingerprint containing invariable bands.

G. Sequence Analysis

RNA sequences and corresponding protein/amino acid sequence translations are analyzed for similarities in the GenBank and dbEST databases using Advanced BLAST from the National Center for Biotechnology Information (NCBI; *http://www.ncbi.nlm.nih.gov*), and BCM Search Launcher from the Human Genome Center, Baylor College of Medicine (BCM, Houston, TX; *http://www.kiwi.imgen.bcm.tmc.edu*).

V. Notes and Comments

A. On the Use of Two Arbitrary Primers

1. The protocol presented here differs from the differential display protocol of Liang and Pardee[11] in that we use an arbitrary primer in both steps of the reaction, instead of an anchored oligo(dT) in the reverse transcription, followed by an arbitrary primer for the PCR. Both protocols

give good fingerprints and have their strengths and weaknesses. However, mRNAs that are not polyadenylated, such as some bacterial mRNAs,[14] can be sampled by RAP-PCR. Moreover, the use of arbitrary primers in both directions allows the sampling of internal RNA fragments, rather than the noncoding 3′ end, increasing the probability of finding open reading frames in the identified transcripts. Sampling the protein-coding region will help to find homologies in the databases or to determine the function of the transcript.

2. The use of two arbitrary primers, one for the RT reaction and another for the PCR, instead of the same arbitrary primer in both steps of the reaction, has the advantage of generating fragments that have the two different primers at the ends. Under the conditions that we use, at least 90% of the PCR products have the second primer at the 5′ end of the sense strand of the transcript and the first primer at the 3′ end, and thus the orientation of the product sequenced is known.

B. Quality of Template RNA

The approach presented here is sensitive to RNA template quality and to small changes in template concentrations. When working with tissues, extra care should be taken to avoid RNA degradation before and during tissue disruption and homogenization. RNA of high quality can be obtained with different commercial kits, allowing the purification of many samples at the same time. Samples must be treated with RNase-free DNase and reverse transcriptase-free control experiments must be carried out for each new set of RNAs. Each RAP-PCR experiment must be performed on three templates (twofold serial dilutions) of each RNA to avoid differential amplification reflecting slight differences in RNA quality or concentration, rather than true differences in transcript abundances. Only those differentially amplified products that are reproducibly present in all three concentrations of each RNA are considered for further analysis.

C. Adaptations of RNA Arbitrarily Primed Polymerase Chain Reaction Fingerprinting

Modifications of the original RAP-PCR and differential display approaches were directed toward the arbitrary nature of the sampling. One adaptation was designed to target the conserved 5′ end of trypanosome mRNA.[15,16,37] The primer used in the reverse transcription is totally arbitrary, but the second primer is complementary to the conserved 5′ end of all mRNAs. The strategy of 5′ anchoring increases the probability of finding

[37] N. B. Murphy and R. Pelle, *Gene* **141,** 53 (1994).

similarities with known functional genes in unrelated organisms because the 5′ ends usually include the N terminus of the open reading frames. Other adaptations attempt to sample particular sequences that are conserved among gene families by fingerprinting with primers derived from conserved amino acid motifs.[25,26] In selecting primers for targeted fingerprinting, another approach involves the use of oligonucleotide sequences that are common and frequent in a list of genes of interest.[27] Two programs have been conceived, one to design primers for *Escherichia coli* transcripts[28] and another to select matched primer pairs for all the sequences in a list of interest, for example, cDNA for genes involved in DNA repair and replication, or genes associated with apoptosis.[29]

D. The Cot Effect

In later cycles of the PCR, self-annealing of abundant products occurs, slowing the amplification efficiency of more abundant products. The rate of amplification of abundant PCR products declines faster than that of less abundant products in the same tube.[31,32] In reference to the dependence of annealing on concentration and time, this self-annealing phenomenon was named the "Cot effect." This effect has two consequences. First, the slow-down in amplification of abundant products will allow less abundant cDNAs to be sampled more efficiently, and rare cDNAs might become visible on a gel. This is an advantage for sampling rarer transcripts that reduces the bias of RAP-PCR fingerprinting toward sampling the most abundant mRNAs. Second, if differences in product abundances decrease as the number of PCR cycles increases, small differences in mRNA abundances will be gradually erased. Thus, the Cot effect causes an underestimate of the true differences in starting template mRNA abundances. These differences in starting templates will be better preserved in the less abundant mRNAs. One way to minimize the Cot effect is to reduce the number of PCR cycles, particularly for the semiquantitative RT-PCR experiments used for confirmation of differential expression.

By reducing the amplification rate of the most abundant products as opposed to the less abundant ones, the Cot effect tends to partially normalize cDNAs. This phenomenon can be exploited for other applications. In fact, abundance normalization at later cycles might be useful for the production of a normalized library.

Acknowledgments

This work was supported in part by the Tobacco-Related Disease Research Program of the University of California, grant No. 6KT-0272, to F.M.-D., and by a generous gift from Sidney Kimmel.

[19] cDNA Representational Difference Analysis: A Sensitive and Flexible Method for Identification of Differentially Expressed Genes

By MICHAEL HUBANK and DAVID G. SCHATZ

Introduction

Control of gene expression is fundamental to the function and diversity of all living organisms. Significant advances have been made in our ability to accurately monitor alterations in gene expression through the development of a number of ingenious and sensitive techniques.[1] The ability to pinpoint which genes are active and which are inactive in any biological context provides us with information critical for understanding living systems and for intervening in dysfunctional states. Here we describe the application, potential modifications and limitations of one of the most sensitive and flexible of these new techniques, cDNA representational difference analysis (cDNA RDA).[2]

cDNA RDA is a polymerase chain reaction (PCR)-based subtractive hybridization technique for identifying genes that differ in their expression between two populations.[3] cDNA RDA employs a positive selection approach in which target cDNA fragments are sequentially enriched by favorable hybridization kinetics and subsequently amplified by PCR, while material common to both populations is eliminated by selective degradation. The exponential degree of enrichment of differences enables the detection of low copy number transcripts by compensating for the rarer annealing events of these species. This generates high sensitivity and allows the application of the technique to small amounts of starting material, including FACS (fluorescence-activated cell sorter)-purified populations of cells. cDNA RDA is rapid and flexible, and has a range of potential modifications that allow the method to be customized to individual experiments. Careful and appropriate usage can result in the successful identification of interesting genes and of the mechanisms by which these genes are regulated. Accumulating literature demonstrates the efficacy of the technique and its wide range of potential applications. To date, these have included the identification of transcriptional targets of hormones[4] and transcription fac-

[1] C. G. Sagerström, B. I. Sun, and H. L. Sive, *Annu. Rev. Biochem.* **66,** 751 (1997).
[2] M. Hubank and D. G. Schatz, *Nucleic Acids Res.* **22,** 5640 (1994).
[3] N. A. Lisitsyn, N. M. Lisitsyn, and M. Wigler, *Science* **259,** 946 (1993).
[4] M-J. Melià, N. Bofill, M. Hubank, and A. Meseguer, *Endocrinology* **139,** 688 (1998).

Copyright © 1999 by Academic Press
All rights of reproduction in any form reserved.

0076-6879/99 $30.00

tors,[5] and of changes in transcription associated with various treatments,[6,7] development,[8] infection,[9] and cancer.[10] Although it can be highly effective, cDNA RDA demands meticulous care, both in the design and execution of experiments, and prior thought should be given to the screening of differentially represented clones. It is also necessary to be aware of the limitations of the technique. The theoretical basis of the method has been described by Lisitsyn *et al.*,[3] and this chapter focuses on the application of RDA to cDNA.

Principles of cDNA Representational Difference Analysis

Genomic Representational Difference Analysis: Reduction of Complexity, Subtraction, and Amplification of DNA

cDNA RDA is a cDNA-specific modification of the PCR-based subtractive hybridization technique of representational difference analysis (RDA), which was first described by Lisitsyn *et al.* as a procedure for the identification of differences between two complex genomes, including deletions, insertions, and translocations.[3] In developing RDA, Lisitsyn and colleagues successfully combined the advantages afforded by both subtractive hybridization and amplification, and achieved a significant advance in the field of subtractive cloning.

Genomic DNA has an extremely high degree of sequence complexity, and requires extensive simplification in order to permit efficient reannealing of rarer species during subtractive hybridization. This simplification is achieved by selecting only the subpopulation of restriction enzyme-cut fragments that will amplify by PCR. DNA is digested with a restriction enzyme having a 6-base pair (bp) recognition site (i.e., a six-cutter) and linkers are ligated onto the cut ends. This material is then used as a template for PCR. The resulting product is limited in its complexity by the ability of each product to amplify within the mixture. Templates that are either too large (>1.2 kb) or too small (<150 bp) do not amplify efficiently under what are effectively competitive conditions, and so the product therefore

[5] B. C. Lewis, H. Shim, Q. Li, C. S. Wu, L. A. Lee, A. Maity, and C. V. Dang, *Mol. Cell. Biol.* **17,** 4967 (1997).

[6] X. Y. Fu and M. P. Kamps, *Mol. Cell. Biol.* **17,** 1503 (1997).

[7] A. Lerner, L. K. Clayton, E. Mizoguchi, Y. Ghendler, W. Vanewijk, S. Koyasu, A. K. Bhan, and E. L. Reinherz, *EMBO J.* **15,** 5876 (1996).

[8] W. P. Zheng and R. A. Flavell, *Cell* **89,** 587 (1997).

[9] M. Dron and L. Manuelidis, *J. Neurovirol.* **2,** 240 (1996).

[10] T. T. Gress, C. Wallrapp, M. Frohme, F. MullerPillasch, U. Lacher, H. Friess, M. Buchler, G. Adler, and J. D. Hoheisel, *Genes Chromosomes Cancer* **19,** 97 (1997).

only "represents" the amplifiable proportion of the digest—approximately 10% of the total material, depending on the six-cutter used. The amplified material is therefore known as a "representation," which then serves as the starting material for successive rounds of subtraction and amplification. The use of several different enzymes is required to achieve coverage of a high percentage of the genome.

cDNA Representational Difference Analysis: Maintenance of Complexity, Subtraction, and Amplification of mRNA

There are two fundamental differences between the RDA of genomic DNA and cDNA RDA. First, in cDNA RDA, the starting material is derived from messenger RNA and not genomic DNA. This therefore targets the subtraction to just those genes that are expressed at the time of isolation. While genomic RDA is geared toward identifying mutations at the DNA level, cDNA RDA aims to find the more dynamic differences between expressed populations of genes. Second, the representation generated by cDNA RDA is as complete as possible. Approximately 15,000 genes are expressed in an average mammalian cell, representing only a small fraction of the total cellular DNA. Simplification of the sample to permit complete hybridization is therefore not obligatory, and whereas in genomic RDA the aim is to reduce complexity, in cDNA RDA the aim is to conserve it. However, to take advantage of the subsequent enrichment of differences afforded by PCR, it is still necessary to generate an amplified population (Fig. 1). To achieve this, double-stranded cDNA is derived from the original mRNA by reverse transcription and second-strand synthesis, and digested with a four-cutter restriction enzyme that maximizes the generation of amplifiable fragments. After digestion, linkers are ligated onto the cut cDNA and PCR is performed. Statistical analysis has shown that the four-cutter enzyme used (*Dpn*II) cuts cDNA to produce fragments with an average length of 286 bp. The resulting representation therefore typically includes one or more fragments from a high proportion of expressed genes. This is the material used in the subsequent subtraction and amplification process.

Polymerase Chain Reaction-Coupled Subtractive Hybridization

In cDNA RDA, the representations are subtracted from one another by hybridization, common material is selectively degraded, and differences are positively enriched by PCR (Fig. 2). This differs from techniques that rely on display techniques in that fragments that are common to both populations should be eliminated from the experiment, simplifying the

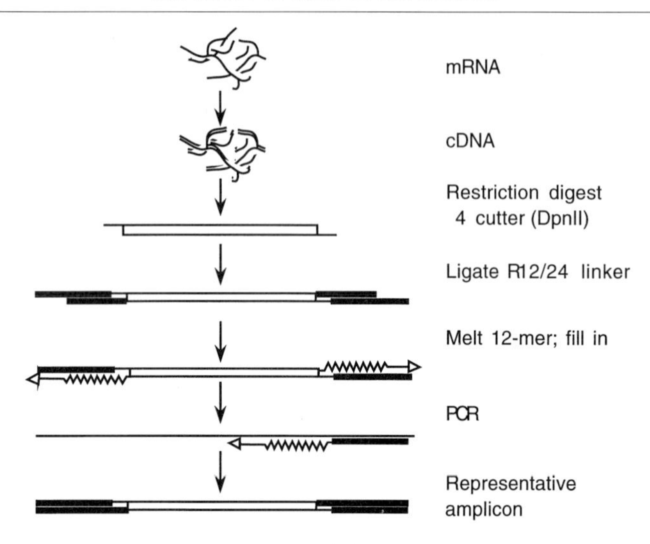

FIG. 1. Generation of representation. Poly(A)$^+$ mRNA is purified from tissues and used as template for the synthesis of double-stranded cDNA. cDNA is digested with *Dpn*II and ligated to the R-12/24-mer linker. The 12-mer is melted away, and a fill-in reaction is performed to generate the primer binding site. This process is the essential basis for tester-specific amplification when generating difference products, and is used in the generation of representations to avoid the problems of concatamerization that would result from the use of phosphorylated oligonucleotides. The representative amplicon is then generated by PCR, using the R-24-mer as the primer.

subsequent identification and analysis of products and greatly reducing the proportion of false positives.

The representation derived from the material containing the target genes is designated the tester, while the control material is known as the driver, which will be used to "drive out" fragments common to both samples. Initially, both representations are digested to remove the linker used to amplify the material. A new linker consisting of unphosphorylated 12- and 24-base oligonucleotides is then ligated to the tester only. In this ligation, as in all cDNA RDA linker ligations, the 12-mer only provides the appropriate end structure to permit ligation of the 24-mer to the digested DNA, and is not itself covalently linked to the DNA. The driver and tester are then mixed at a driver : tester ratio of 100 : 1, melted, and allowed to anneal under optimum conditions.

Three types of hybrids can be formed. The abundant driver : driver hybrids, which form most frequently, lack linkers, cannot generate a primer-binding site during the initial fill-in reaction, and are therefore incapable

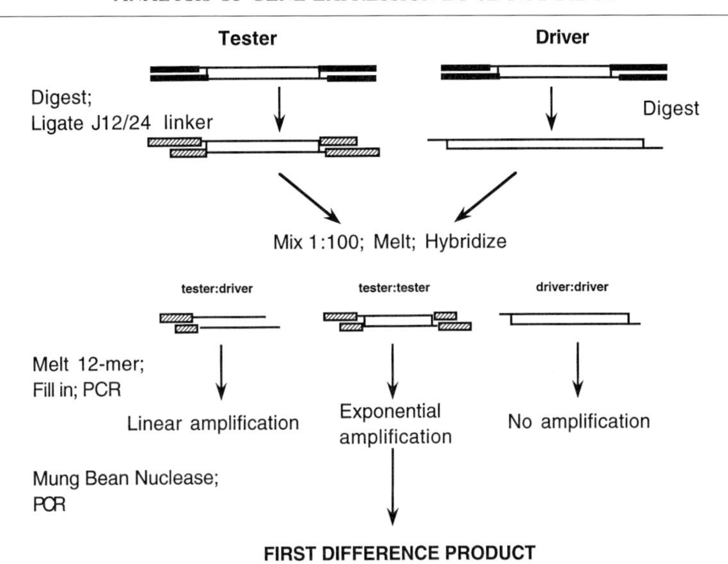

FIG. 2. Subtractive hybridization and amplification. Driver is prepared by removing R-linkers from the representations. The creation of tester requires the ligation of a second set of linkers (J-linkers). Driver and tester are then mixed, melted, and allowed to hybridize for 20 hr. The J-12 oligo is melted away and 5′ ends are filled in. Subsequent amplification results in exponential enrichment of tester : tester hybrids in DP1. For the generation of DP2 the process is identical except that a different set of oligos (N-linkers) is used, and the hybridization is performed at a driver : tester ratio of 800 : 1.

of amplification in the subsequent PCR. Driver : tester hybrids are the next most likely product, with each strand of tester being 100 times more likely to anneal to a complementary strand from the driver than to a complementary strand from the tester. Because the driver strand is unable to generate a primer-binding site, these fragments can amplify only in a linear fashion. If, on the other hand, a target fragment in the tester has no complementary strand in the driver because it is not expressed in that population, then it can anneal only to a complementary strand from the tester itself, generating fragments with 24-mer linkers ligated onto each 5′ end. Residual 12-mers are then melted away, and a fill-in reaction is performed to generate the primer-binding site (complementary to the 24-mer) for the following PCR. The primer is the 24-mer itself, and because only tester : tester hybrids have the primer-binding site on each 5′ end, they alone will amplify exponentially. Part way through the PCR amplification, single-stranded DNA is degraded with mung bean nuclease, thereby eliminating driver (which denatures during the PCR) and unamplified tester DNA. A more extensive explanation of the kinetics involved in hybridization is provided by Lisitsyn and Wigler.[11]

Second and Third Subtractions

Differences may already be apparent in the first product of subtraction and amplification (the first difference product, DP1), but because of random annealing events, many amplified tester : tester hybrids will not represent true differences at this stage. For this reason a second round of subtraction is performed, in a similar manner to the first, generating a second difference product (DP2). However, because of the enrichment in target sequences achieved at DP1, the ratio of driver to tester in the second subtraction is usually increased from 100 : 1 to 800 : 1, increasing the selection against gene fragments present in both populations. Owing to the low complexity of the material compared with genomic DNA, a third round of subtraction is seldom required in cDNA RDA, and in some cases when full stringency is applied (a driver : tester ratio of 40,000 : 1), can actually result in the loss of target sequences. After two rounds of subtraction and amplification, the difference products are visible on agarose gels as clear bands and can be easily subcloned.

Application of cDNA Representational Difference Analysis

Sources of Material, Isolation of RNA, and Preparation of cDNA

cDNA RDA may be applied to virtually any population from which it is possible to generate double-stranded cDNA. Suitable starting materials include whole small organisms (embryos, small metazoans, and unicellular eukaryotes such as yeasts); tissues from animals or plants; *ex vivo* cell populations, for example isolated by FACS purification; and cells cultured *in vitro*. In theory, the method may also be applicable to prokaryotes; however, further work is required to overcome the specific technical difficulties associated with this application, and a reliable protocol is not yet available.

As with any difference cloning technique, results are heavily dependent on the quality of the starting material. Shorter cDNA due to inefficient reverse transcription will reduce the inclusion of potentially amplifiable *Dpn*II fragments and limit detection of target genes, but it is not obligatory for cDNA to be either full length or of exceptional quality. It is, however, absolutely essential that the cDNA that is intended for use as the tester

[11] N. A. Lisitsyn and M. Wigler, *Methods Enzymol.* **254,** 291 (1995).

not be of better quality than the cDNA that will give rise to the driver. A longer average length of tester cDNA will generate fragments (typically from the 5′ end of the gene) that will amplify in the tester representation, but that are absent from the driver. These fragments will then be faithfully enriched by RDA, generating products that are different between the representations, but not between the original mRNA populations. It is therefore important to treat the starting tester and driver material identically, whichever method is chosen for isolation of RNA and production of double-stranded cDNA. If tester and driver are to be derived from cells in culture, it is critical that the two cell populations be grown under identical conditions and harvested at similar densities. RNA should be isolated from similar numbers of cells, using the same procedures, and preferably at the same time. Whenever possible, this process should be monitored—by assessing the quality of either mRNA or double-stranded cDNA. A small aliquot of cDNA from the proposed driver and tester should be compared side by side on an agarose gel (1.2%) to ensure that they are of similar quality. When less material is available, the incorporation of a radiolabeled nucleotide in an aliquot of first- or second-strand cDNA synthesis mix can be used to trace the reaction. Representations themselves can be checked when starting material is in particularly short supply, although this is a less reliable standard.

Ideally, approximately 1–2 μg of double-stranded cDNA should be used for the generation of representations. This is frequently not possible, and so we have developed a protocol for use with much smaller amounts of starting material. mRNA or total RNA can be used as a template for cDNA RDA; however, we have found mRNA to generate more reproducible results, and recommended its isolation prior to making representations. There are several kits capable of preparing adequate mRNA, and choice of a reliable method will likely vary with the source of material. We have found that a modified version of the Oligotex mRNA separation protocol (Qiagen, Chatsworth, CA) generates RNA from which reliable representations can be obtained using as few as 10^4 EL-4 mouse thymic lymphoma cells or 10^5 resting lymphocytes, which are smaller and yield less RNA. While it is possible to create amplicons from smaller quantities of material, problems can arise in maintaining a true representation of starting mRNA, and great care should be taken to confirm the validity of any products generated from small amounts of cDNA.

It is possible to detect transcripts that vary widely in their abundance. Because cDNA RDA combines a high representation of mRNA species, the subtraction of common products, and the amplification of differences, it is therefore highly sensitive. Rare transcripts can be detected (<1 copy per cell) with a low rate of false positives. In most experiments in which absolute differences are desired more than 90% of difference products prove to be genuinely different between the starting populations.

Limitations

cDNA RDA should not be used negatively (i.e., to prove something is not different). The process is unlikely to find point mutation differences or small deletions/insertions unless these affect the restriction site itself. Fragments from nonpolyadenylated transcripts are also undetectable (unless cDNA is prepared from total RNA), as are messages whose cDNA does not contain an amplifiable fragment (more likely with short mRNAs). There can be many reasons for failure of amplification, and occasionally genes containing apparently amplifiable fragments will not be detected. It is not appropriate to use cDNA RDA when a multitude of differences is expected. In these situations, cDNA RDA will not produce a comprehensive library of products, but will enrich just those products that amplify most efficiently. This problem can frequently be avoided by careful choice of driver and tester to avoid cloning irrelevant differences. Tissues should be as closely matched as possible (e.g., derived from close relatives), and when experiments require the processing of cells grown *in vitro* to provide the source of starting material it is best to split a single culture before selected treatment to generate driver and tester, rather than using separate cultures. Similarly, when applicable, the same passage number or frozen aliquot of a cell line should be used for driver and tester production. If representations of mutant lines are being studied to identify the effects of the mutant gene, it is better to subtract them from representations prepared from the same cells complemented with the functional gene rather than from the wild type. For studies in which abundant differences are inevitable, a display-based approach or conventional subtractive hybridization may be more successful. Finally, while cDNA RDA can reproducibly identify small differences in the abundance of transcripts (three- to fivefold and up), detection of all such differences cannot be guaranteed.

cDNA Difference Analysis Protocol (cDNA Representational Difference Analysis)

The first section of the protocol describes the isolation of RNA, reverse transcription to cDNA, and generation of representations from small amounts of starting material (10^4–10^7 cells). If larger amounts of material (tissue or cells) are available, polyadenylated RNA should be prepared and reverse transcribed into double-stranded cDNA, as described in the second section. The third section then describes the difference analysis procedure and is common to any quantity of starting material. Some parts of this procedure are similar or identical to parts of the genomic RDA procedure of Lisitsyn and Wigler,[11] and are included here for convenience. We request that when reporting the use of this protocol, appropriate acknowledgment be given to the originators of the genomic RDA procedure.[3,11]

Only the highest quality reagents should be used, and "PCR-clean" techniques should be observed throughout the protocol, including the use of barrier tips to reduce aerosols when handling any potential templates and the aliquoting of all components. We set up all PCRs in an airflow cabinet. Ensure that all reagents for RNA isolation and processing are RNase free.

Generation of Representations from Low Numbers of Cells (10^4–10^7)

Preparation of Total RNA. Isolate cells or tissues as required. If the cells are in suspension pellet the cells, pour off the medium, disrupt the pellet by flicking, and resuspend it in 0.5 ml of Trizol reagent (Life Technologies, Gaithersburg, MD). Use this volume of Trizol for any number of cells up to 10^7 if the cells are small (e.g., lymphocytes). For larger cells (e.g., fibroblasts) use 1 ml for between 5×10^6 and 10^7 cells and double the following volumes accordingly. For tissues and for adherent cells in culture follow the manufacturer recommendations. Incubate the cells in Trizol at room temperature for 5 min, vortexing to ensure that all cells are completely disrupted. Add 100 μl of $CHCl_3$, and mix well by shaking for 15 sec. Leave at room temperature for 2–3 min, then centrifuge at 12,000 rpm for 15 min at 5° (Eppendorf microcentrifuge). Take the aqueous upper phase to a new tube, add 400 μl of isopropanol, and leave at room temperature for 10 min. Centrifuge at 12,000 rpm for 20 min at 5°. It is possible to include 1 μl of glycogen (5 mg/ml; Boehringer Mannheim, Indianapolis, IN) as a carrier in the precipitation. Do not leave isopropanol precipitations on ice for more than 1 hr at any point in the protocol, and never place isoporopanol precipitations in a freezer, as this produces heavy salt precipitation that will inhibit later reactions. Pour off the supernatant, recentrifuge for 1 min, and remove the residual supernatant with a drawn-out Pasteur pipette. Wash the pellet with 1 ml of 75% ethanol, vortex, and centrifuge at 7500 rpm for 5 min at 5°. Remove the ethanol with a Pasteur pipette, allow to air dry, and resuspend in 75 μl of double-deionized, RNase-free water (ddH_2O).

Removal of Residual Genomic DNA. Even the most carefully prepared RNA contains contaminating genomic DNA, which may compromise the cDNA RDA procedure. This step is particularly important if total RNA, rather than poly(A)$^+$ mRNA, will be the substrate for the generation of cDNA. Genomic DNA should be removed as follows. Mix in order: 10 μl of 0.1 *M* dithiothreitol (DTT); 10 μl of 50 m*M* $MgCl_2$; 3.3 μl of 3 *M* sodium acetate; 0.5 μl of RNase inhibitor (20–40 U/μl, preferably cloned; Promega, Madison, WI); and 1 μl of DNase I (RNase free) (10–50 U/μl; Boehringer Mannheim). Add the mix to 75 μl of RNA and incubate at 37° for 15 min. Extract twice with 100 μl of P/C/I (phenol–$CHCl_3$–isoamyl alcohol; 25:24:1, v/v/v/). Extract once with C/I ($CHCl_3$–isoamyl alcohol; 24:1, v/v) and move the aqueous phase to a new tube.

Preparation of mRNA Using Oligotex mRNA Separation Kit

1. Add 100 μl of 2× binding buffer directly to the aqueous phase from the C/I extraction. Be sure to dissolve any precipitate present in the 2× binding buffer before use. Add 6 μl of Oligotex suspension (Qiagen), mix well, and incubate at 65° for 3 min. Leave the tube at room temperature for 10 min to allow binding of the mRNA and prewarm the elution buffer to 75°. Centrifuge at 14,000 rpm for 3 min and discard the supernatant.
2. Wash the Oligotex twice in 300 μl of wash buffer, each time flicking the tube to resuspend, and centrifuging for 3 min. Remove the wash buffer carefully, and recentrifuge if the Oligotex becomes resuspended during the procedure.
3. Elute the mRNA from the Oligotex with 6 μl of preheated elution buffer by pipetting the Oligotex up and down to resuspend, vortexing, and incubating at 75° for 1–2 min. Centrifuge for 3 min. Remove the supernatant to a fresh tube and repeat. Combine supernatants. Binding, wash, and elution buffers are provided by the manufacturer of Oligotex (Qiagen).

First-Strand cDNA Synthesis

1. Centrifuge the mRNA at 12,000 rpm for 4 min at 4° to pellet any carryover Oligotex. Move 10 μl into a fresh tube to be the +RT sample. Add 8 μl of ddH_2O to the first tube; this will be the −RT sample. Incubate the RNA to disrupt secondary structure (50°, 5 min), then place on ice.
2. For each sample, mix the following components: 4 μl of first-strand buffer (e.g., Superscript II buffer; Life Technologies); 1 μl of 20 m*M* dNTPs (dATP, dCTP, dGTP, TTP); 2 μl of 0.1 *M* DTT; 1 μl of oligo(dT) (Promega; 50 ng/μl); 0.5 μl of RNase inhibitor (20–40 U/μl). Add 8.5 μl of mix to each tube and 1 μl of Superscript II reverse transcriptase (Life Technologies; 200 U/μl) to the +RT tube. Heat to 26° for 8 min, then to 41° for 1 hr 30 min.

Second-Strand cDNA Synthesis. For each reaction, mix the following components: 2 μl of ddH_2O; 6 μl of RT2 buffer [for 1 ml of RT2: 100 μl of Tris-HCl (pH 7.5); 500 μl of 1 *M* KCl; 25 μl of 1 *M* $MgCl_2$; 50 μl of

1 M $(NH_4)_2SO_4$; 50 μl of 1 M DTT; 50 μl of bovine serum albumin (BSA, 5 mg/ml); 225 μl of water]; 0.5 μl of 15 mM β-NAD (Boehringer Mannheim); 0.4 μl of *Escherichia coli* ligase [5 U/μl; New England BioLabs (NEB), Beverly, MA]; 0.3 μl of RNase H (NEB; 2.5 U/μl); 1 μl of *E. coli* DNA polymerase (NEB; 10 U/μl). Add 10 μl of mix to each tube of first-strand synthesis reaction. Incubate at 15° for 2 hr, then at 22° for 1 hr.

Digestion of cDNA with DpnII. Denature the enzymes used to synthesize cDNA by heating to 70° for 10 min, then place on ice. To precipitate cDNA, add to each tube: 1 μl of glycogen (5 mg/ml), 3 μl of sodium acetate, and 90 μl of ethanol, then cool to −20° for 10–20 min. Centrifuge at 14,000 rpm for 30–60 min at 4°. Wash the pellet with 70% ethanol and resuspend in 8 μl of ddH_2O, add 1 μl of 10× *Dpn*II buffer and 1 μl of *Dpn*II (NEB; 10 U/μl). Incubate at 37° for 1–2 hr. Denature the *Dpn*II (65°, 20 min), chill on ice, and spin to collect the sample at the bottom of the tube.

Ligation to R Linkers. The oligonucleotides used for the procedure (particularly the ligations) should be desalted. Many manufacturers of oligonucleotides supply them already deprotected and desalted. Alternatively, a procedure for desalting can be found in Lisitsyn and Wigler.[11]

The entire 10 μl of the cDNA reaction is ligated to the linkers as follows. For each reaction, mix 41 μl of ddH_2O; 6 μl of 10× ligase buffer [10× ligase buffer: 500 mM Tris-HCl (pH 7.8); 100 mM $MgCl_2$; 100 mM DTT; 10 mM ATP]; 1 μl of R-24 (5′-AGCACTCTCCAGCCTCTCACCGCA-3′; 2 mg/ml); and 1 μl of R-12 (5′-GATCTGCGGTGA-3; 1 mg/ml). Add the 49-μl mixture to the 10 μl of digested cDNA. Heat to 50° for 1 min, then cool gradually to 10° at 1°/min to allow annealing of the oligonucleotides. It is convenient to program a PCR machine to perform the gradual cooling. Add 1 μl of T4 DNA ligase (NEB; 400 U/μl) and incubate overnight at 14°. Because the oligodeoxynucleotides (oligos) are not phosphorylated, only the 24-mer (R-24) is covalently linked to the 5′ phosphate group of the *Dpn*II site from the digested cDNA, while the 12-mer remains annealed, but unligated.

Generation of Representative Amplicon. We strongly recommend that a pilot reaction be performed for each representation to be generated and for the corresponding −RT control. This is to establish the optimum amplification conditions, to judge quantities, and to monitor genomic DNA contamination before the entire stock of material is committed.

1. For each pilot reaction to be performed, mix 139.5 μl of ddH_2O; 40 μl of 5× PCR buffer [5× PCR buffer: 335 mM Tris-HCl (pH 8.9 at 25°); 20 mM $MgCl_2$; 80 mM $(NH_4)_2SO_4$; BSA (166 μg/ml)]; 16 μl of 4 mM dNTPs; and 1 μl of R-24 oligonucleotide primer.

2. Add 2.5 μl of the ligation, mix, place in a PCR machine and incubate at 72° for 3 min. During this period the 12-mer (R-12) dissociates, freeing the 3′ ends for subsequent fill-in.

3. Add 1 μl (5 U) of *Taq* polymerase (Life Technologies/Cetus) and incubate at 72° for 5 min to fill in the ends complementary to the 24-mer linkers.

4. Cycle at 95° for 1 min; 72° for 3 min (for various numbers of cycles, see below) with a final extension at 72° for 10 min. Both primer annealing and extension occur at 72°. Use PCR machines equipped with Hot Bonnets or overlay with mineral oil.

5. Pause the reaction after 18, 20, 22, and 24 cycles and remove 10-μl aliquots of the product. To each aliquot add 1 μl of 10× loading buffer and load on a 1.5% agarose gel together with appropriate DNA concentration standards and molecular weight markers.

Concentration standards can be prepared by digesting genomic DNA of known concentration (sheared to an average length of approximately 20 kb) with a restriction enzyme that has a four-base pair recognition site, such as *Sau*3A. Dilutions corresponding to 0.1, 0.5, and 1 μg of concentration standard should be loaded along with the PCR product and electrophoresed until the bromphenol blue has migrated 2 or 3 cm. Stain with ethidium bromide (during or after the run). The yield, which is best judged after short runs, should be assessed on a UV transilluminator, and should ideally be approximately 0.5 μg from the 10-μl aliquot. Reliance on spectrophotometer readings is inadequate. Once a successful representation has been generated it can be used as the standard for other experiments.

The pilot reactions should show a smear in the +RT lane, while the −RT lane should be clear or very faint. Material amplifying from the −RT reaction arises from genomic contamination in the RNA preparation. Remember that the −RT lane contains only 20% of the template in the +RT lane.

Run the gel further to assess the size distribution of products. A smear ranging in size from ~1.5 to 0.2 kb should be obtained (Fig. 3A). Discrete bands may be visible, representing amplified fragments of abundant transcripts, and these will be characteristic of a given sample. Where closely matched, tester and driver samples to be subtracted will not appear different at this stage (Fig. 3A). A product that matches the criteria of yield, size distribution, and clean −RT control indicates the number of cycles necessary for preparing the representation. Underamplification will not provide sufficient material for the subtraction, while overamplification will bias the population and reduce average fragment size. Should more than 24 cycles be required this probably indicates a problem with the mRNA isolation or cDNA synthesis, and it is better to start again. If only two rounds of subtraction are to be performed, a minimum of 100 μg of driver and 20

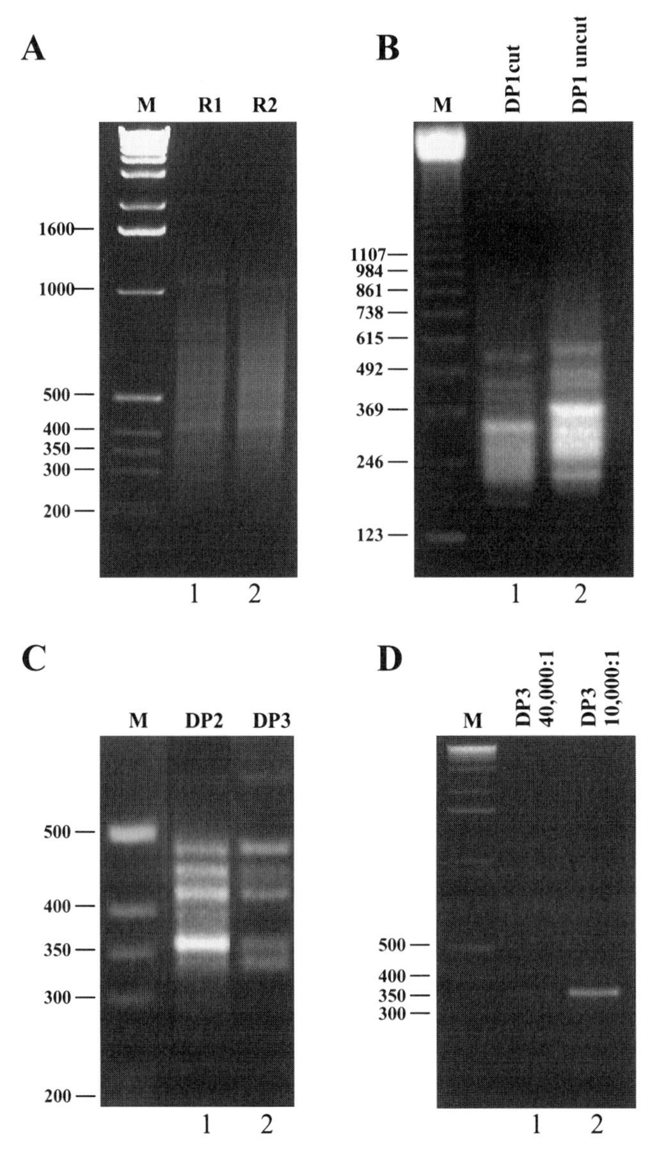

FIG. 3. Representations and difference products. (A) Two representations to be used in a subtraction (R1 as driver and R2 as tester). Note that the concentrations are identical (0.5 μg/μl) and that in this case the representations are indistinguishable. (B) First difference products (DP1) digested with *Dpn*II (lane 1) and undigested (lane 2). Note the shift down in size by 48 bp after digestion. (C) Second and third difference products generated at driver : tester ratios of 800 : 1 and 40,000 : 1, respectively. Note the lower background in DP3, but that most of the differences are already clear in DP2. (D) DP3 from a different subtraction (using the yeast *Schizosaccharomyces pombe*), which produces a single difference product. Note that in this case a driver : tester ratio of 40,000 : 1 is too stringent, and better results are obtained with a lower ratio (10,000 : 1).

μg tester will be required (allowing for the preparation of representation blots). However, it is wise to prepare at least double this quantity to allow for further subtractions and alterations in stringency that may be required later.

6. Usually, a representation will be used as both a tester and a driver, in a reciprocal subtraction. On this assumption, when the correct conditions have been ascertained, prepare twenty 200-μl PCRs in 0.5-ml Eppendorf tubes. This should provide approximately 0.25–0.5 mg of each representative amplicon. If a representation is to be used only as tester, four reactions will suffice. Continue with the difference analysis procedure as described in the next section.

Generation of Representations from Large Amounts of mRNA (>10^7 Cells)

Preparation of cDNA Template. Prepare poly(A)$^+$ mRNA by a reliable procedure, such as a scaled-up version of that described in the preceding section. The DNase step is not essential if the poly(A)$^+$ mRNA is twice selected on an oligo(dT) column or beads. Generate double-stranded cDNA as follows (see the previous section for suppliers of materials).

1. Begin with 5 μg of mRNA dissolved in 20 μl of RNase-free water. Heat to 70° for 10 min. Chill on ice. Perform first-strand synthesis by adding to the mRNA, in order: 4 μl of 10× first-strand buffer; 0.5 μl of 100 m*M* DTT; 0.7 μl of RNase inhibitor; 2.0 μl of oligo(dT) primer (1 mg/ml); 4.0 μl of dNTPs (5 m*M*, for 500 μ*M* final concentration); 1.6 μl of 100 m*M* sodium pyrophosphate; 3.3 μl of Superscript II reverse transcriptase (200 U/μl), and incubate at room temperature for 10 min, then for 1 hr 30 min at 41°.

2. To generate double-stranded cDNA add the following to the 40-μl first-strand reaction: 112.5 μl of water; 40 μl of 5× RT2 (premixed with the water); 2.0 μl of β-NAD (15 m*M*, for 150 μ*M* final concentration); 0.4 μl of *E. coli* DNA ligase (5 U/μl); 4.0 μl of *E. coli* DNA polymerase (10 U/μl); and 1.1 μl of RNase H (2.5 U/μl). Incubate for 2 hr at 15° and then for 1 hr at 22°.

3. Add 50 μl of water. Extract twice with P/C/I, and once with C/I. Add 5 μg of glycogen carrier, 65 μl of 10 *M* ammonium acetate, and 800 μl of 100% ethanol. Precipitate for 15 min on ice and centrifuge (14,000 rpm, 15 min, 4°). Wash with 70% ethanol and resuspend in 30 μl of TE. Run 2 μl on a 0.7% agarose gel. A smear from approximately 8–10 kb down to 0.2 kb should be expected.

4. Digest 2 μg of double-stranded cDNA with *Dpn*II (in 100 μl; 37°; 2–4 hr). Extract with P/C/I, then with C/I. Add 50 μl of 10 *M* ammonium acetate, 650 μl of 100% ethanol (5 μg of glycogen carrier can be used) and

precipitate on ice for 20 min. Centrifuge at 14,000 rpm for 15 min at 4°. Wash the pellet with 70% ethanol. Air dry the pellet and resuspend in 20 μl of TE.

5. Ligate the digested cDNA to R linkers by mixing 12 μl (approximately 1.2 μg) of cut cDNA; 4 μl of R-24 oligo (2 mg/ml); 4 μl of R-12 oligo (1 mg/ml); 6 μl of 10× ligase buffer; 32 μl of water. Anneal the oligos in a PCR machine by heating to 50° for 1 min, then cooling to 10° at 1°/min. Add 2 μl of T4 DNA ligase (NEB; 400 U/μl), and incubate overnight at 14°.

Generation of Representations

1. Dilute the ligations by adding 140 μl of TE. Although quantities are easier to estimate in the large-scale protocol, it is still sensible to carry out a pilot reaction for each sample to determine the optimal cycle number required for generating the representation. For this, follow the procedure described in Generation of Representative Amplicon, above, but use 2 μl of the diluted ligation as template, and take aliquots for assessment at 17, 18, 19, and 20 cycles.

2. Scale up the procedure to generate representations as described in Generation of Representative Amplicon, above. For each PCR, mix in order: 139 μl of water; 40 μl of 5× PCR buffer; 16 μl of 4 m*M* dNTPs; 2 μl of R-24 primer (1 mg/ml); 2 μl of diluted ligation. Place in a PCR machine, and incubate for 3 min at 72° (to melt away the 12-mer). Add 1 μl (5 U) of *Taq* DNA polymerase and incubate for 5 min at 72° (to fill in the ends), then cycle at 95°, 1 min; 72°, 3 min for up to 20 cycles (depending on the pilot), with a final extension for 10 min at 72°. Check the product as previously described on a 1.5% agarose gel.

3. Combine four PCRs into one 1.5-ml Eppendorf tube and extract with 700 μl of P/C/I, then with C/I. Add 75 μl of 3 *M* sodium acetate (pH 5.3), 800 μl of isopropanol, and precipitate on ice for 20 min. Centrifuge at 14,000 rpm for 15 min at 4°. Wash the pellet with 70% ethanol, dry, and resuspend each of the pellets (comprising the products of four PCRs) in approximately 100–120 μl of TE to give a concentration of 0.5 mg/ml. Combine the material. It is important that the concentrations be accurate, and so the combined representation should again be run next to a previously prepared cut standard to allow comparison (see Generation of Representative Amplicon, above).

Difference Analysis

Generation of Driver

1. Digest 240 μg (480 μl) of each representation (20 μg is sufficient if it is to serve as tester only). This will provide sufficient driver for up to

five subtractions, with additional material to prepare representation blots. To digest the representations, mix 480 μl of the representation DNA; 140 μl of 10× *Dpn*II buffer; 50 μl of *Dpn*II (10 U/μl); and 730 μl of water. Incubate at 37° for 4 hr. (Perform tester-only digests in a total volume of 300 μl, with other components scaled down proportionately.)

2. Divide the digests into two tubes and extract with P/C/I, then C/I. Add 70 μl of 3 *M* sodium acetate (pH 5.3), 700 μl of isopropanol, and precipitate on ice for 20 min. Centrifuge at 14,000 rpm for 15 min at 4°. Wash the pellet with 70% ethanol and air dry. Resuspend the representation in a total of 400 μl of TE (200 μl per tube). Combine and assess the concentration by running 1 μl on a 1.5% gel with standards, then dilute with TE to 0.5 mg/ml. This is the cut *driver.*

Generation of Tester

1. To prepare the tester, mix 20 μg (40 μl) of digested representation (see the preceding section) with 50 μl of TE, and 10 μl of 10× loading buffer, and load onto a 1.2% TBE agarose gel prepared in a contamination-free gel box (see Lisitsyn and Wigler[11]). Run the gel until the bromphenol blue has migrated approximately 2 cm. Using a low-energy, long-wave UV transilluminator to visualize ethidium bromide-stained DNA, excise the amplicon-containing gel slice, leaving behind the digested linkers that form a discrete band below the cDNA-derived material. The removal of the R linkers prevents their religation to the amplicon when they are subsequently ligated to J linkers. The driver is not the subject of any further ligations and therefore does not require gel purification. Place the excised band in a preweighed 15-ml Falcon tube, weigh (~1–2 g), and extract the DNA with Qiaex II resin (Qiagen), following the manufacturer protocol. Use 35 μl of resin and extend the incubation time to 20 min at 55°. To maximize yield, dry rapidly but carefully by aspiration, taking care not to overdry the resin, which turns a lighter shade as it dries. Elute the DNA in two 60-μl volumes of TE when the lighter shade occupies most of the pellet but before it is totally white. The first elution should be performed at room temperature, and the second at 37°. Combine eluates to give 120 μl, and estimate the concentration next to standards on a 1.5% gel. It is normal for 2.5–5 μl to correspond to 0.5 μg of cut, gel-purified representation.

2. Ligate the cut, gel-purified representation to J linkers by mixing 2 μg of gel-purified DNA (usually 10–20 μl); 6 μl of 10× ligase buffer; 4 μl of J-24 (5′-ACCGACGTCGACTATCCATGAACA-3′; 2 mg/ml); 4 μl of J-12 (5′-GATCTGTTCATG-3′; 1 mg/ml); bring up to 58 μl with water. Anneal the oligos in a PCR machine by heating to 50° for 1 min, then cooling to 10° at 1° min. Add 2 μl of T4 DNA ligase (400 U/μl), and

incubate overnight at 12–14°. Dilute the ligation to 10 ng/μl by adding 120 μl of TE. This generates the J-ligated *tester*.

Subtractive Hybridization

1. Mix 80 μl of digested *driver* representation (40 μg) with 40 μl of diluted, J-ligated *tester* representation (0.4 μg), at a driver : tester ratio of 100 : 1. Extract with P/C/I, and with C/I, in a 0.5-ml microcentrifuge tube.

2. Add 30 μl of 10 *M* ammonium acetate, 380 μl of 100% ethanol, and precipitate at −70° for 10 min, followed by 37° for 1–2 min to reduce salt precipitation. Centrifuge at 14,000 rpm for 15 min at 4°. Wash twice with 70% ethanol, centrifuging for 2 min each time. Take care not to lose the pellet, which is small and may become difficult to see.

3. Air dry, and resuspend the pellet thoroughly in 4 μl of EE × 3 buffer [EE × 3 buffer: 30 m*M* EPPS (Sigma, St. Louis, MO), pH 8.0 at 20°; 3 m*M* EDTA] by pipetting for at least 2 min, then warming to 37° for 5 min, vortexing, and spinning to the bottom of the tube. Overlay with 35 μl of mineral oil.

4. Denature for 5 min at 98° in a PCR machine. Complete denaturation is essential at this stage, as any tester DNA that fails to denature will remain amplifiable, regardless of its abundance in the driver. Cool to 67°, and immediately add 1 μl of 5 *M* NaCl directly into the DNA. Incubate at 67° for 20 hr to allow complete hybridization.

Generation of First Difference Product

1. Remove as much mineral oil as possible, and dilute the DNA stepwise in 400 μl of TE. The DNA can be quite viscous at this point and should be diluted first by adding 10 μl of TE and repeatedly pipetting, followed by a further 25 μl of TE with more pipetting, and finally made up to 400 μl and vortexed. Residual oil in the dilution does not impair the subsequent amplification.

2. For each subtraction set up four PCRs in 0.5-ml microcentrifuge tubes as follows: mix 122 μl of water; 40 μl of 5× PCR buffer; 16 μl of dNTPs (each 4 m*M*); and add 20 μl of diluted hybridization mix. Incubate in a PCR machine for 3 min at 72°, to melt away the 12-mers. Pause, and add 1 μl (5 U) of *Taq* DNA polymerase; incubate for 5 min at 72°. Pause again, and add 1 μl of J-24 primer (2 mg/ml). Begin PCR (1 min at 95°, 3 min at 70°; 10 cycles) and perform a final extension for 10 min at 72°, then cool to room temperature. The amplification achieved after 10 cycles of PCR guards against potential loss of rare but genuine products during the precipitation steps prior to mung bean nuclease treatment.

3. Combine the four reactions into a single 1.5-ml Eppendorf tube. Extract with P/C/I, and with C/I. Add 10 μg of glycogen carrier, 75 μl of

3 *M* ammonium acetate (pH 5.3), 800 μl of isopropanol, and precipitate on ice for 20 min. Centrifuge at 14,000 rpm for 15 min at 4°, and wash the pellet with 70% ethanol. Resuspend the pellet in 40 μl of TE.

4. Digest the PCR product with mung bean nuclease (MBN) to remove unannealed single-stranded DNA, including linear-amplified products and driver DNA. This is performed as follows: Mix 20 μl of DNA; 4 μl of 10× mung bean nuclease buffer (NEB); 14 μl of water; and 2 μl of mung bean nuclease (NEB; 10 U/μl). Incubate at 30° for 35 min. Stop the reaction by adding 160 μl of 50 m*M* Tris-HCl (pH 8.9), and incubating at 98° for 5 min. Chill on ice. To monitor the success of MBN treatment, a control can be performed in which the MBN is omitted. Ten-cycle products from treated and untreated reactions should then be compared on a 1.5% agarose gel. Most of the ethidium bromide staining should be eliminated in the MBN-treated sample.

5. During the MBN incubation, set up final PCRs on ice (four per subtraction). For each reaction, mix 122 μl of water; 40 μl of 5× PCR buffer; 16 μl of dNTPs (each 4 m*M*); and 1 μl of J-24 (2 mg/ml). On ice, add 20 μl of MBN-treated DNA, and incubate in a PCR machine at 95° for 1 min. Cool to 80°, and add 1 μl (5 U) of *Taq* DNA polymerase. Perform PCR (18 cycles of 95° for 1 min, 70° for 3 min; with a final extension at 72° for 10 mins), then cool to room temperature. Note that the J primer requires an annealing/extension step at 70° rather than 72°. A 10-μl sample (which should contain approximately 0.5 μg of DNA) should be checked on a 1.5% gel next to standards to confirm the presence of products and to estimate the concentration. Should the yield be significantly (more than twofold) lower than this, it is likely that a problem has arisen either during the PCR, or at an earlier stage in the procedure, and it is advisable to return and perform careful controls to determine the source of the difficulty (see Common Problems, below).

6. Combine the four reactions into one Eppendorf tube and extract with P/C/I, and C/I. Add 75 μl of 3 *M* sodium acetate (pH 5.3), 800 μl of isopropanol, and precipitate on ice for 20 min. Centrifuge at 14,000 rpm for 15 min at 4°, and wash the pellet with 70% ethanol. Resuspend the pellet at 0.5 μg/μl in TE (approximately 100 μl, depending on the yield; see previous section). This is the first difference product (DP1).

Change of Adapters on a Difference Product

1. Digest DP1 with *Dpn*II as follows: Mix 20 μl (10 μg) of DP1 DNA; 30 μl of 10× *Dpn*II buffer; 7.5 μl of *Dpn*II (75 U); and 242.5 μl of water, and incubate at 37° for 2–4 hr. This quantity provides ample material for subsequent blots and controls, and may be reduced by 50% if these are not required. Extract with P/C/I, then with C/I. Add 33 μl of 3 *M* sodium acetate (pH 5.3), 800 μl of 100% ethanol and precipitate at −20° for 20

min. Centrifuge at 14,000 rpm for 15 min at 4° and wash the pellet with 70% ethanol. Resuspend the pellet at 0.5 μg/μl in TE (20 μl). Check an aliquot on an agarose gel to ensure that the digest is efficient (Fig. 3B; compare lanes 1 and 2).

2. Dilute 1 μl of the digested DP1 1:10 with TE (final concentration, 50 ng/μl). Dilution of the digested difference product ensures that religation of the free J oligos still present will be insignificant owing to high levels of N oligos in the reaction. Set up ligation to N oligos by mixing 4 μl of diluted DNA (200 ng); 6 μl of 10× ligase buffer; 4 μl of N-24 (5′-AGGCAACTGTGCTATCCGAGGGAA-3′; 2 mg/ml); 4 μl of N-12 (5′-GATCTTCCCTCG-3′; 1 mg/ml); and 40 μl of water. Anneal the oligos in a PCR machine at 50° for 1 min, then cool to 10° at 1°/min. Add 2 μl of T4 DNA ligase (400 U/μl), and incubate overnight at 12–14°.

Generation of Second Difference Product. Dilute the DP1–N linker ligation to 1.25 ng/μl by adding 100 μl of TE. Mix 40 μl (50 ng) of N-ligated DP1 with 80 μl (40 μg) of driver (a driver:tester ratio of 800:1), and proceed through the subtraction and amplification steps as described in the sections Subtractive Hybridization and Generation of First Difference Product (above). Take great care to remember which is the correct driver for the DP1 tester. For amplification, use the N-24 primer and perform the PCR annealing/extension step at 72°. The second difference product (DP2) should look appreciably different from DP1 when run on an ethidium-stained gel. Under most circumstances, DP2 should produce clear bands within a background smear, which will vary between experiments (Fig. 3C, lane 1). These bands are usually real differences (representing either absolutely different or upregulated mRNAs), and can be cut out of the gel and cloned. If, however, a significant background smear remains, proceed directly to a third difference product (Fig. 3C, lane 2).

Generation of Third Difference Product (DP3). Follow the digestion and ligation procedures described in Change of Adapters on a Difference Product above, using J linkers. Dilute J-ligated DP2 to 1 ng/μl, then further dilute 1 μl of this into 39 μl of TE and mix with 40 μg (80 μl) of driver (a driver:tester ratio of 40,000:1). Proceed as described in the sections Subtractive Hybridization and Generation of First Difference Product (above) to generate DP3, performing the final amplification for 23 cycles. If the bands in DP3 are not strong, try a tester:driver ratio of 1:10,000 (Fig. 3D, lane 2).

Common Problems

The identification of many small problems and their possible solutions are covered in the text of the protocol; however, we suggest that special attention be paid to the following points.

Low yields of representations or difference products are typically due to including insufficient template in the PCRs. Insufficient template in the representation can result from excessively degraded mRNA or poor-quality cDNA. Misjudgment of DNA concentrations, failed preliminary reactions (digestion, ligation, and hybridization), or loss of material during the procedure can all account for lack of amplifiable template in the PCRs that generate either the representation or the difference products. A representation or DP1 showing no yield at all is usually due to an error in setting up the PCR, or from loss of the DNA pellet (particularly during the steps leading up to digestion with MBN). The reactions should then be repeated using template from the postligation or post-MBN stages. When a low yield is encountered in the DP1 (or DP2), and the representation was at the correct concentration, the efficiency of the digestion and ligation should be checked. This is most easily accomplished by running uncut, cut, and religated representation (or DP) on an ethidium bromide-stained 1.5% agarose gel and noting how the prominent bands shift down by 48 bp after digestion (Fig. 3B) and back up again after ligation. Failure of these reactions can be caused by the use of poor-quality components or the presence of excessive salt in the DNA. Another potential cause of low yield is failure to properly resuspend the DNA before or after the subtractive hybridization.

The presence of an excessively high background smear after two rounds of subtraction may be due to failure of the MBN digestion, or to incomplete melting of the DNA during the hybridization. In this case, the temperature of the PCR block should be accurately determined.

The cloning of excessive numbers of false positives usually indicates that too low a stringency has been applied, while a lack of any bands in DP2 after obtaining potential products in DP1 probably indicates that too high a stringency has been used. The use of positive controls is advised, at least for first attempts at cDNA RDA, and these can be performed by spiking the original tester mRNA or cDNA with a titrated quantity of an amplifiable *Dpn*II gene fragment known to be absent from the driver. The addition of too many positive control species, however, is counterproductive as it may interfere with amplification of desired unknown species. The isolation of large numbers of clones that have no matches against sequences in the DNA databases can be indicative of genomic DNA contamination and these should be checked for expression at the RNA level.

Screening

It is wise to give careful consideration to the subsequent analysis of difference products before commencement of cDNA RDA. This screening procedure will almost certainly require additional material (RNA, cDNA, or representations) to the samples used for the cDNA RDA procedure itself, and generation of the necessary controls at the outset can save considerable time in the long run. The precise way in which screening is conducted will vary with the experiment being performed, but in almost all cases it is likely that products will need to be cloned, confirmed as true differences, and identified.

The products of cDNA RDA are amplified fragments of cDNA flanked by restriction sites. These are easily excised from agarose gels and cloned into commonly available vectors such as pBluescript. For example, difference products would be digested with *Dpn*II to remove the linkers and ligated into a plasmid digested with a restriction enzyme that generates compatible ends, such as *Bam*HI or *Bgl*II. When the difference product contains a mixture of different-sized fragments, each band should be isolated individually by careful gel purification, as this maximizes product recovery. It is advisable to analyze at least 10 clones for each band isolated, as it is not uncommon for a visible band to contain 2 or more different genuine products. It is also possible that a small fraction of the subclones will contain false products (those that are not real differences between driver and tester), and this can "hide" a more abundant, genuine species if too few colonies are picked and analyzed. A simple way to verify a cloned difference product is to use it as a probe on Southern blots of the original representations. This process has the advantage of being rapid and convenient, as abundant material is already available as a by-product of the cDNA RDA, especially helpful where RNA is in short supply. Results of such "representation blots" should, however, be treated with caution, as a successful subtraction judged in this manner reflects only that particular experiment. As discussed previously, there can be a number of technical reasons why representations may not accurately reflect the mRNA from which they were prepared, and why individual mRNA levels might fluctuate between preparations. It is therefore necessary to generate separate preparations of RNA from identical cell populations so that expression levels of difference products can be independently verified by Northern blot, or by RT-PCR using primers designed from the sequences of the difference products. Other variations on this confirmatory screening have included screening driver- and tester-specific cDNA libraries with products and noting their representations in each.

Identifying products of cDNA RDA is usually straightforward. Unlike differential display, cloned cDNA RDA products are not limited to the extreme 3′ end of the gene. Fragments may be derived from any part of the cDNA, including 5′ and 3′ untranslated regions, but are frequently located in coding regions, so that direct identification of products by sequencing and database searching is often possible. In experiments in which a high driver-to-tester ratio has been used there are likely to be few false products, and so sequencing can often be the most convenient approach to the first round of screening, providing rapid information on products and determining whether the experiment is likely to have been successful. Even clones that do not immediately match with known genes commonly possess homologous expressed sequence tags (ESTs), from which it may be possible to build extensive continuous stretches of sequences for the new genes by sequential database screening.

After verification and identification, difference products can be used as probes for screening cDNA libraries or as the starting fragment for 5′ and 3′ RACE (rapid amplification of cDNA ends). If directionality can be deduced, products may be cloned into expression vectors and may function as antisense constructs. They can also be used to generate antibodies—either as cloned and expressed protein fragments, or as the sequence basis for the synthesis of peptides. Care should be taken with these latter approaches, however, as PCR errors are common after the extensive amplification inherent in generating difference products.

Modifications of cDNA Representational Difference Analysis

cDNA RDA is a flexible approach to cDNA subtraction that can be modified in a number of ways, allowing application to a broad range of questions.

Competition

In many subtractions, one or more products are known in advance to be expressed in the tester but not the driver. These known differences can provide a useful positive control for the cDNA RDA method, but can also compete with and obscure other, more interesting differences. The known differences can be eliminated by adding the cDNA for these genes to the driver. This can be accomplished either with plasmids containing cloned cDNAs, with PCR products, or by including only the amplifiable portions of the genes. Amounts to add for successful competition should be determined experimentally, but up to 50% of the mass of the driver can be composed of specific competitor sequences, with only a twofold drop in overall stringency. By this strategy, even highly abundant species can be successfully eliminated, as the specific competitor : target ratio may far exceed that of the driver : tester ratio. This prevents amplification of the undesired target, increases the chances of identifying new genes, and accelerates the analysis. When multiple differences are expected, it is possible to take advantage of competition to perform successive rounds of cDNA RDA. This "iterative cDNA RDA" can be used to discover genes that are originally concealed by more vigorously amplifying fragments. To perform iterative cDNA RDA, previously cloned differences (and false products) should be combined and ligated to R linkers, which can then be amplified and mixed with the existing driver. Alternatively, R-ligated products can be mixed at an experimentally determined ratio with R-ligated driver and amplified together for subsequent use as the driver. It is sometimes sufficient to include competitor in the second round of subtraction only, although it should be noted that inclusion of 50% competitor reduces the remaining driver : tester ratio to 400 : 1, which may not be adequate to drive out abundant cDNAs.

Combined Driver Populations

On many occasions it may be desirable to combine driver populations to obtain genes that are expressed in one population, but not in several others. For example, it may be interesting to identify genes that are specifically activated by a novel drug but not by any other members of the drug family. To achieve this goal, one or several groups of cells would be treated with drug family members (or left untreated). Representations would then be generated from these populations, and mixed to form the driver. The tester would be derived from cells treated with the novel drug. Application of cDNA RDA would result in the identification of genes that were uniquely expressed in the novel drug-treated population. In another example, it may be desirable to isolate genes that are transcribed at a developmental precursor stage and are not expressed in any of a number of differentiated daughter cells. The same type of combination of representations to generate the driver material would achieve this purpose. These types of studies are possible because virtually any cDNA can be included in the driver without harming the procedure, providing the average length is at least equal to that of the tester cDNA. In such experiments, the driver : tester ratios of individual fragments will vary depending on the composition of the driver, and experimentation with ratios may be necessary to achieve the desired outcome.

Time Course/Multiple Treatments

Representations are relatively easy and rapid to prepare and can serve as the basis for experiments that demand the study of multiple samples, including post "event" or developmental time courses or multiple treatments of the same cells or tissues (e.g., dose responses). In such studies, the zero time or untreated material serves as the driver, while the tester can be composed of representations from either a single time point, or combined from multiple time points or treatments. In the latter case, genes expressed throughout the course of the experiment will be detected. Using these gene fragments as probes on Southern blots of the individual representations that comprise the tester will give an indication of the peak of expression of each product, permitting the convenient identification of acutely regulated genes, and obviating the need to prepare cDNA libraries for each time point or treatment.

Variable Stringency

cDNA RDA has the capability to detect both relative (more than three- to fivefold) and absolute differences in gene expression (although see Limitations, above). The detection of relative differences is favored by the use of a lower stringency, while absolute differences should be present at low stringency, but will also survive high-stringency subtractions. The term *stringency* used in this context refers to the driver : tester ratio, rather than to hybridization conditions of salt concentration and temperature, which remain constant. The driver : tester ratios indicated in the standard protocols are 100 : 1, 800 : 1, and 40,000 : 1 for the first, second, and third subtractions, respectively. In many cases the second difference product (800 : 1) will be enriched for fragments of genes that are upregulated in the tester, but still present in the driver. These are typically detected with a false-positive rate of approximately 10%; however, owing to experimental variation, this rate may occasionally increase to unacceptable levels, and will typically be indicated by a failure to see discrete bands with a low-level background smear on an ethidium bromide-stained gel at DP2. Consequently it is sometimes necessary to increase the stringency to reduce the cloning of false products. This can be achieved by proceeding to generate a third difference product. The prescribed stringency for this hybridization is 40,000 : 1, conditions that generally favor the production of absolute differences only. Although such a high stringency results in a low rate of false positives, many potentially interesting sequences can be lost, and it may be necessary to determine a compromise driver : tester ratio by experimentation. We recommend that this be performed by maintaining the driver quantity constant and varying the amount of tester to generate stringencies between 800 : 1 and 40,000 : 1. Suitable starting ratios of driver : tester would include 5000 : 1 and 10,000 : 1. Bear in mind that when combined drivers are being used, driver : tester ratios may not be identical for all the fragments in the mixture. It should be noted that only a subset of relative differences is likely to be detected, and that if a comprehensive population is desired, other approaches may be more suitable.

Multiple Four-Base Cutters

In the event that a *Dpn*II digest of the tester cDNA does not generate any amplifiable difference products, it is possible to repeat the experiment using a restriction enzyme that cuts at a different 4-bp recognition site, such as *Nla*III. It is preferable to select enzymes that leave 4-bp 5′ overhangs, as these anneal and ligate more efficiently, and it is necessary to redesign the terminal sequence of the 12-mer oligonucleotides to be compatible with the new site.

Prokaryote Applications

cDNA RDA is not necessarily dependent on the presence of polyadenylated mRNA, and therefore in principle is applicable to prokaryotic systems. However, prokaryotes pose particular difficulties in that they synthesize unstable, nonpolyadenylated mRNA. The solution to these problems may lie in the use of random-primed cDNA prepared from total RNA, and an additional step to remove ribosomal RNA is likely to be necessary to generate favorable hybridization conditions and guarantee reproducibility. Although no cDNA RDA protocols for use with prokaryotes are currently available, this possibility is being actively pursued by a number of groups, and it seems likely that a reliable methodology will be available in the near future.

[20] Suppression Subtractive Hybridization: A Versatile Method for Identifying Differentially Expressed Genes

By LUDA DIATCHENKO, SERGEY LUKYANOV, YUN-FAI CHRIS LAU, and PAUL D. SIEBERT

I. Introduction

Subtractive hybridization methods are valuable tools for identifying differentially regulated genes important for cellular growth and differentiation. Numerous subtractive hybridization techniques have been developed and used to isolate significant genes in many systems.[1-5] However, many

[1] S. M. Hedrick, D. I. Cohen, E. A. Nielsen, and M. M. Davis, *Nature (London)* **308,** 149 (1984).
[2] T. D. Sargent and I. B. Dawid, *Science* **222,** 135 (1983).
[3] E. Hara, T. Kato, S. Nakada, S. Sekiya, and K. Oda, *Nucleic Acids Res.* **19,** 7097 (1991).
[4] Z. Wang and D. D. Brown, *Proc. Natl. Acad. Sci. U.S.A.* **88,** 11505 (1991).
[5] M. Hubank and D. G. Schatz, *Nucleic Acids Res.* **22,** 5640 (1994).

Copyright © 1999 by Academic Press
All rights of reproduction in any form reserved.

0076-6879/99 $30.00

of these techniques require either tedious, complicated procedures or large amounts of starting material, thereby reducing their overall utility. Hence, these techniques can be greatly improved if the procedures can be streamlined and/or used with minute amounts of starting material.

We have developed a new polymerase chain reaction (PCR)-based cDNA subtraction method, termed suppression subtractive hybridization (SSH), and demonstrated its effectiveness[6] in generating tissue-specific cDNAs and in identifying functional sequences in a chromosome-specific manner. The central feature of the SSH technique is the suppression PCR effect, previously observed in our laboratories.[7–9] The suppression PCR effect is mediated by long inverted terminal repeats. When attached to the ends of DNA fragments, these inverted repeats form stable panhandle-like loop structures after each denaturation and annealing cycle (Fig. 1). In a PCR with primers derived from the sequences of the long inverted repeats, the panhandle-like structures cannot be amplified exponentially. This is because intramolecular annealing of the long inverted terminal repeats is highly favored and is more stable than intermolecular annealing of the shorter PCR primer to the long inverted repeats. By incorporating this suppression effect in a PCR amplification scheme, undesirable DNA fragments can be eliminated from a mixture of target sequences.

On the basis of this design, we have developed several PCR-based methods, including an earlier version of cDNA subtraction,[10] chromosome walking,[8] rapid amplification of cDNA ends,[9] differential mRNA display,[11] and SSH cDNA subtraction.[6,12] In particular, the subtraction (SSH) method, which we describe here, normalizes (equalizes) sequence abundance among the target cDNA population, eliminates any intermediate step(s) for physical separation of single-stranded and double-stranded cDNAs, requires

[6] L. Diatchenko, Y.-F. C. Lau, A. Campbell, A. Chenchik, F. Moqadam, B. Huang, S. Lukyanov, K. Lukyanov, N. Gurskaya, E. Sverdlov, and P. D. Siebert, *Proc. Natl. Acad. Sci. U.S.A.* **93,** 6025 (1996).

[7] K. A. Lukyanov, G. A. Launer, V. S. Tarabykin, A. G. Zaraisky, and S. A. Lukyanov, *Anal. Biochem.* **229,** 198 (1995).

[8] P. D. Siebert, A. Chenchik, D. E. Kellogg, K. A. Lukyanov, and S. A. Lukyanov, *Nucleic Acids Res.* **23,** 1087 (1995).

[9] A. Chenchik, L. Diachenko, F. Moqadam, B. Tarabykin, S. Lukyanov, and P. D. Siebert, *BioTechniques* **21,** 526 (1996).

[10] S. A. Lukyanov, N. G. Gurskaya, K. A. Lukyanov, V. S. Tarabykin, and E. D. Sverdlov, *Bioorg. Khim.* **20,** 701 (1994). [In Russian]

[11] M. Matz, N. Usman, D. Shagin, E. Bogdanova, and S. Lukyanov, *Nucleic Acids Res.* **25,** 2541 (1997).

[12] N. G. Gurskaya, L. Diatchenko, A. Chenchik, P. D. Siebert, G. L. Khaspekov, K. A. Lukyanov, L. L. Vagner, O. D. Ermolaeva, S. A. Lukyanov, and E. D. Sverdlov, *Anal. Biochem.* **240,** 90 (1996).

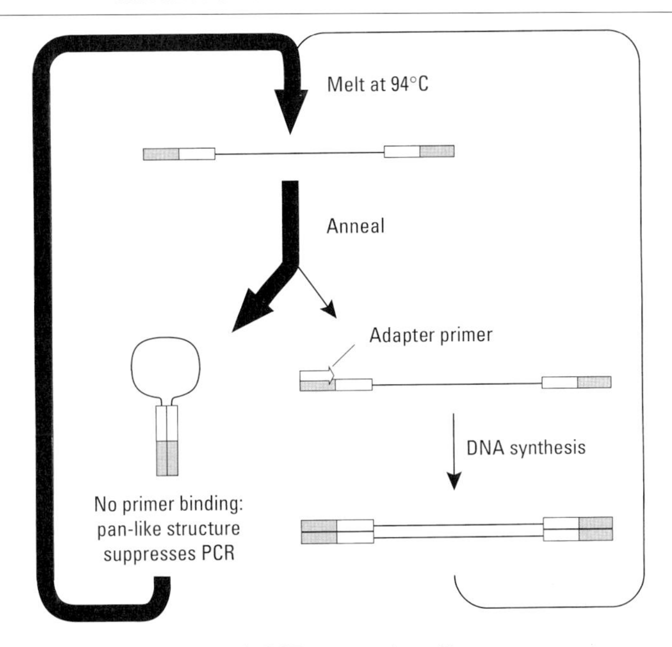

FIG. 1. PCR suppression effect.

only one round of subtractive hybridization, and can achieve greater than 1000-fold enrichment for differentially expressed cDNAs.

Initially, we had demonstrated the effectiveness of the SSH method by generating a testis-specific cDNA library and characterizing selected clones. We showed that the subtracted cDNA mixture can be used directly as a hybridization probe for screening recombinant DNA libraries, such as a human Y chromosome cosmid library, thereby identifying chromosome-specific sequences and tissue-specific transcripts. Furthermore, we demonstrated the feasibility of identifying preferentially expressed genes in both human testis and ovary.[13]

In this chapter we describe an improved version of the SSH technique for generating tissue-specific cDNA libraries. We also describe a modified differential fingerprinting strategy for identifying recombinant DNA clones harboring tissue- and chromosome-specific functional sequences and for rapid isolation of truly differentially expressed cDNA clones.

[13] H. Jin, X. Cheng, L. Diatchenko, P. D. Siebert, and C.-C. Huang, *BioTechniques* **23,** 1084 (1997).

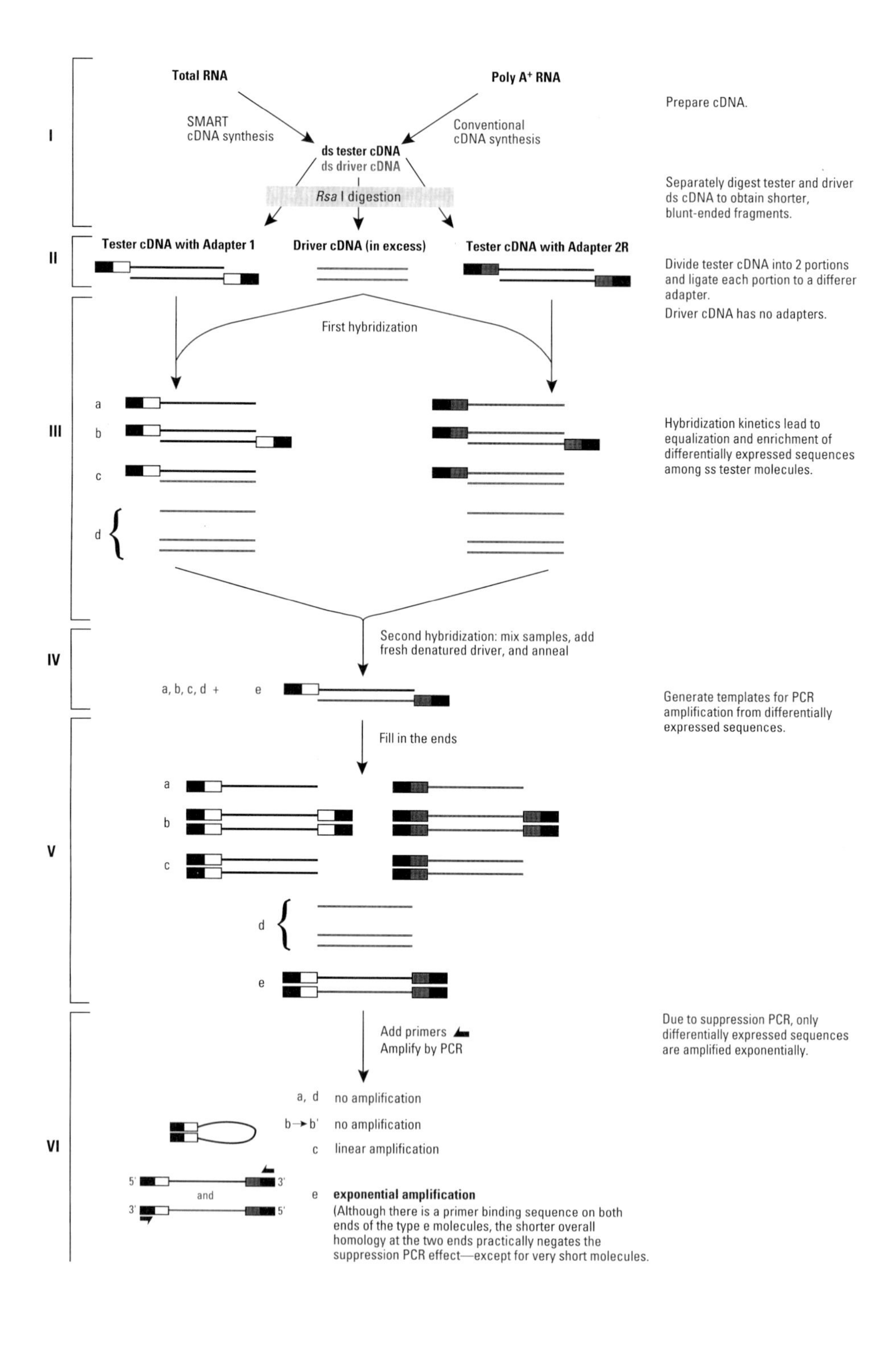
Total RNA
Poly A+ RNA
Prepare cDNA.
SMART
cDNA synthesis
Conventional
cDNA synthesis
ds tester cDNA
ds driver cDNA
Rsa I digestion
Separately digest tester and driver
ds cDNA to obtain shorter,
blunt-ended fragments.
I
II
Tester cDNA with Adapter 1
Driver cDNA (in excess)
Tester cDNA with Adapter 2R
Divide tester cDNA into 2 portions
and ligate each portion to a differer
adapter.
Driver cDNA has no adapters.
First hybridization
III
a
b
c
d
Hybridization kinetics lead to
equalization and enrichment of
differentially expressed sequences
among ss tester molecules.
IV
Second hybridization: mix samples, add
fresh denatured driver, and anneal
a, b, c, d +
e
Generate templates for PCR
amplification from differentially
expressed sequences.
V
Fill in the ends
a
b
c
d
e
VI
Add primers
Amplify by PCR
Due to suppression PCR, only
differentially expressed sequences
are amplified exponentially.
a, d
no amplification
b→b'
no amplification
c
linear amplification
5'
3'
and
3'
5'
e
exponential amplification
(Although there is a primer binding sequence on both
ends of the type e molecules, the shorter overall
homology at the two ends practically negates the
suppression PCR effect—except for very short molecules.

II. Principle of Suppression Subtractive Hybridization

There are six general steps in the SSH strategy (Fig. 2). First, the investigator isolates poly(A)$^+$ RNA for the tissues or cell types being compared. The "tester" RNA contains the differentially expressed sequences, while the "driver" is the reference RNA. Double-stranded cDNAs are synthesized independently from the tester and driver mRNAs, and are digested with a four base-cutting restriction enzyme that yields blunt ends. Second, the tester cDNA fragments are divided into two samples (1 and 2) and ligated with two different adapters (Ad1 and Ad2R), resulting in two populations of tester. The ends of the adapters are designed without phosphate groups, so that only the longer strand of each adapter can be covalently attached to the 5′ ends of the cDNA (Fig. 2, step II). Third, the excess driver cDNA and a small amount of each tester cDNA population are mixed, heat denatured, and allowed to anneal (Fig. 2, step III). During this first hybridization step the single-stranded cDNA tester fraction (a) is normalized, which means that concentrations of high- and low-abundance cDNAs become roughly equal. Normalization occurs because the annealing process generating homohybrid (b) and heterohybrid (c) cDNAs is faster for more abundant molecules, owing to the second order of hybridization kinetics, than annealing of the less abundant cDNAs that remain single stranded (a).[14] By controlling the extent of the hybridization, the single-

[14] R. J. Britten, and E. H. Davidson, *in* "Nucleic Acid Hybridization—A Practical Approach" (B. D. Hames and S. J. Higgins, eds.), pp. 3–15. IRL Press, Oxford, 1985.

Fig. 2. Overall scheme for suppression subtractive hybridization. There are six steps in this procedure. (I) Double-stranded cDNAs are synthesized from tester and driver mRNAs and digested with *Rsa*I to generate optimal fragments for hybridization reactions. (II) Adapters Ad1 and Ad2R are ligated to two separate populations of the tester cDNAs, which are then (III) mixed with 30× excess driver cDNAs. The mixtures are processed in the first hybridization to normalize and enrich differentially expressed sequences among ss tester molecules. (IV) The reactions from the first hybridization are mixed and processed for a second hybridization in the presence of additional driver ss cDNAs, resulting in combination of different hybrid types. (V) The ends of the respective hybrids are filled in to generate cDNA fragments from differentially expressed genes that can be preferentially amplified by PCR using appropriate primers. (VI) Two rounds of PCR are performed to preferentially amplify differentially expressed genes. Solid lines represents the *Rsa*I-digested tester or driver cDNA. Solid boxes represent the outer part of the adapter Ad1 and Ad2R longer strands and corresponding PCR primer P1 sequence. White boxes represent the inner part of the adapter Ad1 longer strand and corresponding nested PCR primers NP1. Shaded boxes represent the inner part of the adapter Ad2R longer strand and corresponding nested PCR primer NP2R.

stranded forms of highly abundant cDNAs can then be reduced to the same levels as those of less abundant ones, thereby normalizing the representation of cDNA population. Furthermore, the single-stranded cDNAs in the tester fraction (a) are significantly enriched in cDNAs for differentially expressed genes, as "common" nontarget cDNAs form heterohybrids (c) with the driver owing to an excess of the driver cDNA.

In the fourth step, the two samples from the first hybridization are combined and annealed further with additional freshly denatured driver cDNAs. Under these conditions, only single-stranded tester cDNAs are able to reassociate and form b, c, and new (e) hybrids (Fig. 2, step IV). Addition of a second portion of denatured driver at this stage increases the extent (C_0t value) of the hybridization, thereby enriching the fraction e hybrids for differentially expressed genes. The entire population of molecules is then subjected to two rounds of PCR to amplify the desired differentially expressed sequences. During the first cycle of the primary PCR, the adapter ends are filled in, creating the complementary primer-binding sites needed for amplification (Fig. 2, step V). This results in several types of hybrids (b, c, d, and e) containing different combinations of adapter sequences at their ends. Among the molecules obtained from the second hybridization and extension reaction, type a and d molecules do not contain primer-binding sites, and type c molecules can be amplified only at a linear rate. Type b molecules contain long inverted repeats on the ends and form stable panhandle-like structures after each denaturation–annealing step and are unsuitable for amplification owing to the PCR suppression effect (see Fig. 1). Only type e molecules have different adapter sequence at their ends: one is from sample 1, and the other is from sample 2. Having different adapters on each end allows them to be exponentially amplified by PCR (there is no suppression PCR effect), thereby generating a cDNA population that is preferentially enriched for genes differentially expressed in the target tissue or cell type. This hybrid (e) is greatly enriched for differentially expressed sequences. The mathematical model for the formation of fraction e molecules, as well as the rate of enrichment, has been described.[12]

III. Detailed Experimental Procedures

Some modifications in the procedure and adapter sequences have been incorporated in the original SSH technique described previously.[6] These modifications improve the overall efficiency of the subtractive hybridization and are described here.

A. Materials

1. Oligonucleotides. The following gel-purified oligonucleotides are used at a concentration of 10 μM:

cDNA synthesis primer: 5′-TTTTGTACAAGCTT$_{30}$-3′
Adapters:
Ad1: 5′-CTAATACGACTCACTATAGGGCTCGAGCGGCCGCCCGGGCAGGT-3′
3′-GGCCCGTCCA-5′
Ad2: 5′-TGTAGCGTGAAGACGACAGAAAGGGCGTGGTGCGGAGGGCGGT-3′
3′-GCCTCCCGCCA-5′
Ad2R: 5′-CTAATACGACTCACTATAGGGCAGCGTGGTCGCGGCCGAGGT-3′
3′-GCCGGCTCCA-5′
PCR primers:
P1: 5′-CTAATACGACTCACTATAGGGC-3′
P2: 5′-TGTAGCGTGAAGACGACAGAA-3′
NP1: 5′-TCGAGCGGCCGCCCGGGCAGGT-3′
NP2: 5′-AGGGCGTGGTGCGGAGGGCGGT-3′
NP2R: 5′-AGCGTGGTCGCGGCCGAGGT-3′
G3PDH 5′ primer: 5′-ACCACAGTCCATGCCATCAC-3′
G3PDH 3′ primer: 3′-TCCACCACCCTGTTGCTGTA-3′
Blocking solution: A mixture of cDNA synthesis primer, nested primers (NP1 and NP2R), and their respective complementary oligonucleotides, at a concentration of 2 mg/ml each

2. Buffers and Enzymes

First-strand synthesis: mouse mammary leukemia virus (MMLV) reverse transcriptase (200 units/μl), 5× first-strand buffer [250 m*M* Tris-HCl (pH 8.3), 30 m*M* $MgCl_2$, 375 m*M* KCl]
Second-strand synthesis: 20× second-strand enzyme cocktail [DNA polymerase I, 6 units/ml (New England BioLabs, Beverly, MA), RNase H, 0.2 units/ml (Epicentre Technologies, Madison, WI), *Escherichia coli* DNA ligase, 1.2 units/ml (New England BioLabs)]; 5× second-strand buffer [500 m*M* KCl, 50 m*M* ammonium sulfate, 25 m*M* $MgCl_2$, 0.75 m*M* β-NAD, 100 m*M* Tris-HCl (pH 7.5), bovine serum albumin (BSA, 0.25 mg/ml)]
T4 DNA polymerase (3 units/μl; New England BioLabs)
Endonuclease digestion: 10× *Rsa*I restriction buffer [100 m*M* Tris-HCl (pH 7.0), 100 m*M* $MgCl_2$, 10 m*M* dithiothreitol (DTT)], *Rsa*I (10 units/μl; New England BioLabs)
Ligation: T4 DNA ligase (400 units/μl; New England BioLabs), 5× DNA ligation buffer [250 m*M* Tris-HCl (pH 7.8), 50 m*M* $MgCl_2$, 10 m*M* DTT, BSA (0.25 mg/ml)], ATP (3 m*M*)
Hybridization: 4× Hybridization buffer [4 *M* NaCl, 200 m*M* HEPES (pH 8.3), 4 m*M* cetyltrimethyl ammonium bromide (CTAB)], dilution buffer [20 m*M* HEPES-HCl (pH 8.3), 50 m*M* NaCl, 0.2 m*M* EDTA]

PCR amplification: Clontech (Palo Alto, CA) Advantage cDNA PCR mix, containing a mixture of KlenTaq-1 and DeepVent DNA polymerases (New England BioLabs) and TaqStart antibody (Clontech); 10× reaction buffer [40 m*M* Tricine-KOH (pH 9.2 at 22°), 3.5 m*M* magnesium acetate, 10 m*M* potassium acetate, BSA (75 mg/ml)]. The TaqStart antibody provides automatic hot start PCR.[15] Alternatively, *Taq* DNA polymerase alone can be used, but five additional thermal cycles will be needed in both the primary and secondary PCR, and the additional cycles may cause higher background. If the Advantage DNA PCR mix is not used, manual hot start or hot start with wax beads is strongly suggested to reduce nonspecific DNA synthesis

General reagents: dNTP mix (10 m*M* each of dATP, dCTP, dGTP, dTTP)

Please note that all cycling parameters have been optimized using a Perkin-Elmer (Norwalk, CT) DNA thermal cycler 480. Perkin-Elmer GeneAmp PCR systems 2400/9600 have also been tested. For the latter, the denaturing time needs to be reduced from 30 to 10 sec. For a different type of thermal cycler, the cycling parameters must be optimized for that machine.

B. Methods

1. Preparation of Subtracted Library

a. FIRST-STRAND cDNA SYNTHESIS. Perform this procedure individually with each tester and driver poly(A)$^+$ RNA sample.

1. For each tester and driver sample, combine the following components in a sterile 0.5-ml microcentrifuge tube. (Do not use a polystyrene tube.)

Poly(A)$^+$ RNA (2 μg)	2–4 μl
cDNA synthesis primer (10 μM)	1 μl
Sterile H_2O (if needed) to	5 μl

2. Incubate the tubes at 70° in a thermal cycler for 2 min.
3. Cool the tubes at room temperature for 2 min and briefly centrifuge the tubes.
4. Add the following to each reaction tube:

First-strand buffer (5×)	2 μl
dNTP mix (10 m*M* each)	1 μl
Sterile H_2O*	1 μl
MMLV reverse transcriptase (200 units/μl)	1 μl

*Optional: To monitor the progress of cDNA synthesis, dilute 1 μl of

[15] D. E. Kellogg, I. Rybalkin, S. Chen, N. Mukhamedova, T. Vlasik, P. Siebert, and A. Chenchik, *BioTechniques* **16,** 1134 (1994).

[α-^{32}P]dCTP (10 mCi/ml, 3000 Ci/mmol) with 9 μl of H_2O, and replace the H_2O above with 1 μl of the diluted label.

5. Incubate the tubes at 42° for 1.5 hr in an air incubator.

6. Place the tubes on ice to terminate first-strand cDNA synthesis and immediately proceed to second-strand cDNA synthesis.

b. SECOND-STRAND cDNA SYNTHESIS

1. Add the following components, previously cooled on ice, to the first-strand synthesis reaction tubes:

Sterile H_2O	48.4 μl
Second-strand buffer (5×)	16.0 μl
dNTP mix (10 m*M*)	1.6 μl
Second-strand enzyme cocktail (20×)	4.0 μl

2. Mix the contents and briefly centrifuge the tubes. The final volume should be 80 μl.

3. Incubate the tubes at 16° (water bath or thermal cycler) for 2 hr.

4. Add 2 μl (6 units) of T4 DNA polymerase to each tube. Mix contents well.

5. Incubate the tubes at 16° for 30 min in a water bath or a thermal cycler.

6. Add 4 μl of 0.2 *M* EDTA to terminate second-strand synthesis.

7. Perform phenol–chloroform extraction and ethanol precipitation. We recommend the use of a 0.5 vol of 4 *M* ammonium acetate rather then sodium acetate for ethanol precipitation.

8. Dissolve the pellet in 50 μl of H_2O.

9. Transfer 6 μl to a fresh microcentrifuge tube. Store this sample at −20° until after *Rsa*I digestion for agarose gel electrophoresis to estimate yield and size range of double-stranded (ds) cDNA products synthesized.

c. *Rsa*I DIGESTION. Perform the following procedure with each experimental ds tester and driver cDNA. This step generates shorter, blunt-ended ds cDNA fragments optimal for subtractive hybridization.

1. Add the following reagents to the tube from step 8 above:

ds cDNA	43.5 μl
*Rsa*I restriction buffer (10×)	5.0 μl
*Rsa*I (10 units/μl)	1.5 μl

2. Mix and incubate at 37° for 2 hr.

3. Use 5 μl of the digest mixture and analyze on a 2% agarose gel along with undigested cDNA (from Section III,B,1,b, step 9) to analyze the efficiency of *Rsa*I digestion.

4. Add 2.5 μl of 0.2 *M* EDTA to terminate the reaction.

5. Perform phenol–chloroform extraction and ethanol precipitation.

6. Dissolve the pellet in 5.5 μl of H_2O and store at −20°.

d. ADAPTER LIGATION. We strongly recommend that subtractions be performed in both directions for each tester/driver cDNA pair. Forward subtraction is designed to enrich for differentially expressed transcripts present in tester but not in driver; reverse subtraction is designed to enrich for differentially expressed sequences present in driver but not in tester. The availability of such forward- and reverse-subtracted cDNAs will be useful for differential screening of the resulting subtracted tester cDNA library.

The tester cDNAs are ligated separately to adapter 1 (tester 1-1 and 2-1) and adapter 2R (tester 1-2 and 2-2). It is highly recommended that a third ligation of both adapters 1 and 2R to the tester cDNAs (unsubtracted tester control 1-c and 2-c) be performed and used as a negative control for subtraction.

Please note: The adaptors are not ligated to the driver cDNA.

1. Dilute 1 μl of each *Rsa*I-digested tester cDNA from the preceding section with 5 μl of sterile H_2O.

2. Prepare a master ligation mix according to the following proportion for each reaction:

Sterile H_2O	2 μl
Ligation buffer (5×)	2 μl
ATP (3 m*M*)	1 μl
T4 DNA ligase (400 units/μl)	1 μl

3. For each tester cDNA mixture, combine the reagents in a 0.5-ml microcentrifuge tube in the order shown. Pipette the solution up and down to mix thoroughly.

Component	Tube 1, tester 1-1 (μl)	Tube 2, tester 1-2 (μl)
Diluted tester cDNA	2	2
Adapter Ad1 (10 μ*M*)	2	0
Adapter AD2R (10 μ*M*)*	0	2
Master ligation mix	6	6
Final volume:	10	10

* This is a new adapter replacing the previous Ad2, which was used in most of the sample applications described in the following section.

4. In a fresh microcentrifuge tube, mix 2 μl of tester 1-1 and 2 μl of tester 1-2. This is the unsubtracted tester control 1-c. Do the same for each tester cDNA sample. After ligation, approximately one-third of the cDNA molecules in each unsubtracted tester control tube will have two different

adapter on their ends, suitable for exponential PCR amplification with adapter-derived primers.

5. Centrifuge the tubes briefly, and incubate at 16° overnight.
6. Stop the ligation reaction by adding 1 μl of 0.2 *M* EDTA.
7. Heat the samples at 72° for 5 min to inactivate the ligase.
8. Briefly centrifuge the tubes. Remove 1 μl from each unsubtracted tester control (1-c, 2-c, etc.) and dilute into 1 ml of H_2O. These samples will be used for PCR amplification (Section III,B,1,h).

Preparation of experimental adapter-ligated tester cDNAs 1-1 and 1-2 is now complete.

e. LIGATION EFFICIENCY TEST. The following PCR experiment is recommended to verify that at least 25% of the cDNAs have adapters on both ends. This experiment is designed to amplify fragments that span the adapter/cDNA junctions of testers 1-1 and 1-2 by adapter-specific P1 primer and a gene-specific primer. The PCR products generated using one gene-specific primer and adapter-specific primer should have about the same intensity as the PCR products amplified using two gene-specific primers, as shown in Fig. 3 for the G3PDH gene. It is important that the amplified

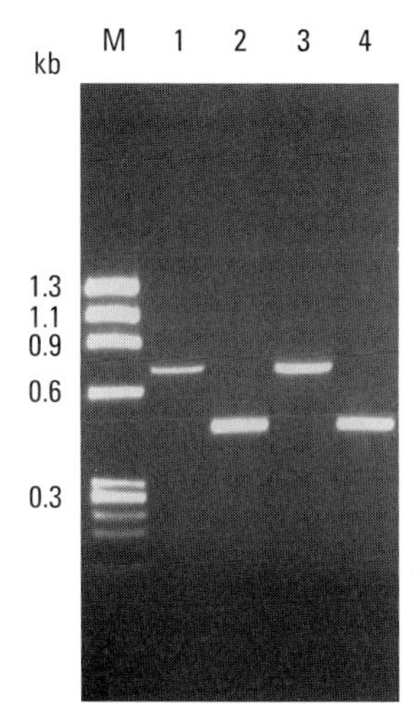

FIG. 3. Typical results of ligation efficiency analysis. The results shown here are for human samples; if mouse or rat samples are used, the PCR product amplified using the G3PDH 3′ primer and PCR primer 1 is ~1.2 kb instead of 0.75 kb. Lane 1, PCR products using tester 1-1 (adapter 1 ligated) as the template and the G3PDH 3′ primer and PCR primer 1; lane 2, PCR products using tester 1-1 (adapter 1 ligated) as the template, and the G3PDH 3′ and 5′ primers; lane 3, PCR products using tester 1–2 (adapter 2R ligated) as the template, and the G3PDH 3′ primer and PCR primer 1; lane 4, PCR products using tester 1-2 (adapter 2R ligated) as the template, and the G3PDH 3′ and 5′ primers. Shown is a 2% agarose–EtBr gel. Lane M, fX174 DNA/*Hae*III-digested size markers.

gene-specific fragment have no *Rsa*I restriction site. The selected G3PDH primers, listed in Section III,A,1, work well for human, mouse, and rat cDNA samples.

1. Dilute 1 μl of each ligated cDNA from Section III,B,1,d (e.g., testers 1-1 and 1-2) into 200 μl of H_2O.
2. Combine the reagents in four separate tubes as follows:

Component	Tube 1 (μl)	Tube 2 (μl)	Tube 3 (μl)	Tube 4 (μl)
Tester 1-1 (ligated to Ad1)	1	1	0	0
Tester 1-2 (ligated to Ad2R)	0	0	1	1
G3PDH 3′ primer (10 μM)	1	1	1	1
G3PDH 5′ primer (10 μM)	0	1	0	1
PCR primer P1 (10 μM)*	1	0	1	0
Total volume:	3	3	3	3

* Primer P1 contains 22 nucleotides corresponding to the 5′-end sequence of both adapters Ad1 and Ad2R. For experiments using adapters Ad1 and Ad2, P1 primer will be used in tube 1 and P2 primer will be used in tube 2.

3. Prepare a master mix for all of the reaction tubes plus one additional tube. For each reaction, combine the reagents in the following order:

Reagent	Amount per reaction tube (ml)	Amount for five reactions (ml)
Sterile H_2O	18.5	92.5
PCR buffer (10×)	2.5	12.5
dNTP mix (10 mM)	0.5	2.5
Advantage cDNA PCR mix (50×)	0.5	2.5
Total volume:	22.0	110.0

4. Mix thoroughly and briefly centrifuge the tubes.
5. Aliquot 22 μl of master mix into each reaction tube from step 2.
6. Overlay with 50 μl of mineral oil. Skip this step if an oil-free thermal cycler is used.
7. Incubate the reaction mixture in a thermal cycler at 75° for 5 min to extend the adaptors. (Do not remove the samples from the thermal cycler.)

8. Immediately commence 20 cycles, each cycle consisting of 94° for 30 sec, 65° for 30 sec, and 68° for 2.5 min.

9. Examine the products by electrophoresis on a 2% agarose–ethidium bromide (EtBr) gel.

Typical results are shown in Fig. 3. If no products are visible after 20 cycles, perform 5 more cycles of amplification, and again analyze the product by gel electrophoresis. The number of cycles will depend on the abundance of the specific gene. The efficiency of ligation is estimated to be the ratios of the intensities of the bands corresponding to the PCR products of tube 2 to tube 1 for adapter Ad1, and of tube 4 to tube 3 for adapter Ad2R. Any ligation efficiency of 25% or less will substantially reduce the subsequent subtraction efficiency; in this case, the ligation reaction should be repeated with fresh samples before proceeding to the next step.

For mouse or rat cDNAs, the PCR products amplified with the G3PDH 3′ primer and PCR primer 1 will be ~1.2 kb instead of the 0.75-kb band observed for human cDNA (because rat and mouse G3PDH cDNAs lack the *Rsa*I restriction site at nucleotide 340). However, for the human cDNA (which contains the *Rsa*I site), the presence of a 1.2-kb band suggests that the cDNAs are not completely digested by *Rsa*I. If a significant amount of this longer PCR product persists, the procedure should be repeated from the *Rsa*I digestion step (Section III,B,1,c).

f. FIRST HYBRIDIZATION

1. For each tester sample, combine the reagents in the following order:

Component	Hybridization 1.1 (μl)	Hybridization 1.2 (μl)
*Rsa*I-digested driver cDNA (Section III,B,1,c)	1.5	1.5
Ad1-ligated tester 1-1 (Section III,B,1,d)	1.5	0
Ad2R-ligated tester 1-2 (Section III,B,1,d)	0	1.5
Hybridization buffer (4×)	1.0	1.0
Final volume:	4.0	4.0

2. Overlay the samples with one drop of mineral oil and centrifuge briefly.

3. Incubate the samples in a thermal cycler at 98° for 1.5 min.

4. Incubate the samples at 68° for 7–12 hr and then proceed immediately to the next section.

g. SECOND HYBRIDIZATION. Repeat the following steps for each experimental driver cDNA.

1. Add the following reagents to a sterile 0.5-μl microcentrifuge tube:

Driver cDNA (Section III,B,1,c)	1 μl
Hybridization buffer (4×)	1 μl
Sterile H_2O	2 μl

2. Place 1 μl of this mixture in a 0.5-ml microcentrifuge tube and overlay it with one drop of mineral oil.
3. Incubate the tube in a thermal cycler at 98° for 1.5 min.
4. Remove the tube of freshly denatured driver from the thermal cycler.
5. To the tube of freshly denatured driver cDNAs, add hybridized sample 1.1 and hybridized sample 1.2 (prepared in Section III,B,1,f) in that order. This ensures that the two hybridization samples are mixed only in the presence of excess driver cDNAs.
6. Incubate the hybridization reaction at 68° overnight.
7. Add 200 μl of dilution buffer to the tube and mix well by pipetting.
8. Incubate in a thermal cycler at 68° for 7 min.

h. Polymerase Chain Reaction Amplification: Selection of Differentially Expressed cDNAs. Each experiment should have at least four reactions: (1) subtracted tester cDNAs, (2) unsubtracted tester control (1-c), (3) reverse-subtracted tester cDNAs, and (4) unsubtracted driver control for the reverse subtraction (2-c).

1. Aliquot 1 μl of each diluted cDNA (i.e., each subtracted sample from Section III,B,1,g and the corresponding diluted unsubtracted tester control from Section III,B,1,d) into an appropriately labeled tube.
2. Prepare a master mix for all of the primary PCR tubes plus one additional tube. For each reaction combine the reagents in the order shown:

Reagent	Amount per reaction (μl)
Sterile H_2O	19.5
PCR buffer (10×)	2.5
dNTP mix (10 mM)	0.5
PCR primer P1 (10 μM)	1.0
Advantage cDNA PCR mix (50×)	0.5
Total volume:	24.0

3. Aliquot 24 μl of master mix into each reaction tube prepared in step 1 above.
4. Overlay with 50 μl of mineral oil. Skip this step if an oil-free thermal cycler is used.

5. Incubate the reaction mixture in a thermal cycler at 75° for 5 min to extend the adapters. Do not remove the samples from the thermal cycler.
6. Immediately commence 27 cycles of the following: 94° for 30 sec, 66° for 30 sec, and 72° for 1.5 min.
7. Analyze 8 μl from each tube on a 2.0% agarose–EtBr gel run in 1× TAE (Tris-acetate electrophoresis) buffer. *Please note:*
   a. The sequence for PCR primer P1 is present at the 5′ ends of both adapters 1 and 2R and hence can be used in a single-primer PCR amplification.
   b. If no PCR product is observed after 27 cycles, amplify for 3 more cycles, and analyze the products again by gel electrophoresis.
   c. If no PCR product is observed in the subtracted or unsubtracted (unsubtracted tester control 1-c) samples, the activity of the *Taq* polymerase needs to be examined. If the problem is not with the polymerase mix, optimize the PCR cycling parameters by decreasing the annealing and extension temperatures in small increments. Lowering the temperature by only 1° can dramatically increase the background. Initially, try reducing the annealing temperature from 66° to 65° and the extension temperature from 72° to 71°.
   d. If PCR products are observed in the unsubtracted (unsubtracted tester control 1-c) samples but not in the subtracted sample, proceed to step 8 and perform more cycles of secondary PCR in step 13.
8. Dilute 3 μl of each primary PCR mixture in 27 μl of H_2O.
9. Aliquot 1 μl of each diluted primary PCR product mixture from step 8 into an appropriately labeled tube.
10. Prepare master mix for the secondary PCR samples plus one additional reaction by combining the reagents in the following order:

Reagent	Amount per reaction (μl)
Sterile H_2O	18.5
PCR buffer (10×)	2.5
Nested PCR primer NP1 (10 μM)	1.0
Nested PCR primer NP2R (10 μM)*	1.0
dNTP mix (10 mM)	0.5
Advantage cDNA PCR mix (50×)	0.5
Total volume:	24.0

* Nested PCR primer NP2 will be used for experiments using adapter 2.

11. Aliquot 24 μl of master mix into each reaction tube from step 9.
12. Overlay with one drop of mineral oil. Skip this step if an oil-free thermal cycler is used.
13. Immediately commence 10–12 cycles of the following: 94° for 30 sec, 68° for 30 sec, and 72° for 1.5 min.
14. Analyze 8 μl from each reaction on a 2.0% agarose–EtBr gel.

Please note: The patterns of secondary PCR products of subtracted samples usually look like smears with or without a number of distinct bands. If no product is observed after 12 cycles, perform 3 more cycles of amplification, and again check the products by gel electrophoresis. Add cycles sparingly; too many cycles will increase background.

i. POLYMERASE CHAIN REACTION ANALYSIS OF SUBTRACTION EFFICIENCY. At this point, it is important to determine the efficiency of the SSH procedure by comparing the abundance of known cDNAs before and after subtraction. Ideally, a nondifferentially expressed gene (e.g., a housekeeping gene) and a gene previously demonstrated to be differentially expressed between the two RNA sources should be used. These comparisons can be performed using either PCR or hybridization techniques.

The test described below uses the G3PDH primers to confirm the reduced relative abundance of G3PDH following the SSH procedure on subtracted samples.

1. Dilute the subtracted and unsubtracted (unsubtracted tester control 1-c and 2-c) secondary PCR products 10-fold in H_2O.

2. Combine the following reagents in 0.5-ml microcentrifuge tubes in the order shown:

Reagent	Tube 1 (μl)	Tube 2 (μl)
Diluted subtracted cDNA (second PCR product)	1.0	
Diluted unsubtracted tester control 1-c (second PCR product)		1.0
G3PDH 5′ primer (10 μM)	1.2	1.2
G3PDH 3′ primer (10 μM)	1.2	1.2
Sterile H_2O	22.4	22.4
PCR buffer (10×)	3.0	3.0
dNTP mix (10 mM)	0.6	0.6
Advantage DNA PCR mix (50×)	0.6	0.6
Total volume:	30.0	30.0

3. Mix and briefly centrifuge the tubes.

4. Overlay with one drop of mineral oil. Skip this step if an oil-free thermal cycler is used.

5. Use the following thermal cycling program for 18 cycles of the following: 94° for 30 sec, 60° for 30 sec, and 68° for 2 min.

6. Remove 5 μl from each reaction, place it in a clean tube, and store on ice. Put the rest of the reaction back into the thermal cycler for five more cycles.

7. Repeat step 6 twice (i.e., remove 5 μl after 28 and 33 cycles).

8. Examine the 5-μl samples (i.e., the aliquots that were removed from each reaction after 18, 23, 28, and 33 cycles) on a 2.0% agarose–EtBr gel.

Figure 4 shows an example of G3PDH reduction in a testis-specific subtracted mixture. In general, for the unsubtracted cDNA, a G3PDH product is observed after 18–23 cycles, depending on the abundance of G3PDH in the particular cDNA. For example, in skeletal muscle and heart poly $(A)^+$ RNA, G3PDH is extremely abundant. However, in the subtracted samples, a product should be observed about 5 to 15 cycles later than it is seen in the unsubtracted samples.

As a positive control for the enrichment of differentially expressed genes, repeat the PCR procedure above using PCR primers for a gene known to be expressed in the tester RNA, but not in the driver RNA. This cDNA should be enriched by the subtraction operation. Do not use PCR primers that amplify a cDNA fragment that contains an *Rsa*I restriction site between the PCR priming sites because it will not be amplified, owing to *Rsa*I digestion prior to the subtraction procedure.

j. GENERATION OF A SUBTRACTED LIBRARY. Once a subtracted sample has been confirmed to be enriched in cDNAs derived from differentially

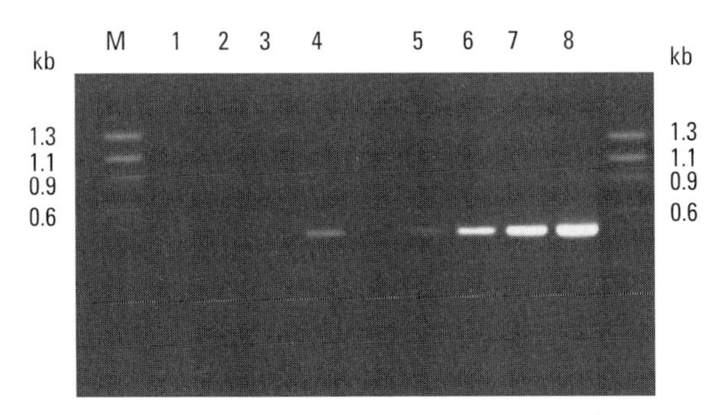

FIG. 4. Reduction of G3PDH abundance by SSH. Tester cDNA was prepared from human testis poly$(A)^+$ RNA and driver cDNA was prepared from a mixture of poly$(A)^+$ RNA samples from 10 different human tissues. PCR was performed on subtracted (lanes 1–4) or unsubtracted (lanes 5–8) secondary PCR product with the G3PDH 5′ and 3′ primers. Lanes 1 and 5, 18 cycles; lanes 2 and 6, 23 cycles; lanes 3 and 7, 28 cycles; lanes 4 and 8, 33 cycles. Lane M, fX174 DNA/*Hae*III-digested size markers.

expressed genes, the PCR products (from Section III,B,1,h) can be subcloned into plasmid vectors using several conventional cloning techniques. The following describes two such methods currently used in our laboratories.

1. T/A cloning: Use 3 μl of the secondary PCR product (from Section III,B,1,h) and a T/A-based cloning system, such as the Advantage PCR cloning kit (Clontech), according to the manufacturer protocol. Typically, 10^4 independent clones from 1 μl of secondary PCR product can be obtained using the preceding cloning kit and electroporation. It is important to optimize the cloning efficiency because a low cloning efficiency will result in a high background derived not from subtracted cDNA colonies.
2. Site-specific or blunt-end cloning: For site-specific cloning, cleave at the *Eag*I, *Not*I, and *Xma* (*Sma*I, *Srf*I) sites embedded in the adapter sequences and then ligate the products to an appropriate plasmid vector. Keep in mind that all of these sites might be present in the cDNA fragments. The *Rsa*I site in the adapter sequences can also be used for blunt-ended cloning. Commercially available cloning kits should be suitable for these purposes, otherwise consult Ref. 16.

Once it has been generated, the subtracted cDNA library can be arrayed into 96- or 384-well microtiter dishes for storage according to established protocols.[16] The number of independent colonies obtained for each library depends on the estimated number of differentially expressed genes, as well as the subtraction and subcloning efficiencies. In general, 500–1000 colonies can be initially arrayed and studied. The complexity of the library can be increased by additional subcloning of the secondary PCR products (from Section III,B,1,h).

2. Differential Screening. Two approaches can be utilized for differential screening of the arrayed subtracted cDNA clones: cDNA dot blots and colony dot blots. For colony dot blots, the arrayed cDNA colonies are spotted on nylon filters, grown on antibiotic plates, and processed for colony hybridization. This method usually is convenient and less expensive, but it is less sensitive and gives a higher background than PCR-based cDNA dot blots. The cDNA array approach is highly recommended and the following protocol is provided.

a. AMPLIFICATION OF cDNA INSERTS BY POLYMERASE CHAIN REACTION. For a high-throughput screening, a 96-well format PCR from one of several thermal cycler manufacturers is recommended. Alternatively, single tubes can be used.

[16] J. Sambrook, E. G. Fritsch, and I. Maniatis, "Molecular Cloning: A Laboratory Manual," 2nd Ed. Cold Spring Harbor Laboratory, Cold Spring Harbor, New York, 1989.

1. Replicate arrayed subtracted cDNA clones in 96-well plates. Alternatively, random colonies can be picked and processed directly without initially arraying the clones.

2. Grow each colony in 100 μl of LB-amp medium in a 96-well plate at 37° for at least 2 hr (up to overnight) with gentle shaking.

3. Prepare a master mix for 100 PCRs:

Reagent	Per reaction (μl)
PCR buffer (10×)	2.0
Nested primer NP1*	0.6
Nested primer NP2R*	0.6
dNTP mix (10 mM)	0.4
H_2O	15.0
Advantage cDNA PCR mix (50×)	0.4
Total volume:	19.0

* Alternatively, primers flanking the insertion site of the vector can be used in PCR amplification of the inserts.

4. Aliquot 19 μl of the master mix into each tube or well of the reaction plate.

5. Transfer 1 μl of each bacterial culture (from step 2, above) to each tube or well containing master mix.

6. Perform PCR in an oil-free thermal cycler with the following conditions: 94° for 1 min (1 cycle), followed by 94° for 30 sec, 68° for 3 min (22 cycles).

7. Analyze 5 μl from each reaction on a 2.0% agarose–EtBr gel.

b. PREPARATION OF cDNA DOT BLOTS OF POLYMERASE CHAIN REACTION PRODUCTS

1. For each PCR, combine 5 μl of the PCR product and 5 μl of 0.6 N NaOH (freshly made or at least freshly diluted from concentrated stock).

2. Transfer 1–2 μl of each mixture to a nylon membrane. This can be accomplished by dipping a 96-well replicator in the corresponding wells of a microtiter dish used in the PCR amplification and spotting it onto a dry nylon filter. Make at least two identical blots for hybridization with subtracted and reverse-subtracted probes (see Section III,B,1,d).

3. Neutralize the blots for 2–4 min in 0.5 M Tris-HCl (pH 7.5) and wash in 2× SSC (1× SSC is 0.15 M NaCl plus 0.015 M sodium citrate).

4. Immobilize the DNA on the membrane using a UV linking device (e.g., UV Stratalinker; Stratagene, La Jolla, CA) or bake the blots for 4 hr at 68°.

c. Preparation of cDNA Probes. Before the forward- and reverse-subtracted cDNA can be used as probes for differential screening, the adapter sequences must be removed. Despite their small size, these sequences cause a high background on the arrayed subtracted library. *Note:* If the adapters were removed for blunt-end or site-specific cloning (Section III,B,1,j, step 2), skip to Section III,B,2,d.

1. Combine two tubes (~40 μl total) of each forward- and reverse-subtracted secondary PCR products (from Section III,B,1,h).
2. Purify the PCR products using a silica matrix-based purification system such as the Advantage PCR-Pure kit (Clontech). The average concentration of each subtracted PCR product is about 10–40 ng/μl. Use ethanol to precipitate the purified cDNA mixture.
3. Adjust the volume after purification to 28 μl with H_2O. Set aside 3 μl for subsequent gel analysis.
4. Digest cDNA mixtures with a combination of *Rsa*I, *Eag*I, and *Sma*I restriction enzymes to remove the adapters as follows:
   a. Add 3 μl of New England BioLabs (NEB, Beverly, MA) 10× restriction enzyme buffer 4 and 1 μl of *Rsa*I restriction enzyme. Incubate for 1 hr at 37°.
   b. Add 1 μl of *Sma*I restriction enzyme and incubate for 1 hr at room temperature. Add 59 μl of H_2O, 10 μl of NEB 10× restriction buffer 3, and 1 μl of *Eag*I restriction enzyme. Incubate for 1 hr at 37°.
5. Analyze 3 μl of digested and undigested cDNA and undigested plasmid DNA on a 2% agarose–EtBr gel. Digested adapter should appear on the gel as a low molecular weight band.
6. Separate the adapters from the cDNA using a silica matrix-based purification resin as before. Adjust the volume of each sample up to 50 μl.
7. Label subtracted probes by random-primer labeling using commercially available kits.

d. Hybridization with Subtracted and Reverse-Subtracted cDNA Probes. Hybridization conditions have been optimized for Clontech ExpressHyb solution. Determine the optimal hybridization for other systems.

1. Prepare a prehybridization solution for each membrane:
   a. Combine 50 μl of 20× SSC, 50 μl of sheared salmon sperm DNA (10 mg/ml), and 10 μl of blocking solution (containing 2 mg/ml of unpurified NP1, NP2R, cDNA synthesis primers, and their complementary oligonucleotides).

   b. Boil the blocking solution for 5 min, then chill on ice.
   c. Combine the blocking solution with 5 ml of ExpressHyb hybridization solution (Clontech).
2. Place each membrane in the prehybridization solution prepared in step 1. Prehybridize for 40–60 min with continuous agitation at 72°.
3. Prepare hybridization probes:
   a. Mix 50 μl of 20× SSC, 50 μl of sheared salmon sperm DNA (10 mg/ml), 10 μl blocking solution, and purified probe (at least 10^7 cpm/100 ng of subtracted cDNA). Make sure the specific activity of each probe is approximately equal.
   b. Boil the probe for 5 min, then chill on ice.
   c. Add the probe to the prehybridization solution
4. Hybridize overnight with continuous agitation at 72°.
5. Prepare low-stringency [2× SSC–0.5% sodium dodecyl sulfate (SDS)] and high-stringency (0.2× SSC–0.5% SDS) washing buffers and warm them to 68°.
6. Wash the membranes with low-stringency buffer (four times, 20 min each) at 68°, then wash with high-stringency buffer (twice for 20 min each) at 68°.
7. Perform autoradiography.
8. If desired, remove probes from the membranes by boiling for 7 min in 0.5% SDS. Blots can typically be reused at least five times.

Note: To minimize hybridization background, we recommend storing the membranes at −20° when they are not in use.

IV. Sample Applications

Since the development of the SSH technique, it has been used in several subtractive hybridization studies to identify differentially expressed genes in our laboratories[6,12,13] and those of others.[17,18] To illustrate the versatility of the SSH technique, the following sample applications are described.

A. Generation and Cloning of Tissue-Specific cDNAs

One of the earliest applications for the SSH technique was in the generation of tissue-specific cDNA libraries. In this strategy, the tester cDNAs are derived from a target tissue, while the driver cDNAs are derived from cDNAs mixed from 10 other tissues. The tester cDNAs are subtracted

[17] T. Yokomizo, T. Izumi, C. Chang, Y. Takuwa, and T. Shimizu, *Nature (London)* **387,** 620 (1997).

[18] O. D. von Stain, W.-G. Thies, and M. Hofmann, *Nucleic Acids Res.* **25,** 2598 (1997).

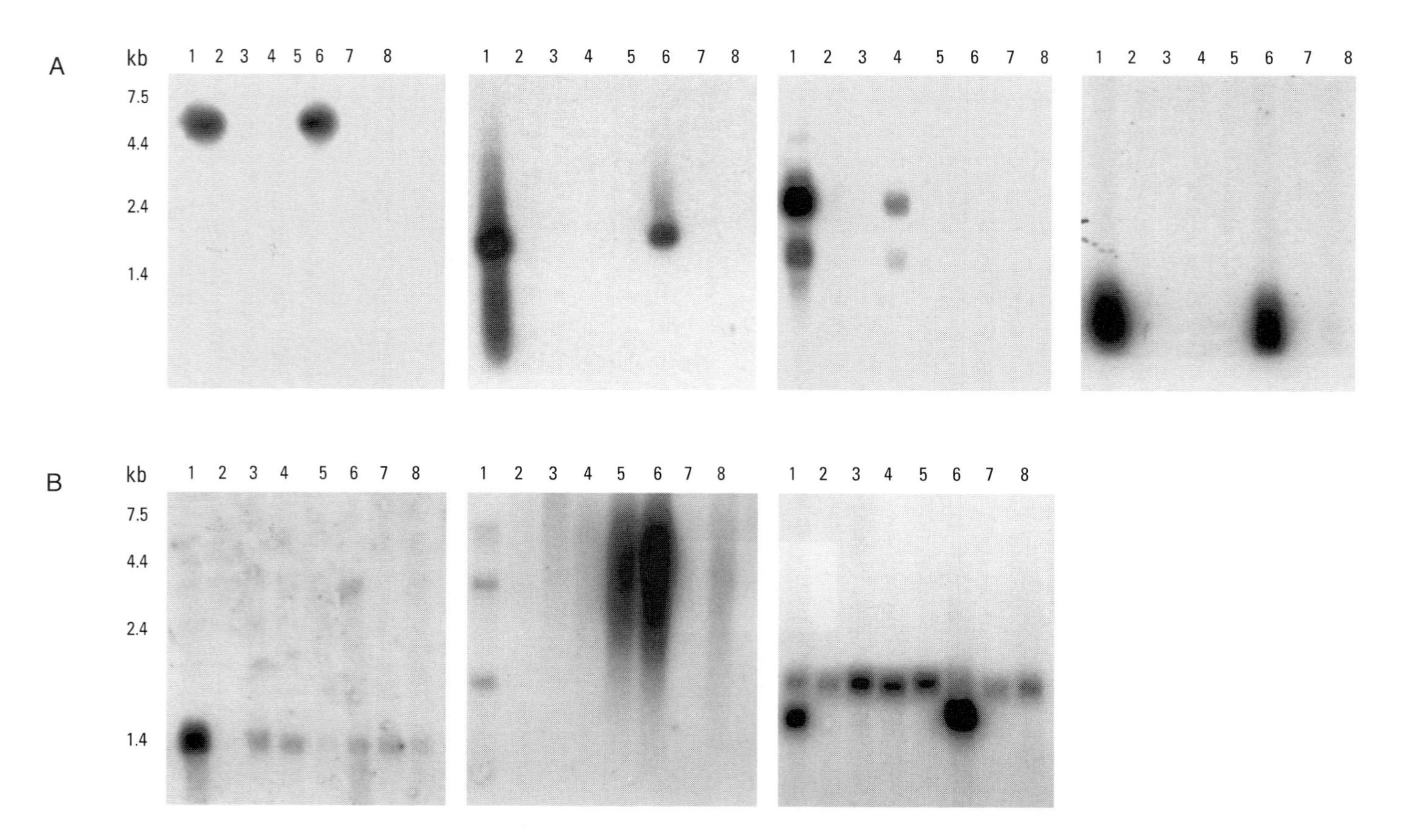
A
kb
7.5
4.4
2.4
1.4
1 2 3 4 5 6 7 8
1 2 3 4 5 6 7 8
1 2 3 4 5 6 7 8
1 2 3 4 5 6 7 8
B
kb
7.5
4.4
2.4
1.4
1 2 3 4 5 6 7 8
1 2 3 4 5 6 7 8
1 2 3 4 5 6 7 8

against the driver cDNAs according to the procedure outlined above, resulting in cDNA clones extremely specific to the target tissue. Such tissue specificity can be demonstrated by random cDNA sequencing and/or probing with multiple tissue Northern (MTN) blots (Clontech). To date, several tissue-specific cDNA libraries (including human testis, ovary, prostate, heart, and brain) have been generated in our laboratories with good to excellent results, as determined by the preceding criteria. Studies of the subtracted testis and ovary cDNA libraries have been reported elsewhere.[6,13] Similar studies of 24 randomly selected clones from the heart-specific cDNA library on MTN blots show that 22 of them hybridized specifically to the heart RNAs (Fig. 5A). Furthermore, cross-hybridization of these cDNA clones was observed in the skeletal muscle RNA sample, confirming the similar gene expression pattern of heart and skeletal muscle. Sequence analysis of 26 randomly selected clones from this subtracted library indicated that they share homology with many genes that are known to be expressed in the heart and 23 of these fragments are derived from independent sequences.

B. Identification of Tissue-Specific Genes on Human Y Chromosome

The specificity of the subtracted tissue-specific cDNA mixtures obtained by SSH technique suggests that they can be used as hybridization probes to detect homologous sequences from a particular human chromosome that may be expressed in the target tissue of interest. Initially, we have used the subtracted testis cDNA mixture as a probe to screen arrayed cosmids constructed from flow-sorted human Y chromosomes (LL0YNC03 "M" constructed by the National Gene Library Project at Lawrence Livermore National Laboratory, Livermore, CA). Figure 6A shows a sample filter that was hybridized with the subtracted human testis cDNA mixture. Several of the positive cosmid clones correspond to those of a testis-specific gene,

---

FIG. 5. (A) Results of MTN blotting of four heart-specific cDNA clones obtained by SSH procedure. Specific hybridization signals are observed in the heart (lane 1) and skeletal muscle (lane 6) RNA samples. Please note that these cDNAs detect mRNAs of different sizes, which suggests that they are derived from different genes. (B) Examples of Northern blot analysis of cosmid clones that specifically hybridized to the heart-specific subtracted probes. Human Y chromosome cosmids were identified by fingerprinting with subtracted heart-specific cDNA mixture. The Y cosmids are individually labeled and used as probes in MTN hybridization. The autoradiograms show a specific hybridization to the heart RNA (*left*), unique bands for the heart RNA, diffuse hybridization to the liver and skeletal muscle RNA (*middle*), and a hybridization pattern similar to that of cardiac actin mRNA (*right*). The MTN blots contained approximately 2 μg of poly$(A)^+$ RNA from human (1) heart, (2) brain, (3) placenta, (4) lung, (5) liver, (6) skeletal muscle, (7) kidney, and (8) pancreas.

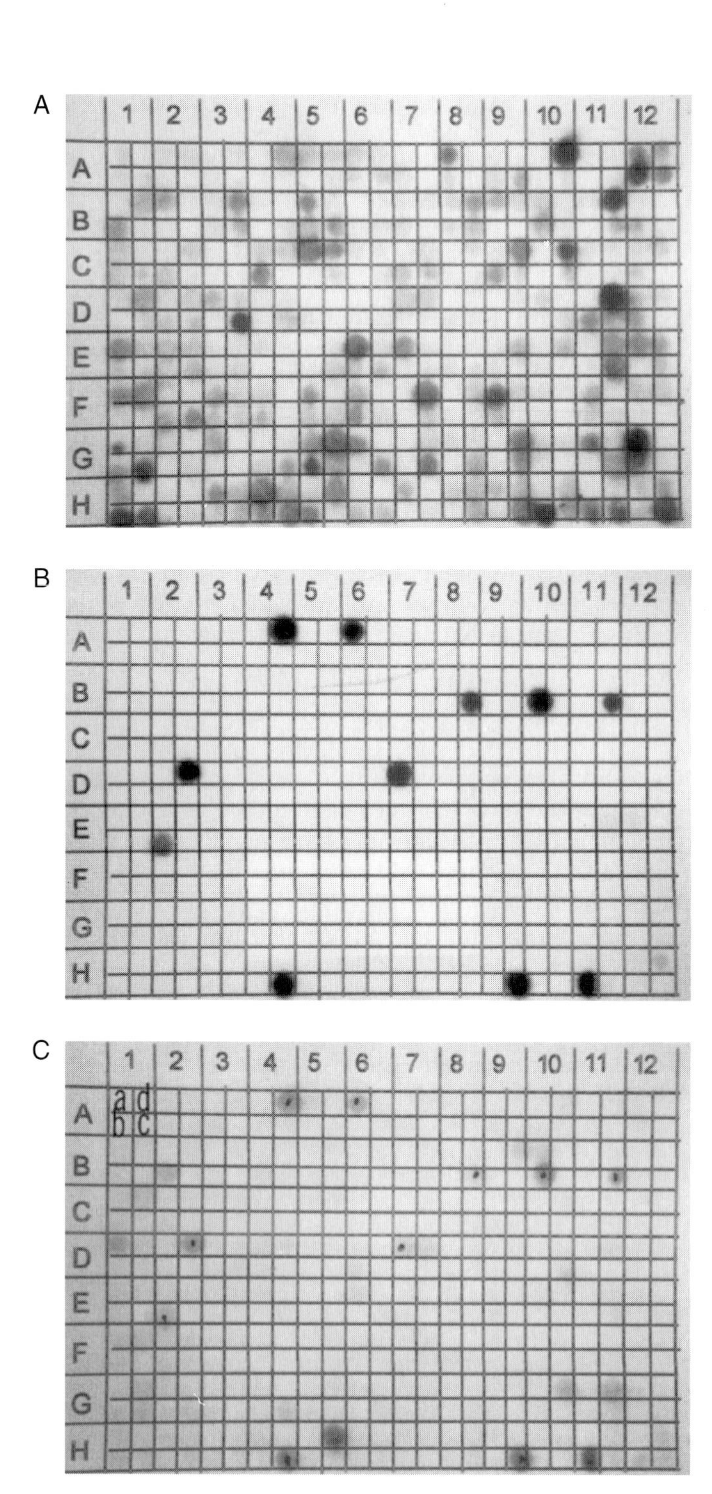

A
1
2
3
4
5
6
7
8
9
10
11
12
A
B
C
D
E
F
G
H
B
C
a
d
b
c

TSPY, which is known to be expressed in the testis (Fig. 6B). Further, hybridization of the same filter with a subtracted human heart cDNA mixture indicated that homologous cosmids were also detected (Fig. 6C). MTN blotting using these Y cosmids indeed detected hybridization to the heart and skeletal muscle RNA samples (Fig. 5B).

We believe that the same technique is applicable to studies designed to isolate potentially functional sequences from recombinant DNA libraries of other human chromosomes, such as those constructed under the National Gene Library Project, or subchromosomal contigs in a tissue-specific manner. This strategy is especially appealing for positional cloning of candidate genes for inherited diseases whose target (affected) tissues and chromosomal locations of the loci have been identified.

C. Identification of cDNAs Preferentially Expressed in Two or More Tissues

We have extended this strategy for differential screening of genes that are expressed in two or multiple tissues, such as the testis and ovary.[13] Although testis and ovary are quite different, they do share some similarities in terms of corresponding cell types, such as Sertoli and granulosa cells. The identification of expressed genes specific for both sex organs may shed light on the functions of these cell types.

To accomplish this differential screening objective, three subtracted cDNAs were obtained: (1) testis-specific cDNA: testis cDNAs subtracted against those derived from nine other human tissues (thymus, brain, liver, placenta, spleen, lung, kidney, heart, and skeletal muscle), (2) ovaray-specific cDNA: ovary cDNAs subtracted against the same nine human tissue cDNAs; and (3) reverse-subtracted cDNA: cDNAs from these nine human tissues subtracted against those of testis and ovary combined. The inserts from the subtracted testis-specific library were amplified by PCR and spotted on filters. The dot blots were screened by hybridization using (1) testis-specific, (2) ovary-specific, and (3) reverse-subtracted cDNA mix-

FIG. 6. Example of sequential screening of a human Y chromosome cosmid library using (A) subtracted testis-specific cDNA, (B) cDNA of a testis-specific Y chromosome gene, *TSPY*, and (C) subtracted heart-specific cDNA as probes. An array of DNAs from four 96-well plates of 384 Y chromosome cosmids were immobilized on a nylon filter and hybridized with the above cDNA probes in the order presented. Testis-specific cDNA probes hybridized more intensely with the Y cosmids than did the heart-specific cDNA probe. Some residual hybridization (indicated by dots) to the *TSPY* cDNA probe was observed in the heart-specific autoradiogram. However, Y cosmids hybridizing to the heart-specific cDNAs probe can be deduced by comparing the two autoradiograms (e.g., H5d, D1a).

tures as probes (Fig. 7). Most of the putative testis-specific cDNAs on the dot blot hybridized to the testis-specific probe (Fig. 7A and unpublished results). Some of the putative testis-specific cDNAs also hybridized to the ovary-specific probe (Fig. 7B), suggesting that they potentially are derived from genes simultaneously expressed in both testis and ovary. However, clones that hybridized to the reverse-subtracted cDNAs (Fig. 7C) represent background clones that were not specific to the testis. This differential screening identified three cDNA clones on this filter that are derived from genes expressed in both the testis and ovary.

By screening about 2000 testis-specific colonies in this manner, we were

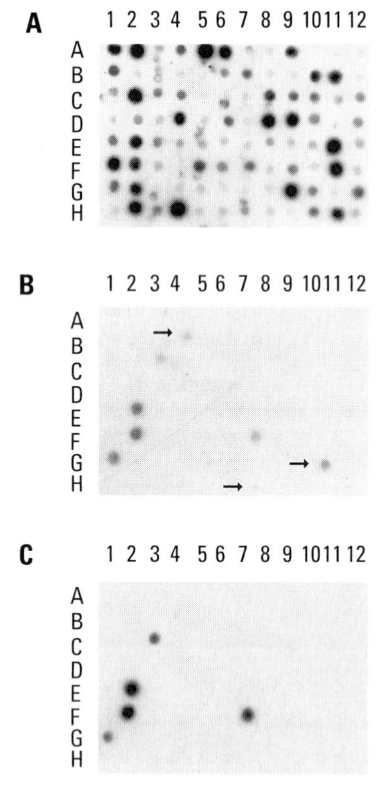

FIG. 7. Identification of specific cDNAs derived from genes expressed in both testis and ovary using a differential screening strategy. PCR-amplified inserts from putative testis-specific cDNAs obtained by SSH were dot blotted onto three identical nylon membranes. The blots were then hybridized with three probes: subtracted testis cDNAs (A), subtracted ovary cDNAs (B), and reverse-subtracted cDNAs (C). Clones corresponding to mRNAs that are preferentially expressed in testis and ovary are marked with arrows (B). Note that the hybridization efficiency of each individual clone is different.

able to identify a total of 14 clones that hybridized to both testis- and ovary-specific subtracted probes, but not to the reverse-subtracted probe. The inserts of these 14 clones were amplified by PCR using NP1 and NP2 primers, labeled with [α-^{32}P]dCTP by random priming, and hybridized to MTN blots containing 8 different human tissues. Twelve of the 14 clones corresponded to RNAs that are either exclusively or highly expressed in testis and ovary. Figure 8 illustrates five examples of MTN blots hybridized with the testis- and ovary-specific cDNA clones.

V. Technical Comments and Considerations

A. *Efficiency of Suppression Subtractive Hybridization*

The SSH technique has been demonstrated to be efficient for generating cDNAs highly enriched for differentially expressed genes of both high and

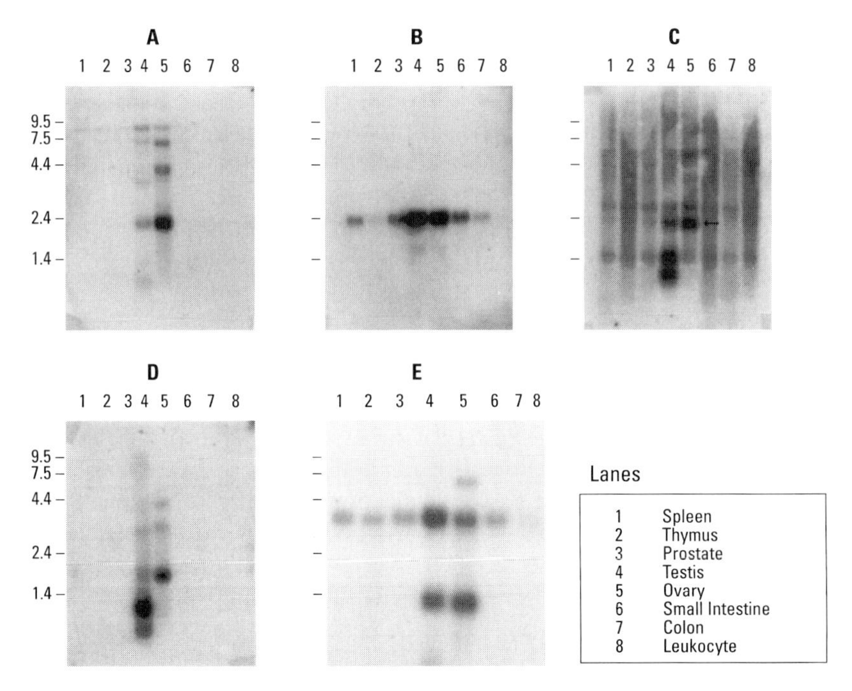

FIG. 8. Northern blot analysis of five representative clones that are preferentially expressed in testis and ovary. MTN blots containing 2 mg of poly(A)$^+$ RNAs isolated from the indicated tissues were probed with DNA inserts from clone TX30E4 (A), TX22A10 (B), TX23C12 (C), TX14G9 (D), or T16E12 (E). The mRNA size markers are indicated on the left. The arrow in (c) indicates the specific band.

low abundance. The high level of enrichment of cDNAs for rare transcripts has been achieved by the inclusion of a normalization step in the subtraction procedure, as evidenced by Northern blot analysis of random cDNA clones from the subtracted testis- and heart-specific cDNA libraries (Ref. 6; and heart-specific study described here). All cDNA clones analyzed gave positive hybridization signals with the respective tissue RNA in Northern blots and were present in different abundance. The same conclusion was also reached by random sequencing analysis. Numerous cDNA sequences showed homology to genes that are expressed in the corresponding tissues, and many of them represent unique and previously uncharacterized sequences.

Using the mathematical model of our subtraction procedure[12,19] and SUBTRACT program,[20] we calculated that the rare specific transcripts can be enriched by more than 1000-fold during one round of subtraction. This conclusion was supported in a model experiment with artificial targets.[6] In practice, the level of enrichment for a particular gene depends greatly on its original abundance, the ratio of its concentrations in driver and tester, and a number of other differentially expressed genes. With the incorporation of a normalization step in our subtraction procedure, the highest enrichment level can be obtained for differentially expressed mRNAs exhibiting low abundance and/or large differences in expression levels in the tester and driver RNA populations. However, as with other subtractive hybridization techniques, the efficiency of SSH is lower in experiments designed to detect mRNAs that show only moderate (e.g., two- to fourfold) differences between the tester and driver populations. Nevertheless, these types of differentially expressed genes can still be identified in our subtracted libraries.[12]

B. Differential Screening of Subtracted Libraries

Although the SSH method greatly enriches for differentially expressed genes, the subtracted sample will still contain some cDNAs that correspond to mRNAs common to both the tester and driver samples, depending somewhat on the quality of RNA purification and the performance of the particular subtraction. But, it mainly arises when few mRNA species are differentially expressed in tester and driver. In general, fewer differentially expressed mRNAs and a less quantitative difference in expression leads to higher background, even if one obtains a good enrichment of differentially expressed cDNAs. In our testis-specific subtracted library, about 95% of

[19] O. D. Ermolaeva and M. C. Wagner, *Cabios* **11,** 457 (1995).

[20] O. D. Ermolaeva, S. A. Lukyanov, and E. D. Sverdlov *in* "Proceedings of the Fourth International Conference on Intelligent Systems for Molecular Biology." AAAI Press, University of St. Louis, Missouri, 1996.

the clones correspond to testis-specific mRNAs. In this case, it was not necessary to prescreen a subtracted library. However, when background is expected to be high, identification of differentially expressed cDNAs by picking random clones from the subtracted library for Northern blot analysis can be time consuming. To circumvent this problem, the incorporation of a differential screening step is an efficient and desirable means to minimize background before embarking on Northern blot analysis.

There are two approaches for differential screening the subtracted library. The first is to hybridize the subtracted library with [α-32P]dCTP-labeled cDNA probes synthesized directly from the tester and driver mRNAs.[1] Clones corresponding to differentially expressed mRNAs will hybridize only with the tester probe, and not with the driver probe. Although this approach is widely used, it has one major disadvantage: only cDNA molecules corresponding to highly abundant mRNAs (i.e., mRNAs that constitute more than about 0.2% of the total cDNA in the probe) can hybridize with the clones.[4,14] Clones corresponding to low-abundance differentially expressed mRNAs will not be detected by this screening procedure.

To avoid missing of low-abundance differentially expressed genes, we recommend a second approach, in which the subtracted library is hybridized with forward- and reverse-subtracted cDNA mixtures. Clones representing mRNAs that are truly differentially expressed will hybridize only with the forward probe; clones that hybridize with the reverse probe may be considered background.[4,21] However, the removal of the adapter sequences from the ends of the cDNA molecules generated by the SSH procedure is critical for reducing the background caused by hybridization of the adapters in the subtracted cDNA probes to those of cDNA clones immobilized on the nylon filters.

C. Starting RNA Materials

Normally, 2–4 μg of poly(A)$^+$ RNA for both the tester and driver will be needed for a comprehensive subtraction scheme using both forward and reverse SSH. The resulting PCR products can be used for subtracted cDNA library construction and differential screening experiments. However, in some cases, such an amount of poly(A)$^+$ RNAs may be difficult to obtain. To circumvent this problem, an amplification step for both the driver and tester poly(A)$^+$ RNA can be incorporated to generate sufficient quantities of both cDNA samples[12] before initiating the SSH procedure. Alternatively, total RNA has been used successfully as starting material for the preampli-

[21] K. A. Lukyanov, M. V. Matz, E. A. Bogdanova, N. D. Gurskaya, and S. A. Lukyanov, *Nucleic Acids Res.* **24,** 2194 (1996).

fication.[22,23] Preamplification of either poly(A)$^+$ or total RNA samples invariably increases the background in the final PCR products and may result in the loss of some sequences. The utilization of preamplification should be minimized whenever possible.

D. Size of cDNA Fragments

For an efficient SSH procedure, the starting tester and driver cDNAs must be cleaved into smaller fragments. A four-base cutter, *Rsa*I, is used for this purpose because it generates optimal fragments (average ~600 bp) for SSH. Although this step may not be a desirable manipulation for obtaining full-length differentially expressed cDNAs, dividing each cDNA into multiple fragments has two important advantages. First, long DNA fragments may form complex networks that prevent the formation of appropriate hybrids needed to position two independent adapters, Ad1 and Ad2R, at the ends of the target molecules. Second, small cDNA fragments provide better representation of individual genes, because cDNAs derived from related but distinct members of gene families may cross-hybridize with each other, thereby eliminating them from the final subtracted cDNA products.[24] Dividing the cDNAs into smaller and different portions increases the possibility that a particular differentially expressed member of a gene family will contain a smaller fragment that is sufficiently different from other homologous members and can be enriched in the final subtracted cDNA mixture.[4,5] Once a small cDNA fragment is cloned and sequenced, numerous approaches, including several PCR-based methods, can be used to quickly obtain corresponding full-length cDNAs (reviewed in Ref. 9).

Another drawback of the SSH technique is the typical small inserts in the final subtracted cDNA libraries (average ~200 bp). This problem is generated by several factors related to the SSH procedure, such as more efficient hybridization of shorter fragments, preferential amplification of these fragments by PCR, and higher efficiency of subcloning of short fragments in plasmid vectors than those of longer ones. To minimize this undesirable selection for shorter fragments, several modifications of the SSH procedure can be incorporated to increase the representation of larger cDNA inserts in the final subtracted products, such as size selection of the subtracted cDNA products before subcloning. To minimize this problem

[22] K. Lukyanov, L. Diatchenko, A. Chenchik, A. Nanisetti, P. Siebert, N. Usman, M. Matz, and S. Lukyanov, *Biochem. Biophys. Res. Commun.* **230,** 285 (1997).

[23] L. Diatchenko, A. Chenchik, and P. D. Siebert, *in* "RT-PCR Methods for Gene Cloning and Analysis" (J. W. Larrick and P. D. Siebert, eds.), pp. 213–238 (1998).

[24] M. S. H. Ko, *Nucleic Acids Res.* **18,** 5705 (1990).

in the current SSH protocol described here, we use adapters with identical sequences for the first 22 nucleotides at their 5′ ends (Ad1 and Ad2R), instead of being completely different in sequence as in the original description (Ad1 and Ad2).[6] These sequence changes introduce short complementary inverted terminal repeats onto the end of the cDNA molecules, which carry different adapters on their ends and allow the primary amplification to be carried out with a single PCR primer. This introduces a weak suppression PCR effect during the primary amplification because the length of the complementary part is equal to the length of the primer. Under this condition the amplification of short (less than 200 bp) DNA fragments significantly diminished[7] and the risk of nonspecific amplification also decreases.[25]

These equalization steps are not intended to eliminate the shorter fragments, which may represent truly differentially expressed cDNAs, but instead are designed to balance representation of different fragment sizes, thereby increasing the complexity of the subtracted cDNA library.

VI. Summary

A new and highly effective method, termed suppression subtractive hybridization (SSH), has been developed for the generation of subtracted cDNA libraries. It is based primarily on a technique called suppression PCR, and combines normalization and subtraction in a single procedure. The normalization step equalizes the abundance of cDNAs within the target population and the subtraction step excludes the common sequences between the target and driver populations. As a result only one round of subtractive hybridization is needed and the subtracted library is normalized in terms of abundance of different cDNAs. It dramatically increases the probability of obtaining low-abundance differentially expressed cDNA and simplifies analysis of the subtracted library. The SSH technique is applicable to many molecular genetic and positional cloning studies for the identification of disease, developmental, tissue-specific, or other differentially expressed genes.

This chapter provides detailed protocols for the generation of subtracted cDNA and differential screening of subtracted cDNA libraries. As a representative example we demonstrate the usefulness of the method by constructing a testis-specific cDNA library as well as using the subtracted cDNA mixture as a hybridization probe. Finally, we discuss the characteristics of subtracted libraries, the nature and level of background nondifferentially

[25] Y. Takarada, *Nucleic Acids Res.* **22,** 2170 (1994).

expressed clones in the libraries, as well as a procedure for the rapid identification of truly differentially expressed cDNA clones.

Acknowledgments

We thank M. Shen and S. Yu for excellent technical assistance, Stephanie Trelogan for critically reading the manuscript, Theresa Provost for the helping in figure preparation, and Jenifer Fishel for helping with manuscript preparation. The work of Chris Lau is partially supported by an NIH grant (HD27392).

[21] Reduced Complexity Probes for DNA Arrays

By THOMAS TRENKLE, FRANÇOISE MATHIEU-DAUDÉ, JOHN WELSH, and MICHAEL MCCLELLAND

Introduction

There have been many advances in methods to manufacture and probe arrays of DNAs (reviewed in Refs. 1 and 2). For example, arrays of thousands of polymerase chain reaction (PCR) products from segments of mRNAs have been attached to glass at high density and addressed using probes from cDNA populations from two sources. Each cDNA or cRNA population is labeled with a different fluorescent dye and hybridization is assessed using fluorescence (e.g., see Refs. 3 and 4). Another class of arrays uses oligonucleotides that are attached to either a glass or silicon surface or manufactured by sequential photochemistry on the chip.[5] Such chips have been manufactured with tens of thousands of different oligodeoxynucleotide (oligo) sequences per square centimeter. Arrays of oligo nucleic acid analogs such as peptide nucleic acids are also being developed.[6]

Sensitivity as high as one copy of a mRNA per cell and a dynamic range of 10,000-fold have been reported when hybridization to arrays is measured

[1] G. Ramsay, *Nature Biotechnol.* **16,** 40 (1998).
[2] A. Marshall and J. Hodgson, *Nature Biotechnol.* **16,** 27 (1998).
[3] J. DeRisi, L. Penland, P. O. Brown, M. L. Bittner, P. S. Meltzer, M. Ray, Y. Chen, Y. A. Su, and J. M. Trent, *Nature Genet.* **14,** 457 (1996).
[4] M. Schena, D. Shalon, R. W. Davis, and P. O. Brown, *Science* **270,** 467 (1995).
[5] M. Chee, R. Yang, E. Hubbell, A. Berno, X. C. Huang, D. Stern, J. Winkler, D. J. Lockhart, M. S. Morris, and S. P. Fodor, *Science* **274,** 610 (1996).
[6] J. Weiler, H. Gausepohl, N. Hauser, O. N. Jensen, and J. D. Hoheisel, *Nucleic Acids Res.* **25,** 2792 (1997).

Copyright © 1999 by Academic Press
All rights of reproduction in any form reserved.

0076-6879/99 $30.00

by fluorescence using a confocal microscope. With these impressive numbers it would seem that the problem of detecting differential gene expression has been solved in cases where sufficient probe can be generated.

However, for many researchers it may be hard to achieve the best levels of sensitivity and dynamic range routinely. In this chapter we present a method to generate reduced complexity probes that, when applied to modern chip arrays, will improve the sensitivity and dynamic range for subsets of cDNAs. The ability to examine simultaneously the level of expression of most mRNAs in a total RNA preparation is sacrificed for improved kinetics and an improved signal-to-noise ratio.

For many researchers the costs of chip arrays remain prohibitive. However, inexpensive arrays of the IMAGE consortium expressed sequence tags (ESTs) are commercially available, such as 18,432 clones in *Escherichia coli,* double-spotted on 22 $\times$ 22 cm membranes, at a cost (at time of writing) of less than $200 per membrane. In addition, membranes with hundreds or even thousands of PCR products from cDNA clones are also commercially available. Any cDNA clones on the array that are from the most abundant few thousand of mRNAs in a cell can easily be detected using a radiolabeled probe containing the full complexity of the mRNA population in the cell[7] (Fig. 1A). The only limitation is background hybridization to all clones. This background hybridization limits the ability to see the rarest mRNAs. If this background hybridization were circumvented then these inexpensive membranes could be of even greater utility. One way to achieve this is to use subtracted probes where most of the complexity is removed, which have shown sensitivity for rare messages comparable to that of confocal scanned chips (e.g., see Ref. 8).

We explain an alternative strategy based on cDNA fingerprints generated by RNA arbitrarily primed PCR (RAP-PCR).[9–12] This method generates different subsets of the mRNA population, depending on the primers used. Some of the mRNAs that are difficult to detect using total cDNA probes are sufficiently represented in these reduced complexity probes to

[7] G. Pietu, O. Alibert, V. Guichard, B. Lamy, F. Bois, E. Leroy, R. Mariage-Sampson, R. Houlgatte, P. Soularue, and C. Auffray, *Genome Res.* **6,** 492 (1996).

[8] T. A. Rhyner, N. F. Biguet, S. Berrard, A. A. Borbely, and J. Mallet, *J. Neurosci. Res.* **16,** 167 (1986).

[9] J. Welsh, K. Chada, S. S. Dalal, R. Cheng, D. Ralph, and M. McClelland, *Nucleic Acids Res.* **20,** 4965 (1992).

[10] J. Welsh and M. McClelland, *Nucleic Acids Res.* **18,** 7213 (1990).

[11] P. Liang and A. B. Pardee, *Science* **257,** 967 (1992).

[12] T. Trenkle, J. Welsh, B. Jung, F. Mathieu-Daudé, and M. McClelland, *Nucleic Acids Res.* **26,** 3883 (1998).

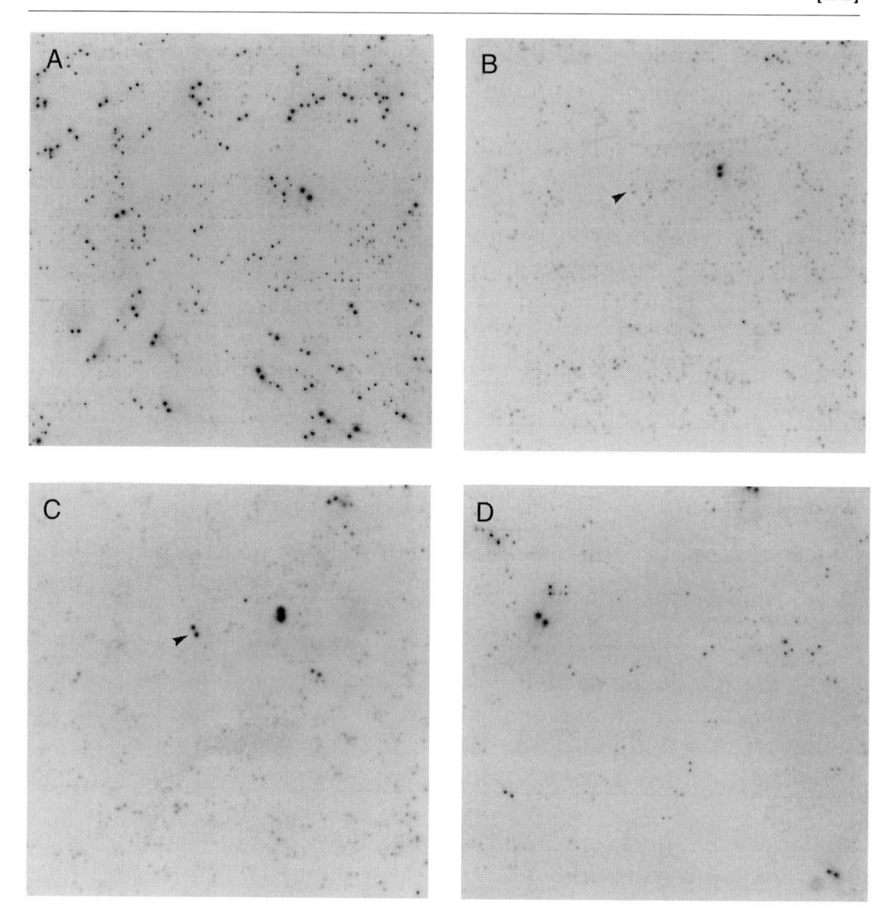

FIG. 1. Differential hybridization to clone arrays. All four images show a closeup of an autoradiogram for the same part of a larger membrane. Each image spans about 4000 double-spotted *E. coli* colonies, each carrying a different EST clone. (A) One microgram of poly(A)$^+$ RNA from confluent human keratinocytes was radiolabeled during reverse transcription. About 500 clearly hybridizing clones can be seen. (B and C) RAP-PCR fingerprint with a pair of arbitrary primers was performed on cDNA from oligo(dT)-primed cDNA of confluent human keratinocytes that were untreated in (B) and treated with EGF in (C). The pattern of hybridizing genes is almost identical in (B) and (C), but entirely different from that seen with total poly(A)$^+$ RNA in (A). The two radiolabeled colonies from one differentially expressed cDNA are indicated by arrows. Differential expression of this gene was subsequently confirmed by specific RT-PCR.[12] (D) A RAP-PCR fingerprint with a different pair of arbitrary primers was performed on RNA from confluent human keratinocytes. Note that this pattern of hybridization is almost entirely different from that found with the previous primer pair, and almost entirely different from that shown with mRNA in (A).

be easily detected on arrays of colonies. This method has the further advantage that it generates a labeled probe using hundreds-fold less RNA than is currently used by any other array probing methods. The use of statistically primed PCR (SP-PCR)[13] to generate probes is also discussed.

It can be envisioned that reduced complexity probes may be useful in other applications such as single nucleotide polymorphism detection on arrays, sequencing by hybridization, diagnostics of infectious disease, and even in phylogenetics. The availability of bacterial and P1 phage artificial chromosome (BAC and PAC) arrays of the genome will also allow reduced complexity probes to detect differences in DNA methylation and retroposable elements.

Equipment and Reagents

The equipment and reagents for purifying RNA and for generating a RAP-PCR fingerprint are described in [18] in this volume.[13a]

Tris-HCl, (pH 8.3), 10 mM
Random hexamer oligonucleotide (Genosys Biotechnologies, The Woodlands, TX), 200 μM
Mix of dATP, dTTP, and dGTP (0.5 mM)
[α-^{32}P]dCTP (ICN Pharmaceuticals, Costa Mesa, CA), 3000 Ci/mmol
dCTP, 2.5 mM
Klenow fragment buffer (10×): 500 mM Tris-HCl (pH 8.0), 100 mM $MgCl_2$, 500 mM NaCl
Klenow fragment (GIBCO-BRL Life Technologies, Gaithersburg, MD): 3.82 U/μl
QIAquick nucleotide removal and PCR purification kit (Qiagen, Chatsworth, CA)
SSC (20×): 3 M NaCl, 0.3 M sodium citrate · $2H_2O$ (pH 7.0)
Sodium dodecyl sulfate (SDS), 10%
Fragmented, denatured salmon sperm (Pharmacia, Piscataway, NJ): 10 mg/ml
Denhardt's solution (50×): 5 g of Ficoll, 5 g of polyvinylpyrrolidone, 5 g of bovine serum albumin in 500 ml of water and sterile filtered
Formamide (Aldrich, Milwaukee, WI)
Fragmented human genomic DNA (1 μg/μl)
cDNA arrays [e.g., arrays purchased from Research Genetics (Huntsville, AL) or Genome Systems (St. Louis, MO)]

[13] G. Pesole, S. Liuni, G. Grillo, P. Belichard, T. Trenkle, J. Welsh, and M. McClelland, *BioTechniques* **25,** 112 (1998).

[13a] F. Mathieu-Daudé, T. Trenkle, J. Welsh, B. Jung, T. Vogt, and M. McClelland, *Methods Enzymol.* **303,** [18], 1999 (this volume).

Hybridization oven with roller bottles (e.g., Techne HB-1D; VWR Scientific, San Francisco, CA), or sealable plastic bags and a water bath, set to 42° and 68°

Horizontal shaker

Flat-bottomed glass or plastic container larger than the membranes to be used

Tabletop microcentrifuge

Methods and Protocols

Generation of Reduced Complexity Probe

One protocol for generating an RNA arbitrarily primed PCR fingerprint is described in [18] in this volume.[13a] In brief, there are three similar protocols to generate these reduced complexity cDNA samples using arbitrary primers. First, reverse transcription uses an oligo(dT), a 3′ anchored primer, or an arbitrary primer. Next, second-strand cDNA is primed with an arbitrary primer. The PCR then takes place between the arbitrary primers or between an arbitrary primer and a 3′ anchor (reviewed in Ref. 14).

Primer selection for statistically primed PCR (SP-PCR) is described in Ref. 13. The main difference between this method and RAP-PCR is that the primers are selected by a computer program, rather than arbitrarily. In brief, a program called GeneUP has been devised. This program uses a greedy algorithm to select primer pairs to sample sequences in the user list of interest (e.g., a list of human mRNAs associated with apoptosis) while excluding sequences in another list (e.g., a list of abundantly expressed mRNAs in human cells and structural RNAs such as rRNAs, *Alu* repeats, and [mitochondrial DNA (mtDNA)].[13] In principle, such a program should make it possible to generate reduced complexity probes in which the messages of interest are all represented. This program is freely available to noncommercial users.

For both of these cDNA sampling methods there is a small subset of cDNA products that are sensitive to whatever conditions are used for amplification.[9,10] To control for variation in these products between RNA samples it is sufficient to generate the probe at two concentrations of RNA, differing by a factor of two or more.

A number of other methods to generate reduced complexity probes are briefly discussed in Note 1 (below).

Labeling of Polymerase Chain Reaction Products

We recommend that the probes developed in the preceding section be prepared using the radiolabeling protocol ([18] in this volume[13a]) and run

[14] M. McClelland, F. Mathieu-Daudé, and J. Welsh, *Trends Genet.* **11,** 242 (1995).

on a denaturing polyacrylamide gel to ensure that the fingerprints are identical at different starting RNA concentrations. Those probes that prove to generate acceptable fingerprints are chosen as probes for an array. However, RAP-PCR does not generate probes that are sufficiently radioactive to be used directly on an array. The following is a protocol for random-primed labeling of the probe.

Up to 10 μg of PCR product from RAP-PCR or SP-PCR is purified using a QIAquick PCR purification kit (*www.qiagen.com*), which removes unincorporated bases, primers, and primer dimers less than 40 base pairs (bp) in length. The DNA is recovered in 100 μl of 10 m*M* Tris, pH 8.3.

Random-primed synthesis with incorporation of ^{32}P from α-labeled dCTP is used under standard conditions. Ten percent of the recovered fingerprint DNA (typically about 100 ng in 10 μl) is combined with 6 μg of random hexamer oligonucleotide primer, and 1 μg of one of the fingerprint primers, in a total volume of 28 μl, boiled for 3 min, and then placed on ice.

The hexamer/primer/DNA mix is combined with 22 μl of reaction mix to yield a 50-μl reaction containing a 0.05 m*M* concentration of three dNTPs (minus dCTP), 100 μCi of [α-^{32}P]dCTP (3000 Ci/mmol; 10 μl), 1$\times$ Klenow fragment buffer, and 8 U of Klenow fragment. The reaction is performed at room temperature for 4 hr. For maximum probe length the reaction is chased by adding 1 μl of 2.5 m*M* dCTP and incubated for 15 min at room temperature, and then for an additional 15 min at 37°. The unincorporated nucleotides and hexamers are removed with the Qiagen nucleotide removal kit and the purified products are eluted twice using 140 μl of 10 m*M* Tris, pH 8.3, each time.

There are many other possible methods for labeling, some of which are briefly discussed in Note 2 (below).

Choice of Array

A number of cDNA arrays from the IMAGE consortium (*www-bio.llnl.gov/bbrp/image/image.html*) are available on nylon membranes, and from a variety of sources including Research Genetics (*www.resgen.com*) and Genome Systems (*www.genomesystems.com*) and from the German Human Genome Project (*www.rzpd.de*). These include clones from various human tissues, stages of development, and diseases. Arrays of mouse and yeast sequences are also available. At present there are two types of array. One type of array contains colonies, for example, 18,432 *E. coli* colonies, each carrying a different IMAGE EST plasmid, and each spotted twice on a 22 × 22 cm membrane, available from Genome Systems. The other kind of array contains more than 5000 PCR products from selected IMAGE clones. The latter are currently more expensive. To date, an array of PCR products is available for every yeast open reading frame (ORF) and for a subset of human ESTs. One can expect a dramatic increase in the number of available arrays, organisms, and accompanying sequence information. Visit the Web sites for current information.

Hybridization to Array

If radioactivity is to be used to label the probe then four membranes will be needed: one membrane for each of two concentrations of RNA for each of the two RNA samples to be compared. If two-color fluorescence is used then only two arrays will be needed, one for each of the two concentrations of starting RNA, because probes from the two RNA samples can be mixed.

Prewash of cDNA Filters. The cDNA filters (Genome Systems) are washed in three changes of 2× SSC–0.1% SDS in a horizontally shaking flat-bottom container to reduce the residual bacterial debris. The first wash is carried out in 500 ml for 10 min at room temperature. The second and third washes are carried out in 1 liter of prewarmed (50°) prewash solution for 10 min each.

Prehybridization. The filters are then transferred to roller bottles and are prehybridized in 60 ml of prewarmed (42°) prehybridization solution containing 6× SSC, 5× Denhardt's reagent, 0.5% SDS, fragmented, denatured salmon sperm (100 μg/ml), and 50% formamide for 1–2 hr at 42°.

Hybridization. The prehybridization solution is exchanged with 7 ml of prewarmed (42°) hybridization solution containing 6× SSC, 0.5% SDS, fragmented, denatured salmon sperm (100 μg/ml), and 50% formamide.

To decrease the background hybridization due to repeats [e.g., *Alu* repeats, long interspersed repetitive elements (LINEs), centromeric repeats], sheared human genomic DNA is denatured in a boiling water bath for 10 min and immediately added to the hybridization solution to a final concentration of 10 μg/ml. See Note 3 (below) for alternatives.

Simultaneously, the labeled probe, in a total volume of 280 μl, is denatured in a boiling water bath for 4 min and immediately added to the hybridization solution. The hybridization is carried out at 42° for up to 48 hr, typically 18 hr.

Wash. For the washes the incubator oven temperature is set to 68°. The hybridization solution is poured off and the membrane is washed twice with 50 ml of 2× SSC–0.1% SDS (room temperature) for 5 min. The wash solution is then replaced with 100 ml of 0.1× SSC–0.1% SDS (room temperature) and incubated for 10 min. For the further washes the wash solution (0.1× SSC–0.1% SDS) is prewarmed to 64°: 40 min in the roller bottles (100 ml), then the filter is transferred to a horizontally shaking flat-bottom container and washed in 1 liter for 20 min under gentle agitation. The filter is transferred back to a roller bottle containing 100 ml of prewarmed 0.1× SSC–0.1% SDS and incubated for 1 hr. The final wash solution is removed and the filter briefly rinsed in 2× SSC at room temperature.

After washing, the membranes are blotted with 3MM paper (Whatman, Clifton, NJ) and the moist membranes are wrapped in mylar TMA-50 (Fralock, Orange, CA), which permits detection of ^{33}P. Ideally, the arrays are read using a PhosphorImager (Molecular Dynamics, Sunnyvale, CA) or a fluorescent scanner. However, if X-ray film is used, the membranes are usually sufficiently radioactive that a 2-day exposure with a screen will clearly reveal the top 2000 products on an array of 18,432 bacterial colonies carrying EST clones. Weaker probes or fainter hybridization events can be seen using a screen at −70° for up to a week.

The number of cDNAs to which a positive hybridization signal can be attributed is limited by the level of the background hybridization and incomplete blockage of *Alu* repeats. In practical terms probably only the top 2000 cDNAs on a colony array of 18,432 clones can be reliably scored by each arbitrarily primed probe derived from a cell line. Arrays of PCR products may yield a higher rate of scorability.

In general, two different reduced complexity probes, generated using different pairs of arbitrary or statistical primers, will hybridize to largely nonoverlapping sets of cDNAs in the array (compare Fig. 1B and D). Typically, fewer than 100 products overlap among the most intensely hybridizing 2000 colonies in two differently primed reduced complexity cDNA samples. The fingerprints are also almost entirely different from those generated by directly labeling the whole mRNA population (compare Fig. 1A and B). As more cDNAs are sampled by more reduced complexity probes, the number of new genes to be sampled will gradually decrease. To some extent the efficiency of coverage of cDNAs can be improved by the use of statistically selected primers that are designed to allow full coverage,[13] but this has yet to be tested.

Confirmation of Differential Expression

The first level of confirmation is the use of two RNA concentrations per sample. Only those hybridization events that seem to indicate differential expression at both RNA concentrations in both RNA samples can be relied on.

One of the advantages of using the arrays from the IMAGE consortium is that more than 80% of the clones have single-pass sequence reads from the 5′ or 3′ end, or both, deposited in the GenBank database. Thus, it is usually not necessary to clone or sequence any DNAs to determine if there is a known gene or other ESTs that share the same sequence. UniGene database clusters human and mouse ESTs that appear to be from the same gene. This database greatly aids in this process (*http://www.ncbi.nlm.nih.gov/UniGene/index.html*). Mapping onto chromosomes at a resolution of a few centimorgans is also available for most of these clusters at the same Web site. The clones on these arrays are all available to be used as probes or to complete the sequencing (*www-biol.llnl.gov*). At that point it is often possible to identify a close homolog in other species. We have used this set of steps many times to generate cDNA sequences of many kilobases.

When cDNA libraries that contain near full-length clones are available on arrays and sequences from the ends of these clones are in the database, then it will be possible to go from a differentially hybridized spot to a full-length cDNA directly. For now, we use 5′ RACE (rapid amplification of cDNA ends)[15] to extend beyond the currently available sequence information.

Confirmation of differential expression does not need a full-length sequence and can be done using reverse transcriptase (RT)-PCR of the known region. In particular, we recommend the use of low-stringency PCR to generate products a few hundred bases in length, as described in [18] in this volume.[13a] This method generates internal "control" PCR products that can be used to confirm the quality of the PCR and the quality and quantity of the RNA used.

When using the array technologies, one is often immediately overwhelmed by too many genes to pursue thoroughly. There are a number of ways to select genes. One strategy is to pick from the list those genes that are already known and for which a new role in the situation of interest can be envisioned. Alternatively, some of the genes may be family members of known genes with known functions for which a plausible role can be hypothesized.

Another approach is to ask a more subtle question, as a result of which less differential gene expression may be observed. For example, the duration of a drug treatment might be shortened to find transcripts that respond early, or the amount of the drug might be lowered to find transcripts that are more sensitive to the treatment. A more ambitious approach is to prepare probes from rather different conditions that will further dissect

[15] Y. Zhang and M. A. Frohman, *Methods Mol. Biol.* **69,** 61 (1997).

the phenomenon of interest.[14,16,16a] For example, one might be interested in apoptosis. When the results of a cDNA probe from apoptotic cells are compared with a cDNA probe from a cell that has been stressed, but which does not undergo apoptosis, then the differentially expressed genes can be divided into apoptosis responses, stress responses, and a large class of genes that respond to both.

It is to be hoped that a public resource will develop in which the transcriptional effect of a growing list of conditions is attached to every gene. Such information will link to the promotors of these genes and to the signal transduction cascades responsible.

Notes and Comments

Note 1: Other Protocols for Generating Suitable Reduced Complexity Probes

Background hybridization and the need for a sufficient mass of probe both lead to limits in detection sensitivity. Complexity reduction methods that allow a few rare mRNAs to contribute significantly to the final mass of the PCR sample can bypass this limitation and enhance the ability to observe differential gene expression among rare mRNAs in the cell.

It should be possible to use almost any method that generates a mixture of products that reliably enriches for only a part of each mRNA and/or only a subset of the mRNA population. Various methods that could generate reduced complexity probes can be found throughout this volume.

It is worth noting that for the purposes of generating reduced complexity probes there are two main classes of methods. One class of methods yields cDNAs representing a subset of the mRNA population that maintains the approximate stoichiometry of the input RNA. Such methods are exemplified by most amplification fragment length polymorphisms (AFLP) and restriction strategies that sample the 3′ end or internal fragments of mRNAs (e.g., see Refs. 17–19). Another example is the use of size-fractionated mRNAs to generate cDNA probes. All of the mRNAs in, say, the 2.0- to 2.1-kb range might be used as a reduced complexity probe. Stoichiometry

[16] M. McClelland, D. Ralph, R. Cheng, and J. Welsh, *Nucleic Acids Res.* **22,** 4419 (1994).

[16a] T. M. Vogt, J. Welsh, W. Stolz, F. Kullmann, B. Jung, M. Landthaler, and M. McClelland, *Cancer Res.* **57,** 3554 (1997).

[17] C. W. Bachem, R. S. van der Hoeven, S. M. de Bruijn, D. Vreugdenhil, M. Zabeau, and R. G. Visser, *Plant J.* **9,** 745 (1996).

[18] Y. Habu, S. Fukada-Tanaka, Y. Hisatomi, and S. Iida, *Biochem. Biophys. Res. Commun.* **234,** 516 (1997).

[19] T. Money, S. Reader, L. J. Qu, R. P. Dunford, and G. Moore, *Nucleic Acids Res.* **24,** 2616 (1996).

among these mRNAs would be mostly preserved in the probe. Rare mRNAs become rare cDNAs in the probe, where they are still difficult to detect although the background is reduced by the reduction in complexity.

A rather different class of complexity reduction strategies does not preserve the stoichiometry of the starting mRNAs, although it does preserve differences among individual RNAs between samples. These methods are exemplified by the two strategies outlined in this chapter, RAP-PCR and SP-PCR. The use of arbitrary or statistical primers of the length we suggest (eight bases or greater) and at the annealing temperatures we suggest (30° to 60°) results in a set of PCR products that reflect a function of both the starting abundance of each mRNA target and the quality of the match. Thus, the final mixture after PCR can include quite abundant PCR products that derive from rarer mRNAs that have a good match with the primers used and are favorable for amplification after the initial priming events. Amplifiability includes effects such as secondary structure and product size. Fortunately, the same mRNA in two different RNA samples experiences an identical combination of "primability" and "amplifiability" so that changes in abundance for particular mRNAs can still be maintained even as the relative abundances between different mRNAs within one sample are profoundly scrambled.

That rarer mRNAs can contribute a significant fraction to the probe in "nonstoichiometric" sampling methods means that nonstoichiometric methods may be better than stoichiometric methods for generating a probe to detect rare mRNAs. In methods that maintain stoichiometry the rare mRNAs are still rare in the probe and so it may be necessary to address more sophisticated arrays of PCR products to detect hybridization by probes derived from these rare mRNAs.

Finally, one can expect that a robust subtraction method would yield an excellent reduced complexity probe. Methods based on representational difference analysis (RDA)[20,21] or suppression subtractive hybridization (SSH)[22] come to mind. In one rather simple subtraction strategy that could be used with RAP-PCR or statistically primed PCR, the PCR products from one RNA sample (A) are labeled and then mixed with a few-fold excess of unlabeled RAP-PCR product from the other RNA sample (B). The whole mixture is denatured and added to the hybridization solution on the membrane. The PCR products that are found in both samples form double-stranded DNA (dsDNA) and the remaining available labeled probe is primarily from the differences between the two samples. The same experi-

[20] N. Lisitsyn and M. Wigler, *Methods Enzymol.* **254,** 291 (1995).

[21] N. Lisitsyn and M. Wigler, *Science* **259,** 946 (1993).

[22] H. Jin, X. Cheng, L. Diatchenko, P. D. Siebert, and C. C. Huang, *BioTechniques* **23,** 1084 (1997).

ment is done with labeled sample (B) and excess unlabeled sample (A) and the two membranes are compared to detect differential gene expression. Incidentally, this procedure also effectively quenches repeats present in the cDNA mixtures.

One remaining limitation is that these methods often require the RNA from thousands of cells, although this is less than is usually needed for total cDNA probes. It would be desirable to develop methods to reduce the amount of probe used so that small samples of RNA could be used as a source.

Note 2: Other Labeling Methods

We have used end labeling of probes but incorporation of radiolabel by random-primed synthesis seems to be superior. The protocol described in this chapter generates approximately equal amounts of randomly primed DNA from both strands of double-stranded PCR products, which will reanneal to some degree during hybridization to the membrane. So far this has not appeared to be a problem but it could be circumvented by various means, such as exonuclease III (exo III) digestion. Another possibility is to incorporate T7, SP6, or other promotors in the primer and use transcription to generate the probe. This has the advantage of generating a single-stranded probe.

In principle, superior results will be obtained by the incorporation of a fluorescent dye into the probe, either in the primer, during cDNA production, or afterward. It is possible to use a different wavelength for each RNA sample and address them all simultaneously on a nonfluorescent substrate, using a fluorescent scanner or confocal microscope. Measuring the relative abundance of two probes simultaneously on the same array rather than on two different arrays eliminates problems that arise owing to differences in the hybridization conditions or the quantity of target PCR product on the glass, thus improving reliability.

Why bother to use a reduced complexity probe if DNA chips and a fluorescence reader are available to the user? One reason is that it is entirely possible that a sensitivity of one RNA molecule per cell is difficult to achieve routinely. A reduced complexity probe will obviate the need to try to achieve that level of sensitivity. Note that most membrane-based arrays do not perform well at most wavelengths owing to a high background. However, infrared dyes are compatible with nylon membranes as long as the membrane is free of protein.

Note 3: Other Quenching Methods

The use of Cot-1 DNA is a standard method of quenching DNA repetitive elements. This genomic fraction is enriched in repeats. Such a quencher could be important for looking at the fainter PCR products, which could be partly quenched by the use of total genomic DNA. However, we have found total genomic DNA adequate for colony arrays. Cot-1 DNA is probably important only for the more sophisticated arrays, such as PCR-based arrays, where the signal-to-noise ratio is sufficiently low to be concerned about relatively poorly amplified fingerprint products.

A simple method for quenching that also acts as a subtraction method is to use the denatured reduced complexity probe from one condition to quench the labeled probe from another condition (as described in Note 2, above).

Acknowledgments

This work was supported in part by a kind donation from Mr. Sidney Kimmel; by NIH Grants CA68822, NS33377, and AI34829 to M.M.; and by a DFG fellowship TR413/1-2 to T.T. This work is licensed to Phenotypics Corporation. We thank Barbara Jung, Steve Ringquist, and Rhonda Honeycutt for helpful discussions.

[22] Targeted Display: A New Technique for the Analysis of Differential Gene Expression

By ALASTAIR J. H. BROWN, CATHERINE HUTCHINGS, JULIAN F. BURKE, and LYNNE V. MAYNE

The phenotype of tissues within an organism is determined by their patterns of gene expression. To this end, several powerful molecular approaches have been developed allowing the simultaneous assessment of gene expression profiles between two populations of cells or tissues under study. We describe in this chapter a new technique, which we have named Targeted Display (TD); this is an extension of a published technique.[1] Targeted display utilizes the surprising finding that there is a nonrandom distribution frequency for octanucleotide sequences throughout randomly selected protein-coding regions of DNA. Statistically designed octanucleotides were therefore adapted for polymerase chain reaction (PCR) and used to investigate the differential expression of genes involved in the acquisition of the differentiated neuronal phenotype in rat PC12 cells. The principle of TD remains, as with similar approaches,[2–4] to identify and isolate genes exhibiting differential patterns of gene expression between

[1] C. E. López-Nieto and S. K. Nigam, *Nature Biotechnol.* **14,** 857 (1996).

Copyright © 1999 by Academic Press
All rights of reproduction in any form reserved.

0076-6879/99 $30.00

two populations of cells. However, this technique offers additional prospects for the specific and highly selective targeting of genomic subsets (gene families), for DNA fingerprinting, phylogenetic analyses, exon trapping, and the elucidation of translocations, insertions, or deletions within genetic lesions.

Introduction

Current methods to distinguish gene expression profiles are all based on the principles of subtractive hybridization,[2,3] PCR-based technologies,[4] expressed sequence tag (EST) approaches,[5] the serial analysis of gene expression[6] (SAGE), or, more recently, large multiplex hybridization assays utilizing microarrays of cDNA clones[7] or oligonuceotides.[8,9] Whatever the chosen methodology, the goal of all of these techniques, within the confines of differential gene expression analysis, remains the same: (1) to target as many as possible of the previously estimated[10] 15,000–20,000 mRNA species per cell, (2) to include both the highly abundant (>1000 copies/cell) and the rare (<10 copies/cell) mRNAs, (3) to be both qualitative and, to an extent, as quantitative as possible, and (4) to incorporate a simple and efficient method for analyzing and cloning the targeted products.

Targeted display belongs to the general class of PCR-based methodologies for both the qualitative and semiquantitative analysis of gene expression. As with differential display,[4] oligonucleotide primers are used in PCRs to reproducibly amplify short DNA fragments, which are then visualized using polyacrylamide or agarose gel formats. Unlike differential display, however, the amplified cDNA fragments primarily represent portions of the protein-coding regions of mRNAs, not the 3′ untranslated regions.

López-Nieto and Nigam[1] first demonstrated the nonrandom distribution frequency of all possible octanucleotide combinations in the protein-coding regions of the sense DNA strands in an analysis of 1000 randomly chosen human mRNAs. A small subset of just 30 octanucleotides was found to be

[2] S. W. Lee, C. Tomasetto, and R. Sager, *Proc. Natl. Acad. Sci. U.S.A.* **88,** 2825 (1991).

[3] N. A. Lisitsyn, N. M. Lisitsyn, and M. Wigler, *Science* **259,** 946 (1993).

[4] P. Liang and A. B. Pardee, *Science* **257,** 967 (1992).

[5] J. Welsh, K. Chada, S. Dalal, R. Cheng, D. Ralph, and M. McClelland, *Nucleic Acids Res.* **20,** 4965 (1992).

[6] V. E. Velculescu, L. Zhang, B. Vogelstein, and K. W. Kinzler, *Science* **270,** 484 (1995).

[7] M. D. Adams, M. Dubnick, A. R. Kerlavage, R. Moreno, J. M. Kelley, T. R. Utterbock, J. W. Nagle, C. Fields, and J. C. Venter, *Nature (London)* **355,** 632 (1992).

[8] M. Schena, D. Shalon, R. W. Davis, and P. O. Brown, *Science* **270,** 467 (1995).

[9] A. Goffeau, *Nature (London)* **385,** 202 (1997).

[10] B. Alberts, D. Bray, J. Lewis, M. Raff, K. Roberts, and J. D. Watson, "Molecular Biology of the Cell." Garland, New York, 1994.

highly overrepresented within 75% of the randomly selected cDNAs. By designing the octanucleotides in both the sense and antisense orientations, with the addition of a short 5-base pair (bp) "tail" to each octanucleotide, the sequences were compatible with a PCR-based format for the subsequent targeted amplification of protein-coding regions of DNA. Computer simulations of PCRs, based on an additional 100 randomly selected cDNAs, suggested a "hit rate" of 75%, which would represent approximately 15,000 mRNA in an average cell.[10]

We have used variants of the original primer sets devised by López-Nieto and Nigam[1] to specifically amplify protein-coding regions, in conjunction with a differential display format, to visualize differences in gene expression between similar populations of cells differing in their disease state or developmental stage. The method has been further developed to target specific families of genes. Using the basic helix–loop–helix transcription factor family (bHLH), 25 octanucleotide sequences were selected on the basis of their overrepresentation in the original gene family and their ability to theoretically target 90% of the sequences used for analysis.

The principle of the technique is simple. Single-stranded cDNA is generated from total RNA from two cell or tissue sources that are likely to express genes differentially in response to environmental or external stimuli. PCR amplifications generate fragments, from the annealing of the antisense primer to the single-stranded cDNA, which produces double-stranded cDNA, to which the sense-orientated primer can anneal and prime the exponential amplification of the fragment. As with any PCR, however, there exists the opportunity for primer mismatches, especially at lower stringency annealing temperatures; however, so long as these occur reproducibly within both populations of cDNA, they do not detract from this technique and may, in fact, form an important part in the amplification of additional products.

Experimental reaction conditions required by this method were optimized, such that the reliable and reproducible amplification of messages from cell or tissue samples were obtained using traditional PCR methods. Subsequent analysis of the amplified fragments was performed using Metaphor agarose gels (Flowgen, Sevenoaks, UK) as opposed to denaturing or native polyacrylamide gels favored in the differential display format. Although the resolution of Metaphor gels is decreased compared with the polyacrylamide variety, this approach does remove the need for radioactivity to be incorporated in the PCRs and decreases both the experimental time and the number of steps involved. Both of these approaches lead to the isolation of bands that represent true difference products on the basis of both qualitative and quantitative analyses.

Applications of this technique are far reaching. Not only does it provide an alternative to the already established differential display method, but the ability to target specific gene families makes it a powerful approach for the elucidation of genes involved in memory, learning, development, and differentiation of all organisms. In addition to cDNA-based approaches, targeted display can be applied to genomic fingerprinting, and owing to its bias toward protein-coding over noncoding regions, the location of exon-containing regions within large genomic subsets can be identified.

As a note of caution, it should be remembered that all applications of TD rely almost exclusively on PCR, and hence the reproducible nature may be comprised by the introduction of any foreign contaminating DNA particles at any stage. In addition, the inherent sensitivity of this method, as with differential display, depends on the uniformity of reagents and reaction conditions, alterations of which within experiments may severely reduce its quantitative nature.

Methodology

Primer Design

For the design of statistically overrepresented primers to be applicable under experimental PCR-based conditions, they must be of optimal length to reproducibly amplify sequences of interest. The frequency of any randomly selected set of 30 hexamers, binding within the genome, is 1 in 136.5 ($4^6/30$); this falls to 1 in 2185.5 for octamers and to 1 in 34,952.5 for decamers. From these figures, one can clearly see that hexamers are of limited use for the design of statistically overrepresented primers, as they would fall in such close juxtaposition as to be relatively ineffective in PCRs. However, for octadecanucleotide combinations, the random frequency is low enough such that any statistical overrepresentation should become relevant, and therefore, applicable to an experimental approach.

An additional problem is that of the annealing temperatures of the primers. Previous work[5,11] has demonstrated that primers shorter that 8 bp do not consistently amplify identifiable products. The addition of identical, short (5 bp), arbitrarily designed 5′ tails is required for all generated statistical primers, in order to increase their effective annealing temperatures. These have a G + C content of either 2 or 3, producing for PCR amplifications a set of 13-mers all having 7 G + C and 6 A + T. It has also been shown[11] that primer specificity in PCRs depends on the sequence at the 3′ end of the primer, therefore the addition of the tail to the 5′ end should

[11] R. Griffais, P. M. Andrew, and M. Thibon, *Nucleic Acids Res.* **19,** 3887 (1991).

not severely compromise primer binding to the corresponding genomic sequences.

The initial set of 30 octanucleotide primers, designed to amplify protein-coding regions throughout the entire human genome, has been previously published.[1] Briefly, these were selected by analyzing the protein-coding regions from 1000 randomly selected mRNAs (1,631,763 bp), using the following criteria: (1) each selected octanucleotide sequence should be of a similar length; (2) they should have a G + C content of 50–60% in order to give similar annealing characteristics to the primers; (3) they should fall within a desired range of 50–600 bp, such that they can be effectively separated by electrophoretic means; (4) they should have a sense gene frequency of 15 ± 6%; and (5) their antisense gene frequency should be <5%. Of all possible combinations of octanucleotides (65,536), 151 sequences met these criteria, from which a final set of 30 was selected on the basis of their high gene yields, when applied to computer-simulated PCR amplifications, using the original 1000 mRNAs. A more useful indicator of their general applicability is shown by the fact that the hit rate remains the same when applied to a further 100 randomly selected human cDNAs, suggesting the apparent nonrandom octamer distribution frequency is not limited to the original subset of 1000 genes. It is also of interest to note that the priming ability of the statistically designed sequences is not restricted to human cDNAs, but when applied to 250 randomly selected mouse, *Drosophila,* and yeast cDNAs, detected 70, 60, and 26% of the genes, respectively.

The approach has been used by both López-Nieto and Nigam[1] and ourselves to target specific gene families. Initially, we analyzed 42 cDNAs (76,743 bp), corresponding to members of the diverse bHLH protein transcription factor family, using a computer program written in the C++ language. Potential primers (of any desired length) are derived on the basis of their distribution frequency in the sense protein-coding strands, using criteria similar to those mentioned above: that is, (1) similar length and G + C contents; (2) antisense frequencies of <5%, but no specified sense frequencies; (3) primers located 50 to 1000 bp apart (this distance is larger than that specified in the original paper[1]; however, we believe that reproducible PCR amplification and analysis are possible over these distances, thereby increasing the effectiveness of targeted display); and (4) the elimination of primer sequences that overlap or are shifted by one or two nucleotides relative to each other (a situation likely to arise within gene families owing to the conservation of important regulatory or catalytic domains) from the primer set if they do not improve the overall coverage of the original gene family (a member of any gene family is "covered" if at least one primer pair exists that satisfies point 3). Using these criteria, a set of

25 octamers was generated showing a theoretical 90% coverage of the bHLH family (see Fig. 1A). The same bHLH gene family was used to demonstrate the theoretical effects of searching for hexamers, decamers and dodecamers, the results of which are shown in Fig. 1B, C, and D. As would be expected from the data mentioned earlier, fewer hexamers are required to give 90% coverage levels; however, these are of limited value

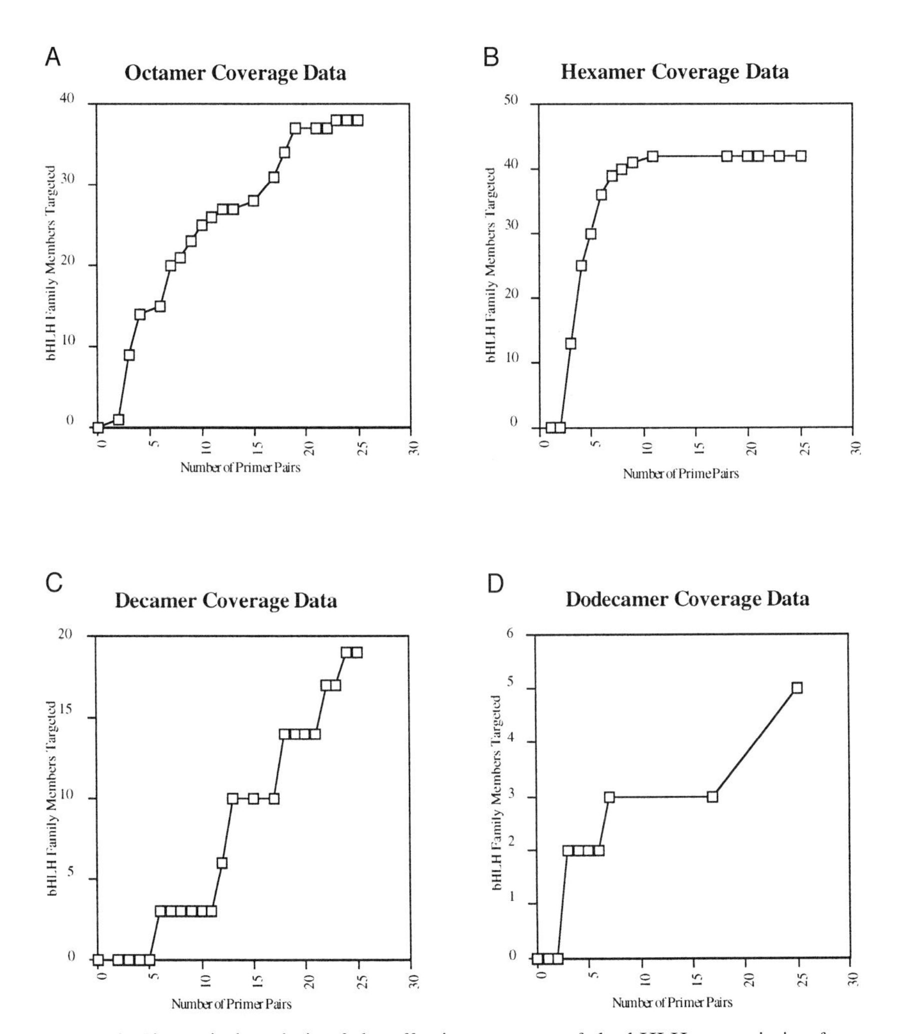

FIG. 1. Theoretical analysis of the effective coverage of the bHLH transcription factor family (maximum number 42) using 25 statistically designed primers of varying lengths in all possible combinations of sense and antisense pairs.

under real experimental conditions, owing to their small size and nonreproducible amplification properties.

Cell Culture

PC12 cells are cultured in suspension at a cell density of 2×10^5 cells per milliliter in Dulbecco's modified Eagle's medium (DMEM; GIBCO, Grand Island, NY) with 10% horse serum and 5% fetal bovine serum at 37°, 5% CO_2. For the long-term induction of the fully differentiated phenotype, PC12 cells are plated onto collagen-coated plates with nerve growth factor (NGF, 100 ng/ml; R&D Systems, Minneapolis, MN). Otherwise, for the short-term time course, NGF (50 ng/ml) is added to flasks containing PC12 cells in suspension. Cells are harvested 7 days (long-term) or 2, 4, and 6 hr (short-term) after NGF induction.

RNA Isolation

Total RNA is the choice starting material for targeted display, owing to its ease of isolation. Following treatment with NGF, cells are pelleted by centrifugation at 4° for 4 min at 400 *g*; the pellet is washed twice with cold phosphate-buffered saline (PBS) and resuspended by flicking the Eppendorf tube (this prevents the cells from clumping in the next step). RNase inhibitor (1 μl; Immunogen International, Newcastle, UK) and 400 μl of RNA lysis buffer [0.15 *M* NaCl, 10 m*M* Tris-HCl (pH 7.4), 1 m*M* $MgCl_2$, 0.5% (v/v) Nonidet P-40 (NP-40)] are added per 10^6 cells and the contents mixed and incubated on ice for 5 min before transferring to cold Eppendorf tubes (~400 μl per tube). These are centrifuged for 5 min at 14,000 rpm in a refrigerated centrifuge to pellet nuclei and unwanted debris. The supernatant is transferred to cold Eppendorf tubes, each containing 200 μl of cold buffered phenol and 50 μl of 10% (v/v) sodium dodecyl sulfate (SDS). The mixture is then vortexed and centrifuged for 2–5 min at 14,000 rpm. The resulting supernatant is removed to a fresh tube and extracted twice with phenol–chloroform and then once with chloroform. RNA is precipitated at −70° by the addition of 40 μl of sodium acetate and 1 ml of RNase-free ethanol per 400 μl of supernatant. The RNA is pelleted by centrifugation for 15 min at 15,000 rpm at 4° and washed in 70% (v/v) ethanol. The pellet is briefly dried before resuspending in 100 μl of RNase-free water and its integrity determined using agarose gel electrophoresis. RNA is routinely stored at −70° under 2 vol of ethanol (2.8×10^8 cells yield approximately 2.4 mg of total RNA). Total RNA from tissue samples is prepared using the method of Chomczynski and Sacchi.[12]

[12] P. Chomczynski and N. Sacchi, *Anal. Biochem.* **162,** 156 (1981).

Genomic DNA Isolation

Genomic DNA for these experiments is isolated as a by-product of the total RNA isolation procedure. After the initial centrifugation step, the nuclei and cell debris are pelleted at the bottom of the Eppendorf tube. The pellet is resuspended in proteinase K (200 μg/ml) in TE buffer [10 m*M* Tris-HCl (pH 7.8), 1 m*M* EDTA] at 37° for 1 hr, phenol–chloroform extracted twice, and resuspended at 100 μg/ml in TE buffer.

DNase I Treatment of RNA

Prior to cDNA synthesis, total RNA is treated with DNase I to prevent contamination with genomic DNA. This step is essential to maximize the reproducibility of subsequent PCRs. RNA aliquots of 99 μl (10–100 μg) are added to 1 μl of RNase-free DNase I (10 U; Boehringer Mannheim, Indianapolis, IN), 10 μl of 50 m*M* $MgCl_2$, 10 μl of 100 m*M* dithiothreitol (DTT), 3.3 μl of 3 *M* sodium acetate (pH 5.3), and 0.5 μl of RNase inhibitor and incubated at 37° for 15 min. The DNase I is inactivated by extracting the supernatant twice with phenol–chloroform and once with chloroform, and the RNA is then ethanol precipitated (2.5 vol of ethanol and a 1/10 vol of 3 *M* sodium acetate) and resuspended in the desired volume of RNase-free water (typically 40 μl).

To check for residual chromosomal DNA contamination, it is strongly recommended that the following negative controls be included: cDNA without the addition of reverse transcriptase and (2) reverse transcriptase enzyme without the addition of any DNA. These should enable one to determine if any contaminating genomic DNA is present in either the original total RNA samples or in the reverse transcriptase enzyme. Such controls become particularly important when the technique is to be used in a quantitative manner.

Reverse Transcription of RNA

First-strand cDNA is prepared from 10-μg aliquots of DNase-free total RNA using Superscript II reverse transcriptase (GIBCO). Freshly precipitated total RNA is added to 500 ng of oligo(dT) on ice [the oligo(dT) is an 18-mer with a degenerate 3′ end, which serves to "lock" it to the end of the poly(A) tail] and the RNA: oligo mixture is denatured to remove secondary structure by heating to 70° for 10 min and chilled on ice. The remaining reagents (4 μl of 5× first-strand buffer, 2 μl of 0.1 *M* DTT, 1 μl of 10 m*M* dNTP mix) are mixed together and chilled on ice and added in 1 vol to the resuspended total RNA. The contents of the tube are mixed gently, the reaction is incubated at 37° for 2 min, and 1 μl (200 U) of

SuperScript II reverse transcriptase is added. Incubation continues at 37° for 20 min, after which the temperature is increased to 42° for 1 hr. At the end of the incubation period, the reaction is terminated by heating the mixture to 70° for 10 min, and then 9 μl of sterile H_2O and 1 μl of RNase H (2 U) are added and the tube is incubated for a further 30 min at 37°. Eppendorf tubes are briefly centrifuged to collect the cDNA samples.

Polymerase Chain Reaction Conditions and Cycling: cDNA

Singe-stranded cDNA, prepared as described above, is used directly in PCR amplifications. PCRs are performed in duplicate to reduce the possibility of misidentification of differentially expressed amplified products. For each individual primer pair combination, the cDNA equivalent of 100 ng of total RNA is used per reaction and added to the following components on ice: 5 μl of 10× buffer (supplied with Taq Express; Genpak, Brighton, UK), 5 μl of dNTP mix (2 m*M* final concentration of dATP, dTTP, dGTP, and dCTP), 10 pmol of sense and antisense primers, and 1 U of Taq Express (1 μl of a 1:25 dilution; Genpak) in a final volume of 50 μl. The reaction mixture is subjected to an initial denaturing step at 94° for 2 min, followed by these cycling conditions: 1 cycle of 94° for 1 min, 42° for 1 min, and 72° for 3 min, and 35 cycles of 94° for 1 min, 42° for 1 min, and 72° for 1 min, with a final extension at 72° for 10 min. When the amounts of total RNA are limited, it is possible to dilute the original PCR 1:100 and use 1 μl of the dilution in a secondary PCR.

Polymerase Chain Reaction Conditions and Cycling: Genomic DNA

In addition to cDNA PCR-based approaches, targeted display can also be used to generate a reproducible array of amplified products from a genomic DNA template. Additional roles can be envisaged for the localization of exon-containing stretches within large regions of noncoding genomic DNA (due to the apparent bias of statistically designed primers toward protein-coding DNA regions) and in phylogenetic studies.

Purified genomic DNA is subjected to PCR cycling conditions using duplicate reactions, each containing 10 or 50 ng of template material in a final volume of 25 μl with the following components: 2.5 μl of 10× buffer (supplied with Taq Express), 2.5 μl of dNTP mix (2 m*M* final concentration of dATP, dTTP, dGTP, and dCTP), 10 pmol of sense and antisense primers, and 1 U of Taq Express (1 μl of a 1:25 dilution; Genpak). An initial denaturation step of 94° for 10 min is followed by 1 low stringency cycle of 94° for 1 min, 38° for 1 min, and 72° for 3 min and 35 cycles at medium stringency of 94° for 1 min, 42° for 1 min, and 72° for 2 min, with a final extension of 72° for 10 min.

Agarose Gel Electrophoresis

As an alternative to denaturing or nondenaturing polyacrylamide gel electrophoresis, PCR products are analyzed using 2% (w/v) Metaphor agarose gels (Flowgen). Agarose gels are precooled for at least 0.5 hr prior to electrophoresis, during which the running buffer (1× TBE) is cooled on ice. This increases the resolution of the bands, allowing higher voltages to be applied across the gel, producing results comparable to those of polyacrylamide gels. DNA bands are visualized by ethidium bromide staining and UV transillumination. Differences in the banding patterns are apparent when applied to amplified cDNA samples (Fig. 2A). As expected, the patterns of amplified fragments originating from genomic DNA templates are more complex (Fig. 2B), which may necessitate the use of polyacrylamide gels to resolve the bands cleanly.

Cloning, Confirmation, and Sequencing of Differentially Expressed cDNAs

Amplified cDNA bands, which are novel or upregulated as judged by their profiles on Metaphor gels, are excised under UV transillumination, recovered from the agarose gel slices using the Qiaex DNA recovery system (Qiagen, Chatsworth, CA) according to the manufacturer instructions, and cloned into the *Eco*RV site of pTT (Genpak). Positive clones are screened by blue/white selection of the colonies and grown in 10 ml of LB broth with ampicillin (final concentration, 100 μg/ml). Plasmid DNA is isolated and the insert is used as a probe against Northern blots to confirm the differential expression status of the clone. Twenty-five nanograms of the digested, cleaned plasmid insert is random-prime labeled using [α-^{32}P]dCTP and hybridized at 65° overnight to a nylon membrane containing 10 μg of each of the original total RNAs per lane. Washes are performed at the following stringencies: twice with 2× SSC (1× SSC is 0.15 *M* NaCl plus 0.015 sodium citrate)–0.1% SDS for 15 min at 65°, once with 1× SSC–0.1% SDS for 15 min at 65°, and once with 0.5× SSC–0.1% SDS for 15 min at 65°, and the membrane is then exposed to X-ray film at −70° for an appropriate time period. Once confirmed, each clone is sequenced on an Applied Biosystems (Foster City, CA) automated sequencer, and sequence homology searches performed using the BLAST network service. Although this method of confirmation remains adequate for both highly and moderately abundant mRNA species, it should be remembered that rare mRNAs may

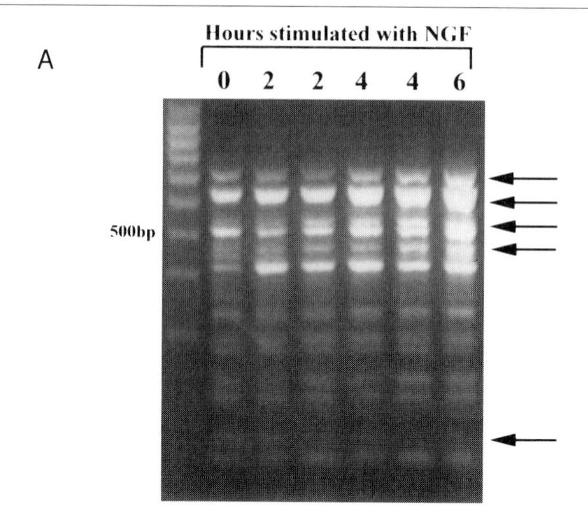

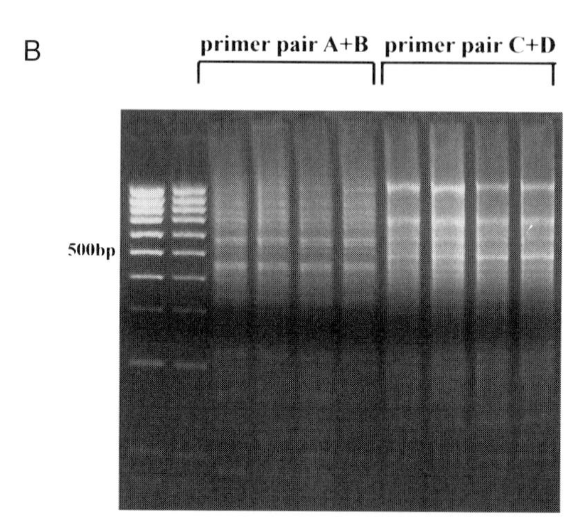

FIG. 2. Examples of amplified targeted display products from NGF-treated PC12 cells analyzed using Metaphor gel techniques. PCR-amplified cDNA, obtained with one primer pair, are shown in (A). This demonstrates the reproducible banding patterns associated with this approach. Potential differentially regulated products are highlighted with arrows. Reactions performed using genomic DNA templates exhibit far more complex banding patterns as shown in (B).

well not be detected in 10-μg total RNA aliquots by this method, in which case alternative confirmation procedures are required.

Results and Discussion

Optimization of Experimental Conditions

The wide-ranging applications of targeted display require certain alterations to the experimental design if the benefits from this new technique are to be maximized. For the original human genome coding primers (HGCP) published by López-Nieto and Nigam,[1] a more degenerate approach is required as it is essential that most, if not all, of the transcribed mRNA species be targeted, whereas for primer sets designed toward specific gene families a more optimized approach is required.

Several parameters were varied in parallel incubations in an attempt to determine the reproducibility of the results: (1) annealing temperatures of 40°, 42°, 44°, and 48°; (2) two different *Taq* DNA polymerases; (3) a range of template concentrations from the equivalent of 1 to 100 ng of starting total RNA per reaction, and 1 to 500 ng of genomic DNA per reaction; and (4) magnesium concentrations between 1.0 and 5.0 μM (data not shown). Using cDNA templates, the effect of the annealing temperature appeared more quantitative than qualitative, decreasing significantly the amount of product amplified as the temperature increased from 40° to 44°. Above this temperature, the number of bands also decreased dramatically. An optimal amount of starting material was adjudged to be between 50 and 100 ng of total RNA per PCR amplification, although this was dependent on the choice of *Taq* polymerase and magnesium concentration. It is, therefore, important that once these conditions have been optimized for a specific enzyme, they remain constant throughout the duration of the experiment.

Primer Performance: Theoretical

Statistical analysis of human genome coding sequences allowed the original authors[1] to develop a set of primers designed to target sequences throughout the genome. In developing a similar computer program capable of the statistical analysis of genomic subsets, it is important to demonstrate that the selected primer subset be specifically enriched for the detection of the family members, in this case the bHLH transcription family. We therefore analyzed the performance of the developed primer set (25 sense and their corresponding antisense primer sequences) against 46 randomly selected genes from the GenBank database, the original 42 bHLH family members, 50 randomly generated strings of sequence (average length, 2000

TABLE I
STATISTICAL ANALYSIS OF BASIC HELIX–LOOP–HELIX PRIMER SET[a]

DNA source	DNA analyzed (bp)	Primer presence (sense primer in sense DNA strand)	Primer presence (sense primer in antisense DNA strand)	Number of theoretical PCR products amplified	Coverage of targeted gene family
Forty-six random cDNAs	93,839	131	88	278	42%
bHLH family (42 members)	76,743	361	56	965	90%
Fifty random strings of sequence	100,000	100	100	70	n/a
Human chromosomal noncoding DNA	86,535	67	72	178	n/a
Human chromosomal coding and noncoding DNA	158,585	100	99	174	n/a
S. pombe chromosomal DNA	104,320	33	31	29	n/a

[a] Theoretical PCR amplification and primer presence calculations were performed using DNAStar, Inc. The GenBank accession numbers for the 46 randomly selected human cDNA sequences are as follows: X87107, X06159, L42324, U71882, X76383, X69256, Z21707, X92426, M90359, M86492, M14564, M65291, M19364, X00383, E01747, Z46376, X94453, X73114, X87843, X84687, Z31367, X73856, M11832, U11700, M54986, M62424, J03248, J00124, L76468, X51408, U16031, D29685, J00299, U36597, M18533, M34664, X61037, X04741, X06705, Y00387, X77549, M55513, X55997, Z80638, and X62071. The Genbank accession numbers for the 42 bHLH family members are as follows: D85845, L35922, M24393, M59764, M65214, M80627, M96739, M96740, S53920, S77532, U18658, U19863, U29086, U36384, D12516, D14029, D16464, D32132, D32200, D82347, D83507, D83674, D85188, S53245, S78079, S79216, U40041, U67776, X14894, X53724, X53725, X56182, X62155, U40039, U58471, U58681, U63841, U63842, U76207, X14894, X17500, and X64840. The GenBank accession numbers for the human and *S. pombe* chromosomal DNA sequences are as follows: Z97988, Z81312, Z54285, and Z68166. The random strings of DNA sequence were generated using a computer program compiled in C++ by N. Jackobi.

bp), 86,535 bp of noncoding chromosomal human genomic DNA,[13] 156,535 bp of human chromosomal DNA containing 15,137 bp of transcriptionally active DNA, and 104,320 bp of *Schizosaccharomyces pombe* genomic DNA (which contains transcriptionally active DNA). The results of this analysis are shown in Table I. These results show a clear overrepresentation of the statistically designed bHLH primer set in the sense versus antisense DNA strands of the bHLH gene family, and an almost equal ratio within all of

[13] Sanger Centre Genome Project at *http://www.sanger.uk*.

the other gene groupings tested. Using computer-modeled PCR simulations, this translates into 965 amplifiable PCR products from the bHLH gene set (producing a gene family coverage figure of 90%) and demonstrates a significant enrichment for this primer set within the original gene family over both the number of products generated from the randomly selected human cDNAs (278) and the random strings of sequence (70). As previously suggested,[1] there appears to be some enrichment for protein-coding regions in general, as demonstrated by the targeting of 42% of the 46 randomly selected cDNAs, and the targeting of 88% of the 15,137 bp of coding sequence located within the 158,585 bp of human chromosomal DNA. Using the same DNA data sets, we tested the original HGPS[1] to act as a comparison (Table II). As predicted,[1] they targeted 72% of the randomly selected human cDNAs and, interestingly, 70% of the bHLH family members, although only 21% of the total number of PCR products were generated for this gene family compared with the bHLH primer set. These results demonstrate both a significant enrichment for the bHLH primer set within the bHLH gene family, and the general ability of the HGPS to target transcriptionally active sequences within the genome.

TABLE II

STATISTICAL ANALYSIS OF ORIGINAL HUMAN GENOME PRIMER SET[a,b]

DNA source	DNA analyzed (bp)	Primer presence (sense primer in sense DNA strand)	Primer presence (sense primer in antisense DNA strand)	Number of theoretical PCR products amplified	Coverage of targeted gene family
Forty-six random cDNAs	93,839	183	77	486	72%
bHLH family members	76,743	111	78	207	70%
Fifty random strings of sequence	100,000	120	120	100	n/a
Human chromosomal noncoding DNA	86,535	175	145	320	n/a
Human chromosomal coding/noncoding DNA	158,535	113	134	275	n/a
S. pombe chromosomal DNA	104,320	44	52	67	n/a

[a] As devised by Lopez-Nieto and Nigam.[1]

[b] The analysis was as described for Table I.

Primer Performance: Experimental

Initial investigations into this technique utilized the well-studied phenomenon of nerve growth factor-induced neuronal differentiation in rat pheochromocytoma PC12 cells.

Primer efficiency can be analyzed experimentally on three levels: (1) the number of bands amplified by each primer pair; (2) the number of mismatches tolerated in each primer binding site, and (3) the number of products correctly amplified by sense and antisense primers annealing to either end of the fragment. Together, these data give an indication as to the annealing and amplification characteristics of the primer. Initial analysis was limited to 30 primer pair combinations, selected from the HGPS (i.e., 3.5% of the total number of 870 possible combinations), each amplified in separate PCRs and analyzed electrophoretically on 2% Metaphor agarose gels. These conditions generated, on average, 18 amplified bands per lane, ranging in size from 50 bp to approximately 900 bp. It was, however, noticeable that certain primer pairs performed far more effectively than others, probably owing to the formation of primer dimers. Complete forward and reverse sequencing of each clone revealed the correct priming at both ends of the fragment in 26 of 32 differentially expressed samples. Seven of the 32 clones were revealed to be known genes of rodent origin, with exact database matches, and these were used to analyze the degree of permitted mismatch within each primer-binding site. Seven of 14 primers annealed exactly to their complementary sequence. Of the remaining seven octamer annealing sites, mismatches were tolerated at one or two positions, generally being found toward the 5′ end. These results demonstrated that, while mismatches and mispriming are tolerated, the specific octamers are still able to reproducibly amplify mRNAs. In fact, nonexact amplification characteristics may even increase the ability of the statistically designed primer pairs to target the transcriptionally active genome.

One final point, regarding mismatches and mispriming, is the increased specificity of PCR conditions required for gene family subsets. In this scenario, increased annealing temperatures are required to help decrease the degree of misprimed products, which may otherwise detract from the specific targeting ability of these primer subsets. Routinely, we have used annealing temperatures of 52° and 54°, when amplifying from the bHLH gene family.

Genome Coverage

The ability to amplify as many as possible of the estimated 10,000 to 20,000 mRNAs in a mammalian cell is a prerequisite of any technique attempting to analyze differential gene expression. Initial data suggested an average of 18 amplified bands per primer pair. Assuming no duplications in the targeting of mRNAs, for the full set of 870 primer pairs, a total of 13,920 mRNAs would be recognized. Although duplications in the targeting of certain mRNAs are possible, and in fact might even be expected within the larger mRNAs, this estimate still correlates well with the proposed number of mRNAs per cell.

A second important aspect in assessing the fidelity of this new technique is its ability to target both abundant and rare mRNA species within the cell. Confirmation by Northern blot analysis of 56% of the experimentally isolated differentially expressed clones from PC12 cells suggested that these sequences represented abundant mRNA species. Extrapolating from this, it is possible that the remaining 44% of unconfirmed genes represent rare mRNAs. Few data are available for the transcript prevalence of specific genes within the cell; however, previous estimates suggest ratios of 25% : 50% : 25% for high-, medium-, and rare-abundance mRNAs, within a typical mammalian cell, suggesting targeted display reproducibly amplifies both rare and abundant mRNAs with an almost equal ratio.

Prospects and Limitations

There are many potential advantages with targeted display over other differential display techniques, most notably the speed and ease of use, and the ability to specifically target gene families, a problem not yet tackled by other display techniques. However, it also has its limitations. As with all PCR-based techniques, the inherent problems of background noise and the reproducibility of experimental results, especially if the reaction conditions are altered, will contribute significantly to the possibility of false positives (i.e., the isolation of nondifferentially expressed genes) and difficulties in the quantitative determination of gene expression. Although we found the percentage of false positives to be only 6%, this would probably depend on the system being studied and, therefore, special attention should be paid to performing the correct controls for each experiment. The importance of controls increases if the technique is to be used in a semiquantitative way, where the PCR amplification must become limiting and the number of products, as a result, may drop. If one were to use all 870 primer pair combinations, then the amount of starting material may also become limiting (an estimated 40 to 80 μg of total RNA would be required for the full primer set). This can be reduced at least 100-fold using dilutions and secondary PCRs, however, at the expense of doubling the number of PCRs required. As with differential display, there may also be questions concerning the ability to target rare mRNA species, which may comprise only 1–10 copies per cell. However, our preliminary data suggest we have isolated

44% of rare mRNAs. Finally, and in a fashion similar to the subtractive hybridization techniques, targeted display is unlikely to be able to detect point mutations or nontranscriptional differences between genes that may be responsible for changes in the levels of gene transcription.

Acknowledgments

We thank Genpak Limited for providing the complete human coding primer set, Nicholas Jackobi for kindly providing the computer expertise required, and Dr. Helen Stewart (UCL, London) for collaboration on the bHLH family. This work was supported by the Trafford Renal Trust (L.V.M.) and an MRC Ropa Award to L.V.M. and J.F.B. L.V.M. was an R.M. Phillips Senior Research Fellow. [For further information, e-mail *J.F.Burke@sussex.ac.uk,* and for primer information *info@genpakdna.com.*

Section IV

Functional Relationship among cDNA Translation Products

[23] RNA Polymerase III-Based Two-Hybrid System

By MARIE-CLAUDE MARSOLIER and ANDRÉ SENTENAC

Introduction

Identifying protein–protein interactions and their role in biological processes such as signal transduction, cell cycle regulation, or the control of gene expression has become a major aim for biologists in today's postgenomic era. This endeavor has been facilitated by the emergence of new *in vivo* strategies called "two-hybrid systems."

The concept of the two-hybrid system was developed by Fields and Song[1] on the basis of the modularity of RNA polymerase II (Pol II) transcriptional activators in yeast. Since then, other two-hybrid systems have been invented, which rely again on the functional independence of transcriptional activator DNA-binding domains[2] or of plasma membrane-targeting signals,[3] but they have not yet proved to be of such wide use as the original strategy of Fields and Song. These systems can be used to analyze interactions between two known proteins or to search libraries for proteins that interact with a target protein. We present a two-hybrid system based on yeast RNA polymerase III (Pol III), which is more specifically designed for the study of Pol II activators or of nuclear proteins that spuriously activate Pol II reporter genes. We first briefly present the yeast Pol III transcription system and its genetic modification to derive a Pol III-based two-hybrid system.

RNA Polymerase III-Based System

The Pol II-based two-hybrid system of Fields and Song made use of the fact that Pol II transcriptional activators such as Gal4p contain two independent, separable domains, a DNA-binding domain and an activating domain,[4] which do not need to be covalently attached to be functional.[5] *In vivo* interactions between a protein X, fused to the DNA-binding domain, and a protein Y, fused to the activating domain, reconstitute a functional transactivator, whose activity can be readily monitored by a reporter gene harboring the corresponding upstream binding sites.

[1] S. Fields and O. Song, *Nature* (*London*) **340,** 245 (1989).
[2] S. L. Dove, K. Joung, and A. Hochschild, *Nature* (*London*) **386,** 627 (1997).
[3] A. Aronheim, A. Zandi, H. Hennemann, S. J. Elledge, and M. Karin, *Mol. Cell. Biol.* **17,** 3094 (1997).
[4] R. Brent and M. Ptashne, *Cell* **43,** 729 (1985).
[5] J. Ma and M. Ptashne, *Cell* **55,** 443 (1988).

Copyright © 1999 by Academic Press
All rights of reproduction in any form reserved.

0076-6879/99 $30.00

We have extended this concept of functional modularity to the transcription factor C (TFIIIC) of the yeast Pol III transcription system. The transcription of yeast class III genes (*SNR6*, tRNA, or 5S RNA genes) is relatively simple. Activation of the *SNR6* gene, for example, relies on only two general transcription factors: TFIIIB and TFIIIC. TFIIIC, or τ, is a large, multisubunit factor that recognizes the promoter elements (the A and B blocks) and assembles the second factor, TFIIIB, on the DNA, upstream of the transcription start site. TFIIIB is the initiation factor that recruits RNA polymerase III[6] (Fig. 1A). No transactivators, similar to Pol II transactivators, are known to regulate the Pol III transcription rate in yeast. We have shown that the specific interaction of TFIIIC with the B block element is dispensable if TFIIIC is provided with another means of anchorage on the DNA. If the essential B block element of *SNR6* is replaced by Gal4p-binding sites (UAS_G sites; Fig. 1B), the wild-type TFIIIC cannot bind the UAS_G–*SNR6* template, but its DNA-binding function can be restored artificially by fusing one of its subunits to the Gal4p DNA-binding domain[7] (one-hybrid system). TFIIIC–DNA interaction is in fact sufficiently flexible, as in the case of Pol II transactivators, to tolerate a loose protein–protein interaction that connects the factor to the DNA. Hence, TFIIIC remains functional when one of its subunits is fused to a protein X that interacts with a protein Y, which is itself fused to a Gal4p DNA-binding domain[8] (two-hybrid system). Several subunits of TFIIIC have been tested for their efficiency in triggering UAS_G–*SNR6* transcription in the one-hybrid or two-hybrid systems (τ138,[7,8] τ131,[7,9] τ95,[9,10] and τ91[9,10]), and τ138 has proved to be the most efficient.

Practically, two plasmids encoding the fusions to τ138 and to the Gal4p DNA-binding domain are introduced together into a yeast strain that contains a UAS_G–*SNR6* template. We have developed two methods for detecting interactions between two hybrid proteins. The first assay is simply the quantification of the transcripts produced from a modified UAS_G–*SNR6* reporter gene. The second assay is based on the fact that *SNR6* is an essential gene. Activation of the UAS_G–*SNR6* template via the two-hybrid interaction allows the loss of a *URA3*-marked plasmid harboring the wild-type *SNR6* gene, and the cells then become 5-fluoroorotic acid (5-FOA) resistant. These two techniques are discussed sequentially.

[6] G. A. Kassavetis, B. R. Braun, L. H. Nguyen, and E. P. Geiduschek, *Cell* **60,** 235 (1990).

[7] M.-C. Marsolier, N. Chaussivert, O. Lefebvre, C. Conesa, M. Werner, and A. Sentenac, *Proc. Natl. Acad. Sci. U.S.A.* **91,** 11938 (1994).

[8] M.-C. Marsolier, M.-N. Prioleau, and A. Sentenac, *J. Mol. Biol.* **268,** 243 (1997).

[9] M.-C. Marsolier, unpublished (1996).

[10] R. Arrebola, N. Manaud, S. Rozenfeld, M.-C. Marsolier, C. Carles, P. Thuriaux, C. Conesa, and A. Sentenac, *Mol. Cell. Biol.* **18,** 1 (1998).

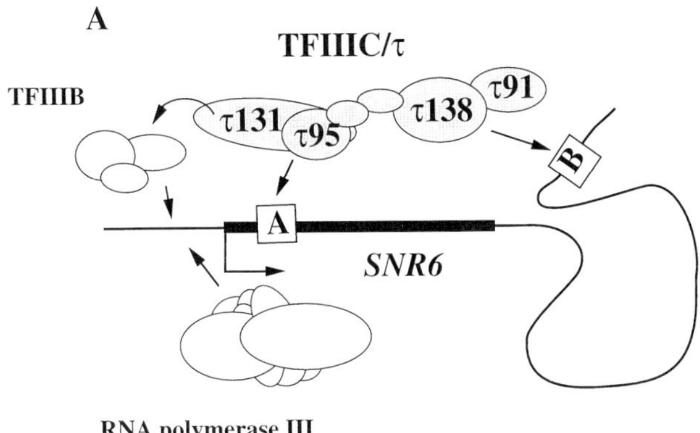

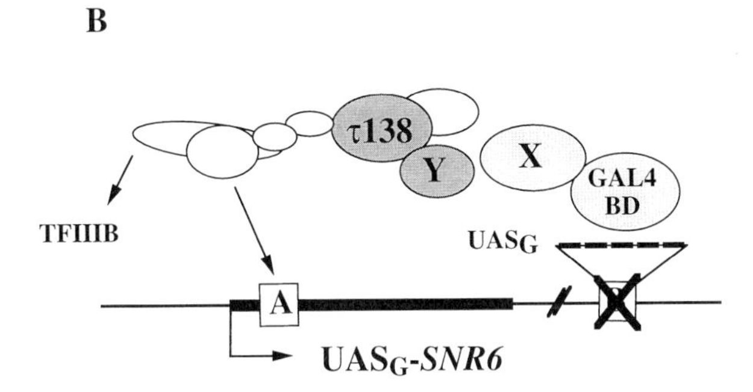

FIG. 1. The Pol III-based two-hybrid system. *SNR6* A and B blocks are represented by open boxes marked A and B. (A) Transcriptional activation of wild-type *SNR6*. The TFIIIC subunit τ138 binds the downstream B block as the first step of gene activation, triggering the successive recruitment of TFIIIB and of RNA polymerase III. (B) The UAS_G–*SNR6* templates have five head-to-tail UAS_G sequences (small black rectangles) inserted at the location of the *SNR6* B block (which has been destroyed by a 2-bp deletion). τ138 is recruited to the UAS_G–*SNR6* template indirectly, through the interaction between proteins X and Y, fused to Gal4p DNA-binding domain (GAL4 BD) and τ138, respectively.

One of the advantages of this two-hybrid strategy is that it makes possible the analysis of proteins that activate Pol II transcription and that cannot be studied by the original two-hybrid technique. We have found that neither Gal4p nor the activation domains of VP16 or p53 fused to the Gal4p DNA-binding domain could by themselves activate the transcription of the UAS_G–*SNR6* gene.[8] Other proteins can also activate Pol II reporter genes spuriously without being involved in transcription. This promiscuity

TABLE I
COMPARISON OF RELATIVE ACTIVATION LEVEL IN RNA POLYMERASE II AND III TWO-HYBRID SYSTEMS

	Pol II two-hybrid system	Transcriptional activation (%)	Pol III two-hybrid system	Transcriptional activation (%)
Reporter gene	USA_G–*lacZ*		UAS_G–*SNR6*	
DNA-binding domain	*GAL4*(1–147)		*GAL4*(1–147)	
Activating component	*GAL4*(768–881)		τ138	
Quantification	β-Galactosidase activity		*SNR6* RNA level	
Activation through direct interaction	*GAL4*(1–147)–*GAL4* (768–881)	100[a]	*GAL4*(1–147)–τ138	100[a]
Activation through indirect interaction	*GAL4*(1–147)–Prp9 with *GAL4*(768–881)–Prp21;	4	*GAL4*(1–147)–Prp9 with τ138–Prp21;	70
	GAL4 (1–147)–Prp11 with *GAL4*(768–881)–Prp21	6	*GAL4* (1–147)–Prp11 with τ138–Prp21	80

[a] The transcriptional activation obtained through direct interaction between the UAS_G sites and the *GAL4*(1–147)–activator construct of each system is assigned the value 100%.

is likely due to the relative lack of specificity of Pol II-activating sequences.[11] Because the UAS_G–*SNR6* gene is insensitive to all the bona fide Pol II transcriptional activators that we tested, there is no theoretical reason why it should be activated by analogous sequences unrelated to transcription.

Another advantage of the system is that it has probably a large domain of applications. Indeed, only two factors, TFIIIC and TFIIIB, are involved in the activation of *SNR6*-like genes, which makes it likely that few sequences will activate the UAS_G–*SNR6* gene by themselves.

Finally, this Pol III system is sensitive. In one set of experiments,[9] we have been able to compare directly the activation of the reporter genes of the Pol II and Pol III two-hybrid systems, when the same pairs of interacting proteins (Prp9p/Prp21p and Prp11p/Prp21p) are involved. The results are shown in Table I. The UAS_G–*SNR6* template is almost as strongly activated by an indirect activation as it is by a direct one,[9] whereas indirect activation of the UAS_G–*lacZ* gene via the two hybrid proteins amounts to only a few percent of the activation obtained when the Gal4p activating domain is fused to the Gal4p DNA-binding domain.[11,12]

Materials and Methods

The vectors, the construction of hybrid genes, and the protocol of yeast transformation are common to both assays and are described first. The

[11] J. Ma and M. Ptashne, *Cell* **51,** 113 (1987).

[12] P. Legrain, M.-C. Dokhelar, and C. Transy, *Nucleic Acids Res.* **22,** 3241 (1994).

yeast strains, the UAS_G–*SNR6* templates, and the protocols specific to each assay are addressed thereafter.

Vectors

pGEN-τ138 is specifically designed for the efficient expression of τ138-fusion proteins. This vector derives from the pGEN plasmid[13] (2μ, *TRP1*), and contains the full-length τ138-coding sequence under the regulation of the phosphoglycerate kinase (PGK) promoter.[13] τ138 is a nuclear protein, therefore, no nuclear localization signal needs to be added to the vector. The map and polylinker sequence of pGEN-τ138 are shown in Fig. 2.

The present two-hybrid system makes use of the Gal4p DNA-binding domain for one of the fusion proteins. All of the vectors that have been constructed for the Pol II two-hybrid system developed by Fields and collaborators and that contain this sequence can thus be used. These vectors are typical yeast/*Escherichia coli* shuttle vectors containing the bacterial and yeast sequences and markers required for plasmid replication and selection in *E. coli* and in yeast. The most widely used are pMA424,[11] pGBT9,[14] pAS1,[15] and pAS2.[16] These vectors contain the gene fragment encoding the Gal4p DNA-binding domain controlled by the *ADH1* promoter, and followed by unique restriction sites for in-frame fusions. No nuclear localization signal needs to be added because one is found in the Gal4p DNA-binding domain. All of these vectors are multicopy plasmids and possess the *TRP1* marker, except pMA424, which harbors the *HIS3* gene. A comparative study[12] indicates that expression of fusion proteins from the pGBT9 vector is much lower than from pMA424, resulting in a significant reduction in the reporter gene expression of the conventional two-hybrid assay. While lower production of fusion proteins may be advantageous to overcome their potential toxic effects, the ensuing reduced sensitivity of the two-hybrid assays could lead to overlooking weak interactions.

The construction of hybrid genes between the proteins of interest and the Gal4p DNA-binding domain or the τ138 subunit requires in-frame fusions. These constructs can be made using either standard molecular biology techniques or polymerase chain reactions (PCRs) in order to generate the proper cloning sites.

[13] G. V. Shpakovski, J. Acker, M. Wintzerith, J.-F. Lacroix, P. Thuriaux, and M. Vigneron, *Mol. Cell. Biol.* **15,** 4702 (1995).

[14] P. L. Bartel, C.-T. Chien, R. Sternglanz, and S. Fields, *in* "Cellular Interactions in Development: A Practical Approach" (D. A. Hartley, ed.), p. 153. Oxford University Press, Oxford, 1993.

[15] T. Durfee, K. Becherer, R.-L. Chen, S. H. Yeh, Y. Yang, A. E. Kilburn, W. H. Lee, and S. J. Elledge, *Genes Dev.* **7,** 555 (1993).

[16] J. W. Harper, G. R. Adami, N. Wei, K. Keyomarsi, and S. J. Elledge, *Cell* **51,** 805 (1993).

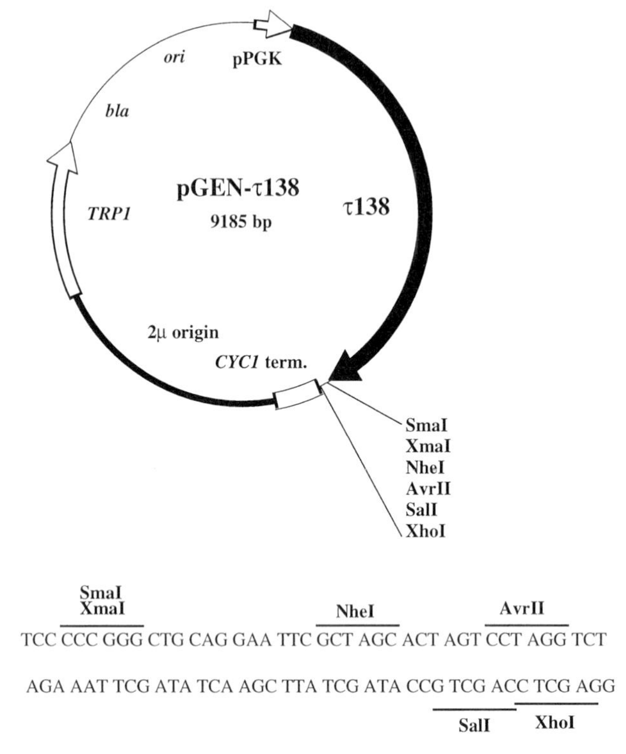

FIG. 2. Map of pGEN-τ138. Important features include the PGK promoter (pPGK), which drives transcription of the τ138 gene, the *CYC1* terminator (*CYC1* term), the yeast origin of replication (2μ origin), the *E. coli* origin of replication (*ori*), the *E. coli* selectable marker for ampicillin resistance (*bla*), and the yeast auxotrophic marker *TRP1*. The restriction sites diagrammed below the map are unique and hence suitable for the insertion of heterologous genes. The nucleotide sequence is arranged in codons continuing in the same reading frame as the upstream τ138 gene.

Yeast Transformation

Plasmids can be introduced into yeast cells in a variety of ways. We use a simplified version of the lithium acetate method developed by Ito *et al.*[17] and improved by Schiestl *et al.*[18] Yeast cells are grown in liquid medium, resuspended in lithium acetate along with the plasmid DNA, heat shocked, and plated onto minimal media lacking the amino acids or the nucleotides corresponding to the auxotrophic markers harbored by the plasmids to be

[17] H. Ito, Y. Fukuda, K. Murata, and A. Kimura, *J. Bacteriol.* **153,** 163 (1983).
[18] R. H. Schiestl, P. Manivasakam, R. A. Woods, and R. D. Gietz, *Methods* **5,** 79 (1993).

selected. The description of standard yeast media can be found in any general book about yeast genetics.[19]

Protocol

1. Inoculate 20 ml of selective medium with yeast cells. Incubate at 30° overnight.
2. Dilute the cells into 40 ml of YEPD to give a final OD_{660} of 0.2. Incubate at 30° until the culture reaches an OD_{660} between 0.4 and 0.8.
3. Pellet the cells by centrifugation (4000 *g*, 5 min, room temperature).
4. Wash the pellet with 4 ml of sterile water.
5. Resuspend the pellet in 1 ml of TE–lithium acetate made fresh from 10× sterile stocks: 10× TE (100 m*M* Tris-HCl, 10 m*M* EDTA, pH 7.5), 10× lithium acetate (1 *M* $C_2H_3O_2Li$). Transfer to a 1.5-ml microcentrifuge tube and centrifuge for 1 min at 3000 rpm.
6. Resuspend in 400 μl of TE–lithium acetate.
7. For one transformation, mix 1 μg of plasmid DNA, 50 μg of Clontech (Palo Alto, CA) carrier DNA (Yeastmaker), 100 μl of cell suspension, 700 μl of 40% polyethylene glycol 4000 (PEG 4000)–TE–lithium acetate. Mix gently by pipetting up and down.
8. Incubate at 30° for 30 min or longer (up to 3 hr).
9. Heat shock at 42° for 15 min.
10. Pellet the cells by centrifuging for 1 min at 5000 rpm in a microcentrifuge.
11. Resuspend the cells in sterile water and plate them onto the appropriate medium.

The plasmids harboring the fusion constructs can be introduced sequentially into the yeast reporter strain, or simultaneously, with a transformation frequency of about 10^5 transformants per microgram of plasmid DNA for a single transformation, and 10^4 transformants per microgram of plasmid DNA for a double transformation.

Quantifying UAS_G–SNR6 Transcripts

A straightforward way to measure the intensity of the interaction between two hybrid proteins consists in quantifying the level of the UAS_G–*SNR6* transcripts generated through this interaction. Because *SNR6* is an essential gene, a wild-type copy must be present in the cell, and the UAS_G–*SNR6* construct must be modified so that its transcripts can be distinguished

[19] C. Guthrie and G. R. Fink, "Guide to Yeast Genetics and Molecular Biology." Academic Press, San Diego, California, 1991.

from the ones derived from the wild-type gene. In order to do so, we have introduced a 24-bp sequence into the transcribed region of the gene, at position +73 relative to the *SNR6* transcription start site. The UAS_G–*SNR6* maxigene construct thus generates transcripts 24 nucleotides longer than the wild-type gene, which is used as an internal standard in Northern blots.

The UAS_G–*SNR6* maxigene, with 5 head-to-tail Gal4p-binding sites at the location of its B block and a 24-bp insertion in its transcribed sequence, has been cloned into 2μ vectors whose high copy number provides a greater sensitivity to the system: YEp352[7,20] (*URA3*), pRS423[9,21] (*HIS3*), and pRS425[9,21] (*LEU2*). The strains to be used for this assay do not have any other requirement than the auxotrophic markers necessary for the selection of the plasmids harboring the UAS_G–*SNR6* template and the two hybrid fusion constructs (in particular, they do not need to be deleted for the *GAL4* gene).

In our hands, the easiest way to quantify the amount of UAS_G–*SNR6* maxigene transcripts has been the Northern blot. Total RNA is extracted according to a modified version of the method described by Schmitt *et al.*[22] Small RNA species are separated by electrophoresis on a denaturing polyacrylamide gel, blotted onto a membrane, then hybridized with a ^{32}P-end labeled oligonucleotide to probe both wild-type *SNR6* and UAS_G–*SNR6* transcripts.

Protocol

RNA EXTRACTION. Wear gloves throughout the preparation to avoid ribonuclease contamination.

1. Grow yeast cells overnight in 40 ml of selective medium, until they reach an OD_{660} of 0.3–0.4. It is important to harvest the cells at that stage and not later, otherwise the steady state level of the transcripts derived from the UAS_G–*SNR6* maxigene will drop compared with the level of the wild-type transcripts (this is probably because the maxigene transcripts are less stable than the wild-type *SNR6* RNA, so that their steady state level decreases when the cells enter the stationary phase and slow down their Pol III transcription). Pellet the cells by centrifugation (4000 *g*, 5 min, room temperature). Use 50-ml Falcon tubes with conic ends, so that the pellet is tightly packed.
2. Resuspend the cells in 400 μl of AE buffer [50 m*M* sodium acetate

[20] J. E. Hill, A. M. Myers, T. J. Koerner, and A. Tzagoloff, *Yeast* **2,** 163 (1986).
[21] T. W. Christianson, R. S. Sikorski, M. Dante, J. H. Shero, and P. Hieter, *Gene* **110,** 119 (1992).
[22] M. E. Schmitt, T. A. Brown, and B. L. Trumpower, *Nucleic Acid Res.* **18,** 3091 (1990).

(pH 5.3), 10 m*M* EDTA, autoclaved]. Transfer the cells to a 1.5-ml microcentrifuge tube. *Note:* The protocol can be interrupted at this step and cells can be kept at −80°.

3. Add 50 μl of 10% sodium dodecyl sulfate (SDS).
4. Vortex.
5. Add 550 μl of phenol previously equilibrated in AE buffer.
6. Vortex.
7. Incubate at 65° for 4 min.
8. Quickly chill the mixture in a dry ice–ethanol bath.
9. Centrifuge in a microcentrifuge at 15,000 rpm for 6 min at room temperature.
10. Transfer the aqueous phase (about 600 μl) to a fresh tube and extract with phenol–chloroform for 5 min at room temperature.
11. Take 400 μl of the aqueous phase; add 40 μl of 3 *M* sodium acetate (pH 5.3) and 1.1 ml of ethanol.
12. Centrifuge at 4° for 30 min at 15,000 rpm. The pellets should be clearly visible.
13. Wash once with 80% ethanol. Centrifuge for 5 min at 4° at 15,000 rpm.
14. Dry the pellets in a Speed-Vac evaporator (Savant, Hicksville, NY) (15 min maximum).
15. Resuspend in 20 μl of sterile distilled water. Store at −80°.
16. Quantify the amount of RNA by determining the OD_{260} of 1 μl of RNA preparation diluted in 1 ml of sterile H_2O. The RNA concentration is usually around 4–10 μg/μl.

Gel Electrophoresis, Blotting, and Hybridization

Small-size RNAs are separated on a denaturing polyacrylamide gel [8% acrylamide (acrylamide–bisacrylamide, 20 : 1), 8 *M* urea, 1× TBE (90 m*M* Tris-borate, 2 m*M* EDTA)]. Ten to 15 μg of RNA is loaded per lane. Four volumes of loading buffer [90% (v/v) formamide, 10% (v/v) 10× TBE, 0.1% (w/v) bromophenol blue, 0.1% (w/v) xylene cyanol blue] are added to the RNA samples, which are then heated in boiling water for 5 min and cooled on ice for 5 min before loading.

After electrophoresis, the gel is blotted onto a positively charged nylon membrane (Boehringer Mannheim, Indianapolis, IN) using the Bio-Rad (Hercules, CA) Trans-Blot cell (60 V, constant voltage, 1 hr, in 0.5× TBE).

The membrane is hybridized with a probe consisting of a fragment of *SNR6* transcribed sequence. We have used either the entire *SNR6* sequence labeled by nick translation with [α-^{32}P]dCTP or an oligonucleotide (5′-

AATCTCTTTGTAAAACGGTTCATCC-3′) kinased with [γ-^{32}P]dATP. The hybridization protocol is derived from the method described by Church and Gilbert.[23]

Protocol

1. Heat the hybridization buffer [0.5 *M* sodium phosphate (pH 7.2), 10 m*M* EDTA, 7% SDS] to 65° and add to the membrane containing the fixed nucleic acids.
2. Add the denatured probe.
3. Hybridize overnight at 65°.
4. Wash four to six times with the washing buffer [40 m*M* sodium phosphate (pH 7.2), 1% SDS] initially at 65°, for 5–10 min each time.

Assaying for Reporter Gene Expression by 5-Fluoroorotic-Acid-Resistance Test

A quicker and easier way to assess potential interactions between two given hybrid proteins consists in testing whether enough transcripts can be generated from a UAS_G–*SNR6* template to support cell growth in the absence of the wild-type *SNR6* gene. We constructed a strain, yMCM616,[8] whose chromosomal *SNR6* gene has been inactivated, and which survives with a wild-type *SNR6* gene harbored by a *URA3*-marked plasmid (pRS316-*SNR6*). This strain cannot grow on a medium containing 5-fluoroorotic acid (5-FOA) because it cannot lose the vector harboring the wild-type *SNR6* copy (*SNR6* is an essential gene) and therefore the *URA3* gene, whose product metabolizes 5-FOA to a toxic compound.[24] yMCM616 is then transformed with plasmids harboring *GAL4*(1–147)- and τ138-fusion constructs, and the UAS_G–*SNR6* reporter gene. If the interactions between the hybrid proteins trigger the transcription of the UAS_G–*SNR6* template to a sufficient extent, the corresponding *SNR6* transcripts can support cell growth, hence the cells can lose the *URA3*-marked pRS316–*SNR6* plasmid and grow on 5-FOA.

This growth assay is easy to perform: transformants of the reporter strain containing the two-hybrid constructs are selected on the appropriate medium, and several independent clones are streaked on 5-FOA plates. The positive clones usually take 2–3 days to grow, and their growth rate (reflecting the strength of the interaction between the hybrid proteins) can be compared with that of a positive control containing the *GAL4*(1–147)–τ138 construct.

[23] G. M. Church and W. Gilbert, *Proc. Natl. Acad. Sci. U.S.A.* **81,** 1991 (1984).
[24] J. D. Boeke, F. Lacroute, and G. R. Fink, *Mol. Gen. Genet.* **197,** 345 (1984).

This assay must be performed in a yeast strain presenting the characteristics of yMCM616: it must have auxotrophic markers for the selection of the required plasmids, its *SNR6* chromosomal copy must be inactivated, and a wild-type copy of *SNR6* must be present in the cell, harbored by a plasmid that can be both selected and counterselected. yMCM616 is derived from YPH500α[25] and possesses the following genotype: *MAT*α *ura3-52 lys2-801*^{amber} *ade2-101*^{ochre} *trp1-*Δ*63 his3-*Δ*200 leu2-*Δ*1 snr6-1* (pRS316-*SNR6*:*CEN ARS URA3 SNR6*). (*snr6-1* is a mutant allele of *SNR6* with a 2-bp deletion in its B block that inactivates the gene).[26]

The UAS_G–*SNR6* template consists of the *SNR6* gene with five head-to-tail Gal4p-binding sites inserted at the location of its B block (its transcribed sequence is not modified because the corresponding transcripts must be functional). This construct has been cloned in the *LEU2* centromeric plasmid pRS315.[8,25]

Protocol

1. Prepare 500 ml of a 2× 5-FOA mix containing 7 g of yeast nitrogen base without amino acids, 1 g of 5-FOA, 20 g of glucose, and 50 mg of uracil. Sterilize by filtration.
2. Prepare 500 ml of a 4% agar solution. Autoclave to sterilize.
3. Melt the agar preparation and place it along with the 2× 5-FOA mix into a water bath at 60°.
4. Mix the two solutions and add the appropriate nutrients corresponding to the auxotrophic markers of the strain.

Because this assay is a genetic test, it could be used to screen libraries for τ138-fused proteins interacting with a bait fused to the Gal4p DNA-binding domain. This could be performed either by replicating on 5-FOA plates transformants which have been previously selected for the plasmid auxotrophic markers, or by plating the cells directly after transformation on 5-FOA plates.

Troubleshooting

This system is based on the interactions between two hybrid proteins in the nucleus of yeast cells. The two hybrid proteins must be expressed at a sufficient level (which does not occur if they are toxic to the cell or unstable), they must be correctly folded and targeted to the nucleus, and their Gal4p or τ138 moiety must not hamper their interacting properties.

[25] R. S. Sikorski and P. Hieter, *Genetics* **122,** 19 (1989).

[26] M.-C. Marsolier, S. Tanaka, M. Livingstone-Zatchej, M. Grunstein, and A. Sentenac, *Genes Dev.* **9,** 410 (1995).

A problem with one of these requirements could possibly be solved by making different constructs including only part of the proteins of interest, but one cannot rule out the possibility that important physiological interactions could not be observed with this system.

Other Applications

This system lends itself to further development. First, a conditional expression system is being developed to facilitate library screening procedures. Once an interaction has been observed between two proteins, their interacting domains can be mapped using the same procedures. DNA-binding proteins that bind to a given site can be identified by screening a τ138-fusion library with a modified *SNR6* reporter gene in which the Gal4p-binding sites have been replaced by the relevant DNA sequences. Three-hybrid systems can be developed in which the overexpression of a third partner enhances the interactions between two hybrid proteins. Conversely, components can be searched whose expression reduces a given protein–protein interaction.

Acknowledgments

This research was supported by grants from the European Community (Science and Biotechnology programs).

[24] Screening for Protein–Protein Interactions

By F. JOSEPH GERMINO and NEAL K. MOSKOWITZ

Introduction

A variety of molecular approaches have been developed to assay for protein–protein interactions. These assays for protein–protein interactions include the yeast two-hybrid assay[1] (along with a number of modifications), the phage display technique,[2–4] the screening of bacterial expression libraries with a labeled protein of interest (the filter lift assay),[5] and the double-tagging assay.[6,7] Because the phage display technique is most useful for

[1] S. Fields and O. Song, *Nature (London)* **340,** 245 (1989).
[2] J. Scott and G. Smith, *Science* **249,** 386 (1990).
[3] J. Devlin, L. Panganiban, and P. Devlin, *Science* **249,** 404 (1990).
[4] S. Cwirla, E. Peters, R. Barrett, and W. Dover, *Proc. Natl. Acad. Sci. U.S.A.* **87,** 6378 (1990).
[5] M. Blanar and W. Rutter, *Science* **256,** 1014 (1992).
[6] F. Germino, Z. Wang, and S. Weissman, *Proc. Natl. Acad. Sci. U.S.A.* **90,** 933 (1993).
[7] Z. Wang, A. Bhargava, R. Sarkar, and F. Germino, *Gene* **173,** 147 (1996).

Copyright © 1999 by Academic Press
All rights of reproduction in any form reserved.

0076-6879/99 $30.00

identifying relatively small peptides (or epitopes) that can interact with a protein of interest, this chapter concentrates on the three other techniques.

Two-Hybrid Assay

Perhaps the most widely used approach for identifying the binding partner(s) of a protein of interest is the yeast two-hybrid assay. This assay is based on the observation that the 881-amino acid yeast GAL4 protein can be divided into a 147-amino acid amino-terminal domain ($GAL4_{bd}$), which binds to specific DNA sequences known as upstream activating sequences (UAS_G), and an acidic, carboxy-terminal domain (amino acids 768–881), the transcription-activating domain ($GAL4_{ad}$). In order for this protein to serve as a functional transcriptional activator for promoters containing UAS_G, these two separate domains must be in close proximity, although not necessarily physically joined.

A similar approach, using the DNA-binding domain of the *Escherichia coli* LexA protein and the *lexA* operator instead of the $GAL4_{bd}$ and UAS_G,[8,9] respectively, also has been employed, indicating that different operators and DNA-binding proteins can be used in place of the GAL_{bd} and the UAS_G. Moreover, other transcriptional activators can replace the GAL_{ad}: the activation domain of the bacterial BA42 protein is also functional and has been used in the LexA-based system. The LexA-based system has some advantages in that yeast proteins are not utilized and thus it can be used in $GAL4^+$ yeast. The fusion "bait" genes can be inserted downstream of a GAL4-inducible promoter so that their expression can be regulated. In addition, there are a number of strains that have been engineered so that testing for the production, nuclear transport, and DNA binding of the hybrid "bait" protein is relatively simple. On the other hand, because the "bait" and "fish" hybrid proteins are not yeast proteins, the $GAL4_{bd}$ may be more stable for some fusions and may be transported to the nucleus more efficiently.

For this assay, the gene or portion of a gene of interest ("bait") is fused to the $GAL4_{bd}$ and transformed into the cells of an appropriate strain of yeast with a cDNA library cloned downstream of the $GAL4_{ad}$ ("fish"). In those cells in which an interaction between the bait and fish has occurred, transcription from reporter genes containing UAS_G will be initiated and the yeast will survive on the appropriate selective plates (see Fig. 1). Because this is an *in vivo* genetic assay, a large number of clones can be screened

[8] J. Ma and M. Ptashne, *Cell* **55,** 443 (1988).

[9] J. Gyuris, E. Golemis, H. Chertkov, and R. Brent, *Cell* **75,** 791 (1993).

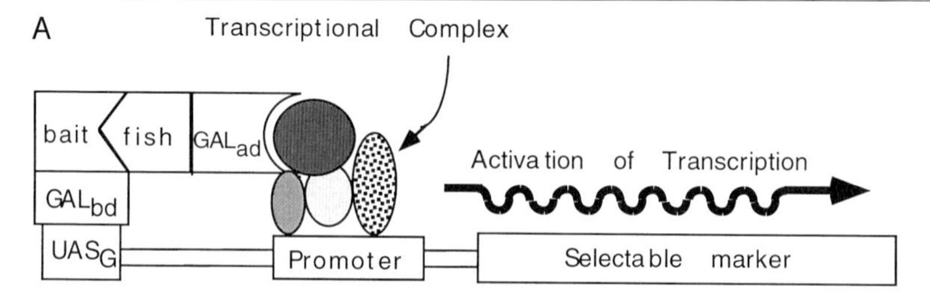

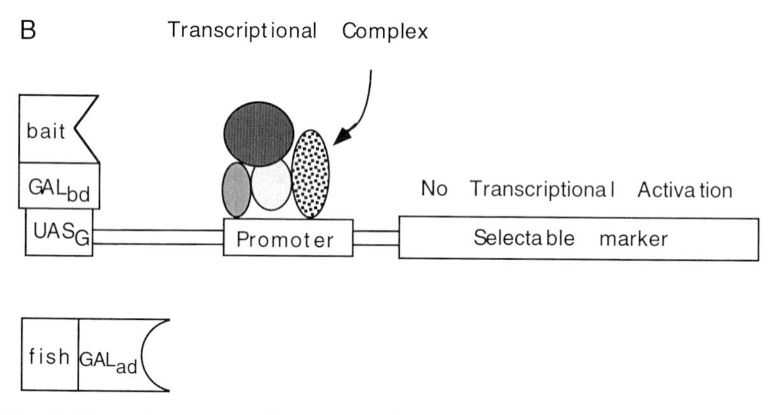

FIG. 1. Schematic representation of transcriptional activation induced by the interaction of two proteins ("bait" and "fish") with the yeast two-hybrid assay. The hybrid protein, consisting of the yeast GAL4 binding domain (GAL_{bd}) and the bait protein, binds to DNA sequences containing GAL4 activating sequences (UAS_G). Interaction of the bait hybrid protein with another hybrid consisting of the GAL4 activation domain (GAL_{ad}) and a fish protein (usually that encoded by a random cDNA) leads to activation of transcription of a downstream selectable marker or reporter gene such as *LacZ* (A). However, $GAL4_{ad}$ hybrids, which do not interact with the bait hybrid protein, fail to activate transcription (B).

at one time under conditions that may approximate that inside a mammalian cell and has been used successfully for a wide variety of different genes.

Choice of Yeast Strains

To undertake a yeast two-hybrid screen using the GAL4-based system, the yeast host strains must carry deletions of the *gal4* gene (to avoid induction of the selectable marker irrespective of whether an interaction occurs between the protein of interest and one from the cDNA library) and the *gal80* gene (which binds to GAL4 and blocks its transcriptional activation). The yeast cells also must harbor one or more selectable markers or reporter genes, the expression of which is regulated by the GAL4 protein. In most systems, the selectable markers are either *lacZ, HIS3,* or *LEU2.* In general, these markers contain either the natural *GAL1* UAS (which contains four GAL4-binding sites) or multimers (multiple copies generally lead to higher levels of transcription than a single copy) of the 17-mer GAL4 consensus binding site upstream of a minimal promoter. Ideally, the level of transcription of these marker genes is undetectable in the absence of a protein–protein interaction that generates a functional GAL4 protein but becomes high when there is such an interaction. However, in some strains, there can be "leaky" expression of one of these markers even in the uninduced state (see below). It is often desirable to use a strain that contains two different GAL4-inducible selectable markers but with different promoters. One can then select for further study only those clones that activate transcription in the yeast two-hybrid assay from both GAL4-inducible promoters. This strategy would minimize the likelihood of some artifactual activation of transcription from a single promoter. However, it is important to realize that unless the level of inducible expression from both promoters is similar, one might obtain positive results with one marker and not the other. A list of some commonly used strains is provided in Table I, along with the relative promoter strengths of the reporter genes.

TABLE I
COMMONLY USED YEAST STRAINS FOR THE YEAST TWO-HYBRID SCREEN

Strain	Cycloheximide	Reporter gene	Promoter (UAS-TATA)[a]	Expression level (basal/induced)
Hf7c	Resistant	*lacZ*	$(UAS_G)_3$-*CYC1*	Negative/low
		HIS3	GAL1-*GAL1*	Negative/high
Y190	Resistant	*lacZ*	GAL1-*GAL1*	Negative/high
		HIS3	GAL1-*HIS3*	Leaky/high
CG1945	Resistant	*lacZ*	$(UAS_G)_3$-*CYC1*	Negative/low
		HIS3	GAL1-*GAL1*	Low/high
YRG-2	Sensitive?	*lacZ*	$(UAS_G)_3$-*CYC1*	Negative/low
		HIS3	GAL1-*GAL1*	Negative/high
Y187	Sensitive	*lacZ* (2 copies)	GAL1-*GAL1*	Negative/very high
SFY526	Sensitive	*lacZ*	GAL1-*GAL1*	Negative/high

[a] UAS, Upstream activating sequence; TATA, promoter TATA sequence.

Choice of Plasmids

For GAL4-based systems, the gene of interest must be inserted downstream of and in frame with the $GAL4_{bd}$. A number of commercially available plasmids are available (see Table II). For some of these plasmids (e.g., pGBT9), a truncated *ADH1* promoter directs the transcription of the hybrid gene containing the $GAL4_{bd}$. This leads to only low levels of hybrid protein expression. Nonetheless, a sufficient amount of fusion protein is synthesized to detect many known interactions. However, for some interactions this level of expression is not sufficient to yield a positive result in the yeast two-hybrid screen. For higher levels of expression, plasmids containing the full-length *ADH1* promoter should be used. Once the appropriate hybrid has been constructed, it is important to test whether this "autoactivates," i.e., causes transcriptional activation from GAL4-inducible promoters in the absence of any fish protein. For obvious reasons, transcription factors or proteins that interact with transcription factors often autoactivate in the yeast two-hybrid screen. If the level of transcription or autoacti-

TABLE II
COMMONLY USED VECTORS FOR YEAST TWO-HYBRID ASSAY

Plasmid	Fusion	Promoter[a]	Expression level	Commercial source	Marker[b]
pGBT9	$GAL4_{bd}$	*ADH1**	Low	Clontech (Palo Alto, CA)	Trp
pAS2-1	$GAL4_{bd}$	*ADH1*	High	Clontech	Trp, Cyc^S
pGAD424	$GAL4_{ad}$	*ADH1**	Low	Clontech	Leu
pACT2	$GAL4_{ad}$	*ADH1*[1]*	Moderate	Clontech	Leu
pGADGH	$GAL4_{ad}$	*ADH1*	High	Clontech	Leu
pBD-GAL4 Cam	$GAL4_{bd}$	*ADH1*	High	Stratagene (La Jolla, CA)	Trp
pAD-GAL4	$GAL4_{ad}$	*ADH1*	High	Stratagene	Leu
pEG202	$LexA_{bd}$	*ADH1*	High	Origene (Rockville, MD)	His
pJG-5	B42 activation domain	*GAL1*	High Inducible	Origene	Trp
pHybLex/Zeo	$LexA_{bd}$	*ADH1*	High	Invitrogen (San Diego, CA)	Zeocin
pLexA	$LexA_{bd}$	*ADH1*	High	Clontech	His
pB42AD	B42 activation domain	*GAL1*	High Inducible	Clontech	Trp

[a] *ADH1**, Truncated *ADH1* promoter; *ADH1*[1]*, truncated *ADH1* promoter with pBR sequence, which acts as transcriptional activator.
[b] Cyc^S, cycloheximide sensitivity.

vation is relatively low and the *HIS3* marker is used to detect positive interactions, the addition of relatively low concentrations (5–20 m*M*) of 3-amino-1,2,4-triazole (3-AT; Sigma, St. Louis, MO) to the yeast plates may suppress the growth of yeast cells lacking interacting hybrid proteins. One can then proceed with the screen using a suppressive concentration of 3-AT. If this strategy is unsuccessful, modifying the fusion so that a smaller domain of the gene of interest is used may eliminate the autoactivation and permit screening (although this may reduce the sensitivity of the assay). Alternatively, a two-hybrid screen based on RNA polymerase III (see [23] in this volume[9a]) may be attempted. If these approaches fail, an alternative method of screening should be considered.

To complete the screen, the "query" genes or "fish" must be fused in frame with the $GAL4_{ad}$. Several plasmids for generating these fusions are commercially available. Some of these generate relatively low levels of fusion protein (e.g., pGAD424) because they are under the control of a truncated *ADH1* promoter. For some interactions, a negative result may be obtained with these poorly expressed hybrids even when the proteins being tested are known to interact, so the choice of the plasmid in which to generate a cDNA library should be considered carefully. A number of cDNA libraries from a wide variety of tissues are commercially available in several different plasmids. Because only one in three clones of directionally constructed libraries will be in frame with the $GAL4_{ad}$ gene, a library with at least 3×10^6 independent clones is necessary to screen 1×10^6 $GAL4_{ad}$ fusion proteins. Should one construct a library, it is generally advisable to construct the library in a λ vector (such as λACT or Hybrizap II; Stratagene, La Jolla, CA) to minimize the biasing of the library toward smaller inserts. Many of the plasmids used for the construction of fusions with the activation domain are rather large, potentially resulting in the underrepresentation of large cDNA inserts if the cDNAs were ligated directly with these plasmids and transformed into *Escherichia coli.* The use of a λ phage vector that harbors a yeast expression plasmid that can be excised subsequently from the phage vector minimizes the likelihood of such biasing. Many of the available λ vectors for the yeast two-hybrid assay permit one-step mass excision of the yeast expression plasmid from the parent phage vector.

If a commercially available cDNA library is used, it may need to be converted into plasmid form according to the manufacturer recommendations. However, if it is provided in plasmid form in an *E. coli* strain, the library must be amplified so that a sufficient amount of plasmid DNA is obtained for screening (50–500 μg, depending on the transformation

[9a] M.-C. Marsolier and A. Sentenac, *Methods Enzymol.* **303,** [23], 1999 (this volume).

efficiency). Although plating the library on multiple (up to 100) large (150-mm) plates at a density of about 30,000 colonies per plate is often recommended to prevent biasing the library (some cells may grow more quickly than others and the plasmids contained within these cells will be overrepresented relative to the other clones), we have found little difference in the quality of the plasmid library whether it is prepared from cells scraped from plates versus from cells grown directly in liquid culture (Terrific broth). Because the latter approach is considerably easier and faster, it is routinely employed in our laboratory.

Reagents

The following solutions should be prepared and autoclaved prior to the screen.

YPD: Difco (Detroit, MI) peptone (20 g/liter), yeast extract (10 g/liter), agar (20 g/liter), and 2% dextrose, which is added separately after autoclaving

1× M9 salts: Na_2HPO_4 (6 g/liter), KH_2PO_4 (3 g/liter), NH_4Cl (1 g/liter), NaCl (500 mg/liter), $CaCl_2$ (3 mg/liter); after autoclaving, add sterile 1 *M* $MgSO_4$ to yield a final concentration of 1 m*M*, add sterile 20% dextrose solution (10 ml/liter) and sterile 0.5% thiamine solution (1 ml/liter), along with any required amino acids

LB: Bacto-Tryptone (1 g/liter), Bacto-Yeast extract (5 g/liter) and NaCl (5 g/liter)

Synthetic dropout plates: Difco nitrogen base (6.7 g/liter) without amino acids, agar (20 g/liter), 850 ml of sterile water, 50 ml of a 40% dextrose solution, and 100 ml of 10× dropout solution (the latter two of which are autoclaved separately and added after autoclaving). If 3-AT is required, a 1 *M* stock solution should be filter sterilized and added to the SD medium once it has cooled to less than 55°

10× Dropout solution: This should contain all of the following components except for the selectable markers encoded by the bait and fish plasmids (e.g., Trp and Leu) as well as that activated by GAL4 protein if a positive interaction is observed (e.g., His): L-adenine hemisulfate salt (200 mg/liter), L-arginine (200 mg/liter), L-histidine (200 mg/liter), L-isoleucine (300 mg/liter), L-leucine (1000 mg/liter), L-lysine (30 mg/liter), L-methionine (200 mg/liter), L-phenylalanine (500 mg/liter), L-threonine (2000 mg/liter), L-tryptophan (200 mg/liter), L-tyrosine (300 mg/liter), L-valine (1500 mg/liter), and L-uracil (200 mg/liter). Prepared dropout supplements with the appropriate components omitted are commercially available.

Fresh PEG–LiAc solution, made immediately prior to the transformation step: 40% polyethylene glycol 4000 (PEG 4000), 100 m*M* lithium acetate, 10 m*M* Tris, and 1 m*M* EDTA (pH 7.5), made by mixing 48 ml of a 50% PEG 4000 solution with 6 ml of a sterile 1 *M* lithium acetate solution (10× lithium acetate buffer), adjusted to pH 7.5 with acetic acid, and 6 ml of a 10× TE solution [100 m*M* Tris-HCl, 10 m*M* EDTA (pH 7.5)]

Z buffer (for the β-galactosidase assays): Prepare by mixing $Na_2HPO_4 \cdot 7H_2O$ (16.1 g/liter), $NaH_2PO_4 \cdot H_2O$ (5.5 g/liter), KCl (0.75 g/liter), and $MgSO_4 \cdot 7H_2O$ (0.25 g/liter). Adjust pH to 7.0 and add 2-mercaptoethanol after autoclaving to a final concentration of 10 m*M*

Transformation

The transformation efficiency of yeast cells is considerably poorer than that of bacterial cells, and can vary significantly depending on the transformation conditions as well as the yeast strains used. It should be calculated for each screen so that one can assess whether an adequate number of transformants have been screened. The two most widely used protocols for transforming plasmid DNA into yeast cells are electroporation[10] and lithium acetate–PEG.[11] While the former protocol permits efficient transfer of DNA into yeast cells, the efficiency drops off rapidly as the amount of input DNA increases beyond 100 ng. Thus, for a typical library screen, many electroporations would be required. A lithium acetate transformation protocol can yield an acceptable transformation efficiency for library screening. A detailed protocol and set of reagents is available from Clontech (Palo Alto, CA); a modified lithium acetate protocol routinely used in our laboratory is provided below. A third approach that may offer a higher efficiency of introducing the plasmid library into the yeast cells is to transform the bait and fish plasmids into yeast strains of opposite mating type and then to mate the cells.

One can simultaneously transform both plasmids (the bait and the fish plasmids) into the yeast cells or perform a sequential transformation. First, transform the cells with the $GAL4_{bd}$ plasmid containing the gene of interest (the bait). On recovery of the yeast cells transformed with this plasmid, a second transformation is performed with the $GAL4_{ad}$ plasmid library. Simultaneous transformation results in a lower transformation efficiency (typically, 10^4 to 10^5 colonies per microgram of DNA) than sequential transformation (approximately 10^5 to 10^6 colonies per microgram of DNA)

[10] D. Becker and L. Guarente, *Methods Enzymol.* **194,** 182 (1991).

[11] D. Gietz, A. St. Jean, R. Woods, and R. Schiestl, *Nucleic Acids Res.* **20,** 1425 (1992).

but is somewhat quicker and avoids any effects the $GAL4_{bd}$ fusion protein may have on yeast growth and transformation efficiency. Should simultaneous transformations be chosen, the bait plasmid needs to be checked for autoactivation prior to the screen. The appropriate yeast strain should be transformed with the bait plasmid containing the gene of interest (e.g., pAS2-1) and the fish plasmid lacking any insert (e.g., pACT2 as indicated below) and plated on an agar selective medium plate lacking tryptophan and leucine. Once colonies grow (typically 3–4 days), several should be streaked onto −Trp, −Leu, −His selective medium plates. Growth on these plates indicates autoactivation.

Another critical component of the yeast transformation is the carrier DNA used. Optimal transformation efficiencies are obtained when the salmon sperm DNA has an average size of 5–10 kb (range, 2–15 kb). It is best prepared by dissolving dried salmon sperm DNA in 1× TE buffer at a concentration of 10 mg/ml. Overnight stirring at 4° may be required to completely dissolve the DNA. Once dissolved, the DNA should be sonicated briefly to generate an average fragment size of 5–10 kb. Check the size of the sonicated DNA on an agarose gel between sonication bursts (20–30 sec). Once the DNA has been sonicated to the appropriate size range, phenol–chloroform extract and then ethanol precipitate the DNA. Wash the DNA pellet with 70% ethanol, dry briefly, and dissolve in 1× TE buffer to a final concentration of 10 mg/ml. Yeast total RNA can also be used as a carrier instead of salmon sperm DNA. While more difficult to prepare, it may yield more reproducible results (S. Elledge, personal communication, Baylor University).

One day prior to the transformation, inoculate the appropriate yeast strain (e.g., Hf7c) into 100 ml of YPD and incubate at 30° with shaking. The next morning, this culture should be added to 1 liter of YPD in a 4-liter flask and incubated at 30° for an additional 3–4 hr (to an OD_{600} of ~0.4–0.8) to allow the cells to undergo approximately two divisions. While the cells are incubating, prepare 10 ml of sterile 1× TE–LiAc buffer [made fresh by diluting 1 ml of 10× TE and 1 ml of 10× lithium acetate (LiAc) buffers in 8 ml of sterile water]. Pellet the cells by centrifugation (1000 *g*) for 5 min at room temperature in a GS3 or equivalent rotor. Wash the cells with 100 ml of sterile water and repellet the cells by centrifugation. Discard the supernatant and resuspend the cell pellet in 5 ml of 1× TE–LiAc buffer. Store on ice until ready for transformation (allowing the cells to incubate on ice for more than 30–60 min may decrease the transformation efficiency).

To a 250-ml sterile flask, add 250–500 μg of each plasmid along with 1–2 mg of salmon sperm carrier DNA, which has been boiled for 20 min prior to use and rapidly cooled on ice (alternatively, 5 mg of total RNA can be used as a carrier), and the yeast competent cells in TE–LiAc buffer. Mix well and add 51 ml of freshly prepared sterile PEG–LiAc solution. Vortex briefly and incubate at 30° for 30 min with shaking (200 rpm). Longer incubation times can significantly decrease the transformation efficiency. After 30 min, add dimethyl sulfoxide to a final concentration of 10% and heat shock the cells for 10–15 min at 42°. The cells should be swirled gently during the heat shock period.

At this stage, the cells can be pelleted by centrifugation and plated directly onto the selective plates. Alternatively, they can be added to 500 ml of the appropriate dropout media (usually −Trp, −Leu) and incubated with shaking for 3–4 hr at 30° and then pelleted by centrifugation. This step allows the cells to recover from the transformation and dilutes out the PEG solution (which is toxic to the cells) prior to plating. The pelleted cells should be resuspended in 3–4 ml of sterile 1× TE buffer, plated on 10–15 large (150-mm) agar plates containing the appropriate selective medium, and incubated at 30°. An aliquot of the cells (100 μl) should be plated on −Trp, −Leu selective medium plates (corresponding to the markers on the bait and fish plasmids) to determine the transformation efficiency.

Positive colonies generally appear within 3–5 days. We have found that most colonies that appear after 5 days, or those that appear as small colonies prior to that time but that do not enlarge even after an additional 1–2 days, turn out to be negative after further testing so we generally ignore these colonies. The transformation efficiency can be calculated by counting the number of colonies on the −Trp, −Leu selective plate and using the following formula:

$$\frac{\text{No. of colonies on } -\text{Trp}, -\text{Leu plate} \times \text{total volume of transformed cells } (\mu\text{l})}{\text{Volume plated } (100\ \mu\text{l}) \times \mu\text{g of ``bait'' or ``fish'' plasmid used (choose lesser amount)}}$$

A low transformation efficiency (less than 10^3 colonies per microgram of DNA) will yield too few transformants for an adequate screen. If so, it may be necessary to perform sequential transformations. One should determine the total number of colonies screened. It is desirable to screen at least 3×10^6 colonies.

β-Galactosidase Assays

If the screen has been performed with a strain that also contains the β-galactosidase gene under the control of the Gal4 protein (e.g., Hf7c, CG1945, Y190), all positive colonies should be checked for the production of β-galactosidase. Pick each colony onto duplicate plates, both containing the appropriate selective medium and one containing a nylon filter that has been sterilized by exposure to a UV light source for 10 min. Incubate the plates for 1–2 days at 30°. Once colonies have appeared on the nylon filter, remove the filter and soak it in liquid nitrogen for at least 10 sec. Allow the cells and filter to thaw at room temperature.

Prepare an X-Gal plate by boiling 0.2 g of agarose in 25 ml of Z buffer. After cooling to 55°, add 250 μl of 2% 5-bromo-4-chloro-3-indolyl β-D-galactopyranoside (X-Gal) in *N,N'*-dimethyl formamide, pour the solution into a 150-mm petri plate, and allow it to gel. Once firm, the thawed filter containing the colonies can be overlaid onto the agar gel and incubated at room temperature. A positive signal will generally appear within 2–12 hr. A positive and negative control should be added to each filter.

For more quantitative results, the β-galactosidase activity can be measured in liquid cultures. However, the level of β-galactosidase activity is not necessarily an indication of the strength of the interaction.[12] The sensitivity of the assay is dependent on the substrate used. *o*-Nitrophenol galactopyranoside (ONPG) is less sensitive than chlorophenol red galactopyranoside (CPRG; Boehringer Mannheim, Indianapolis, IN), which is less sensitive than a chemiluminescent substrate for β-galactosidase (although the latter may have a higher false-positive or background rate). For the ONPG assay, grow the yeast cells to be tested overnight in selective medium. Dilute 2 ml of the overnight culture into 8 ml of YPD and incubate at 30° for 3–5 hr until the OD_{600} is 0.5–0.8. Record the OD_{600} and then harvest 1.5 ml of the cells by centrifugation. Multiple aliquots can be prepared for more accurate results. Remove the supernatant and resuspend the cells in 1.5 ml of Z buffer. Pellet the cells again by centrifugation and replace the supernatant with 300 μl of fresh Z buffer. After resuspending the cells, take 100 μl of the suspension and freeze in liquid nitrogen for 1–2 min. Thaw the cells at 37° in a water bath. Repeat the freeze–thaw cycles at least two more times to ensure adequate cell lysis.

Begin the assay for β-galactosidase activity by mixing 200 μl of ONPG solution (4 mg/ml dissolved in Z buffer) with 700 μl of Z buffer. Add 100 μl of the lysed cell extract to the ONPG–Z buffer mixture. For a negative control, prepare a sample from the host strain that lacks β-galactosidase activity. Incubate the reaction at 30° and record the time of incubation. A yellow color should develop within minutes to hours but may take up to 24 hr. If the reaction proceeds rapidly (a deep yellow color develops within minutes), it will be necessary to stop the reaction using 0.4 ml of 1 *M* Na_2CO_3 in order to obtain an accurate result. If the yellow color takes several hours to develop, reliable measurements can be obtained without stopping the reaction. Because cellular debris may affect the accuracy of the results, centrifuge each sample to pellet any cellular debris prior to

[12] J. Estojak, R. Brent, and E. Golemis, *Mol. Cell. Biol.* **15,** 5820 (1995).

measuring the absorbance. Measure and record the OD_{420} of each sample. The negative control should be used as a blank to zero the spectrophotometer. The linear range of the assay is between OD_{420} values of 0.2 and 1.0; if the OD_{420} falls outside this range, the incubation time or amount of cellular extract should be altered accordingly. Short incubation times (less than 2 min) should be avoided. The number of β-galactosidase units (micromoles of ONPG hydrolyzed per minute) present in the culture can be calculated according to the following formula[13]:

$$\text{No. of units} = \frac{1000 \times OD_{420}}{\text{Incubation time (in min)} \times V \times C \times OD_{600}}$$

where V is the volume (in ml) of the cell suspension used (0.1 ml) and C is the enrichment of cell suspension from the original culture (1.5/0.3 = 5).

A similar protocol can be used for CPRG except that the OD_{578} should be measured instead of the OD_{420}. The linear range of this assay is between OD_{578} 0.25 and 1.8.

Confirmation of Positives

All colonies that grow on selective medium and express β-galactosidase activity (if the strain used contains the β-galactosidase gene under the control of a GAL4 responsive element) are potential positives. While the recovery of multiple different isolates of the same fish protein in a screen is suggestive of a "real" interaction, it is necessary to confirm that the observed interaction is dependent on the presence of both interacting pairs and is specific for both. Some clones can yield positive results irrespective of the protein bait. To verify that the interaction is dependent on the presence of both plasmids, it is necessary to recover the isolate cells containing the fish plasmid without the bait plasmid. If the yeast strain is cycloheximide resistant (e.g., CG1945) and the bait plasmid confers cycloheximide sensitivity (e.g., pAS2-1), one can select for cells that have been cured of the bait plasmid by growing the cells in cycloheximide. Cycloheximide resistance is conferred by a mutation in the yeast ribosomal L29 protein, but the wild-type protein is dominant to the mutant protein. Thus, cells containing both wild-type and mutant L29 protein will be sensitive to cycloheximide. Because pAS2-1 encodes a wild-type L29 protein, yeast cells harboring this plasmid will be sensitive to cycloheximide. If, however, these yeast cells encode a chromosomal L29 mutation, then on loss of pAS2-1 they will become resistant to cycloheximide. To select for cells that

[13] J. Miller, "Experiments in Molecular Genetics." Cold Spring Harbor Laboratory Press, Cold Spring Harbor, New York, 1972.

have lost the pAS2-1 bait plasmid, pick a small colony (<3 mm in diameter) and resuspend it in 300 μl of sterile YPD. Spread 10–100 μl of the resuspended cells onto an appropriate selective medium plate (the amino acid corresponding to the selective marker on the plasmid that is to be lost should be included in the medium to avoid selective pressure for the presence of that plasmid) containing cycloheximide (the concentration of cycloheximide is strain dependent: 1 μg/ml is sufficient for CG1945 while 10 μg/ml is needed for Y190). The cycloheximide solution can be prepared as a 1000× solution by dissolving 10 mg in 10 ml of sterile water. It should be filter sterilized and added to the selective medium after autoclaving. Once colonies appear (it generally requires 3–5 days of incubation at 30°), loss of the bait plasmid can be confirmed by streaking the cells on selective medium plates lacking tryptophan (for pAS2-1 and related plasmids). The cells should fail to grow in the absence of tryptophan.

If cycloheximide counterselection cannot be employed, growth of the yeast in medium lacking selective pressure for the plasmid will result in the gradual loss of that plasmid from the cells. Thus, if the bait plasmid encodes a marker for tryptophan, growth of the yeast in selective medium containing tryptophan will result in the loss of the bait plasmid but not the fish plasmid if selective pressure is maintained for it. To employ this strategy, pick individual colonies and grow the cells for 2 days at 30° in the appropriate dropout medium that does not maintain selection pressure for the plasmid that is to be lost (i.e., the bait plasmid). Once the cells have reached stationary phase, dilute them in fresh selective medium and plate on selective medium agar plates that do not maintain selection for the bait plasmid. Incubate the plates at 30° until colonies form (generally 2–3 days). Pick 50–100 colonies onto the appropriate selective medium plates to verify that the desired plasmid is lost (e.g., onto selective medium plates which are Trp^+ and Trp^- if pAS2-1 is used; the desired cells are those that grow on the former but not on the latter).

A third approach that can be used if there are a relatively small number of positive clones is to isolate the fish plasmids directly from the yeast cells. Add a colony to 2 ml of YPD and incubate overnight at 30°. Transfer 1.5 ml of the culture to a microcentrifuge tube the next morning and pellet the cells. Resuspend the cells in 200 μl of lysis solution (2% Triton X-100, 1% SDS, 100 m*M* NaCl, 10 m*M* Tris, 1 m*M* EDTA, pH 8.0). After the cells have been thoroughly resuspended, add 200 μl of phenol–choroform–isoamyl alcohol (25:24:1) solution and 300 mg of acid-washed glass beads (Sigma). Vortex the mixture vigorously for 2 min. Centrifuge the mixture at 14,000 rpm for 5 min at room temperature. Transfer the supernatant to a clean microcentrifuge tube and ethanol precipitate the DNA. After incubation at −20° for 1–2 hr, pellet the precipitate by centrifugation. Wash the pellet with 70% ethanol and dry briefly. Resuspend the pellet in 10 μl of sterile 1× TE.

Because the *Leu2* gene can be expressed in *E. coli* and can rescue *leuB E. coli,* it is possible to select for fish plasmid containing the *Leu2* gene. HB101 is a commonly used *E. coli* strain that is *leuB*. To isolate the fish plasmid in this strain, add 1 μl of the previously prepared plasmid DNA mixture containing the bait and fish plasmids to 20 μl of electrocompetent HB101 cells. Perform the electroporation (the transformation efficiency of chemically competent HB101 cells may not be high enough to yield any transformants). Transfer the cells to 5 ml of LB medium and incubate at 37° for 1 hr. Pellet the cells by centrifugation at 5000 rpm for 5 min and wash the cells with 5 ml of M9 minimal medium. Spread the washed cells on M9 agar minimal medium containing ampicillin (50 μg/ml), proline (40 μg/ml), and 1 m*M* thiamine-HCl and incubate at 37° until colonies appear (about 2–3 days). Prepare DNA from three or four colonies and verify that the same fish plasmid has been isolated from each yeast colony picked. For fish plasmids encoding the *Trp1* gene, a similar strategy can be employed if *E. coli* cells harboring a *trpC* mutation (e.g., KC8) and M9 minimal medium plates lacking tryptophan are used for the selection. An aliquot of this plasmid DNA obtained from bacterial cells can be retransformed into an appropriate yeast strain.

Once yeast have been isolated that contain only the fish plasmid, they can be transformed with the original bait plasmid and a bait plasmid containing an irrelevant $GAL4_{bd}$ gene fusion. Alternatively, they can be mated with a yeast strain of alternate mating type but containing an unrelated fusion in the bait plasmid. Any fish clones that are positive with the unrelated or irrelevant GAL_{bd} fusion should be discarded.

There are some genes that are recovered from a yeast two-hybrid screen rather frequently. It sometimes may be difficult to determine whether these represent "real" interactions that occur *in vivo* or simply represent some limited, nonspecific interaction and are therefore of less interest. Some of these genes include those encoding heat shock proteins, ribosomal proteins, cytochrome oxidase, mitochondrial proteins, cytoskeletal proteins, elongation factors, ubiquitin-conjugating proteins, proteasome subunits, ferritin, tRNA synthases, collagen-related proteins, vimentin, and zinc finger proteins. Should one of these genes be recovered in a screen, one must entertain the possibility that it is a false positive.

All positive results can be confirmed by switching the genes in the bait and fish plasmids. While many false positives will not be positive when switched, some true positives also may fail to activate GAL4 transcription by this test, so a negative result may be ambiguous.

Because yeast two-hybrid screenings frequently yield a significant number of false-positive clones, it is important to confirm all positives by at least one other assay. If antibodies are available for both proteins, immunoprecipitate one of the two interacting proteins from mammalian cells followed by Western blot analysis for the other interacting protein in the precipitate. Suitable negative controls should be performed simultaneously. If antibodies are available to only one protein, the other protein can be epitope tagged and expressed in mammalian cells. Immune precipitation can be performed using an antibody against the epitope tag followed by Western blot analysis with the other available antibody. If antibodies are not available to either protein, one of the proteins can be expressed in mammalian cells as a fusion with the widely used glutathione-*S*-transferase (GST) protein, while the other protein can be epitope tagged and coexpressed in the same cells.[14] The GST fusion protein can be purified from cells by using glutathione–agarose affinity chromatography. Because this purification does not require harsh conditions, any associated proteins may purify with the GST fusion protein. After elution with glutathione, Western blot analysis using the antibody against the epitope tag can confirm the presence of an interaction between the two proteins. Again, suitable negative controls must be performed.

In vitro binding assays also can be used to confirm the interaction. One of the two interacting proteins can be fused to a "tag" such as GST, which permits it to be rapidly and gently purified from *E. coli* or a eukaryotic cell extract. If posttranslational modifications (such as phosphorylation) are important for the interaction, the protein should be synthesized in eukaryotic cells. The other protein of the interacting pair can be synthesized and labeled *in vitro* with [^{35}S]methionine or epitope tagged and expressed in bacterial or eukaryotic cells. Passage of this tagged protein over a column containing the other immobilized protein (or a control protein) should result in retention of the tagged protein if there is a significant interaction.

For some yeast two-hybrid screens, no true positive clones will be recovered. This may be a technical limitation, such as owing to the absence of the desired clone in the library used (which may require the preparation of a new library with more independent clones or the use of a library prepared from a different source) or the screening of too few clones. Under such circumstances, repeating the screening may be fruitful. It also is possible that one or more of the hybrid clones may be toxic to the yeast cells when expressed at high levels. If so, the use of an inducible system, such as the LexA-based system, may overcome this problem. Another possibility is that the GAL4 domains cause steric hindrance and interfere with the desired interaction, particularly if the relevant binding domains are near

[14] B. Chatton, A. Bahr, J. Acker, and C. Kedinger, *BioTechniques* **18,** 142 (1995).

the GAL4 domains. The use of a "spacer" region may relieve the steric hindrance and permit the interaction to occur. If the interacting proteins are highly conserved through evolution, it is possible that one or more endogenous yeast proteins compete with the fish test protein for binding to the bait hybrid protein. Overexpression of both hybrid proteins may overcome this competition and permit the recovery of a positive clone. Although many mammalian proteins can be synthesized and are fully functional in yeast cells, one or both of the hybrid proteins may not fold properly in the yeast cells outside their normal milieu (e.g., membrane proteins), or they may not localize to the nucleus. Without nuclear localization, no transcriptional activation can occur. The former problem may be resolved by choosing different fragments of the test protein (although this may eliminate a potential binding domain) while the latter might be corrected by adding a simian virus 40 (SV40) nuclear localization signal to the hybrid construct. (*Note:* the GAL4 domains already possess a nuclear localization signal.) Nonetheless, none of these changes may help and no positive clones may be recovered using the yeast two-hybrid assay. Indeed, it is possible that some complexes may form in yeast cells but still not generate a transcriptionally active GAL4 protein. If a protein–protein interaction is still suspected, an alternative method of screening should be considered.

It should be noted that various modifications to the basic yeast two-hybrid assay have been developed, including adaptation of the system to mammalian cells,[15,16] the reverse two-hybrid screen,[17–19] the triple-hybrid assay,[20] and the two-hybrid kinase screen.[21] The reverse two-hybrid screen can be used to select for mutations in one protein of an interacting pair that interrupts its binding to its interacting partner, while the triple-hybrid assay can be used to select for the binding partner of various small drugs or bioactive molecules. The two-hybrid kinase screen was developed to isolate proteins that interact only with bait proteins phosphorylated on key residues.

[15] H. Vasavada, S. Ganguly, F. Germino, Z. Wang, and S. Weissman, *Proc. Natl. Acad. Sci. U.S.A.* **88,** 10686 (1991).

[16] E. Fearon, T. Finkel, M. Gillison, S. Kennedy, J. Casella, G. Tomaselli, J. Morrow, and C. Van Dang, *Proc. Natl. Acad. Sci. U.S.A.* **89,** 7958 (1992).

[17] H. Shih, P. Goldman, A. DeMaggio, S. Hollenberg, R. Goodman, and M. Hoekstra, *Proc. Natl. Acad. Sci. U.S.A.* **93,** 13896 (1996).

[18] M. Vidal, P. Braun, E. Chen, J. Boeke, and E. Harlow, *Proc. Natl. Acad. Sci. U.S.A.* **93,** 10321 (1996).

[19] M. Vidal, K. Brachmann, A. Fattaey, E. Harlow, and J. Boeke, *Proc. Natl. Acad. Sci. U.S.A.* **93,** 10315 (1996).

[20] E. Licitra and J. Liu, *Proc. Natl. Acad. Sci. U.S.A.* **93,** 12817 (1996).

[21] K. Keegan and J. Cooper, *Oncogene* **12,** 1537 (1996).

Filter Lift Assay

For those proteins for which the yeast two-hybrid assay or one of its derivatives is not suitable, one can attempt a protein–protein interaction assay using the filter lift assay.[5] For this assay, the fish cDNAs are cloned in a λ phage expression. After bacterial cells are infected with this phage library, the expression of the fish proteins is induced. The phage eventually causes lysis of the cells, exposing the proteins to capture on a nylon or nitrocellulose filter. After processing, the fish proteins are incubated with bait protein labeled to high specific activity. Once the filters have been washed and developed, localization of the bait protein on the filters identifies the original plaque(s) that synthesized an interacting protein. This plaque can be recovered by secondary and tertiary screenings, if necessary.

Preparation of Labeled Protein Probe

The protein of interest can be labeled to high specific activity with ^{32}P. The gene of interest is cloned downstream of a glutathione-*S*-transferase gene containing a heart muscle kinase recognition site (Arg-Arg-Ala-Ser-Val) and the fusion protein is expressed and labeled directly from *E. coli* cell extracts or after its purification by affinity column chromatography.

Dissolve 250 units of heart muscle kinase (Sigma) in 25 μl of 40 m*M* dithiothreitol (DTT) to yield a final concentration of 10 units/μl. It is best to use this within several hours of preparation although the kinase activity is usually maintained for 2–3 days if it is stored at 4°. Prepare 10× HMK buffer [200 m*M* Tris-HCl (pH 7.5), 10 m*M* DTT, 1 *M* NaCl, and 120 m*M* $MgCl_2$]. For each 30-μl reaction, add 3 μl of 10× HMK buffer, 2–5 μl of [γ-^{32}P]ATP with the highest specific activity available (preferably >7000 Ci/mmol), 1–10 μl of purified bait–GST fusion protein or protein extract containing this bait hybrid (several different amounts should be tried to determine which yields the highest specific activity; probes with higher specific activity tend to yield better results), 1 μl of the freshly prepared HMK solution (10 units/μl), and sterile water to a final volume of 30 μl. Incubate at 37° for 60 min and then pass through a G-50 column as described below to remove the unincorporated [α-^{32}P]ATP. Failure to eliminate the unincorporated radionucleotide will result in high background levels on the filters.

Wash 1 ml of preswollen G-50 Sepharose (Pharmacia, Piscataway, NJ) with 5 ml of buffer A [25 m*M* HEPES-KOH (pH 7.7), 12.5 m*M* $MgCl_2$, 20% glycerol, 100 m*M* KCl, filter-sterilized bovine serum albumin (BSA, 3 mg/ml), and 1 m*M* DTT]. Resuspend the G-50 in 1 ml of buffer A and rotate at room temperature for 1 hr. After the G-50 has been washed three times with 1 ml of buffer A, pack it in a sterile 2-ml plastic pipette. Wash the column with 10 ml of buffer A. Immediately prior to loading the column with the labeled protein, add 70 μl of cold buffer A to the 30 μl of the HMK reaction. Load and run the column with buffer A. Collect 1-drop fractions (store on ice) and measure the number of counts in 2 μl of each fraction. Pool the excluded peak fractions (generally 300–500 μl), which contain the labeled protein of interest. To test the quality of the labeled protein, run 1 μl of the pooled fractions (or each individual fraction) on a sodium dodecyl sulfate (SDS)–polyacrylamide gel and perform autoradiography. A positive signal can be detected usually after a brief exposure (less than 1 hr).

Screening the Library

The λ phage expression library should be plated in the standard manner at a density of about 20,000–30,000 PFU/150-mm plate. Phage that produce high levels of the cloned gene tend to yield a stronger signal than those that produce only low levels of the cloned gene. If the expression of the cloned genes is inducible with isopropyl-β-D-thiogalactopyranoside (IPTG), overlay an IPTG-saturated filter (the filters can be slightly moist prior to use but should not be excessively wet, which can cause smearing of the plaques) onto the plates and incubate at 37° for an additional 6–8 hr. This permits the synthesis and capture of relatively large amounts of protein on the filter. Once the plaques have barely begun to touch one another, transfer the plates and filters to 4°. This step hardens the top agar and makes is easier to remove the filters without disturbing the top agar containing the plaques. It is important to mark the filters relative to the plates so that they can be oriented later. Remove the filters and allow to air dry at room temperature briefly. The remaining steps are all performed at 4° to prevent protein degradation.

Because some proteins may not fold properly when synthesized and captured in this way or may precipitate with irrelevant proteins, a denaturation–renaturation step is used. First, the filters pulled from the plates are incubated for 5 min in HBB buffer [250 m*M* HEPES-KOH (pH 7.7), 50 m*M* $MgCl_2$, 250 m*M* NaCl, and 1 m*M* DTT]. The proteins bound to the filters are denatured by two 10-min incubations in 6 *M* guanidine-HCl in HBB buffer, followed by slow renaturation using sequential 10-min washes in the following solutions: 3 *M* guanidine-HCl in HBB, 1.5 *M* guanidine-HCl in HBB, 0.75 *M* guanidine-HCl in HBB, 0.38 *M* guanidine-HCl in HBB, and 0.19 *M* guanidine-HCl in HBB. Two final washes of 10 min each are performed using HBB.

The filters are prehybridized for 60 min with 5% nonfat dried milk in HBB containing 0.05% Nonidet P-40 (NP-40) followed by 30 min with 1%

nonfat dried milk in HBB containing 0.05% NP-40. Following prehybridization, the filters should be transferred to hybridization buffer [20 m*M* HEPES-KOH (pH 7.7), 75–200 m*M* KCl, 0.1 m*M* EDTA, 2.5 m*M* $MgCl_2$, 1% nonfat dried milk, 1 m*M* DTT, and 0.05–0.10% NP-40; the KCl and NP-40 concentrations may need to be optimized for each protein]. Use approximately 20–25 ml of hybridization buffer per large (150-mm) filter and 7–10 ml per small (100-mm) filter. Enough labeled protein should be added to the solution to yield 250,000 cpm per milliliter of solution. Incubate the filter with the labeled probe overnight at 4°.

Remove and save the hybridization solution with the labeled probe (it can be used for secondary and tertiary screenings, if necessary). Three 10-min washes at 4° with hybridization buffer should eliminate nearly all background counts. Dry the filters briefly at room temperature and expose them overnight to X-ray film with intensifying screens. Positive plaques can be recovered for secondary and tertiary screening by aligning the filters with the plates. *In vitro* binding assays can be used to confirm the observed interaction.

Limitations and Uses

The filter lift assay has several obvious limitations. Because the fish proteins are produced in *E. coli,* those interactions that are dependent on posttranslational modifications in the fish protein will not be recovered. In addition, while many proteins will refold properly after the graded denaturation–renaturation steps outlined above, some may not and therefore will not react with their interacting partner. Interactions that require the cotranslation of both interacting partners also will not be detected using this assay. Finally, the preparation of the labeled protein and the denaturation–renaturation steps are somewhat time consuming and one must work with relatively large amounts of radioactivity for each screen. Labeling the protein probe with biotin instead of ^{32}P or detecting any protein probe bound to the filter with primary and secondary antibodies can be used to eliminate exposure to radioactivity.

On the other hand, this assay has several advantages. Because it does not involve transcriptional activation, proteins that bind to transcriptional activators or are activators themselves can nonetheless be used as probes without modification. The filter lift assay probes for direct protein–protein interactions and does not rely on a secondary event to detect the interaction, so false positives may be less common. Because bacteria and eukaryotic cells are so divergent, competition with endogenous host proteins is less likely to be encountered. Finally, one can use the protein labeled in this way as a probe in Western blots (with the proteins denatured–renatured as outlined above) prior to the screening. Although this may be useful only for relatively abundant target proteins, a positive result would not only confirm the presence of an interacting protein and indicate its size but would also establish appropriate wash conditions.

Double-Tagging Assay

A third assay that maintains some of the advantages of the filter lift assay but overcomes three of its limitations—no radioactivity is used, the bait and fish proteins are cotranslated within the same cells, and no denaturation–renaturation step is required—is the double-tagging assay.[6] For this assay, the protein of interest (the bait) is fused with a protein that can be specifically captured while the fish proteins are fused in-frame with a reporter protein (β-galactosidase). All cells in the screen produce the bait hybrid protein while different cells produce different fusion fish proteins. The bait hybrid protein can be captured specifically from the cell lysate via its tag while the presence of a fish protein can be detected by assaying for β-galactosidase activity.

Generation of Bait Hybrid

The plasmid pTrc.EZZ.BCCP[7] can be used to generate fusions between the protein of interest and the EZZ domain (a synthetic, protein A-like, 13-kDa protein that binds to IgG) at its amino-terminal end and the 13-kDa carboxy-terminal portion of the biotin carboxylase carrier protein (BCCP). This latter protein fragment is biotinylated *in vivo* in *E. coli* even when proteins are fused to its amino-terminal end[22] and can bind to avidin with high avidity. We have shown that similar results are obtained with either tag. However, because IgG is considerably less expensive than avidin or streptavidin, we generally use the former to capture the bait protein.

There is a unique *Eco*RI site (reading frame: GAA TTC, identical to that in λgt11) available for cloning between the EZZ and BCCP genes on pTrcEZZ.BCCP to generate a fusion with EZZ. We generally synthesize two oligonucleotides for every gene of interest: a sense primer that can hybridize to the 5′ end of the gene of interest but that introduces an *Eco*RI site immediately upstream of the start codon and in frame with the EZZ domain, as well as an antisense oligonucleotide that is complementary to the 3′ end of the gene and inserts a *Sal*I site immediately upstream of the stop codon (in the reading frame XXG TCG ACX if an in-frame fusion

[22] J. Cronan, *J. Biol. Chem.* **265,** 10327 (1990).

with the BCCP moiety is desired). The gene of interest is then polymerase chain reaction (PCR) amplified to generate *Eco*RI and *Sal*I sites at the 5′ and 3′ ends of the coding region, respectively. If the gene of interest has an internal *Eco*RI or *Sal*I site, the oligonucleotides can be synthesized with *Mun*I (CAATTG) or *Xho*I (CTCGAG) sites, respectively.

The *Eco*RI and *Sal*I sites on these PCR-amplified genes often cut poorly unless the amplified material is first washed. Apply 200–300 μl of the amplified PCR product to a Centricon-100 concentrator (Amicon, Danvers, MA) and centrifuge in a SS-34 rotor at 2500 rpm for 15 min. The concentrated product is then washed three times, each time with 1 ml of sterile water and centrifugation at 2500 rpm for 15 min. After washing, the concentrated product is collected and diluted to a final volume of 200 μl. Digest 90 μl of the PCR product in a total volume of 100 μl in the appropriate buffer with 100 units of *Eco*RI and 100 units of *Sal*I overnight at room temperature. Gel purify the digested DNA, ligate with *Eco*RI–*Sal*I digested pTrc.EZZ.BCCP, and transform into *E. coli* cells that are $LacZ^-$ and λ^- (e.g., LE392).

After verifying that the appropriate construct has been retrieved, confirm that the proper fusion protein is being synthesized in the cells. Start a small (10 ml of LB) overnight culture of the cells containing this plasmid. The next morning, add to 200 ml of sterile LB containing ampicillin (100 μg/ml) and incubate with shaking at 37° until the cells reach an OD_{600} of ~0.4. Add IPTG to a final concentration of 0.6 m*M* and allow the cells to incubate for an additional 2–3 hr. Harvest the cells by centrifugation and resuspend them in 5 ml of Z buffer containing lysozyme (1 μg/ml). Incubate on ice for 15 min and then rapidly freeze in liquid nitrogen for 5 min. Thaw the cells at 37°. If the solution has become viscous, sonicate briefly to reduce the viscosity. If not, repeat the freeze–thaw cycle and then sonicate. Transfer the lysate to microcentrifuge tubes and centrifuge at 4° for 15 min. Recover the supernatant and discard the pellet.

Run 1, 5, 10, and 25 μl of the supernatant on SDS–polyacrylamide gels and transfer the proteins to a nitrocellulose or nylon filter. Production of the desired fusion protein can be confirmed by probing the filter with rabbit IgG and a secondary, goat anti-rabbit IgG antibody coupled to a reporter protein (alkaline phosphatase or horseradish peroxidase). If the protein is fused to the BCCP moiety, the filter can also be probed with avidin–alkaline phosphatase to verify that full-length protein is being produced (some eukaryotic proteins are cleaved within *E. coli* cells). A positive signal should be easily detected with the 5- and 10-μl supernatant aliquots; if only a barely detectable signal is present in these lanes, there may not be sufficient production of the bait protein to detect an interaction.

Construction of Fish Fusions

The cDNAs for the fish proteins are ligated with *Eco*RI–*Hin*dIII- or *Eco*RI–*Not*I-restricted λFJG2. This λ vector has a unique *Eco*RI site (reading frame: GAA TTC) at the extreme 3′ end of the *LacZ* gene and downstream unique *Hin*dIII and *Not*I sites. The *Eco*RV endonuclease gene has been cloned between the *Eco*RI and *Not*I sites. This latter feature permits the positive selection of inserts. When genes are cloned between the *Eco*RI and *Not*I or *Hin*dIII sites, the *Eco*RV gene is lost and the phage can be propagated in cells lacking *Eco*RV methylase. On the other hand, if the ligation results in the reinsertion of the *Eco*RV gene, these phage cannot be propagated in cells lacking *Eco*RV methylase. Cloning into the *Eco*RI site generates fusions with active β-galactosidase protein.

Preparation of Filters

A critical aspect of the assay is to have a sufficient amount of the bait fusion protein captured on a filter so that detectable levels of β-galactosidase are present. We have found that the Ultrabind US450 filters (Gelman, Ann Arbor, MI) yield a much better signal than do nitrocellulose or nylon filters. In addition, because the capture protein (IgG or avidin) is covalently linked to the filter, it is extremely stable and will not wash off.

To covalently link the capture protein to the filter, make a 1-mg/ml solution of IgG or avidin in binding buffer [0.5 *M* potassium phosphate buffer (pH 7.4), 200 m*M* NaCl, 1 m*M* KCl]. It is extremely important to avoid the use of any buffers containing free primary amines (e.g., Tris) in the binding buffer because it will compete with the capture protein for binding to the filter. For every large (150-mm) filter, add 10 ml of solution and allow to incubate with gentle rocking overnight at room temperature. In the morning, the filters should be blocked at room temperature with 5% NFDM-TBST [10 m*M* Tris (pH 8.0), 150 m*M* NaCl, 0.05% sodium azide, 0.05% Tween containing nonfat dried milk (50 g/liter)] for several hours. The blocking solution can be rinsed away with sterile water and the filters are stable for several months if stored at 4°.

Screening

Once the fish genes have been ligated with λFJG2 DNA and packaged *in vitro,* use the packaged phage to infect cells harboring the desired pTrc.EZZ.BCCP fusion. Grow the cells overnight in 50 ml of LB containing ampicillin (100 μg/ml) and 0.5 ml of sterile 20% maltose. For each 150-mm plate, add approximately 20,000–30,000 PFU to 600 μl of cell culture and incubate at room temperature for 15 min. Add 10 ml of melted top agar (LB broth containing 0.6% agarose, cooled to 65°) to the cells, followed by 36 μl of 100 m*M* IPTG. Pour the mixture on a 150-mm LB plate and incubate at 37°. After 4–6 hr, tiny plaques will appear. When the plaques reach an average diameter of 1–2 mm, overlay the IgG or avidin-coated filter onto the plates and incubate at 37° for an additional 1–2 hr. Be certain to make orientation marks on the filters and plates so that they can be realigned later. Remove the filters and wash briefly three times at room temperature for 10 min each with TBST [10 m*M* Tris (pH 8.0), 150 m*M* NaCl, 0.05% Tween].

The filters can be assayed directly for β-galactosidase activity using CPRG or X-Gal. The former substrate is water soluble and yields a positive result within minutes; the latter substrate yields an insoluble product (which is permanent) and takes several hours to develop. For the CPRG assay, dissolve 1–2 μg in 10 ml of Z buffer (the solution should appear red-orange) and spread 1 ml over the filter. A positive interaction is indicated by a purple spot. Because the spot diffuses rapidly, a more permanent signal can be generated by adding 100 μl of 2% X-Gal (in dimethyl formamide) to 10 ml of Z buffer containing 0.6% agarose. Cool to 55° and pour directly on top of the filter. A deep blue color should develop within several hours at the site of positive interactions. The positive plaques can be recovered from the original plate for secondary and tertiary screenings.

Localization of Cyclin- and Cdk-Binding Domains of p21

One use of the double-tagging assay is to localize the binding domains of two interacting proteins. To illustrate this, we present double-tagging assay results in localizing the cyclin- and Cdk-binding domains of p21.

The p21 protein, also known as Cip1 and Waf1, is a Cdk inhibitor that was isolated by S. Elledge and co-workers[23] using the yeast two-hybrid assay. It also was isolated by the Vogelstein laboratory as a p53-induced gene.[24] Thus, these early studies suggested two important features of p21—that it is transcriptionally activated by p53 and that it interacts with and inhibits Cdk–cyclin complexes. These observations were consistent with earlier data indicating that DNA damage leads to elevated p53 levels and cell cycle arrest in normal cells. On repair of the DNA damage, p53 and p21 levels would fall, relieving the cell cycle block. Consistent with this

[23] J. Harper, G. Adami, N. Wei, K. Keyomarsi, and S. Elledge, *Cell* **75,** 805 (1993).

[24] W. El-Deiry, T. Tokina, V. Velculescu, D. Levy, R. Parsons, J. Trent, D. Lin, W. Mercer, K. Kinzler, and B. Vogelstein, *Cell* **75,** 817 (1993).

model, fibroblasts from mice lacking p21 fail to arrest properly following DNA damage.[25]

Further insight into the function of the p21 family of Cdk inhibitors came from the discovery of additional family members p27 (Kip1)[26,27] and p57 (Kip2).[28,29] All three family members share a region in their amino-terminal halves that is highly conserved. It was subsequently shown that a portion of this conserved region was sufficient for binding to Cdks and was essential for the Cdk inhibitory effect of p21.[7,30] Because one of the most highly conserved stretches of amino acids in the conserved region of p21 (amino acids 1–39 in Fig. 2) could be deleted without affecting Cdk binding, it seemed likely that this amino-terminal portion of the conserved region had another important function, perhaps in protein–protein interaction.

Several observations suggested that this amino-terminal segment of the conserved region might bind to cyclins. First, the related protein p27 was isolated by Toyoshima and Hunter using a yeast two-hybrid screen with cyclinD1.[26] They found that p27 could interact with cyclinD1 but not with a somewhat shortened Cdk4 protein. However, the interpretation of this result was complicated by the presence of homologous yeast cyclins and the yeast cyclin-dependent kinase Cdc28. Hall *et al.*[31] reported *in vitro* binding assay results that indicated that p21 could bind to a number of cyclins but not to Cdks. Finally, Zhu *et al.*[32] found that the retinoblastoma-related protein known as p107, which also binds to cyclin A–Cdk2 complexes, has a region highly homologous with a segment of the amino-terminal portion of the conserved region of p21.

The double-tagging assay is a useful technique for evaluating interactions between eukaryotic cell cycle regulatory proteins, because bacteria do not contain any homologous cyclin or Cdk proteins that might interfere with the binding studies. Demonstration of independent binding to cyclin and Cdk2 is complicated by the fact that Cdk2 can associate with several human and yeast cyclins, making *in vitro* or eukaryotic cell-derived proteins possibly contaminated with unwanted cell cycle proteins.

[25] C. Deng, P. Zhang, J. Harper, S. Elledge, and P. Leder, *Cell* **82,** 675 (1995).

[26] H. Toyoshima and T. Hunter, *Cell* **78,** 67 (1994).

[27] K. Polyak, M. Lee, H. Erdjument-Bromage, A. Koff, J. Roberts, P. Tempst, and J. Massague, *Cell* **78,** 59 (1994).

[28] M.-H. Lee, I. Reynisdottir, and J. Massague, *Genes Dev.* **9,** 639 (1995).

[29] S. Matsuoka, M. Edwards, C. Bai, S. Parker, P. Zhang, A. Baldini, J. Harper, and S. Elledge, *Genes Dev.* **9,** 650 (1995).

[30] M. Nakanishi, R. Robertorye, G. Adami, O. Pereira-Smith, and J. Smith, *EMBO J.* **14,** 555 (1995).

[31] M. Hall, S. Bates, and G. Peters, *Oncogene* **11,** 1581 (1995).

[32] L. Zhu, E. Narlow, and B. Dynlacht, *Genes Dev.* **9,** 1740 (1995).

```
EcoRI                                                  HincII         PvuII

GAA TTC GCC TGC CGC CGC CTC TTC GGC CCA GTC GAC AGC GAG CAG CTG

 E   F   A   C   R   R   L   F   G   P   V   D*  S   E   Q*  L

         _______________________________________________________

                                              NaeI

AGC CGC GAC TGT GAT GCG CTA ATG GCC GGC TGC ATC CAG GAG GCC CGT

 S*  R   D   C   D*  A   L   M   A   G   C   I   Q*  E   A*  R

_____________________________________  =========================

                                                       XhoI

GAG CGC TGG AAC TTC GAC TTT GTC ACC GAG ACA CCA CTC GAG GGT GAC

 E   R*  W   N   F   D   F   V   T   E   T   P   L   E   G   D

================================================================


TTC GCC TGG GAG CGT GTG CGA GGC CTT GGC CTG CCC AAG CTC TAC CCA

 F*  A   W   E*  R   V   R   G   L   G   L   P   K   L   Y   P

================================================================


HindIII

AGC TTG

 S   L
```

FIG. 2. Sequence of the conserved region of p21. DNA and protein sequence of the conserved region of p21 used in this study. The underlined amino acids represent conserved residues between p21 and p27; asterisks represent homologous amino acids. Several unique restriction sites were introduced by creating silent mutations in the DNA sequence. The single underline denotes the cyclin-binding domain whereas the double underline spans the Cdk-binding domain of p21 as determined by the double-tagging assay.

The cDNAs for human Cdk2 and Cdk7 (which does not bind to p21), as well as murine cyclinD1, were cloned into plasmid pTrc.EZZ.BCCP in-frame with the EZZ domain. Western blot analysis indicated that adequate levels of all three proteins were synthesized in *E. coli*. The entire p21 gene was cloned into λFJG2 and tested for binding by the double-tagging assay. A control phage (λFJG2 with a mutant E1a gene) was mixed with the test phage to serve as the "negative" background level of binding. As shown in Figure 3A, p21 could bind to both Cdk2 and cyclinD1 but not to human Cdk7. When the *Hin*cII-*Hin*dIII fragment of p21 was cloned into λFJG2 and tested for binding, significant binding was observed to Cdk2 but not to cyclinD1 or Cdk7 (Fig. 3A). Similar results were obtained with the *Nae*I–*Hin*dIII fragment of p21 (data not shown), indicating that the carboxy-terminal portion of the conserved region was involved in Cdk2 but not cyclin binding. When the 3′ end of this fragment was truncated (the *Hin*cII–*Xho*I fragment) and tested for binding, no binding above background level was observed to any of the test proteins (Fig. 3A). Thus, one or more of these deleted residues were important for Cdk2 binding. In summary, the double-tagging assay results confirmed that the Cdk2-binding domain of p21 was encoded within the carboxy-terminal half of the conserved region (see Fig. 2).

We next tested the binding properties of the amino-terminal half of the conserved region of p21. First, the entire conserved region of p21 was cloned into λFJG2 and mixed with the control phage λFJG2.*Hin*cII–*Xho*I p21 (because it yielded only background levels of binding to the test proteins). By the double-tagging assay, positive binding was observed to both Cdk2 and cyclinD1, but not to Cdk7, consistent with earlier results (see Fig. 3B). Positive binding to cyclinD1 but not to either Cdk2 or Cdk7 was observed when the *Eco*RI–*Xho*I fragment of p21 was tested. A similar pattern of binding was evident when the *Eco*RI–*Nae*I fragment of p21 was used (Fig. 3B). Further 3′ deletions (*Eco*RI–*Hin*cII and *Eco*RI–*Pvu*II) caused a significant decrease in cyclinD1 binding to a level barely above background (data not shown). The double-tagging assays results are summarized in Table III and indicate that the conserved region of p21 contains separate, nonoverlapping domains that can bind to Cdk2 and cyclin. These data are consistent with the crystal structure work of Russo *et al.*[33] and the results of Fotedar *et al.*[34] Thus, the

[33] A. Russo, P. Jeffrey, A. Patten, J. Massague, and N. Pavletich, *Nature* (*London*) **382,** 325 (1996).

[34] R. Fotedar, P. Fitzgerald, T. Rousselle, D. Cannella, M. Doree, H. Messier, and A. Fotedar, *Oncogene* **12,** 2155 (1996).

A

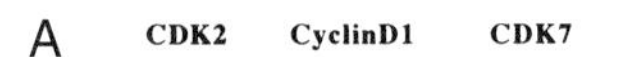

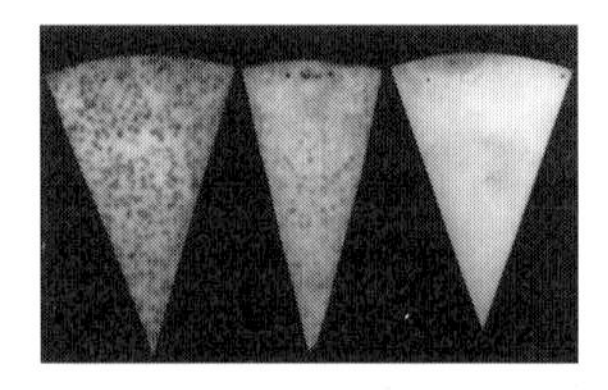

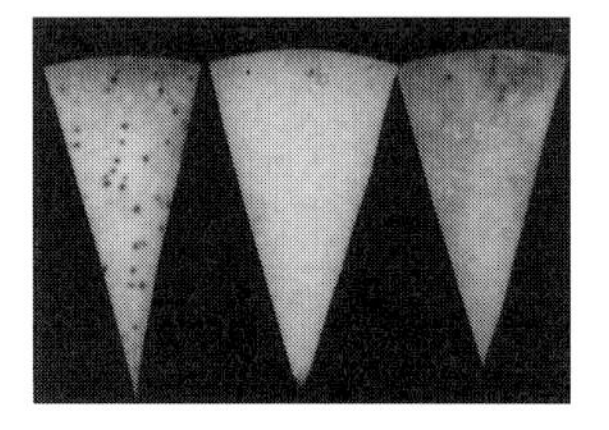

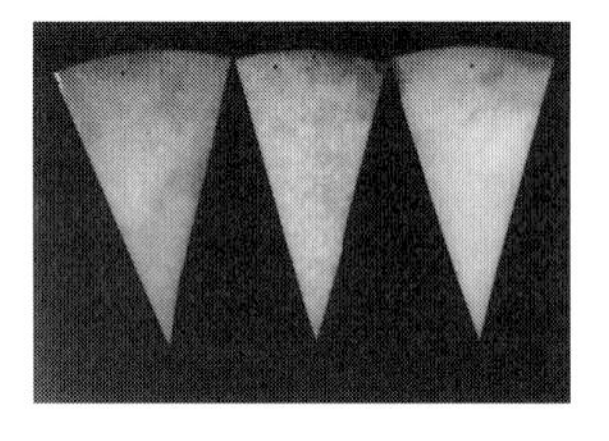

B

CDK2 CyclinD1 CDK7

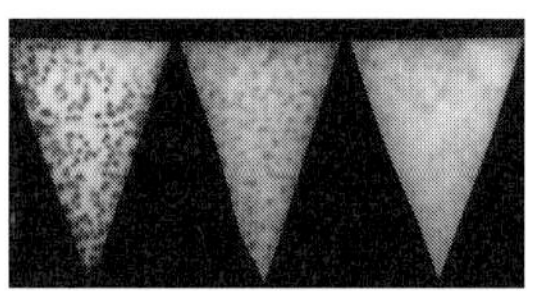

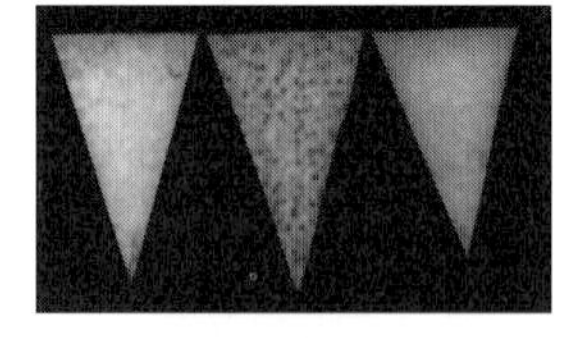

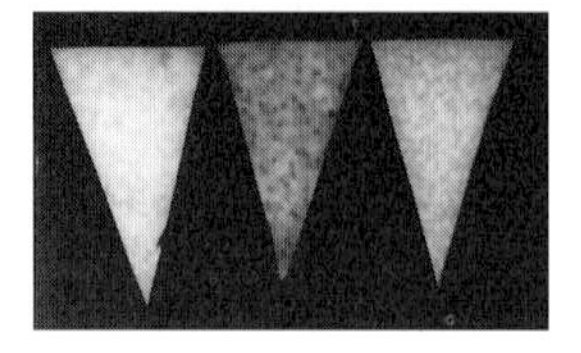

TABLE III
LOCALIZATION OF Cdk- AND CYCLIN-BINDING DOMAINS OF p21 BY THE DOUBLE-TAGGING ASSAY

p21 fragment[a]	Cdk2 binding	CyclinD1 binding	Cdk7 binding
Entire p21	+	+	–
Conserved p21 (aa 1–78)	+	+	–
*Hinc*II–*Hin*dIII (aa 25–78)	+	–	–
*Nae*I–*Hin*dIII (aa 39–78)	+	–	–
*Hinc*II–*Xho*I (aa 25–59)	–	–	–
*Eco*RI–*Xho*I (aa 1–59)	–	+	–
*Eco*RI–*Nae*I (aa 1–39)	–	+	–
*Eco*RI–*Pvu*II (aa 1–29)	–	+ (weak)	–
*Eco*RI–*Hinc*II (aa 1–25)	–	+ (weak)	–

[a] Refers to the amino acids in Fig. 2.

double-tagging assay was able to predict accurately the cyclin- and Cdk-binding domains of p21.

Limitations

While relatively rapid, the double-tagging assay suffers from the same limitations inherent in all bacterially based screens, namely, that essential posttranslational modifications do not occur in the bacterial cells and that some eukaryotic proteins may not fold properly within the bacterial cells. The double-tagging assay is best suited for screening a relatively limited

FIG. 3. Localization of the cyclin and Cdk2 binding domains of p21 by the double-tagging assay. The cDNAs for Cdk2, cyclinD1, and Cdk7 were cloned into pTrc.EZZ.BCCP. The indicated fragments of p21 were cloned into λ FJG2, mixed with a control phage (either λFJG2.mutant E1a or λFJG2.*Hinc*II-*Xho*I p21), and used to infect cells expressing EZZ–Cdk2–BCCP, EZZ–cyclinD1–BCCP, or EZZ–Cdk7–BCCP fusion proteins. Filters containing immobilized human IgG were overlaid onto the plates, incubated at 37° for 1 hr, and then removed. After three brief washes, the filters were developed for β-galactosidase activity as indicated. (A) Localization of the Cdk-binding domain of p21. Positive binding to Cdk2 was observed with the entire p21 molecule and the region encoded by the *Hinc*II–*Hin*dIII fragment of p21. In contrast, only the entire p21 fusion bound to cyclinD1 and no binding was observed to Cdk7. (B) Localization of the cyclin-binding domain of p21. Positive binding to cyclinD1 was observed with the entire conserved region of p21, and the region of p21 encoded by the *Eco*RI–*Xho*I and *Eco*RI–*Nae*I fragments of p21 as indicated in Fig. 2. In contrast, only the entire conserved region of p21 bound to Cdk2 and no binding was observed to Cdk7.

set of proteins (such as mutations in a single protein) rather than an entire cDNA library. We have found that between 0.1 and 1% of the cDNA library proteins expressed as a fusion with β-galactosidase in λFJG2 yield a false-positive result with the double-tagging assay—a positive signal is generated regardless of the bait protein used. These false positives are most likely due to the precipitation of insoluble β-galactosidase fusion proteins on the filter along with the captured bait hybrid protein. They are not removed with more stringent or prolonged washings. This remains the most serious obstacle to library screening with the double-tagging assay. Modifications of the screening protocol have not yet been unable to reduce significantly the frequency of these false positives.

Another use of the double-tagging assay is as a secondary assay for all the positive clones from a yeast two-hybrid screen. While secondary yeast screens often require 4–5 days for preparing the competent cells and allowing the cells to grow, the double-tagging assay can yield results within 24 hr of constructing the appropriate clones. Inserts from the fish clones are PCR amplified from the yeast vectors and ligated with *Eco*RI–*Hin*dII (or *Not*I)-digested λFJG2. The packaged phage are plated directly on cells synthesizing the EZZ–bait fusion protein and tested for binding as described above. An interaction that is positive by both the yeast two-hybrid and double-tagging assays is likely to be significant and should be explored further.

Summary

The assays described above can be used to screen for cellular proteins that can interact with a protein of interest, to screen for mutant proteins that retain the ability to bind to its partner, and to identify the domains and amino acids involved in known protein–protein interactions. Nonetheless, the biological significance of some of these interactions needs to be confirmed by appropriate cellular studies.

[25] Using the Lac Repressor System to Identify Interacting Proteins

By NICOLE L. STRICKER, PETER SCHATZ, and MIN LI

Introduction

Protein–protein interactions are essential to all biological processes. The physical associations between two proteins are generally achieved by one of two mechanisms. First, two interacting partners can form a complex via elaborate surface to surface contacts, which often require a proper tertiary structure of both partners. Alternatively, the two molecules can associate by way of a "lock-and-key" mode, as has been found in many enzyme–substrate interactions.[1] In the latter case, one interacting partner needs to be properly folded, while the other could be as simple as a short linear peptide. A large number of such protein interaction modules have been identified, and many interact by binding to short linear regions of target proteins. This class of protein interaction module includes SH2, SH3, PTB, and PDZ domains. These domains have been found in diverse proteins including many with important roles in signal transduction. Proteins with one or multiple copies of these interaction modules are thought to bind multiple target proteins because each module presumably has distinct substrate specificity. Thus determination of the optimal binding sequence for an orphan protein interaction module would provide a probe to electronically identify proteins with these binding sequences from either protein or DNA sequence databases. In this chapter we describe a series of experimental protocols that have been optimized for determination of peptide-binding sequences of PDZ domains. These protocols should also be applicable to other protein interaction modules.

PDZ domains are modular protein–protein interaction domains that were first identified in the postsynaptic density protein PSD-95, the *Drosophila* tumor-suppresser protein discs-large (*dlg*), and the tight junction protein ZO1.[2,3] There are currently more than 150 distinct PDZ domains present in a diverse array of proteins. Many of these proteins contain multiple protein interaction modules, either tandem PDZ domains or in combination with other interaction modules. However, the physiological binding partners for most PDZ domains are currently unknown. Studies

[1] E. M. Phizicky and S. Fields, *Microbiol. Rev.* **59,** 94 (1995).

[2] M. B. Kennedy, *Trends Biochem. Sci.* **20,** 350 (1995).

[3] M. Sheng, *Neuron* **17,** 575 (1996).

Copyright © 1999 by Academic Press
All rights of reproduction in any form reserved.

0076-6879/99 $30.00

have suggested that a number of PDZ domains play a role in the assembly of protein complexes involved in signal transduction cascades. Both biochemical and crystallographic studies have shown that the PDZ domain contacts only the four C-terminal residues of the target protein.[4,5] A strategy for *in vitro* selection of random peptides for optimal PDZ-binding sequences may provide an important entry point to the function studies of orphan PDZ domains and their parent proteins.

There are a number of methods currently used to perform peptide library screenings, including phage-displayed and chemically synthesized random peptide libraries. Because PDZ domains demonstrate a preference for C-terminal residues, we chose to apply the "peptides-on-plasmids" system[6] because it displays the C termini of the library peptides. Like the phage display system, the peptides-on-plasmids random peptide library is an oriented, biologically amplified library. To ensure amplification fidelity, both systems must have a way of linking each random peptide with the DNA encoding it. In the phage display system, this is easily accomplished because each peptide is displayed on the phage coat that encloses the encoding DNA. The peptides-on-plasmids system uses a DNA-binding protein to accomplish this linkage. Oligonucleotides are cloned into a plasmid, 3′ to the lac repressor gene (*lacI*). This same plasmid contains the DNA sequence recognized by LacI, the *lac* operator (*lacO*). Thus, on protein expression, each plasmid will express a LacI–random peptide fusion protein; the repressor will immediately bind to the *lacO* sequence on that plasmid, maintaining a physical link between each random peptide and its encoding DNA (Fig. 1).

Because this system displays peptides fused to the C terminus of the LacI protein, it is an ideal system for identifying PDZ-interacting peptides. In addition, because each binding peptide is individually amplified and sequenced, it is possible to differentiate among several preferred sequences. Once binding peptides have been identified, binding specificity can be assessed via binding assays such as an enzyme-linked immunosorbent assay (ELISA). This protocol describes the methods that apply a peptides-on-plasmids library to identify binding peptides specific for the nNOS and PSD-95 PDZ domains and uses of this information to identify potential *in vivo* interacting proteins.

Screening the Library

We screened a random 15-mer library with the PDZ domain of nNOS (the library was kindly provided by P. Schatz of Affymax Research Institute;

[4] N. Stricker, K. Christopherson, B. Yi, P. Schatz, R. Raab, G. Dawes, D. Bassett, D. Bredt, and M. Li, *Nature Biotechnol.* **15,** 336 (1997).

[5] D. Doyle, A. Lee, J. Lewis, E. Kim, M. Sheng, and R. MacKinnon, *Cell* **85,** 1067 (1996).

[6] M. G. Cull, J. F. Miller, and P. J. Schatz, *Proc. Natl. Acad. Sci. U.S.A.* **89,** 1865 (1992).

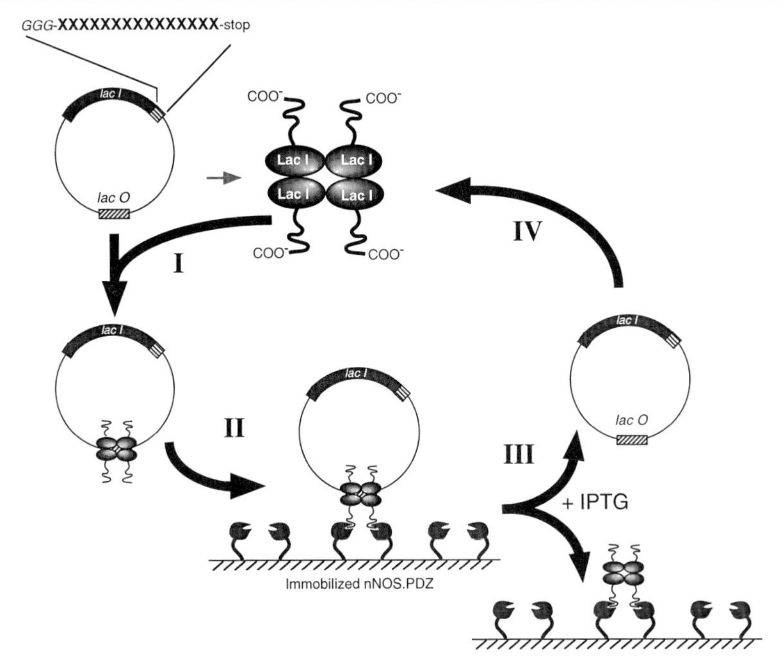

FIG. 1. Schematic diagram showing affinity selection from a C-terminal peptide library. Step I: A pool of oligonucleotides encoding 15 random amino acids (X_{15}) was cloned in frame, C terminal to *lacI*. Protein expression from each plasmid of the library yields a LacI fusion with a distinct peptide sequence. The recombinant LacI binds the *lacO* sites present on the same plasmid, yielding LacI–plasmid complexes that are purified from *E. coli*. Step II: Affinity panning selects peptides that interact with target receptor, e.g., the PDZ domain of nNOS. Step III: The bound plasmid DNA can be specifically recovered by addition of isopropyl β-D-thiogalactopyranoside (IPTG). Step IV: The recovered plasmids are retransformed, amplified, and used for subsequent rounds of panning.

the nNOS construct was provided by D. Bredt of University of California). This library has an initial complexity of 10^9. The screening consists of four repeated rounds of affinity panning against the immobilized glutathione-*S*-transferase (GST) fusion protein. After each round of panning, the progress of enrichment is monitored by comparing the number of input and recovered clones to estimate the percentage of complexes that are bound to the PDZ domain.

Protein Purification and Immobilization

The protein used for panning should be relatively pure to avoid isolation of peptides that may be specific for impurities in the sample. To complete the library screening, and subsequent analysis, at least 0.5 mg of purified protein is required. We expressed nNOS PDZ as a GST fusion protein and affinity purified with glutathione–Sepharose (Pharmacia Biotech, Piscataway, NJ). There are a number of ways to conveniently express and purify fusion proteins. These include histidine or monoclonal tagging, and maltose-binding protein fusion. In principle, other types of fusion proteins should also work well. The major consideration is to minimize binding noise in the target preparation, such as isolation of peptides that interact with the purification tag or other impurities. For example, we have found several C-terminal peptides capable of binding to GST (K. Loeffert and M. Li, unpublished results, 1998).

The GST–nNOS PDZ fusion proteins are expressed in *Escherichia coli* BL21. After a 4-hr induction with isopropyl-β-D-thiogalactopyranoside (IPTG), cells are washed once with cold water and lysed by sonication and the GST fusion proteins are harvested as described.[4] Purified protein is stored at a concentration of 3 mg/ml in a buffer containing 10 m*M* glutathione, 50 m*M* Tris-HCl (pH 8.0) at −20° until use. Purified protein is immobilized by overnight incubation on an Immulon 4 microtiter plate (Dynatech Laboratories, McLean, VA).

Reagents

Phosphate-buffered saline (PBS, for 1 liter): 8 g of NaCl, 0.2 g of KCl, 1.44 g of Na_2HPO_4, 0.24 g of KH_2PO_4, pH 7.4, in H_2O (filter sterilize or autoclave and store at 4°)

Procedure

1. Dilute protein to 0.05 mg/ml in PBS.
2. *Round 1:* Add 100 μl of diluted protein to 24 Immulon microtiter wells (5 μg/well). *Rounds 2–4:* Coat six wells with 100 μl of diluted protein (5 μg/well). Add PBS alone to a further six wells.
3. Cover all filled wells with tape (Scotch No. 845 BookTape; Stark Office Supplies, Baltimore, MD, to guard against evaporation during the immobilization.
4. Incubate the plate at 4° overnight (or for 2–3 hr at room temperature).

Because the diversity of a library is maximal during the first round of panning, we usually use 24 nNOS-PDZ-coated wells in this round to provide more binding substrates for higher recovery of interacting peptides. For subsequent rounds, six coated wells are sufficient to extract a significant number of plasmids for interacting peptides. It is also useful in later rounds (2, 3, and 4) to coat an additional six wells with PBS alone as a control; the recovered plasmids from these wells are used to calculate the background level of recovery (described below).

Affinity Panning

The panning procedure consists of blocking the wells containing the immobilized protein, then exposing the coated surface to the random peptide library. The peptide–LacI–plasmid complexes are released from the bacteria by gentle cell lysis, and the crude lysate is incubated in the nNOS PDZ-coated wells, allowing the random peptides to interact with the nNOS PDZ domain. Unbound clones are washed away, and plasmids encoding peptides that interact with the PDZ domain ("binders") are eluted with IPTG, which disrupts the LacI/*lacO* interaction (Fig. 1). These plasmids are amplified by transforming *E. coli,* and the amplified peptide–repressor–plasmid complexes are used as input for the next round of panning.

Reagents

For the reagents listed below, the volumes given are enough for five panning rounds; store all reagents at 4°.

Bovine serum albumin (BSA, 1.0% in PBS): 25 ml
Carnation dry milk (1.0% in PBS): 25 ml
BSA (10 mg/ml in HE): 5 ml
HE (100 ml): 35 m*M* HEPES (pH 7.5), 0.1 m*M* EDTA
HEK (50 ml): 35 m*M* HEPES (pH 7.5), 0.1 m*M* EDTA, 50 m*M* KCl
HEKL (50 ml): 35 m*M* HEPES (pH 7.5), 0.1 m*M* EDTA, 50 m*M* KCl, 0.2 *M* α-lactose
HEKL–BSA (500 ml): 1% BSA in HEKL
Bulk DNA (30 ml): Salmon sperm DNA (0.1 mg/ml in HEKL–BSA)
Dithiothreitol (DTT), 0.5 *M* in double-distilled H_2O (ddH_2O): 100 μl
Phenylmethylsulfonylfluoride (PMSF, 0.1 *M* in isopropanol): 0.5 ml
Lysis buffer: Four 15-ml Falcon tubes, each containing 4.2 ml of HE and 1 ml of 50% glycerol
Lysozyme: Prepare four Eppendorf tubes, each containing 5 mg of lysozyme; add 0.5 ml of HE just before use
KCl (2 *M*): 5 ml
α-Lactose (20%): 20 ml
IPTG–KCl (5 ml): 1 m*M* IPTG and 0.2 *M* KCl in HE

Procedure

1. Wash the protein-coated wells four times with room temperature PBS. Slap the plate on a bed of paper towels to remove any residual liquid.
2. *Rounds 1 and 4:* Block the wells by adding 200–300 μl/well of 1% BSA in PBS. Cover with tape and incubate at 37° for 1–2 hr. *Rounds 2 and 3:* Block the wells by adding 200–300 μl/well of 1% dry milk in PBS. Cover with tape and incubate at 37° for 1–2 hr.

3. Wash the wells twice with HEKL–BSA. Slap the plate to dry as before.
4. Preblock with bulk DNA (50 μl/well). Cover with tape and shake at 4° for 10–60 min.
5. While wells are preblocking, prepare the library lysates.
   a. Add 750 μl of BSA (10 mg/ml), 10 μl of DTT, and 12.5 μl of PMSF to one HE–glycerol "lysis buffer" tube.
   b. Thaw the library (naive or amplified) and transfer it into the lysis buffer.
   c. Add 0.5 ml of HE to an Eppendorf tube containing 5 mg of lysozyme; pipette up and down to dissolve the lysozyme. Add 10 μl of PMSF to the centrifuge tube that will be used to centrifuge the lysed cells.
   d. Add 150 μl of the lysozyme (10 mg/ml) to the lysis buffer only when everything else is prepared (i.e., be ready to add the α-lactose and KCl, and have a prechilled rotor ready for the centrifugation), as lysis will proceed quickly and it is important not to overlyse the cells.
   e. Invert the tube several times. Let the cells lyse for 1–2 min. Lysis can be observed by the slow migration of bubbles to the top of the tube.
   f. Quickly add 2 ml of 20% α-lactose and 250 μl of 2 *M* KCl. Pour the lysed cells into the centrifuge tube and spin at 4° for 15 min at 12,000 rpm (~10,000 *g*).
   g. While the lysate is centrifuging, prepare a 15-ml conical tube by adding 10 μl of PMSF and labeling with the type of library (i.e., Lac-15mer), the protein used for screening (i.e., nNOS PDZ), the round of panning for which the lysate will be used (i.e., round 3 input), and the date.
   h. When the centrifugation is finished, gently pour the supernatant into this 15-ml conical tube (6 to 9 ml of clear supernatant should be obtained). Save a sample of the crude lysate (labeled "Pre 1, 2, 3, or 4") to determine the number of clones used as "input" for that panning round. We usually save 1% of the amount used for panning, as this facilitates calculations later (i.e., *Round 1,* between 48 and 60 μl; *round 2,* 6 μl; *rounds 3 and 4,* 6 μl of the diluted sample).
6. Without removing the bulk DNA, add crude lysate to the preblocked wells:

Round 1: 200–250 μl of lysate per well
Round 2: 100 μl/well

Rounds 3 and 4: 100 μl/well of lysate diluted 1:10 in HEKL–BSA. Cover with tape and shake at 4° for 1 hr. Store any extra lysate at −80°.

7. Remove the tape and discard (it may have nonbinding clones stuck to it, which can contaminate later preparations). *Round 1:* Skip to step 9. *Rounds 2–4:* Wash the wells four times with HEKL–BSA, slap plate to dry as before.
8. Add 200 μl of bulk DNA. Shake uncovered at 4° for 30 min.
9. Wash the wells four times with HEKL. Wash twice with HEK, slap the plate to dry as before.
10. Elute binding clones by adding 50 μl of IPTG–KCl. Shake at room temperature for 30 min.
11. *Round 1:* Remove all eluates to a microcentrifuge tube by holding the plate at a 45° angle to collect as much of the liquid as possible. Divide the sample into two tubes to proceed with the DNA precipitation; label these tubes "Pan" (indicating recovered clones) and note the round number. *Rounds 2–4:* Remove eluates from receptor-coated wells to the Pan tube, remove eluates from the PBS wells to a microcentrifuge tube marked NC (negative control), and note the round number.

The library lysing step is critical, especially for the first round of panning. Under- or overlysing of the library should be avoided, as incomplete lysis of the library will compromise library complexity, while overlysing will prevent the formation of a compact cell pellet during centrifugation, resulting in a viscous supernatant that cannot be used for panning. First, it is crucial to use only fresh lysozyme; enzyme prepared as early as the night before may not lyse properly and result in loss of that library aliquot. Second, the lysate should be centrifuged as quickly as possible after the lysis (within 5–10 min), despite the slowing in lysozyme activity that results from KCl and α-lactose addition (to increase LacI solubility and prevent LacI/*lacO* dissociation, respectively). Finally, it is important to be careful when labeling the tube containing this supernatant. The crude lysates can be useful for repeating a round of panning or performing a "biased" screen with a related protein, and a clear label is critical for their reuse. A common downfall is the label "round 3," which is ambiguous as to whether the lysate contains peptide complexes to be used during the third round of panning (round 3 input), or if they are lysates from cells used to amplify round 3 binders (round 3 recovery/round 4 input). Explicit labels are the best way to avoid future confusion.

All panning washes are performed with a repeating pipette. If a standard pipettor is used, extreme care should be taken to avoid contamination of the wash solutions with the peptide–repressor–plasmid complexes that may remain in the wells at all steps subsequent to the panning. The blocking reagent is changed for panning rounds 2 and 3 to avoid isolation of BSA-sticky peptides and to enhance isolation of only those complexes that interact with the nNOS PDZ fusion protein. The long wash (step 8) is skipped during the first round of panning to avoid loss of weak binders; the subsequent amplification of binding clones makes this precaution unnecessary in the later panning rounds. Similarly, because later rounds involve panning with an amplified population of binders, it is no longer necessary to maintain high input levels in later rounds, hence the lysate dilution in rounds 3 and 4.

Amplification of Interacting Plasmids

After the first panning round, the complexity of the library will have been reduced significantly (from $\sim 10^8$ to $\sim 10^4$), and many nonbinding complexes will have been eliminated. However, because there are relatively few copies of any one binding species in this population, the eluted plasmids are amplified in bacteria before a second round of panning is carried out. This amplification step ensures that all clones are present in equal numbers for the subsequent panning round, and prevents binding complexes from being lost owing to a weak interaction or a low copy number. Binding clones are amplified after each round of panning by transforming bacteria with the eluted plasmid DNA via electroporation. The transformed bacteria are grown to late log phase, then centrifuged and washed to prepare the membranes for the gentle lysis that will release the peptide–repressor–plasmid complexes for the subsequent round of panning.

Reagents

Glycogen (Boehringer Mannheim, Indianapolis, IN)
Ethanol (80%): 50 ml
NaCl (5 *M*): 1.5 ml
WTEK (500 ml): 50 m*M* Tris (pH 7.5), 10 m*M* EDTA, 100 m*M* KCl
TEK (250 ml): 10 m*M* Tris (pH 7.5), 0.1 m*M* EDTA, 100 m*M* KCl
ARI814 electrocompetent cells (8- to 100-ml aliquots)[6]: Transformation frequency should be $\geq 2 \times 10^{10}$ colonies per microgram of DNA

DNA Precipitation

1. Add IPTG–KCl to the Pre sample to bring it to the same volume as the Pan and NC samples.
2. Add 1 μl of glycogen and a 1/10 vol of 5 *M* NaCl to each Eppendorf tube. Add an equal volume of isopropanol and invert several times to mix.

3. Centrifuge for 10 min to pellet the DNA (a pellet may or may not be visible).

4. Carefully aspirate the supernatant or remove with a pipettor. Gently add 0.5 ml of cold 80% ethanol, taking care not to disturb the pellet. Let the pellet sit for 2–5 min at room temperature in the ethanol.

5. Centrifuge for 2 min then repeat with a second ethanol wash.

6. *Round 1:* Resuspend the "pre" DNA in 200 μl of ddH_2O. Resuspend each Pan pellet in 3 μl of DNA and pool them into one tube. *Rounds 2–4:* Resuspend the Pan and NC pellets in 6 μl each, and resuspend the Pre DNA in 200 μl. Store the DNA at 4°.

Careful washing of the pellet is important, as insufficient washing will leave salt in the DNA, causing the electroporation to "pop." The Pan DNA should be treated with extra care; it represents plasmids encoding all binders from the previous panning round, and loss of either the pellet, or the DNA due to a pop, will force a repetition of the entire panning round.

Electroporation

1. Thaw 75 to 100 μl of ARI814 electrocompetent cells and chill three or four 0.1-cm-gap cuvettes on ice (Bio-Rad, Hercules, CA).

2. Label three (or four) 14-ml round-bottom tubes as Pre, Pan A, and Pan B (and NC for later rounds). Add 2 ml of LB to the Pre (and NC) tube, add 1 ml of LB to each of the Pan tubes. Place all tubes in a 37° heat block or water bath.

3. Set the Bio-Rad Gene Pulser apparatus to 1.8 kV, 25 μF capacity, and the pulser controller unit to 200 Ω.

4. Put 1.5 μl of Pan DNA into a sterile Eppendorf tube and chill it on ice. Add 25–30 μl of competent cells to the tube and pipette up and down to mix.

5. Transfer the cells to a chilled gap cuvette and apply the pulse. If the time constant (τ) is between 4 and 5 msec, use the same volumes of DNA and competent cells for the remaining electroporations.

6. If the sample pops, adjust the volume of input DNA in the electroporation. Prepare a fresh DNA–competent cell mixture using 0.7 μl of DNA (or 1.5 μl of a 1:2 dilution) and 25–30 μl of cells.

7. After a successful electroporation, immediately add 0.5–1.0 ml of the prewarmed LB from the appropriate tube to the cuvette to resuspend the cells. Transfer the cell suspension into the round-bottom tube and incubate at 37° without shaking for 1 hr.

8. Complete steps 4–7 for the Pre (NC), and two Pan samples. (Pan DNA is electroporated twice to obtain a higher number of transformed bacteria and to prevent loss of all eluted DNA if one sample pops.)

9. During the recovery period, label 10 LB-Amp plates. These will be used to plate the cells to determine the level of enrichment for each round of panning (as described in the following section). For round 1, it is useful to prewarm the 200 ml of LB growth medium supplemented with ampicillin (100 μg/ml) (denoted as LBA).

10. After the recovery period, pool the two 1-ml Pan cultures and remove 100 μl to an Eppendorf tube for plating.

Amplification

11. *Round 1:* Inoculate the 200 ml of LBA with the 2-ml Pan recovery culture. Grow at 37° with shaking to an A_{600} of 0.5–1.0 (this usually takes 7 to 10 hr, depending on the competency of the cells). *Rounds 2–4:* Add 2 ml of LB and 8 μl of ampicillin (50 mg/ml; final concentration, 10 μg/ml) to the 2-ml Pan recovery culture. Grow the cultures overnight with shaking at 37°. Use this 4-ml overnight culture to inoculate 200 ml of LBA. Grow at 37° with shaking to an A_{600} of 0.5–1.0 (this usually takes 2 to 2.5 hr).

12. Chill the flask in an ice–water bath for 10 min.

13. Pour the culture into a 250-ml centrifuge tube and centrifuge at 6000 rpm (3000 *g*) for 6 min.

14. Resuspend the cells in 100 ml of cold WTEK.

15. Centrifuge at 6000 rpm for 6 min.

16. Resuspend the cells in 50 ml of cold TEK.

17. Centrifuge at 6000 rpm for 6 min.

18. Resuspend the cells in 4 ml of cold HEK. Divide the cells into two 2-ml Nalgene screw-top vials. Again, be sure that the vials are carefully labeled with respect to the round number (i.e., whether the cells contain plasmids that are the input or recovery products for a particular round).

19. Quickly freeze the cells in dry ice powder and store at −80° until use. One aliquot will be lysed for input in the next panning round, the other can be stored indefinitely and eventually used to repeat a round, or to conduct a biased library screen. It is best to freeze both aliquots, even if one is to be used the same day, as the freeze–thaw process aids cell lysis.

Overnight growth of the transformation recovery culture is discouraged for the first round of amplification to prevent fast-growing clones from overgrowing slower ones. This is an important consideration because the library is still relatively diverse at this stage; there will be fewer than 100 copies of any particular binding clone in the eluted population. The amplification step is necessary to enhance the competition among different binding moieties by increasing the number of ligands to a level significantly higher than the available number of receptors. If fast-growing clones are allowed to overtake slow-growing clones during this first amplification step, then the relative levels of the different binding complexes will be unequal for the second panning round, creating a biased library that may lead to loss of certain binding sequences. In later rounds, the diversity of the library will have been reduced enough so that each distinct peptide will be present at higher levels before amplification, rendering this precaution unnecessary.

Monitoring Panning Efficiency and Determining Enrichment

At the end of each round of panning, it is useful to compare the complexity of the library with that at the beginning. This information helps evaluate whether there are peptides that interact with the target PDZ domain, and whether an appropriate level of binding clone enrichment is taking place. "Enrichment" describes the degree to which the complexity of the library is being reduced by the panning process and gives an indication of the diversity of the binding population. Too little enrichment precludes determination of a binding preference, while too much enrichment eventually results in the isolation of a single or a few dominant clones.

To monitor the progress, the number of clones input into each round is compared with the number eluted from the wells. These values are determined by plating dilutions of Pre, Pan, and NC transformed cells, then counting colonies and extrapolating to determine the total number of clones present in each sample. Because each eluted plasmid participates in a single binding interaction, and subsequently transforms a single *E. coli* cell, each colony on the plate represents a single binding clone. Several dilutions of the transformed cells are plated so that the number of colonies per microliter of DNA can be accurately determined.

Enrichment of a binding population can be monitored in several ways. First, "percent recovery" compares the number of clones eluted from PDZ-coated wells with the total number of clones panned during that round; this describes the fraction of total clones that bind to the coated plate in a given round. Second, "enrichment" compares the number of clones eluted from PDZ-coated wells with the number of clones eluted from buffer-coated wells and indicates whether binding to the PDZ domain is occurring at a level significantly above the background peptide binding level with the Immulon plate. Both percent recovery and enrichment should increase significantly with each round of panning (see Table I).

1. Label LBA plates with the transforming DNA (i.e., Pre, Pan, or NC), the round number, and the number of microliters of cells plated. Transform the DNA samples by electroporation as described above. For round 1, plate 100, 10, 1, 0.1, and 0.01 μl from the 2 ml of recovered cells for both Pre and Pan samples. In later rounds, a good starting point is 10,

TABLE I
SUMMARY OF nNOS PDZ PANNING[a]

Round	Input (Pan)	Output (Pan)	Output (NC)	Percent recovery (Pan)	Enrichment
1	6.0×10^9	1.7×10^5	—	2.9×10^{-5}	—
2	3.2×10^9	1.4×10^5	4.4×10^4	4.4×10^{-5}	3
3	1.2×10^8	1.1×10^6	4.1×10^3	5.9×10^{-3}	270
4	8.4×10^7	4.8×10^6	2.8×10^3	5.7×10^{-2}	1700

[a] NC, Negative control.

1, 0.1, and 0.01 μl for Pre and Pan samples, and 1 and 0.1 μl of NC cells. However, adjustments may need to be made if these dilutions yield plates with too few or too many colonies to count.

2. Use a disposable Falcon 3911 flexible assay plate (Becton Dickinson Labware, Lincoln Park, NJ) to make serial dilutions. Add 100 μl of LB to each of four wells. Add 10 μl of culture to the first well and pipette up and down to mix. Use a new tip to add 10 μl of the 1:10 dilution to the next well and pipette up and down to mix; continue for all four wells. Failure to change the tips between dilutions will compromise the dilution fidelity because cells from more concentrated wells will stick to the tip and end up in more dilute wells. This practice can result in inconsistent colony numbers among one set of dilutions.

3. After the dilutions are made, plate the cells and spread them evenly, using sterilized glass beads or a glass spreader. Incubate the plates overnight at 37°.

4. Count the colonies on each plate. If there are too few or too many colonies, try to obtain data from at least two plates (dilutions) of the same cell culture. After counting the colonies, plates from the initial rounds can be discarded. However, plates containing clones from round 3 and 4 pannings should be retained for future examination of individual clones by ELISA (described in the next section).

5. Use the average number of colonies per microliter of cells plated to calculate the number of clones in the entire cell culture, and thus the number of clones in the total DNA sample (6 μl for Pan and NC, 200 μl for Pre). For Pan and NC samples, this number equals the total number of recovered and background clones, respectively. For Pre samples, recall that DNA was precipitated from only 1% of the total lysate used for the panning, so this number needs to be multiplied accordingly to obtain the total number of input clones for the round of panning.

6. Determine the percent recovery by taking the ratio of recovered (Pan) clones to input (Pre) clones. Determine the enrichment by taking

the ratio of recovered clones to background (NC) clones. Representative values for these numbers are listed in Table I.

Identification of PDZ-Specific Clones

After affinity purification of a population of binding clones, it is necessary to identify those clones that bind specifically to the nNOS PDZ domain. Because the GST fusion protein is used as the panning substrate, a subset of the isolated clones may be specific for the GST portion of the protein, not the PDZ domain. PDZ-specific clones are identified using an enzyme-linked immunosorbent assay (ELISA). This procedure involves screening of a number of individual binding clones to determine binding specificity. Each clone is incubated in a set of three wells, which are coated with PBS alone, a GST control protein, and the GST–nNOS fusion protein, respectively. The binding of an antibody to the LacI repressor followed by binding of alkaline phosphate-conjugated secondary antibody allows for visual detection of a binding interaction. A strong ELISA signal in PDZ-coated wells, but not in GST-coated or empty wells, indicates a PDZ-specific interaction (Fig. 2).[4] After a subset of PDZ-specific clones have been identified, the plasmids can then be sequenced to deduce the amino acid sequence of each binding peptide.

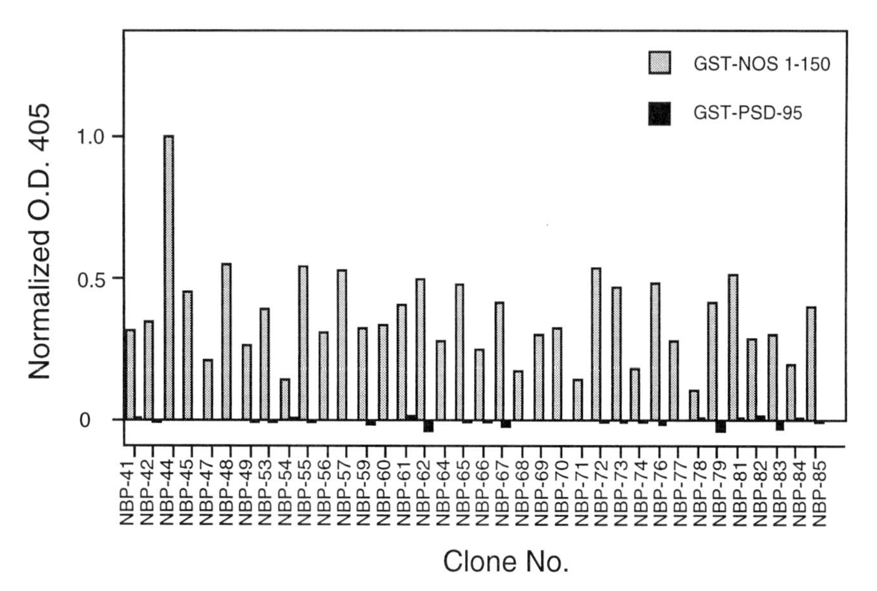

FIG. 2. nNOS-binding peptides (NBPs) bind specifically to nNOS PDZ. ELISA results of 36 randomly chosen NBP clones are shown. Horizontal axis, NBP clone number; vertical axis, ELISA signal normalized against the clone with strongest binding (NBP-44).

Lysis of Individual LacI Clones for ELISA

Before the ELISA can be performed, individual clones must first be amplified and lysed to release the peptide–repressor–plasmid complexes. Individual colonies are picked from the Pan plates used to determine recovery data (previous section). These clones are then amplified, washed, and lysed in the same manner as the 200-ml cultures used for panning. For the initial ELISA screen, examination of 16 to 24 individual clones should be sufficient to evaluate the percentage of PDZ-specific clones and the relative level of peptide diversity at a given round of panning. Larger scale ELISA can be performed later if necessary.

Reagents

Lysis buffer (make fresh just before use): 42 ml of HE, 5 ml of 50% glycerol, 3 ml of BSA (10 mg/ml in HE), 125 μl of 0.1 *M* PMSF, 750 μl of lysozyme (10 mg/ml in HE).

L-Arabinose, 20% in ddH_2O, sterile

1. Pick 12 individual colonies from the Pan plate and inoculate 2-ml LBA cultures and shake at 37° overnight. These clones do not need meaningful names at this point, but should be numbered so that they can be kept track of and used as the source to prepare plasmid DNA for sequencing.
2. Inoculate 3 ml of LBA with 300 μl of the overnight cultures. Save the remaining 1.7 ml of the overnight cultures for minipreps of interesting clones.
3. After incubation at 37° for 1 hr induce each culture with 33 μl of 20% L-arabinose (final concentration, 0.2%) for 3 hr at 37°.
4. Chill the tubes on ice and centrifuge at 2000 *g* for 5 min (keep the tubes chilled for the rest of the procedure).
5. Pour off the supernatant and add 3 ml of cold WTEK buffer. Resuspend the cells by vortexing.
6. Centrifuge at 2000 *g* for 5 min.
7. Pour off the supernatant and add 1 ml of cold TEK buffer. Resuspend the cells and transfer the suspension to an Eppendorf tube.
8. Centrifuge at 12,000 *g* in a microcentrifuge tube for 2 min.
9. Aspirate the supernatant and resuspend the cells in 1 ml of lysis buffer. Incubate on ice for 1 hr.
10. Add 110 μl of 2 *M* KCl, and invert to mix.
11. Centrifuge at 12,000 *g*, 4°, for 20 min. A longer centrifugation is sometimes required for the production of a pellet compact enough to allow removal of at least 500 μl of clear supernatant.
12. Transfer 500–900 μl of the clear crude lysate to a fresh, clearly

labeled Eppendorf tube and store at 44° for immediate use or at −80° for later use.

LacI ELISA

The ELISA allows visual detection of binding interactions so that clones that bind to the GST portion of the fusion protein can be distinguished from those specific for the nNOS PDZ domain. In addition, if there are related proteins with known peptide-binding partners, such as PDZ3 of PSD-95, these can serve as useful controls. The lysis procedure described above produces enough crude lysate for many ELISAs; thus, a large number of possible binding interactions can be tested for each individual peptide complex. The standard controls used are (1) a PBS-coated well to detect an interaction with the plastic or the BSA, (2) a well coated with GST alone to detect an interaction with the GST portion of the fusion protein, (3) a well coated with the nNOS PDZ fusion protein used in the pannings to detect an nNOS-specific binding interaction, and (4) related PDZ domains to provide clues regarding the degree of nNOS PDZ specificity.

Reagents

PBS–Tween (500 ml): 0.05% Tween 20 in PBS
PBT (250 ml): 1% BSA, 0.05% Tween 20 in PBS
TBS (100 ml): 10 m*M* Tris (pH 7.5), 150 m*M* NaCl
Rabbit anti-LacI antibody
Alkaline phosphatase (AP)-conjugated goat anti-rabbit antibody (Sigma, St. Louis, MO)
p-Nitrophenyl phosphate tablets (Sigma)
Development buffer (50 ml): 10% diethanolamine (w/v), 0.5 m*M* $MgCl_2$, pH 9.8 with HCl. Store in the dark at 4°

1. For each clone to be tested, coat one Immulon well with PBS alone, one with GST control protein, and one with GST–nNOS fusion protein. Dilute the GST control and GST–nNOS PDZ proteins to 5 μg/ml and add 100 μl to each well (0.5 μg protein per well). Immobilize the protein by incubating the covered plate at 4° overnight or at room temperature for 2–4 hr.
2. Wash the wells four times with PBS.
3. Block the wells by adding 200–300 μl of 1% BSA. Incubate for 1 hr at 37°.
4. Wash the wells four times with PBS–Tween.
5. Add 100 μl of the diluted crude lysates. This is most easily achieved by adding 95 μl of PBT to all wells, then carefully adding 5 μl of the lysate. It is not necessary to pipette up and down, as the samples will be mixed

with shaking during the incubation. Extreme care must be taken to assure that the correct lysate is added to each well.

6. Incubate the plate at 4°, with shaking, for 30 min.
7. Wash the wells four times with PBS–Tween. Add 100 μl of anti-LacI antibody to each well. The antibody is diluted 1:15,000 with PBT or according to manufacturer instructions.
8. Incubate the plate at 4°, with shaking, for 30 min.
9. Wash the wells four times with PBS–Tween. Add 100 μl of AP-conjugated goat anti-rabbit antibody (diluted 1:3000 in PBT) to each well.
10. Incubate the plate at 4°, with shaking, for 30 min.
11. During the incubation, prepare the development solution in a foil-covered conical tube. For every 45 wells in the assay, dissolve one 5-mg *p*-nitrophenol tablet in 5 ml of development buffer. Vortex until the tablets are completely dissolved. This solution should be made freshly each time, and it is not necessary to keep this solution on ice.
12. Wash the wells four times with PBS–Tween, then twice with TBS.
13. Develop the ELISA by adding 100 μl of the development solution to each well. Remember that the color of these wells will be compared, so it is important to perform this step as quickly as possible to avoid significant differences in incubation times among the wells.
14. Read the plate in a microtiter plate reader, at A_{405}. Take readings of several time points (about one every minute) to assure a reading in the linear range. The signal is no longer linear above an A_{405} of 1.0.

When adding the individual lysates to the wells, it is important to assure that each lysate is added to the correct well. Losing track of which lysate has been added to which well can ruin an entire plate and force repetition of the experiment. Also, changing the tip after each 5-μl addition will ensure that each well receives the same amount of lysate. Because ELISA results are interpreted by comparing the signals from adjacent wells, it is important to establish binding conditions in these wells that are as similar as possible.

Washes are easiest with Nunc-Immuno wash 12 (Fisher, Pittsburgh, PA), but can also be carried out with a multichannel pipette. If using a multichannel pipette, it is acceptable to use one set of tips for each of the four washes at any given step (aerosol-resistant tips are recommended to prevent liquid from entering the pipette). While washing the wells with a multichannel pipette, be sure that all tips are filling and dispensing equally.

Sequence Analysis and Search for Candidate Interacting Proteins

After ELISA and DNA sequencing, the corresponding peptide sequences for individual binding peptides can be aligned either visually or by DNA sequence analysis software. The amino acid abundance at each

>gi|244607|bbs|79586 cleaved prolactin-1, cIPRL-1=fragment A [rats, Peptide Partial, 20 aa] (Match **DRV**)
>gi|497021 (U05699) cytochrome c oxidase subunit Va [Mus spretus] (Match **DKV**)
>gi|505029 (D14849) meiosis-specific nuclear structural protein 1 [Mus musculus] (Match **DGV**)
>gi|531881 (U12877) vascular cell adhesion molecule-1 [Mus musculus] (Match **DTV**)
>gi|191913 (M11895) A-1 alpha-amylase [Mus musculus] (Match **DKV**)
>gi|191919 (M11896) B-1 alpha-amylase [Mus musculus] (Match **DKV**)
>gi|192098 (M18187) B144 protein A [Mus musculus] (Match **DYV**)
>gi|196056 (M34984) Ig H-chain [Mus musculus] (Match **DTV**)
>gi|554244 (K03547) myb protein [Mus musculus] (Match **DSV**)
>gi|1363194|pir||A53202 MAMA protein precursor - mouse gi|297033 (X67809) mama gene product [Mus musculus] (Match **DMV**)
>gi|423447|pir||S35792 glutamate receptor GluR6C - mouse gi|312494 (X66117) glutamate receptor subunit GluR6C [Mus musculus] (Match **DTV**)
>gi|117099|sp|P12787|COXA_MOUSE CYTOCHROME C OXIDASE POLYPEPTIDE VA PRECURSOR. gi|90420|pir||S05495 cytochrome-c oxidase (EC 1.9.3.1) chain Va precursor - mouse gi|50527 (X15963) cytochrome c oxidase subunit Va preprotein [Mus musculus] (Match **DKV**)
>gi|805000 (X83536) MT-MMP [Mus musculus] (Match **DKV**)
>gi|939951 (X73037) partial paired box; pid:e74985 [Mus musculus] (Match **DGV**)
>gi|1184877 (U46562) MHC class II transactivator CIITA [Mus musculus] (Match **DMV**)
>gi|1215666 (U17267) T cell receptor-Zeta [Mus musculus] (Match **DEV**)
>gi|1326151 (U52222) Mel-1a melatonin receptor [Mus musculus] (Match **DSV**)

FIG. 3. A list of mouse proteins that contain C termini that match the binding consensus of the nNOS PDZ domain.

position can be calculated and a binding consensus can then be obtained. Depending on the length and degeneracy of the binding consensus, one can chose suitable computer software or an appropriate URL on the Internet to search for a potential interacting protein. If the consensus is relatively long, e.g., more than five residues, one can try a simple BLAST similarity search (*URL:http://www.ncbi.nlm.nih.gov/BLAST/*). If the consensus is short, fewer than four residues, one reasonable place to start would be to submit the sequence as a "Pattern Match" query using the GCG sequence analysis software package. An example of a search for nNOS binding proteins is shown in Fig. 3.

Summary

The present protocol describes a series of procedures to identify peptides interacting with PDZ domains. It is conceivable that the procedures can be applied to other purified protein modules or intact proteins without substantial modifications. With the deduced consensus combined with sequence information, it is possible to identify proteins present in the database with compatible sequences. If the expression of target protein and potential interacting candidate overlap temporally and spatially, biochemical and molecular experiments can be designed to study their physical and functional interactions.

Acknowledgments

We thank David Bredt for providing the nNOS constructs. This work is supported in part by grants from NIH (NS33324) and the American Heart Association.

[26] A Genetic Selection for Isolating cDNA Clones that Encode Signal Peptides

By KENNETH A. JACOBS, LISA A. COLLINS-RACIE, MAUREEN COLBERT, MCKEOUGH DUCKETT, CHERYL EVANS, MARGARET GOLDEN-FLEET, KERRY KELLEHER, RONALD KRIZ, EDWARD R. LAVALLIE, DAVID MERBERG, VIKKI SPAULDING, JEN STOVER, MARK J. WILLIAMSON, and JOHN M. MCCOY

Introduction

Secreted and membrane-bound proteins mediate many of the signals of intercellular and interorgan communication in metazoans. Identifying the functions of these proteins requires that their genes be identified. This is not currently feasible if traditional bioassay-dependent approaches are applied. The bioassay-independent approach of random sequencing of

Copyright © 1999 by Academic Press
All rights of reproduction in any form reserved.

0076-6879/99 $30.00

cDNA clones is an alternative but is labor intensive and becomes more so for rarely expressed genes. Because secreted and membrane-bound proteins are synthesized with signal sequences, another approach is to isolate these clones by using a bioassay that identifies signal sequences. Because all signal sequences perform the same function, directing the entry of proteins into the endoplasmic reticulum,[1] one assay can be used to isolate a variety of proteins.

Several techniques that identify cDNA clones encoding signal sequences have been published. The first, by Tashiro *et al.*[2] and others that followed,[3,4] used mammalian cells for the assay. This was followed by papers by Klein *et al.*[5] and Jacobs *et al.*,[6] who described the use of yeast cells for the assay. The yeast cell approaches use a genetic selection to identify signal sequences and consequently are more efficient and more sensitive. The selection is for secretion of invertase, a *Saccharomyces cerevisiae* protein that must be secreted for yeast to utilize sucrose and raffinose as carbon sources.[7] In our application mammalian cDNA fragments that are ligated adjacent to a truncated invertase gene provide the signals for protein synthesis and secretion.

Invertase is a propitious choice for a reporter protein for this application. Studies by Kaiser *et al.*[8] show that invertase secretion and activity are tolerant of many different amino-terminal extensions and that secretion of as little as 2.6% of wild-type levels gives growth detectable above background. Thus even cDNA invertase protein fusions that produce small amounts of invertase activity can be detected.

Description of Yeast Signal Sequence Trap

The signal sequence trap involves a process of several discrete steps that begins with the construction of a cDNA library and ends with identification of clones of interest. To facilitate the isolation and characterization

[1] P. Walter and A. E. Johnson, *Annu. Rev. Cell Biol.* **10,** 87 (1994).

[2] K. Tashiro, H. Tada, R. Heilker, M. Shirozu, T. Nakano, and T. Honjo, *Science* **261,** 600 (1993).

[3] M. Yokoyama-Kobayashi, S. Sugano, T. Kato, and S. Kato, *Gene* **163,** 193 (1995).

[4] W. C. Skarnes, J. E. Moss, S. M. Hurtley, and R. S. P. Beddington, *Proc. Natl. Acad. Sci. U.S.A.* **92,** 6592 (1995).

[5] R. D. Klein, Q. Gu, A. Goddard, and A. Rosenthal, *Proc. Natl. Acad. Sci. U.S.A.* **93,** 7108 (1996).

[6] K. A. Jacobs, L. A. Collins-Racie, M. Colbert, M. Duckett, M. Golden-Fleet, K. Kelleher, R. Kriz, E. R. LaVallie, D. Merberg, V. Spaulding, J. Stover, M. J. Williamson, and J. M. McCoy, *Gene* **198,** 289 (1999).

[7] M. Carlson, R. Taussig, S. Kustu, and D. Botstein, *Mol. Cell. Biol.* **3,** 439 (1983).

[8] C. A. Kaiser, D. Preuss, P. Grisafi, and D. Botstein, *Science* **235,** 312 (1987).

of large numbers of clones, we developed a novel invertase selection and adapted several protocols from the literature. These are presented in detail or referenced. Rationales and reagents are described in the accompanying comments.

Vector

The vector pSUC2T7M13ORI (Fig. 1) is designed to identify sequences in cDNA clones that mediate synthesis and transport of invertase into the

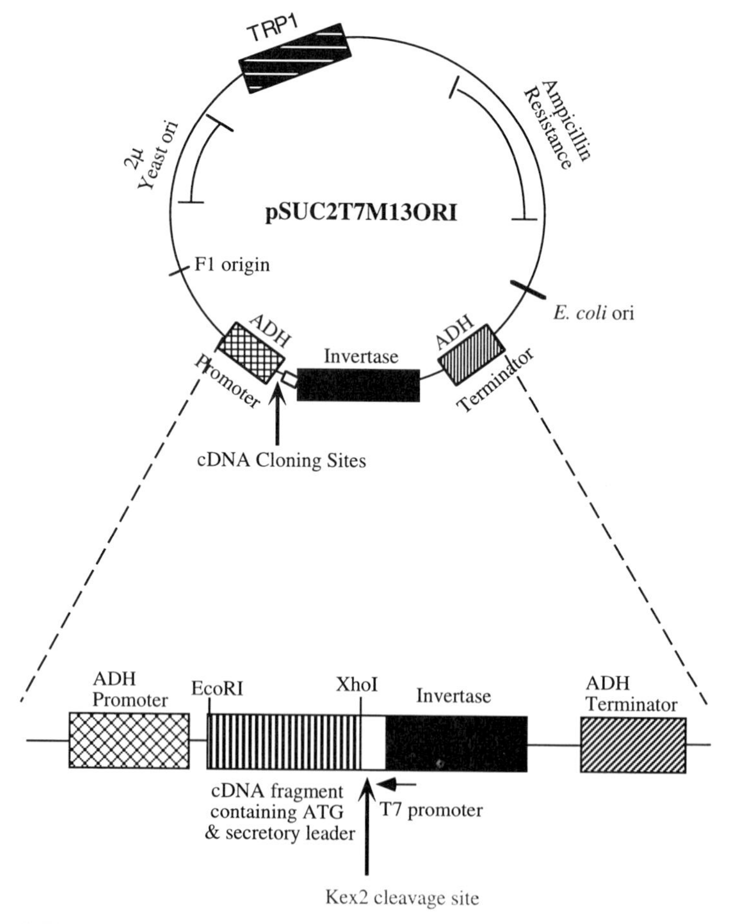

FIG. 1. The yeast signal sequence trap vector. [Reprinted from K. A. Jacobs, L. A. Collins-Racie, M. Colbert, M. Duckett, M. Golden-Fleet, K. Kelleher, R. Kriz, E. R. LaVallie, D. Merberg, V. Spaulding, J. Stover, M. J. Williamson, and J. M. McCoy, A genetic selection for isolating cDNAs encoding secreted proteins. *Gene* **198,** 289 (1999). Copyright 1997, with kind permission from Elsevier Science-NL, Sara Burgerhartstraat 25, 1055 KV Amsterdam, The Netherlands.]

endoplasmic reticulum and hence into the secretion pathway. It is derived from a shuttle vector that replicates episomally in both *Escherichia coli* and yeast and carries β-lactamase and tryptophan (Trp1)-selectable markers for *E. coli* and yeast, respectively.[9] The invertase gene is transcribed from the alcohol dehydrogenase promoter. The protein, however, is not expressed because it lacks a methionine to initiate translation and is not secreted because it lacks a signal sequence. These elements must be provided by inserted cDNA fragments. The selection for both a methionine and a signal sequence is necessary to minimize the frequency of false positives.[8]

cDNA Library Construction

Protocol. A random primed, directional cDNA library is synthesized from 5 μg of poly(A)$^+$ mRNA and size fractionated prior to cloning into the invertase selection vector.

We follow the protocol provided with the GIBCO-BRL (Gaithersburg, MD) Superscript cDNA synthesis kit, with the noted exceptions. Prime first-strand synthesis with random ninemers. After second-strand synthesis, ligate the cDNA to *Eco*RI–*Sfi*I adapters and phosphorylate the ligated cDNA with T4 polynucleotide kinase as described in the kit.

At this point the protocols diverge. Heat the phosphorylated, adapted cDNA at 65° for 10 min. Add 75 μl of water, 15 μl of New England BioLabs (Beverly, MA) restriction enzyme buffer 2, 80 units of *Xho*I, and incubate for 2 hr at 37°. Phenol–chloroform extract the reaction; ethanol precipitate from 2.5 *M* ammonium acetate, and resuspend the pellet in 18 μl of water.

Add 2 μl of 10× gel loading buffer [25% (v/v) Ficoll, 0.1% (w/v) bromphenol blue, 0.1% (w/v) xylene cyanol FF] and heat at 65° for 5 min. Purify 200- to 800-base pair (bp) cDNA following nondenaturing polyacrylamide gel electrophoresis.[10] Phenol–chloroform extract, ethanol precipitate, and ligate the size-fractionated cDNA into *Eco*RI- and *Xho*I-digested invertase selection vector.

Transform this material into *E. coli* by electroporation.[11] (We use ElectroMAX DH10B *E. coli* from GIBCO-BRL, but any high-efficiency cells should work.) Aim for libraries of greater than or equal to 2×10^7 independent clones (see Comments). Grow 1-liter cultures of transformed *E. coli*

[9] J. Gyuris, E. Golemis, H. Chertkov, and R. Brent, *Cell* **75,** 791 (1993).

[10] J. Chory, *in* "Current Protocols in Molecular Biology" (F. M. Ausubel, R. Brent, R. E. Kingston, D. D. Moore, J. G. Seidman, J. A. Smith, and K. J. Struhl, eds.), p. 2.7. Wiley Interscience, New York, 1995.

[11] C. E. Seidman, J. Sheen, and T. Jessen, *in* "Current Protocols in Molecular Biology" (F. M. Ausubel, R. Brent, R. E. Kingston, D. D. Moore, J. G. Seidman, J. A. Smith, and K. J. Struhl, eds.), p. 1.8. Wiley Interscience, New York, 1995.

in ampicillin. Isolate and purify the plasmid by alkaline lysis and CsCl–ethidium bromide equilibrium centrifugation.[12]

Comments. Using a directional cDNA library reduces the frequency of false positives. Random ninemers that contain a *Xho*I site to facilitate cloning into the vector are used to prime the first strand. These oligodeoxynucleotides (oligos) show no apparent priming bias.[6] The oligos are synthesized separately and mixed to give a solution of 25 μM of each oligo. Add 3 μl of the mixture for first-strand synthesis. The sequences (5′ to 3′) of the oligos are

CGATTGAATTCTAGACCTGCCTCGAGNNNNNNNNa
CGATTGAATTCTAGACCTGCCTCGAGNNNNNNNNc
CGATTGAATTCTAGACCTGCCTCGAGNNNNNNNNg
CGATTGAATTCTAGACCTGCCTCGAGNNNNNNNNt

The sequences (5′ to 3′) of the *Eco*RI–*Sfi*I adapters are

AATTCGGCCAAAGAGGCCTA
pTAGGCCTCTTTGGCCG

where "p" represents a phosphate. Anneal the adapters by mixing 2 nmol of each adapter, 2 μl of 1 M Tris (pH 8.0), 4 μl of 1 M $MgCl_2$, and water to a volume of 60 μl. Heat the mixture to 70° and cool it slowly to room temperature. Add 10 μl of the annealed adapters to the ligation reaction in place of the adapters that come with the GIBCO-BRL Superscript cDNA synthesis kit.

The cDNA is size fractionated to allow for 5′ untranslated regions of mRNA and to reduce the possibility of having the translation termination codon at the end of protein-coding regions prevent the expression of invertase. Because the functional size of the library is at least 15-fold smaller than the numerical size, a library of 2×10^7 or more clones is preferred. Large quantities of plasmid DNA are necessary because yeast transform poorly compared with *E. coli.*

Yeast Transformation and Invertase Selection

Protocol. To facilitate the identification of large numbers of signal sequences, we devised a new invertase selection. The selection is performed with the intent of isolating a single yeast colony that secretes invertase.

Transform the amplified cDNA library into yeast strain YTK12[6] by lithium acetate.[13] [Other strains that are available from the American Type

[12] J. S. Heilig, K. Lech, and R. Brent, *in* "Current Protocols in Molecular Biology" (F. M. Ausubel, R. Brent, R. E. Kingston, D. D. Moore, J. G. Seidman, J. A. Smith, and K. J. Struhl, eds.), p. 1.7. Wiley Interscience, New York, 1995.

[13] D. M. Becker and L. Guarante, *Methods Enzymol.* **194,** 182 (1991).

Culture Collection (ATCC, Rockville, MD), ATCC 96100 or ATCC 96099, may work equally well.] We obtain approximately 20,000 Trp^+ yeast per microgram of DNA. Spread 300,000 Trp^+ yeast per 100-mm plate directly onto tryptophan dropout plates containing sucrose (CMS/loD-W); grow the yeast for 3 days and then replicate onto rich medium plates containing raffinose and antimycin A (YPRAA) (Accutran, Schleicher & Schuell, Keene, NH). Grow the replicas for 7 days. Pick and streak single colonies onto YPRAA plates, and grow for an additional 7 days. Patch isolated colonies onto tryptophan dropout plates containing dextrose (CMD-W) and grow overnight. All incubations are at 30°.

Comments. Substituting sucrose for dextrose in the CMS/loD-W plates allows preferential growth of yeast that secrete invertase. The absence of tryptophan permits a higher plating density. The plating conditions also further reduce the growth of yeast that express only cytoplasmic invertase. Plating higher numbers of transformants suppresses the recovery of yeast that secrete invertase.

Our recipes for yeast media are derived from those presented in Treco and Lundblad.[14] CMD-W medium is complete minimal medium lacking tryptophan. CMS/loD-W medium is identical to CMD-W medium with two exceptions: the dextrose concentration is reduced to 0.1% (w/v) from 2% (w/v), and filter-sterilized sucrose is added to 2% (w/v).

YPRAA medium is identical to YPD medium with two exceptions: filter-sterilized raffinose, 2% (w/v) final concentration, replaces dextrose, and antimycin A is used to mimic anaerobic growth conditions to inhibit the growth of yeast that do not secrete invertase. To prepare 1 liter of YPRAA autoclave the other components, dissolved in 900 ml of water, with a stirring bar in the flask. Cool the medium to 60°. Add 100 ml of a 20% (w/v) filter-sterilized solution of raffinose. Add 1 ml of the antimycin A solution (below), final concentration 2 μg/ml, dropwise with vigorous stirring to prevent premature precipitation.

To prepare 100 mg of a 1000-fold concentrated stock of antimycin A (Sigma, St. Louis, MO), add 40 ml of isopropanol directly to the vial. Mix until it is in solution. Adjust to a final volume of 50 ml with ethanol. Aliquot to sterile tubes and store at −20°. *Note:* Antimycin A is toxic. Take proper safety precautions when handling. Wear a laboratory coat, gloves, particle mask, and safety glasses. Handle the powder and solution in a fume hood with proper precautions.

[14] D. A. Treco and V. Lundblad, *in* "Current Protocols in Molecular Biology" (F. M. Ausubel, R. Brent, R. E. Kingston, D. D. Moore, J. G. Seidman, J. A. Smith, and K. J. Struhl, eds.), p. 13.1. Wiley Interscience, New York, 1995.

Plasmid Isolation and Amplification

Protocol. Plasmid DNA is isolated from the yeast colonies and partially sequenced to identify cDNA clones of interest.

To each well of a deep-well microtiter plate (Beckman, Fullerton, CA) add 300 μl of CMD-W medium containing 15% (v/v) glycerol. Transfer yeast from the individual patches described above with sterile pipette tips into the wells. Seal the plates with pressure-sensitive film (Falcon, Becton Dickinson Labware, Lincoln Park, NJ) and incubate overnight at 30°. After removing an aliquot for isolating plasmid DNA, store the yeast overnight cultures at −80°.

Isolate plasmid DNA as described by Baldini *et al.*[15] Add 20 μl of lysis buffer [1% Triton X-100, 20 m*M* Tris (pH 8.9), 2 m*M* EDTA] to 96-well polymerase chain reaction (PCR) plates (Robbins Scientific, Sunnyvale, CA). Transfer 5 μl of each overnight culture to the PCR plate. Cover the plate with a rubber mat (Perkin-Elmer, Norwalk, CT) and incubate on a PCR thermal cycler for 10 min at 100°. Centrifuge the plates at 1500 rpm for 7 min.

To amplify the DNA add 38 μl of PCR cocktail to each well of a clean 96-well PCR plate (Robbins Scientific). Transfer 2 μl of DNA template from the lysis plate to the corresponding wells. Cover the plate with a rubber mat (Perkin-Elmer). Amplify for 30 cycles: 94° for 1 min, 55° for 45 sec, 68° for 5 min. Hold at 4°. These are the conditions for an M.J. Research (Watertown, MA) Tetrad thermal cycler or a Perkin-Elmer 9600.

Comments. To amplify 100 yeast cultures prepare a PCR cocktail as follows:

Reagent	Volume (μl)
10× PCR buffer (Perkin-Elmer)	400
10 m*M* Tris (pH 8.0), 0.1 m*M* EDTA	230
dNTPs, 1.25 m*M* (Pharmacia, Piscataway, NJ)	400
5′ Primer (3 μM)	400
3′ Primer (3 μM)	400
H_2O	1950
AmpliTaq polymerase (Perkin-Elmer)	20
	3800

The sequence (5′ to 3′) of the primer for the PCR and DNA sequencing reaction from the 3′ side of the cDNA insert is GGTGTGAAGTGGAC CAAAGGTCTA.

[15] A. Baldini, M. Ross, D. Nizetic, R. Vatcheva, E. A. Lindsay, H. Lehrach, and M. Siniscalco, *Genomics* **14,** 81 (1992).

The sequence (5′ to 3′) of the primer for the PCR and DNA sequencing reaction from the 5′ side of the cDNA insert is CCTCGTCATTGTTCTC GTTCCCTT.

Plasmid Purification and DNA Sequencing

Purify the DNA by solid-phase reversible immobilization (SPRI).[16] Transfer the PCR products to a 96-well flex plate (serocluster "v" vinyl plate; Costar, Cambridge, MA). Transfer 10 μl of beads to each well. Add 50 μl of 20% polyethylene glycol (PEG) 8000–2.5 *M* NaCl. Shake the plate gently and incubate at room temperature for 10 min. Place the PCR plate on a magnet (Dynal, Great Neck, NY) for 2 min. Aspirate off the liquid. Wash the particles twice with 150 μl of 70% ethanol. Air dry for 2 min. Resuspend the beads in 20 μl of 10 m*M* Tris-acetate, pH 7.8. Place the PCR plate on the magnet for 2 min and transfer the liquid to another 96-well plate for DNA sequencing. Store the plates at $-20°$.

Sequence the purified DNA templates using standard methods.[17–20] We sequence 2 μl of each template and analyze the reactions on an Applied Biosystems (Foster City, CA) 377 DNA sequencer.

Comments. To purify DNA templates from 100 PCRs, prepare the magnetic beads (PerSeptive Diagnostics, Framingham, MA) as follows: Transfer 1 ml of beads to a 1.5-ml centrifuge tube. Wash three times with 0.5 *M* EDTA. Resuspend in 1.0 ml of 10 m*M* Tris-acetate, pH 7.8.

For the first sequencing reaction we sequence the PCR products from the 3′ side, using the same primer that was used in the PCR. This DNA sequence is commonly referred to as the "first-pass" sequence. For complete sequence or "second-pass" sequence we sequence from both sides using the PCR primers.

Computational Analysis

First-pass sequence is accurate enough to determine if the clone is identical, homologous, or distinct from sequences in the public databases. Compare the clone sequences with sequences in GenBank and GenPept successively, utilizing the programs BLASTN,[21] BLASTX,[21] and FASTA.[22]

[16] M. M. DeAngelis, D. G. Wang, and T. L. Hawkins, *Nucleic Acids Res.* **23,** 4742 (1995).

[17] F. Sanger, S. Niklen, and A. R. Coulson, *Proc. Natl. Acad. Sci. U.S.A.* **92,** 6339 (1995).

[18] V. Murray, *Nucleic Acids Res.* **17,** 8889 (1989).

[19] M. Craxton, *in* "Methods: A Companion to Methods in Enzymology" (J. N. Abelson and M. I. Simon, eds.), Vol. 3, p. 20. Academic Press, San Diego, California, 1991.

[20] T. Hawkins, Z. Du, N. D. Halloran, and R. K. Wilson, *Nucleic Acids Res.* **23,** 4742 (1995).

[21] S. F. Altschul, W. Gish, W. Miller, E. W. Myers, and D. J. Lipman, *J. Mol. Biol.* **215,** 403 (1990).

[22] W. R. Pearson and D. J. Lipman, *Methods Enzymol.* **183,** 63 (1990).

Obtain second-pass sequence for clones of interest. Repeat the database searches. Analyze the clones for a reading frame into invertase and for a signal peptide with the program SigCleave.[23] In most cases these are readily identified. False positives are identified by the presence of open reading frames encoding fewer than 15 amino acids, homology to a nonsecreted protein, or by identity or homology with a repetitive sequence. *Alu* sequences are mostly present in noncoding regions of genes.

Applications

Partial lists of genes isolated by this approach have been published.[5,6] We have isolated cDNA clones encoding secreted proteins, type 1 transmembrane proteins, type 2 transmembrane proteins, multipass transmembrane proteins, and proteins located in intracellular organelles. The individual proteins have different functions and structures, attesting to the versatility of the approach.

A discussion of the parameters that could limit the scope of the approach has been presented.[6] The true positive rate is approximately 80% and varies with the source of RNA. A more relevant number is the false-negative rate—signal peptides that should be isolated but are not. One way to approach this question is to determine if the yeast signal sequence trap has isolated all of the cDNAs known to encode signal sequences. We assembled this list by searching release 34.0 of the SWISS-PROT database[24] for proteins of human origin annotated as having signal sequences. We then compared proteins on this list to our database, using the program TBLASTN.[21] To date we have isolated approximately 25% of the proteins on this list.

The relevance of this percentage is not readily apparent. Failure to isolate a clone may be a consequence of events unrelated to yeast biology; for example, the mRNA may not be expressed in the sources we have sampled. We expect this percentage to increase as we screen more RNA sources and incorporate cDNA subtraction techniques into the process. This list does permit useful inquiries about the fidelity and biases of the yeast signal sequence trap. In one inquiry we asked whether the yeast clones contained the signal sequences of their cognate SWISS-PROT entries or whether another part of these cDNAs mediated secretion of invertase

[23] Program Manual for the Wisconsin Package, version 8 (August, 1994), Genetics Computer Group, 575 Science Drive, Madison, WI 53711. Program Manual for the EGCG Package, Peter Rice, The Sanger Centre, Hinxton Hall, Cambridge, CB10 1RQ, England.

[24] A. Bairoch and R. Apweiler, *Nucleic Acids Res.* **24,** 21 (1996).

(Fig. 2). The minimum aligned residue is the most amino-terminal amino acid in the yeast clone that aligned with the SWISS-PROT entry in the TBLASTN search. For 75% of these proteins this was amino acid 62 or less, suggesting that these clones did contain the corresponding signal sequence. The analysis is not precise because clones are sequenced from the 3′ end to generate first-pass sequence data that often does not include the entire insert.

The data (Fig. 2) also show that for some clones the insert clearly does not contain the natural signal sequence. These are represented by clones with a minimum aligned residue of 600, for example. Analysis of a representative sample of these revealed the presence of a methionine apposed to a transmembrane segment that presumably mimicked a natural signal sequence. These clones are isolated much less frequently than are clones encoding natural signal sequences.

We also compared the SigCleave scores of the SWISS-PROT entries identified by the invertase selection with the remaining SWISS-PROT entries not yet identified but annotated as described above. The distribution of SigCleave scores for both sets is presented in Fig. 3. They are superimposable, suggesting that the properties measured by SigCleave are not a source of bias.

Because genes encoding secreted proteins represent only a fraction, perhaps 10%, of the total genome and a smaller fraction of expressed

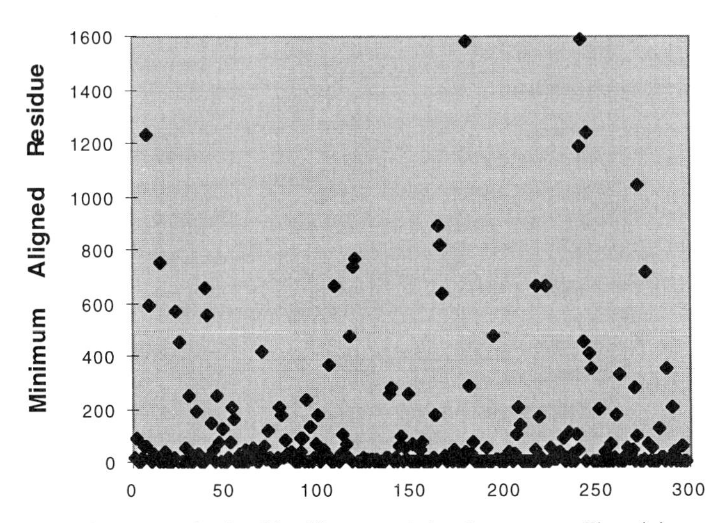

FIG. 2. The invertase selection identifies natural signal sequences. The minimum aligned residue (◆) of each identified gene is plotted against the individual genes. For 25% of the genes the minimum aligned residue is 1; for 50% it is 14 or less; for 75% it is 62 or less.

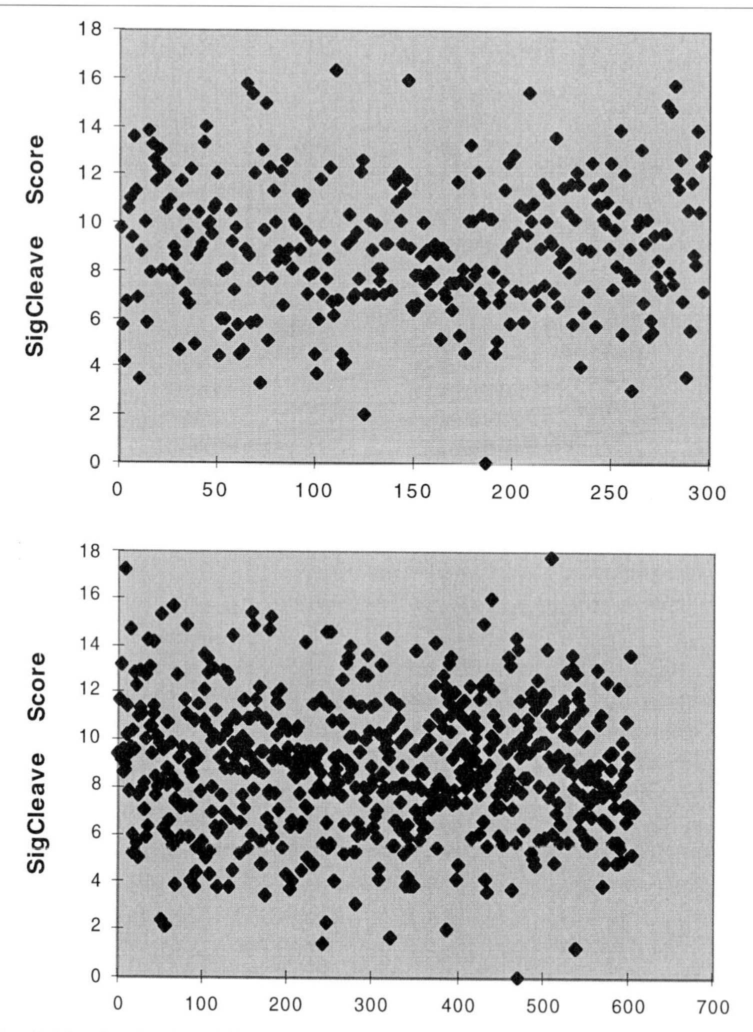

FIG. 3. The distribution of SigCleave scores of proteins identified by the invertase selection (top) is identical to the scores of proteins not yet identified (bottom). The SigCleave score of individual proteins (◆) is plotted against the individual genes.

mRNAs, the yeast signal trap is more efficient at identifying them than is random sequencing. In addition, because the yeast signal trap isolates the 5′ end of mRNAs, it provides an important advantage in isolating full-length cDNAs. The technique is particularly useful at isolating genes for which no bioassays exist. Having these genes in hand should make it easier to determine function. In addition, because signal sequences possess features conserved throughout evolution, we anticipate that the procedure can be used to isolate genes encoding signal sequences from most extant organisms.

Acknowledgments

This work was supported by Genetics Institute, Inc.

[27] The Signal Sequence Trap Method

By KEI TASHIRO, TOMOYUKI NAKAMURA, and TASUKU HONJO

Introduction

To understand the mechanisms for development and maintenance of multicellular organisms, it is of great importance to identify proteins involved in the intercellular signal trunsduction. Most of the molecules involved in intercellular signaling or cell adhesion are secreted or are membrane-anchored proteins. Many of these proteins contain a signal sequence or leader peptide at the N-terminus of the premature form.[1] The traditional strategy for cloning these genes requires a functional assay to detect the specific function of each molecule. Most cytokines have been cloned or purified using assay systems depending on growth-stimulating activities because the growth-stimulating activities tend to be more easily detected than are other activities such as differentiation and growth arrest.

To understand the roles of each type of cell in multicellular organisms, it is also of great importance to identify cell surface proteins, which can be utilized as specific surface markers for each cell lineage. To establish a general cDNA cloning method for cytokines, hormones, neuropeptides, their receptors, adhesion molecules, transporters, and so-called surface markers, we have developed a cloning strategy for selecting cDNA fragments encoding N-terminal signal sequences.[2,3]

This method, named the signal sequence trap method, has turned out to be an efficient method for isolating 5′ cDNA fragments from secreted

[1] D. D. Sabatini, G. Kreibich, T. Morimoto, and M. Adesnik, *J. Cell Biol.* **92,** 1 (1982).
[2] K. Tashiro, H. Tada, R. Heilker, M. Shirozu, T. Nakano, and T. Honjo, *Science* **261,** 600 (1993).
[3] K. Tashiro, T. Nakano, and T. Honjo, *Methods Mol. Biol.* **69,** 203 (1996).

Copyright © 1999 by Academic Press
All rights of reproduction in any form reserved.

0076-6879/99 $30.00

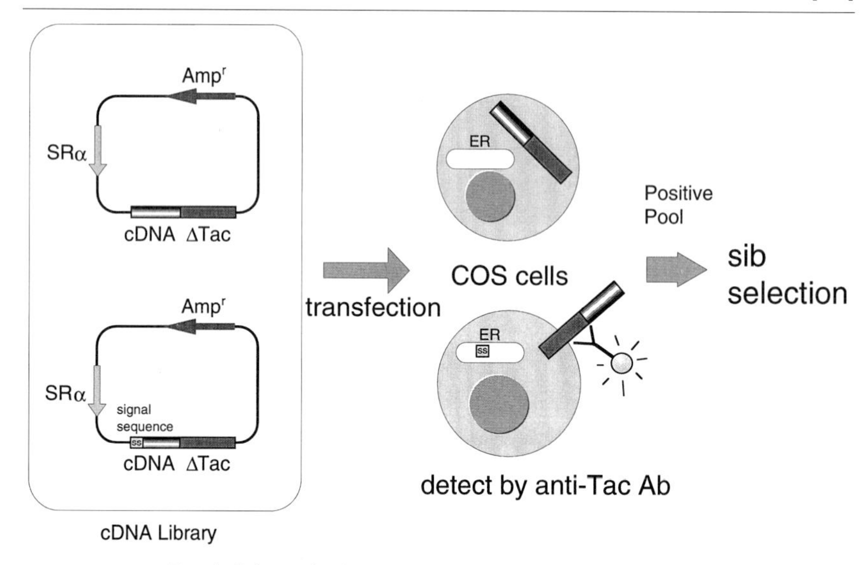

FIG. 1. Schematic view of the signal sequence trap method.

or transmembrane proteins[2–7] (see Fig. 1). We have already obtained a number of cDNA clones of putative cytokines, receptors, or adhesion molecules such as SDF-1 (stromal cell-derived factor 1),[2] SDF-5,[6] and lymphotoxin-β receptor.[4] Interestingly, many of them seem to bear no (or, if any, weak) growth-stimulating activities. For example, SDF-1 is a potent chemoattractant factor for T lymphocytes[8] and hematopoietic stem cells,[9] and has strong anti-HIV-1 (human immunodeficiency virus type 1) virus infection activity as SDF-1 is the ligand for the HIV-1 coreceptor, named

[4] T. Nakamura, K. Tashiro, M. Nazarea, T. Nakano, S. Sasayama, and T. Honjo, *Genomics* **30,** 312 (1995).

[5] T. Hamada, K. Tashiro, H. Tada, J. Inazawa, M. Shirozu, K. Shibahara, T. Nakamura, T. Nakano, and T. Honjo, *Gene* **30,** 112 (1996).

[6] M. Shirozu, H. Tada, K. Tashiro, T. Nakamura, N. D. Lopez, M. Nazarea, T. Hamada, T. Sato, T. Nakano, and T. Honjo, *Genomics* **37,** 273 (1996).

[7] K. Mori, Y. Ogawa, M. Tamura, K. Ebihara, T. Aoki, S. Muro, S. Ozaki, I. Tanaka, K. Tashiro, and T. Honjo, *FEBS Lett.* **401,** 218 (1997).

[8] C. C. Bleul, R. C. Fuhlbrigge, J. M. Casasnovas, A. Aiuti, and T. A. Springer, *J. Exp. Med.* **184,** 1101 (1997).

[9] A. Aiuti, I. J. Webb, C. Bleul, T. Springer, and J. C. Gutierrez-Ramos, *J. Exp. Med.* **185,** 111 (1997).

CXCR-4.[10,11] However, SDF-1 exerts rather weak growth-stimulating activity on pre-B cells, which can be measured only when another cytokine, interleukin 7 (IL-7) is added.[12] Some clones obtained by the signal sequence trap method, such as SDR-1, seem to be good surface markers for limited cell lineages.[12a] Not only secreted proteins and plasma membrane proteins, but also proteins located in the endoplasmic reticulum (ER), the Golgi apparatus (GA), lysosomes, and glycosyl-phosphatidylinositol (GPI)-anchored proteins can be cloned by the signal sequence trap method.[13]

Perspectives for Signal Sequence Trap

pcDLSRα-Tac(3′) Vector

The epitope-tagging expression plasmid vector used in the signal sequence trap method is the pcDLSRα-Tac(3′) vector,[2,14] which directs the cell surface expression of fusion proteins with Tac (human CD25, α chain of the human interleukin 2 receptor) when cDNAs with N-terminal signal sequences are cloned in-frame, in the correct orientation.[15] The Tac epitope-tagged fusion protein expressed on the plasma membrane is easily detected with anti-human CD25 antibodies[16] (see Fig. 1).

Unidirectional cDNA Library Construction for Signal Sequence Trap

The first strand of cDNA is synthesized with a *Sac*I linker primer that contains random nonamers; deoxyadenine (dA) tails are added at the 3′ end of the first-strand cDNA, and second-strand synthesis is carried out with an *Eco*RI linker primer that contains polydeoxythymine [poly(dT)] (see Fig. 2). The product is size fractionated to 350–600 bp by agarose gel

[10] C. C. Bleul, M. Farzan, H. Choe, C. Parolin, I. Clark-Lewis, J. Sodroski, and T. A. Springer, *Nature (London)* **382,** 829 (1996).

[11] E. Oberlin, A. Amara, F. Bachelerie, C. Bessia, J.-L. Virelizier, F. Arenzana-Seidedos, O. Schwartz, J.-M. Heard, M. Loetscher, and M. Baggiolini, B. Moser, *Nature (London)* **382,** 833 (1996).

[12] T. Nagasawa, H. Kikutani, and T. Kishimoto, *Proc. Natl. Acad. Sci. U.S.A.* **91,** 2305 (1994).

[12a] N. D. Lopez, A. Kinoshita, M. Taniwaki, H. Tada, M. Shirozu, T. Nakano, K. Tashiro, and T. Honjo, *Biomed. Res.* in press.

[13] D. Yabe, T. Nakamura, N. Kanazawa, K. Tashiro, and T. Honjo, *J. Biol. Chem.* **272,** 18232 (1997).

[14] Y. Takebe, M. Seiki, J. Fujisawa, P. Hoy, K. Yokota, K. Arai, M. Yoshida, and N. Arai, *Mol. Cell. Biol.* **8,** 466 (1988).

[15] T. Nikaido, A. Shimizu, N. Ishida, H. Sabe, K. Teshigawara, M. Maeda, T. Uchiyama, J. Yodoi, and T. Honjo, *Nature (London)* **311,** 631 (1984).

[16] T. Uchiyama, S. Broder, and T. A. Waldman, *J. Immunol.* **126,** 1393 (1981).

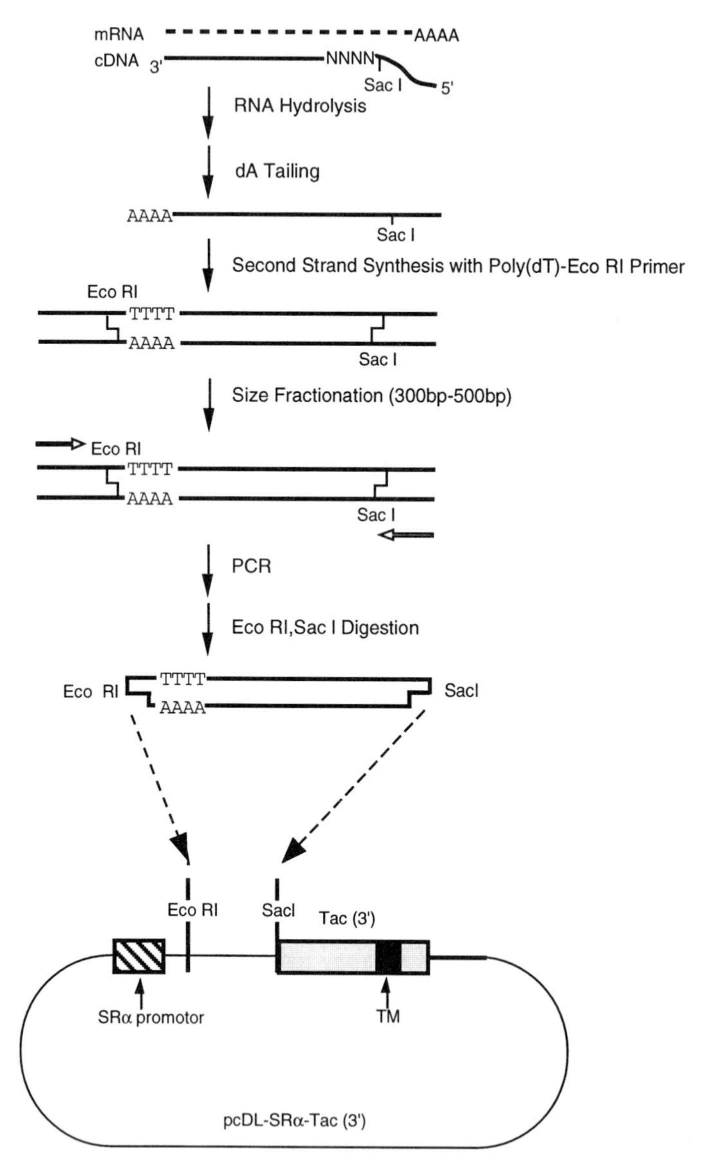

FIG. 2. cDNA library construction for the signal sequence trap method.

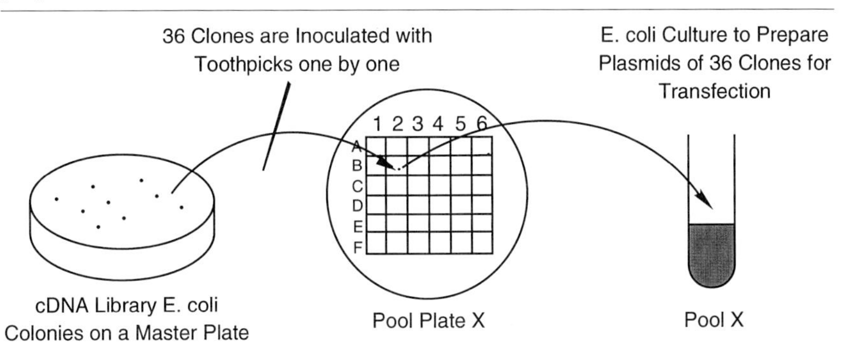

FIG. 3. Preparation for first screening in sib-selection.

electrophoresis, and amplified by the polymerase chain reaction (PCR).[17] The amplified fragments are digested with *Eco*RI and *Sac*I, purified by agarose gel electrophoresis again, and inserted into the pcDL-SRα-Tac(3′) vector in the same orientation with the Tac cDNA.

Assay Procedures

Using the resulting expression cDNA library, sib screening[18] should be done (see Fig. 3). After *Escherichia coli* transformation, 36 individual colonies are plated on a 9-cm LB–agar plate in a matrix format (6 rows by 6 lines) and assigned to one pool. COS-7 cells[19] are transfected with plasmid DNAs of each pool, and fusion proteins expressed on the cell surface are microscopically detected by immunostaining with anti-human CD25 antibodies. If a pool contains any positive clone plasmid, 12 smaller pools consisting of 6 individual clones in each row or line of the matrix are tested until a single positive clone can be determined. The nucleotide sequence information from positive clones is used to identify hydrophobic amino acid residues that are the core of the N-terminal signal sequence. As well, a homology search of the database is done to compare the sequence with known proteins. Full-length cDNAs of positive clones can be obtained from an oligo(dT)-primed cDNA library under stringent hybridization conditions, using trapped cDNA fragments that encode the N-terminal signal sequence as a probe.

[17] R. K. Saiki, S. Scharf, F. Faloona, K. B. Mullis, G. T. Horn, H. A. Erlich, and N. Arnheim, *Science* **330,** 1350 (1985).

[18] J. Sambrook, E. F. Fritsch, and T. Maniatis, "Molecular Cloning: A Laboratory Manual." Cold Spring Harbor Laboratory Press, Cold Spring Harbor, New York, 1989.

[19] Y. Gluzman, *Cell* **23,** 175 (1981).

Signal Sequence Trap cDNA Library Construction

First-Strand Synthesis and RNA Hydrolysis

The priming procedure using the *Sac*I linker primer that contains random nonamer lessens the probability of failure in cDNA library construction, compared with the older version using *Sac*I adapter ligation,[4] probably because the *Sac*I adapters are added to both ends of the cDNA in the latter method, which might interrupt the PCR in the next step. cDNA library construction should be started with 100 ng to 5 μg of poly(A)$^+$ RNA. Contamination with rRNA results in an increase in the number of clones that must be screened. The probability of positive clones is typically 1 in 150 to 1 in 300; it can be 1 in 100 to 1 in 1000, depending on the quality of poly(A)$^+$ RNA, the random hexamer : poly(A)$^+$ RNA ratio, the elongation of first-strand cDNA, the rate of nonspecific priming in second-strand synthesis, the size fraction, and the cell source.

Protocol

1. Mix, with caution, 1 μg of poly(A)$^+$ RNA and 10 pmol of *Sac*I linker primer that contains random nonamers (5′-GAG ACG GTA ATA CGA TCG ACA GTA GGA GCT CNN NNN NNN N) in diethylpyrocarbonate (DEPC)-treated H_2O in a final volume of 11 μl. Stir gently and centrifuge (15,000*g*, 5 sec, RT). Incubate at 70° for 10 min, and quickly chill on ice. Add 4 μl of 5× reverse transcription buffer [0.25 *M* Tris-HCl (pH 8.3), 0.375 *M* KCl, 15 m*M* $MgCl_2$], 2 μl of 0.1 *M* dithiothreitol (DTT), and 1 μl of 10 m*M* dNTP mix solution (1 m*M* dATP, 1 m*M* dTTP, 1 m*M* dCTP, 1 m*M* dGTP), and mix gently.
2. Immediately add 2 μl of SuperScript II reverse transcriptase (GIBCO-BRL, Gaithersburg, MD) (200 U/μl), gently stir, and centrifuge (15,000*g*, 5 sec, RT). Incubate at 42° for 60 min.
3. Add 1 μl of 0.5 *M* EDTA and gently mix. Add 38 μl of H_2O, 15 μl of 1 *N* NaOH, gently mix, and centrifuge (15,000*g*, 5 sec, RT). Incubate at 70° for 30 min. Add 4 μl of 2 *M* Tris (pH 7.4), 14.7 μl of 1 *N* HCl, and 109 μl of TE [10 m*M* Tris-HCl (pH 8.0), 0.1 m*M* EDTA].
4. Add 600 μl of NaI solution (GeneClean II; Bio 101, La Jolla, CA) and 10 μl of glass milk. Put the mixture on ice for 10 min, and then centrifuge at full speed for 5 sec. Wash the pellets with 700 μl of New Wash three times. Elute the single-stranded cDNA in 50 μl of TE. Add 25 μl of 7.5 m ammonium acetate, 3 μl of glycogen (10 mg/ml), and 150 μl of cold ethanol. Put the mixture on dry ice for 30 min, and centrifuge at full speed for 20 min. Dissolve the pellet in 16 μl of TE. One microliter of first-strand cDNA

can be checked by alkaline agarose gel electrophoresis, if a tracing amount of [^{32}P]dCTP is added.

dA Tailing, Second-Strand Synthesis, and Size Selection

dA tailing gives us better results than dC tailing. If dC tailing is too long, there will be some difficulties in determination of base sequences, because poly(dG) longer than 25 bases prevents sequencing reaction. The tailing reaction, using 1/2× reverse transcription buffer, which contains as little as 1.5 m*M* $MgCl_2$, tends to result in a rather constant (15–40 nucleotides) length of the dA tail.[20] Cacodylate has been recommended for the tailing reaction so far.[18,21,22] However, it is difficult to control the number of nucleotides added, using the classic tailing buffer containing cacodylate.

To enrich 5′ ends of cDNAs, the frequency of nonspecific priming during the second-strand synthesis should be decreased. That is why we chose the most heat-durable polymerase, reverse transcriptase, and adopted a so-called hot start PCR.

The 350- to 600-bp fraction should be tested first for the following two reasons: (1) longer cDNAs, which contain the whole coding regions, do not generate fusion proteins owing to the appearance of stop codons; and (2) fractions shorter than 250 bp give many pseudopositive clones. When we tested short fractions of cDNAs, pseudopositive clones such as cDNA of ribosomal RNA were trapped.

Protocol

1. Heat the cDNA at 70° for 5 min and chill it on ice. Mix 15 μl of first-strand cDNA with 2 μl of 5 m*M* dATP and 2 μl of 5× reverse transcription buffer, which makes the final 1/2× reverse transcription buffer. Add 1 μl of terminal deoxynucleotidyl transferase (20 U/μl), gently stir, and centrifuge (15,000*g*, 5 sec, RT). Incubate at 37° for 10 min, then at 70° for 10 min.

2. Keep the tube at 46° during the assembly of the reaction. Add 2 μl of ESTN primer (5′-CCG CGA ATT CTG ACT AAC TGA TTT TTT TTT TTT TTT TTN N) (12.5 μ*M*), 8 μl of 5× reverse transcription buffer, which makes a final 4/5× reverse transcription buffer. Add 2.5 μl of 0.1 *M* DTT, 5 μl of 10 m*M* dNTP, and 10.5 μl of H_2O, mix gently, and centrifuge

[20] F. Grosse and A. Manns, *Methods Mol. Biol.* **16,** 95 (1993).

[21] D. M. Schuster, G. W. Buchman, and A. Rashtchian, *Focus* **14,** 46 (1993).

[22] F. U. Ausubel, R. Breut, R. E. Kingston, D. D. Moore, J. G. Seidman, J. A. Smith, and K. Struhl, "Current Protocol in Molecular Biology." Greene Publishing Associates and Wiley-Interscience, New York, 1987.

(15,000*g*, 5 sec, RT). Add 2 μl of SuperScript II (200 U/μl), mix gently, and centrifuge again. Incubate at 42° for 60 min.

3. Add 50 μl of TE and 100 μl of phenol–CIAA [phenol equilibrated with 0.1 *M* Tris-HCl, (pH 8)–chloroform–isoamyl alcohol (25:24:1, v/v/v)]; mix on the vortexer. Centrifuge (15,000*g*, 10 min, RT) to separate the phases, then add 50 μl of 7.5 *M* ammonium acetate and 2.5 vol of cold ethanol to the aqueous phase. Set on dry ice for 30 min, and centrifuge at full speed for 20 min. Rinse the pellets with 950 μl of cold 75% ethanol, and air dry. Dissolve the DNA in 10 μl of H_2O. Prepare a 1.4% agarose–TAE (0.04 *M* Tris acetate, 1 m*M* EDTA) gel containing ethidium bromide (30 ng/ml) and load the 10 μl of DNA solution with 4 μl of 6× loading buffer (0.25% bromphenol blue, 30% glycerol in H_2O). Load the DNA size markers, keeping them at least one slot away from the cDNA. Run the DNAs on the gel.

DNA Recovery from Agarose Gel Blocks with Ultrafree-MC Filter Unit

The DNA recovery method using an Ultrafree-MC filter unit (Millipore, Bedford, MA) is recommended, because the loss of ~400-bp DNA fragments and the contamination of small DNA fragments, some of which are generated by dA tailing onto the *Sac*I linker primer, are minimized.

Protocol

1. Cut out agarose gel blocks containing the required fractions (e.g., 350–600, 600–900, and 900–1200 bp), and put each block into the upper cup of an Ultrafree-MC 0.45-μm pore size filter unit (Millipore). The weight of blocks in one tube should be less than 400 μg. If the weight of a block is more than 400 μg, divide it into two or more blocks. Freeze it at −20° for 20 min, and thaw it at room temperature. Centrifuge at 10,000 rpm for 10 min.

2. Transfer the liquid dropped into the lower tube to the upper cup of a new Ultrafree-MC 30,000NMWL filter unit (Millipore). Centrifuge at 4000 rpm for 15 min. Add 200 μl of H_2O and centrifuge at 4000 rpm for 10 min, or until the remaining liquid in the upper cup comes to be less than 3 μl. Add 40 μl of TE and try to dissolve all of the DNA on the membrane. Transfer all of the liquid on the membrane to a new tube. This DNA solution is referred to as the original PCR template solution.

Polymerase Chain Reaction Amplification and Insert cDNA Digestion with SacI and EcoRI

The Mg^{2+} concentration is critical in this step. When the Mg^{2+} concentration is too low, only a limited amount of DNA can be amplified, whereas if the Mg^{2+} concentration is too high, nonspecific DNA amplification occurs. Thus, optimization is recommended between 1 and 3 m*M* Mg^{2+}. Ex Taq DNA polymerase (Takara Shuzo, Kyoto, Japan) is recommended because of its high elongation efficiency and relatively high fidelity. *Pfu* DNA polymerase (Stratagene, La Jolla, CA) and Deep Vent DNA polymerase (New England BioLabs, Beverly, MA) are also worth trying, although elongation efficiency is lower. There is no necessity to focus on achieving the highest accuracy in DNA amplification in this step, because the full-length cDNA for corresponding signal sequence trap (SST) clones must be cloned later.

Protocol

1. Make a series of dilutions of the original PCR template solution (e.g., 1:10, 1:100, 1:1000) to check the quality of the cDNA library.
2. Mix 1 μl of the original or a diluted PCR template solution, 2.5 μl of 10× PCR buffer [100 m*M* Tris-HCl (pH 8.3), 500 m*M* KCl, 20 m*M* $MgCl_2$, 0.01% (w/v) gelatin], 4 μl of 2.5 m*M* dNTP, 1 μl of RSP primer (5′-GAC GGT AAT ACG ATC GAC AGT AGG) (25 μ*M*), 1 μl of ETP primer (5′-CCG CGA ATT CTG ACT AAC TGA TT) (25 μ*M*), 17.2 μl of H_2O, and 0.3 μl of Ex Taq DNA polymerase (5 U/μl) for a total volume of 25 μl. Mix on the vortexer and centrifuge (15,000*g*, 5 sec, RT). Overlay a drop of mineral oil.
3. Start the thermal cycle: 94° for 4 min, followed by 25 cycles of denaturation for 30 sec at 94°, annealing for 1 min at 56°, and synthesis for 3 min at 72°, and then followed by synthesis for 5 min at 72°.
4. Prepare a 1.6% agarose–TAE gel. Load 5 μl of each PCR product with 2 μl of 6× loading buffer. Run the electrophoresis, and observe the result under UV. Make sure smears of the DNA appear not only in the reaction using the original PCR template solution but also in the reaction using the diluted solutions. When smearing DNA fragments are observed in the 1:1000 diluted reaction, even if it is faint, there will be little chance to trap exactly the same clones generated by PCR. If DNA appears in the 1:100 diluted reaction, the experiment can proceed.
5. Prepare four tubes of reactions generated from the original PCR template solution; combine them into 100 μl of PCR product. Add 100 μl of phenol–CIAA and mix on the vortexer. Centrifuge as described above to separate the phases, then add 50 μl of 7.5 *M* ammonium acetate and 2.5 vol of cold ethanol to the aqueous phase. Set it on dry ice for 30 min, and centrifuge it at full speed for 20 min. Rinse the pellets with 950 μl of 75% ethanol, and air dry. Dissolve the DNA in 85 μl of H_2O.
6. Mix 85 μl of the recovered DNA solution, 10 μl of 10× low-salt restriction enzyme buffer [100 m*M* Tris-HCl (pH 7.5), 100 m*M* $MgCl_2$, 10 m*M* dithiothreitol], and 5 μl of *Sac*I (10 U/μl). Incubate at 37° for 3 hr.

Add 12 μl of 10× medium salt restriction enzyme buffer [100 m*M* Tris-HCl (pH 7.5), 100 m*M* $MgCl_2$, 10 m*M* dithiothreitol, 500 m*M* NaCl], 3 μl of H_2O, and 5 μl of *Eco*RI (10 U/μl). Incubate at 37° for 3 hr. Add 120 μl of phenol–CIAA and vortex. Centrifuge as described above to separate the phases, then add 60 μl of 7.5 *M* ammonium acetate and 2.5 vol of cold ethanol to the aqueous phase. Set it on dry ice for 30 min, and centrifuge at full speed for 30 min. Rinse the pellets with 950 μl of 75% ethanol; air dry. Dissolve the DNA in 20 μl of H_2O.

7. Prepare a 1.4% agarose–TAE gel containing ethidium bromide (30 ng/ml), and load the 20 μl of DNA solution with 5 μl of 6× loading buffer. Load the DNA size markers, and be sure to keep them at least one slot away from the cDNA. Run the DNA on the gel and cut out the required fractions (e.g., 350–600, 600–900, and 900–1200 bp). Cut out the agarose gel containing the DNA smear and recover the DNA with Ultrafree-MC filters as described above. Dissolve the DNA in 20 μl of H_2O.

Vector Preparation

For vector preparation, negative control plasmid pcDLSRα-hRAR(5′)-Tac(3′) should be used. If the positive control plasmid pcDL SRα-hG-CSF(5′)-Tac(3′) were used, there might be the potential risk that the small amount of undigested plasmid would be enriched during the selection procedure.

Protocol

1. Assemble 3 μl of 10× low-salt restriction enzyme buffer, 3 μg of negative control plasmid pcDLSRα-hRAR(5′)-Tac(3′), and H_2O to 27 μl; mix and centrifuge as described above. Add 3 μl of *Sac* I (10 U/μl), mix, and centrifuge again. Incubate at 37° for 3 hr.

2. Add 4 μl of 10× medium-salt restriction enzyme buffer, 3 μl of H_2O, and 3 μl of *Eco*RI (10 U/μl); mix and centrifuge as before. Incubate at 37° for 3 hr, then incubate at 65° for 5 min and quickly chill on ice.

3. Run the mixture on a 1.0% agarose–TAE gel. Two bands should appear: the 291-bp band representing the retinoic acid receptor insert, and the 4.5-kb digested vector band. Recover the digested vector DNA with an Ultrafree-MC filter unit as described above. Dissolve in 15 μl of TE.

4. Run 1 μl of the preceding DNA solution on a 0.9% agarose–TAE gel and estimate the DNA concentration.

Vector–Insert Ligation and Transformation

When the number of clones that contain reasonable insert DNA is small, try to remove the small DNA fragments, generated by dA tailing onto the *Sal*I linker primer, using the Ultrafree-MC filter unit.

Protocol

1. Add 1 μl of the insert DNA, 100 ng of the digested vector, 1 μl of 10× ligation buffer [500 m*M* Tris-HCl (pH 7.8), 100 m*M* $MgCl_2$, 100 m*M* dithiothreitol, 10 m*M* ATP, bovine serum albumin (BSA, 250 μg/ml)], H_2O to 9 μl, and 1 μl of T4 DNA ligase (400 U/μl; New England BioLabs); mix and centrifuge as described above. As a control without the insert DNA, add 100 ng of digested vector, 1 μl of 10× ligation buffer, H_2O to 9 μl, and 1 μl of T4 DNA ligase; mix and centrifuge again.

2. Incubate at 16° for 4–16 hr, then store the ligation reaction solutions at 4° until required for transformation.

3. Thaw a 100-μl aliquot of XL-10 Gold (Stratagene) competent cells on ice and transfer to a Falcon 2159 tube. Add 1 μl of the ligation solution, and set it on ice for 30 min. Heat shock at 42° for 30 sec, then immediately place on ice. Add 1.1 ml of SOC medium, and shake at 37° for 50 min.

4. Spread 100 μl of the preceding mixture on 10-cm LB–agar plates containing ampicillin (100 μg/ml). The transformation efficiency usually ranges between 0.5×10^5 and 5×10^6 transformants per microgram of DNA. Check the insert size by miniprep DNA, *Sac*I–*Eco*RI digestion, and 1.6% agarose–TAE gel electrophoresis.

Sib Screening with Anti-Tac Immunostaining

Making Pool Plates and DNA Preparation

In our hands, the sib screening method described here is faster than fluorescence-activated cell sorting (FACS) followed by plasmid recovery from the positive cells.

Protocol

1. Draw six rows by six lines in a matrix format on the bottom surface of LB–agar plates containing ampicillin (100 μg/ml), as shown in Fig. 3, and assign pool plate numbers. It is advisable to check the system by screening at a small scale, such as 1000 clones. When 1000 clones are going to be screened, prepare 28 pool plates and 28 two-ml cultures of Terrific broth containing ampicillin (100 μg/ml) in Falcon 2159 tubes (see Fig. 3). One thousand to 3000 clones can be inoculated on LB–agar plates in the 6 × 6 format by one person in 6 hr, and 20–60 pools of plasmids can be prepared on the next day.

2. Thirty-six individual colonies are picked up and inoculated onto both individual squares on the pool plate and in a 2-ml Terrific broth culture, and assigned to one pool as shown in Fig. 3. Set the pool plates at 37°

overnight and let a single colony grow in a square. Thirty-six individual colonies should appear on each pool plate. Shake the 2-ml culture tube at 37° overnight. Transfer 1.2 ml of the full growth bacterial culture to an Eppendorf tube, and centrifuge at full speed for 30 sec. The remaining cultures should be kept at 4°.

3. Plasmid DNAs are prepared with the Wizard Minipreps DNA purification system (Promega, Madison, WI). Resuspend the cells in 180 μl of cell resuspension solution. Add 180 μl of cell lysis solution and invert several times. Add 180 μl of neutralization solution; invert five times. Centrifuge at 14,000 rpm for 5 min. Decant cleared supernatant to a clean tube. Add 800 μl of DNA purification resin, and invert five times. Assemble an empty syringe and column on the vacuum manifold. Decant the slurry into the syringe barrel and wash columns with 2 ml of column wash solution. Set the column in a tube and centrifuge at 14,000 rpm for 2 min. Place the columns in fresh tubes, apply 50 μl of warmed TE to the column, and centrifuge at 14,000 rpm for 2 min.

4. Using 2 μl of preceding DNA solution, check the DNA concentration by running it on a 1% agarose–TAE gel. At least 200 ng of DNA is needed in one transfection procedure. Prepare pcDLSRα-hRAR(5′)-Tac(3′) and pcDLSRα-hG-CSF(5′)-Tac(3′) DNA, using the same procedure as described above.

Transfection into COS-7 Cells

In our hands, the lipofection method gives lower numbers of pseudopositive clones than does the DEAE-dextran method, and higher transfection efficiency than the calcium phosphate method, so long as COS-7 cells are chosen as the target cell line. The protocol with LipofectAMINE (Life Technologies, Gaithersburg, MD) is as follows.

1. Mix 500 ng of pcDLSRα-hRAR(5′)-Tac(3′) and OptiMEM (Life Technologies) to 100 μl and designate it the negative control. Mix 10 ng of pcDLSRα-hG-CSF(5′)-Tac(3′), 490 ng of pcDLSRα-hRAR(5′)-Tac(3′), and OptiMEM to 100 μl and designate it as the positive control. Twenty-four hours before transfection, harvest exponentially growing COS-7 cells by trypsinization and replate a six-well culture plate at 7×10^4 cells/well. Add 2 ml of DMEM(+) [Dulbecco's modified Eagle medium supplemented with fetal calf serum (FCS, 10%, v/v)], and incubate at 37° in a humidified incubator in an atmosphere of 5% CO_2. Remove the medium from the cells by aspiration, and rinse with 2 ml of OptiMEM.

2. Prepare the following solutions in Eppendorf tubes.

Solution A: Approximately 500–1000 ng of DNA in 100 μl of OptiMEM

Solution B: 10 μl of LipofectAMINE reagent diluted with 100 μl of OptiMEM.

Combine the two solutions, mix gently, and incubate for 15–45 min. For each transfection, add 0.8 ml of OptiMEM to the tube containing the mixture, mix gently, and overlay onto the rinsed cells. Return the cells to the incubator and incubate them for 6 hr. Add 1 ml of DMEM supplemented with 20% FCS; return the cells to the incubator and incubate them for 48–72 hr.

Cell Surface Immunostaining with Anti-Human CD25

The optimal concentrations of fluorescein isothiocyanate (FITC)-conjugated anti-human CD25 should be titrated before use. Alternatively, the combination of anti-Tac monoclonal antibodies [anti-Tac ascites can be obtained from T. A. Waldman (NIH, Bethesda, MD) or from T. Uchiyama (Kyoto University, Kyoto, Japan)] and FITC-conjugated AffinPure goat anti-mouse IgG(H+L)(1 mg/ml) (Jackson ImmunoResearch Laboratories, West Grove, PA) can be used.

Protocol

1. Mix 50 μl of FITC-conjugated anti-human CD25 (Dako, Carpinteria, CA) and 950 μl of PBS(–) supplemented with 1% (v/v) FCS. Set on ice. Remove the DMEM(+) from the COS-7 cells and wash with 2 ml of PBS(–) twice. Add 1 ml of 0.02% (w/v) EDTA. Scrape off the COS-7 cells with a cell scraper, suspend the cells with a Pipetman p1000 (Gilson, Villiers-le-Bel, France), and transfer the cells into Eppendorf tubes. Centrifuge at 10,000 rpm for 5 sec and remove the supernatant.

2. Add 20 μl of diluted FITC-conjugated anti-human CD25 prepared in step 1, and suspend well. Set the tubes on ice for 20 min (shield them from light). Tap each tube, add 800 μl of PBS(–) supplemented with 1% (v/v) FCS, centrifuge at 10,000 rpm for 5 sec, and remove the supernatant. Tap each tube, add 800 μl of PBS(–) supplemented with 1% (v/v) FCS, suspend the cells with a Pipetman p1000, centrifuge at 10,000 rpm for 5 sec, and remove the supernatant. Resuspend the COS-7 cells in 6–10 μl of PBS(–) supplemented with 1% (v/v) FCS.

3. Observe with a fluorescence microscope. In COS-7 cells transfected with positive control DNA, the margins of some of the cells are glittering intensely in a ring shape. The frequency is 1 in 20–500 cells. In other words, more than 10 surface-staining positive cells are detected in 1 μl of positive control DNA-transfected cell suspension. Only COS-7 cells that show surface fluorescence as strong as positive control DNA-transfected cells should be judged as those containing positive pools. For unknown reasons, a limited

number of weakly surface-stained cells can sometimes be observed in cells transfected with negative control DNA. Thus, comparison with controls in terms of intensity and numbers is necessary to determine if a pool or clone is positive or negative. Typically, 3–8 positive clones are trapped in 1000 clones (28 pools).

Making Smaller Pools and Identifying a Single Positive Clone

Among approximately 10% of positive pools obtained in the first screening, no positive clones appear in the secondary screening for unknown reasons. Sometimes more than one clone appears from one positive pool in the secondary screening.

Protocol

1. Pick six individual colonies from a row or line on the positive pool plate and inoculate into 2 ml of Terrific broth, containing ampicillin (100 μg/ml). Twelve sets of smaller pools consisting of 6 colonies should be prepared from 1 positive pool. Prepare DNAs from 12 sets of 2-ml cultures with the Wizard miniprep system as described above.
2. Transfect COS-7 cells with the smaller pools and stain the cells with FITC-conjugated anti-human CD25 as described. A single positive clone can be obtained by determining which one in the smaller pool of six rows and which one in the smaller pool of six lines is positive.
3. Prepare DNA of one clone assumed to be positive, transfect COS-7 cells with this DNA, and stain the transfected COS-7 cells with anti-human CD25.

Analysis

Criteria for Positive Clones

The three criteria for the positive clones are as follows.

1. Check if the open reading frame is fused with Tac (3′) in the proper reading frame.
2. Check if (a) the base sequence near the ATG fits with Kozak's rule[23] or if (b) there is one or more in-frame stop codons upstream of the start codon. Either one of (a) or (b) is enough.
3. Draw the hydropathy profiles of deduced amino acid sequences of positive clones, compare the shape of hydropathy profiles with that of authentic N-terminal signal sequences, and make sure the putative N-termi-

[23] M. Kozak, *J. Cell Biol.* **108,** 229 (1989).

nal regions are as hydrophobic as authentic ones.[24,25] Calculate by von Heijne's method[25] using the Gene Works program (IntelliGenetics, Mountain View, CA) or the PSORT program (*http://psort.nibb.ac.jp/*)[26] and make sure there is a reasonable cleavage site for the signal peptidase.

Criterion 1 is a prerequisite for criteria 2 and 3. Therefore check criteria 2 and 3 only when criterion 1 is fulfilled.

In the cDNA fraction (350–600 bp), 10–25% of the anti-Tac surface-stained positive clones will not match the three criteria described above. We call such clones "pseudopositive." The ratio of pseudopositive clones among trapped clones depends on the stringency of the judgment of cell surface anti-Tac immunostaining on microscopic observation. Results of computer homology search or cloning of full-length cDNA show that 10–30% of clones that match the three criteria do not encode the N-terminal region but middle regions. Some of them encode putative transmembrane regions, and our interpretation is that the system worked as a so-called transmembrane trap. Clones matching criteria 1 and 3, and not matching criterion 2, can also be clones trapped owing to their transmembrane regions. Such a transmembrane region can be either that of type 1 or type 2 transmembrane proteins. With the signal sequence trap method, we trapped not only secreted proteins and type 1 transmembrane proteins, but also type 2 transmembrane proteins and GPI-anchored proteins.

Hydropathy Analysis and Database Search

1. Determine base sequences of positive clones in both directions, using SRA primer (5′-TTT ACT TCT AGG CCT GAC G) and Tac primer (5′-CCA TGG CTT TGA ATG TGG CG). Check whether the positive clones match the three criteria described above. Compare their sequence information with that registered in both DNA and protein databases, using searching programs such as BLAST[27] or FASTA.[28]

2. Using obtained cDNA fragments as probes, check for RNA expression, and screen an oligo(dT)-primed cDNA library to obtain full-length clones. As cDNAs encoding complete coding regions can never be trapped by the signal sequence trap method, the screening for full-length clones is always necessary.

3. Draw the hydropathy profiles of deduced amino acid sequences of

[24] J. Kyte and R. F. Doolittle, *J. Mol. Biol.* **157,** 105 (1982).

[25] G. von Heijne, *Nucleic Acid Res.* **14,** 4683 (1986).

[26] K. Nakai and M. Kanehisa, *Genomics* **14,** 897 (1992).

[27] S. F. Altschul, W. Gish, W. Miller, E. W. Myers, and D. J. Lipman, *J. Mol. Biol.* **215,** 403 (1990).

[28] W. R. Person and D. J. Lipman, *Proc. Natl. Acad. Sci. U.S.A.* **85,** 2444 (1988).

full-length clones, and check if each hydrophobic region trapped by the signal sequence method is the N-terminal portion of the full-length clone. Compare the sequence information with that registered in both DNA and protein databases, using searching programs such as BLAST or FASTA. Check whether the deduced proteins have ER retention signals or putative GA retention signals[13,29] or lysosome transport signals.[30]

4. After the first round of screening, about 10 clones that were trapped frequently should be eliminated by colony hybridization, in order to decrease the chance of trapping the same molecule repeatedly.

Concluding Remarks

The signal sequence trap method can serve to identify secreted proteins, type 1 and type 2 transmembrane proteins, proteins with multiple membrane-spanning regions, GPI-anchored proteins, and proteins in the ER, GA, lysosomes, and other reticular structures downstream of the ER. Many trials to isolate those molecules have been performed by us and other researchers using not only the original Tac(3′)-COS7 reporter system, but also other reporter proteins and host cells such as CD4 without the signal sequence,[31,32] the protease domain of urokinase-type plasminogen activator,[33] and yeast invertase.[34,35] Each generic system has some advantages and disadvantages. With the yeast system, more clones can be trapped in a short period of time. Each system has its own biases in the kinds of proteins trapped, because each reporter system and host cells have some preferences in the kinds of proteins allowed to enter the ER, although in every system short fragments of DNAs without hydrophobic regions are trapped for unknown reasons. So long as reasonable-length DNAs are tested, the signal sequence trap method is a potent tool for identifying surface markers and intercellular signal-transducing proteins. In particular, candidate molecules involved in the signal transduc-

[29] T. Nilsson and G. Warren, *Curr. Opin. Cell Biol.* **6,** 517 (1994).

[30] T. Ludwig, R. Le Borge, and B. Hoflack, *Trends Cell Biol.* **5,** 202 (1997).

[31] T. Yoshida, T. Imai, M. Kakizaki, M. Nishimura, and O. Yoshie, *FEBS Lett.* **360,** 155 (1995).

[32] T. Imai, T. Yoshida, M. Baba, M. Nishimura, M. Kakizaki, and O. Yoshie, *J. Biol. Chem.* **271,** 21514 (1996).

[33] M. Yokoyama-Kobayashi, S. Sugano, T. Kato, and S. Kato, *Gene* **163,** 193 (1995).

[34] R. D. Klein, Q. Gu, A. Goddard, and A. Rosenthal, *Proc. Natl. Acad. Sci. U.S.A.* **93,** 7108 (1996).

[35] K. A. Jacobs, L. A. Collins-Racie, M. Colbert, M. Duckett, M. Golden-Fleet, K. Kelleher, R. Kriz, E. R. LaVallie, D. Merberg, V. Spaulding, J. Stover, and M. J. Williamson, J. M. McCoy, *Gene* **198,** 289 (1997).

tion can be isolated without setting up biological assay systems for each molecule.

Acknowledgments

We are grateful to T. Uchiyama for providing anti-Tac ascites; and to Y. Takebe for the pcDLSRα plasmid. We are very grateful to all of our colleagues who responded generously to our request for published and unpublished information on the signal sequence trap method.

[28] Solid-Phase Differential Display and Bacterial Expression Systems in Selection and Functional Analysis of cDNAs

By STEFAN STÅHL, JACOB ODEBERG, MAGNUS LARSSON, ØYSTEIN RØSOK, ANNE HANSEN REE, and JOAKIM LUNDEBERG

Introduction

There are two main categories of techniques that can be applied to identify the differences in gene expression between two samples of different biological origin: the global and the selection techniques. The global approach involves quantitative estimation of the number of expressed genes in a defined biological material. Expressed sequence tag (EST) techniques, in which tags from individual clones from a cDNA library are sequenced, belong to this category. Sequencing 1000–5000 clones of a library generates a transcript image of the most commonly expressed genes.[1] By parallel EST sequencing of two or more cDNA libraries, "*in silico*" subtraction is possible to identify differentially expressed genes. Considering the immense personnel and hardware resources required for such analyses alternative strategies have been sought. The use of chip technology offers one attractive possibility, although it has not yet matured into an established, widespread technique. A conceptually attractive EST sequencing approach in which polymers of short gene-specific tags (9 bp) were assembled in contigs and sequenced, improving the through-put by 20- to 40-fold, was demonstrated in 1995.[2]

[1] M. D. Adams, J. M. Kelley, J. D. Gocayne, M. Dubnick, M. H. Polymeropoulos, H. Xiao, C. R. Merril, A. Wu, B. Olde, R. F. Moreno, A. R. Kerlavage, W. R. McCombie, and J. C. Venter, *Science* **252,** 1651, (1991).

[2] V. E. Velculescu, L. Zhang L., B. Vogelstein, and K. W. Kinzler, *Science* **270,** 484 (1995).

Copyright © 1999 by Academic Press
All rights of reproduction in any form reserved.

0076-6879/99 $30.00

The second category of techniques applicable in the identification of unique genes includes the so-called selection techniques. These methods strive to identify the differentially expressed genes in a more cost-effective and direct manner. The selection has traditionally been achieved by subtractive hybridization and differential screening of cDNA libraries. However, by the introduction of *in vitro* amplification techniques (polymerase chain reaction, PCR), several new strategies have been designed that speed up the gene discovery process, and, more importantly, allow for analysis of small amounts of biological material. Two comparative PCR-based techniques have been described, arbitrary primed cDNA[3] and differential display,[4] which involve visual comparison of genetic fingerprints from two (or more) biological conditions. Unique fragments are physically isolated and characterized, and results obtained by these methods require confirmation by Northern blotting, reverse transcriptase (RT)-PCR or nuclease protection assays. Although the methods conceptually are attractive, reports have shown problems with contaminating rRNA, cDNA, or DNA and with reproducibility.[5]

The use of solid supports, such as magnetic carriers, has led to improved reproducibility and handling in numerous molecular biology applications.[6] Several commercial vendors now offer a variety of specific ligands (antibodies, nucleic acids, streptavidin, etc.) coupled to a solid support. The obvious reason for using solid-phase methodology lies in the improved reproducibility and efficiency, as clearly demonstrated in peptide and nucleic acid synthesis. Furthermore, in efforts to develop closed systems, the use of magnetic particles has circumvented the difficult steps in automation, such as centrifugation or precipitation, because the particles allow for a rapid change of reaction buffers and reagents by simply applying a magnetic field. In a previous study we have developed a solid phase-based differential display technology using magnetic carriers for analysis of genes found in early hematopoietic stem cells.[7] The differential display technique developed was successfully used in the analysis of a few stem cells, although that study clearly showed that at least 1000 cells would be required for reliable interpretations using the arbitrary primer approach.

To understand fully the function of the identified cDNA and the encoded cognate protein, its specific activity and role in the complex context of a cell,

[3] J. Welsh, K. Chada, S. S. Dalal, R. Chang, D. Ralph, and M. M. McClelland, *Nucleic Acids Res.* **20,** 4965 (1992).

[4] P. Liang and A. B. Pardee, *Science* **257,** 967 (1992).

[5] C. Debouck, *Curr. Opin. Biotechnol.* **6,** 6231 (1995).

[6] J. Lundeberg and F. Larsen, *Biotechnol. Annu. Rev.* **1,** 373 (1995).

[7] Ø. Røsok, J. Odeberg, M. Rode, T. Stokke, S. Funderud, E. Smeland, and J. Lundeberg, *BioTechniques* **21,** 114 (1996).

tissue, or organism, is indeed a challenge. The work required to decipher the function of the 80,000 or more proteins delivered by the human genome project is not even possible to overview with the techniques available at present. Obviously, several different approaches must be used in parallel in order to elucidate the function of a gene product. A common first approach is to perform a sequence homology analysis with the help of available databases. However, this approach gives valuable information only if a truly homologous protein with known function is characterized. A protein containing a common motif does not guarantee that the function would be completely the same. Analysis of mRNA levels by various techniques, including *in situ* hybridization, will give information as to whether a gene is transcribed in a certain tissue or cell, but it will give only limited information concerning protein function. Because most proteins probably exert their activity by interaction with other proteins or in protein complexes, procedures allowing the identification of protein interactions are of significant importance.[8] Powerful analysis techniques, such as mass spectrometry coupled to two-dimensional gel electrophoresis,[9] represent yet other approaches for the identification and characterization of new proteins.

The genome approaches so far have concentrated on the accumulation of various sequence data, with interpretations relying on analogies to known genes, proteins, and structures. Because the final understanding of a gene function needs to be verified experimentally, gene expression of the isolated cDNAs or ESTs is required.[10] The function of a certain protein could then be investigated directly by full-length cDNA expression or by transgene/gene knockout experiments. These approaches are elaborate and time consuming, but are most probably needed for a complete understanding of the protein function.

One highly interesting feature is the protein localization within a cell. The localization to the nucleus, a certain cellular organelle or membrane, perhaps together with homology analyses, can indeed also give valuable information about protein function. High-throughput approaches toward the elucidation of gene product localization have hitherto been lacking, but are discussed further below.

Taken together, a multitude of approaches exist to identify and characterize new proteins, and this chapter describes a protocol to perform solid-phase differential display for samples containing small amounts of target mRNA. In addition, we present a strategy using bacterial expression and antibodies to elucidate protein function by localization experiments.

[8] R. K. Brachmann and J. D. Boeke, *Curr. Opin. Biotechnol.* **8,** 561 (1997).

[9] P. Roepstorff, *Curr. Opin. Biotechnol.* **8,** 6 (1997).

[10] G. Gross and H. Hauser, *J. Biotechnol.* **41,** 91 (1995).

Selection of Differentially Expressed cDNAs Using Solid-Phase Differential Display

The rationale with the outlined protocol was to simplify sample preparation and handling and to increase the throughput in the differential display methodology by taking advantage of magnetic bead technology. The overall scheme is outlined in Fig. 1. Biotinylated differential display probes [5′-$(T)_{11}$VA-3′, 5′-$(T)_{11}$VG-3′, 5′-$(T)_{11}$VC-3′, 5′-$(T)_{11}$VT-3′] are immobilized onto streptavidin-coated superparamagnetic beads for use in capture of a subset of mRNAs present in a cell lysate. For hematopoietic cells a simple lysis in a LiCl–sodium dodecyl sulfate (SDS)-based buffer is used that achieves a quick disintegration and inactivation of RNases. Selective target mRNA capture is achieved, during a 5-min incubation, by the differential display oligo(dT) probe having a poly(T) stretch and two additional partly degenerated positions at the 3′ end (as described above). Noncaptured mRNA (noncomplementary) and potentially interfering DNA and ribosomal RNA are removed from the immobilized mRNA by subjecting the beads to a magnetic field for magnetic separation followed by repeated washing cycles. Hereby the beads (with the selected mRNA template) can be conditioned into a suitable and optimal buffer for solid-phase cDNA synthesis. After completed reverse transcription, the supernatant-containing reaction components and degradation products can be removed by magnetic separation, leaving the beads with immobilized first-strand product. This again allows for a swift change into a suitable buffer for the next enzymatic step, being PCR. The PCR is essentially performed as described previously,[11,12] using a primer pair consisting of an arbitrary decamer and the identical biotinylated oligo(dTVN) primer used in the previous mRNA capture step. The difference with this protocol is that the template consists of magnetic beads with immobilized first-strand cDNA. Prior to the first cycle of PCR we perform a hot start at an elevated temperature. This incubation at 75° (and the ramping to the first denaturation step at 94°) not only renders a sensitive and specific PCR but also releases the first-strand cDNA product into solution, allowing for efficient amplification. Differentially expressed fragments, labeled with [^{35}S]dATP during PCR, can be identified by conventional comparative gel electrophoresis and can be isolated from the polyacrylamide gel eluate either by magnetic beads or by ethanol precipitation. The reamplified material can be either directly sequenced using solid-phase sequencing[13] with dye-labeled

[11] P. Liang, L. Averboukh, K. Keyomarsi, R. Sager, and A. B. Pardee, *Cancer Res.* **52,** 6966 (1992).

[12] P. Liang, L. Averboukh, and A. B. Pardee, *Nucleic Acids Res.* **21,** 3269 (1993).

[13] T. Hultman, S. Ståhl, E. Hornes, and M. Uhlén, *Nucleic Acids Res* **17,** 4937 (1989).

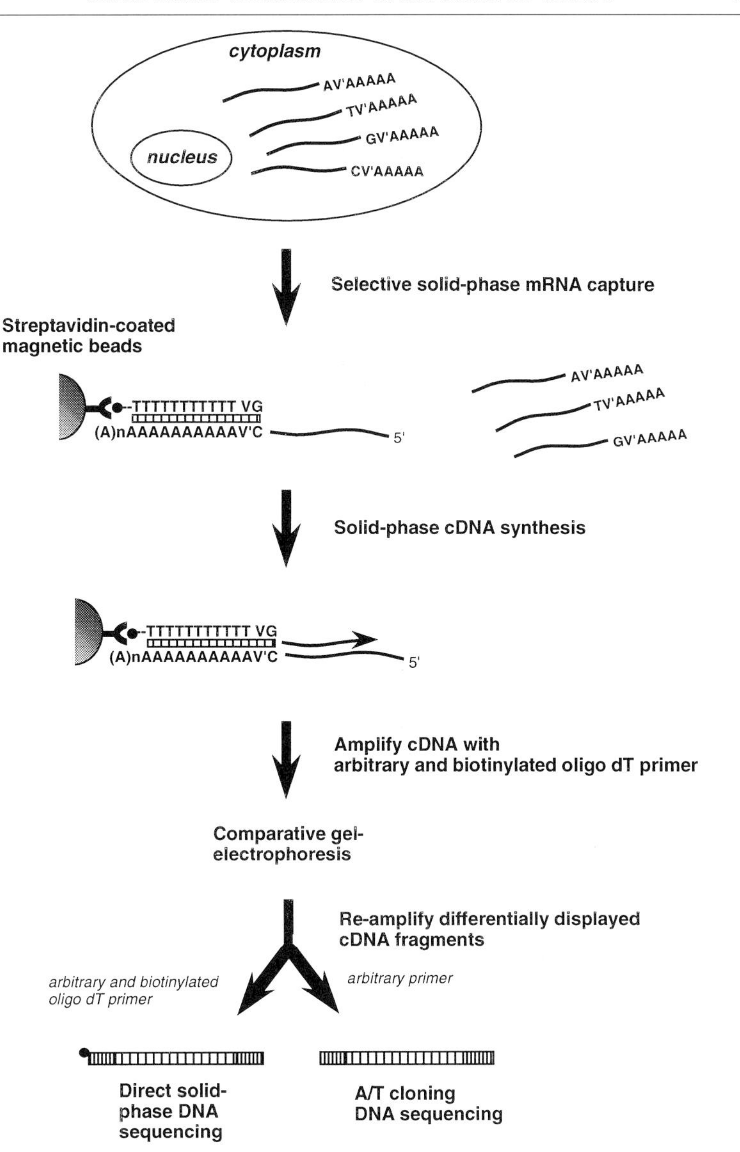

FIG. 1. Schematic outline of the solid-phase differential display.

terminators or cloned into a pGEM-T vector to allow amplification with vector-specific primers followed by solid-phase sequencing.

In the initial report the solid-phase cDNA synthesis was investigated in detail.[7] Yield and selectivity were analyzed by variation of probe length and different reverse transcriptases. It was possible to investigate the effect of these modifications, without affecting the subsequent PCR, because the cDNA primer remain coupled to the support after cDNA synthesis. Among the investigated experiments were tests with an extended oligo(dT) probe (with up to 25 thymine nucleotides) to allow for higher and more stringent temperatures during mRNA hybridization and cDNA synthesis. Surprisingly, it was found that only the oligo$(dT)_{11}$ primers resulted in the specificity required in differential display, and it was also shown that the reproducibility generally was higher using mesophilic reverse transcriptases (such as SuperScript; Life Technologies, Gaithersburg, MD) compared with thermostabile alternatives (recombinant *Thermus thermophilos,* rTth). In addition, the reproducibility was markedly lower for decreasing amounts of target mRNA. For the analysis of bone marrow cells a minimum limit was established, corresponding to approximately 10 ng of total RNA.

This solid-phase differential display protocol was primarily developed for gene expression analysis in samples having a limited amount of target mRNA, such as stem cells and lymph node metastases (see example in Fig. 2). The optimized solid-phase differential display protocol uses oligo$(dT)_{11}$VN-coated beads for capture and cDNA priming combined with Superscript RT (or a derivative) in first-strand synthesis for the highest possible reproducibility. In Fig. 2, two sets of experiments are depicted comparing micrometastatic cells from bone marrow (BM) and lymph node (N+) from one and the same patient having breast cancer. Figure 2 shows the reproducibility at different PCR annealing temperatures as well as differentially expressed fragments with alternating arbitrary primers. The average length of the solid-phase differential display fragments is approximately 450 bp. An interesting feature using this approach is that the majority (approximately 65%) of the obtained fragments are primed only by the arbitrary primer during amplification, which benefits the subsequent bacterial cDNA expression system because these fragments are more likely to be within the coding region. Although a focus in our studies has been to optimize the protocol for small amounts of mRNA, the method can easily be used in other situations when target amounts are not limiting. A detailed description of the procedure is given below.

Preparation of Oligo(dT) Probe-Coated Beads for Differential Display

1. Use 20 μl (200 mg) of resuspended Dynabeads M-280 Streptavidin (10 mg/ml) (Dynal AS, Oslo, Norway) per template; pipette the suspended

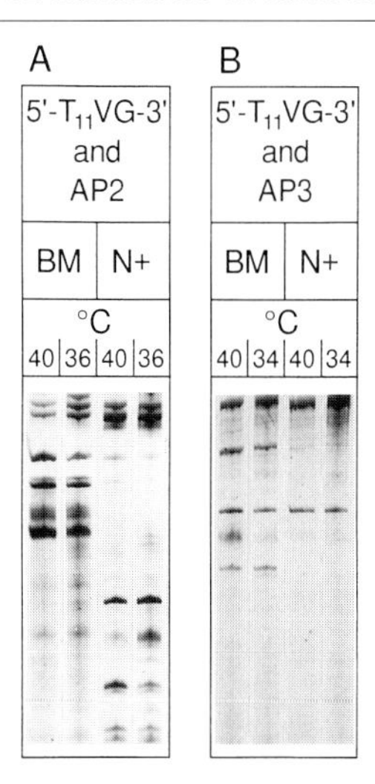

FIG. 2. Solid-phase differential display on axillary node metastases. The display pattern of samples (BM, micrometastatic cells from bone marrow; N+, micrometastatic cells from lymph node) for each set of conditions demonstrates the reproducibility and specificity of the protocol for one differential display oligo(dT) primer and two different arbitrary primers (A, B) used at alternating PCR annealing temperatures as indicated.

beads into an RNase-free 1.5-ml microcentrifuge tube. The beads are suitably washed in bulk for the total number of templates.

2. Place the tube in the neodymium–iron–boron magnet tube holder (MPC; Dynal AS) and allow the beads to adhere to the magnet at the side of the wall. After approximately 20 sec remove the supernatant, using a pipette (without removing it from the magnetic holder).

3. Wash the beads by adding an equal volume of binding buffer [10 m*M* Tris-HCl (pH 8.0), 2 *M* LiCl, m*M* EDTA (pH 8.0)] and gently pipette to resuspend. Repeat once using the magnetic holder.

4. Immobilize the differential display oligo(dT) probe by resuspending the beads in 100 μl of binding buffer and add 4 μl of the chosen biotinylated oligo(dT) probe (25 μM). The four HPLC-purified probes are 5′-biotin-$(T)_{11}$VG-3′; 5′-biotin-$(T)_{11}$VT-3′; 5′-biotin-$(T)_{11}$VC-3′; and 5′-biotin-$(T)_{11}$VA-3′; V denotes a degenerate position: A, G, and C. Incubate at room temperature for 15 min and mix occasionally.

5. Remove unbound probe by sedimenting the beads in the magnetic holder and remove the supernatant. Wash the beads once with 100 μl of binding buffer and twice with 100 μl of lysis buffer [100 m*M* Tris-HCl (pH 8.0), 500 m*M* LiCl, 10 m*M* EDTA (pH 8.0), 1% SDS, 5 m*M* dithiothreitol (DTT)], and remove the supernatant.

6. Finally, prepare ready-to-use beads by resuspending the beads in 100 μl of lysis buffer and then put the tube on ice.

Preparation of RNA Lysate

Here, one specific example of sample preparation for differential display analysis of hematopoietic cells is given, although other protocols for solid tissues can easily be adapted.

1. Wash the isolated hematopoietic cells with ice-cold phosphate-buffered saline [PBS: 137 m*M* NaCl, 2.7 m*M* KCl, 4.3 m*M* $Na_2HPO_4 \cdot 7H_2O$, 1.4 m*M* KH_2PO_4 (pH 7.5)].
2. After sedimentation by centrifugation (1000–2000 rpm for 5 min), resuspend the cell pellet in PBS and aliquot into tubes to contain the desired number of starting cells in the following experiments.
3. Sediment the cells, resuspend the cell pellet in 100 μl of lysis buffer, and vortex immediately. Store the lysed cells, preferably at −130° (or −80°) until analyzed.

Solid-Phase Capture of mRNA

1. Thaw the RNA sample (cell lysate) slowly on ice (~3 min) and add the lysate (100 μl) to the prepared differential display beads with coupled oligo(dT) probe (100 μl).
2. Incubate on ice for 5 min and mix gently during the hybridization to allow for selective mRNA capture.
3. Remove nonbound RNA by sedimenting the beads, using the magnetic holder, and wash the beads twice with 100 μl of washing buffer [10 m*M* Tris-HCl (pH 7.5), 0.15 *M* LiCl, 1 m*M* EDTA, 0.1% SDS] and once with 200 μl of RT washing buffer [10 m*M* Tris-HCl (pH 8.3), 50 m*M* KCl, 5 m*M* $MgCl_2$].
4. Sediment the beads with captured mRNA with the magnetic holder, remove the supernatant, and resuspend the beads in 100 μl of RT washing buffer; transfer the solution to a new tube. Repeat the washing once with 200 μl of RT washing buffer and remove the supernatant.

Solid-Phase cDNA Synthesis

1. Prepare a master mixture for first-strand cDNA synthesis by taking 4 μl of 5× first-strand buffer, 2 μl of 0.1 *M* DTT, 1 μl of dNTPs (10 m*M*), 200 U of SuperScript (or equivalent) reverse transcriptase, and sterile water to 20 μl per reaction according to the manufacturer recommendations (Life Technologies). Prewarm the cDNA mixture at 37° for 1–2 min just prior to adding to the beads with captured mRNA.

2. Initiate solid-phase cDNA synthesis by resuspending the beads in 20 μl of prewarmed cDNA master mixture. Incubate for 1 hr at 37°, using a rotator to keep the beads in suspension.

3. Remove the reaction components, using the magnetic holder. The first-strand cDNA products will remain immobilized on the magnetic particles.

4. Wash the beads twice with 100 μl of PCR washing buffer [20 m*M* Tris-HCl (pH 8.3), 50 m*M* KCl, 0.1% Tween 20].

5. Sediment the beads with the magnetic holder, remove the supernatant, and resuspend the beads with 25 μl of PCR washing buffer. Use directly in PCR or store at −80°.

Hot Start-Mediated Polymerase Chain Reaction

1. Prepare a master mixture for PCR by taking 2.5 μl of 10 × PCR buffer, 2.5 μl of dNTPs (20 m*M*), 1.25 μl of [^{35}S]dATP (1200 Ci/mmol; Amersham, Arlington Heights, IL), 1 U of AmpliTaq DNA polymerase, and sterile water to 17.5 μl per reaction according to the manufacturer recommendations (Perkin-Elmer, Norwalk, CT).

2. Prepare a primer master mixture for hot start PCR by taking 2.5 μl of biotinylated oligo(dT) primer (25 μ*M*) and 2.5 μl of arbitrary primer (5 μ*M*, D388: 5′-GCAGATGATG-3′; D389: 5′-GATCTCCTCA-3′; D395: 5′-CTTGATTGCC-3′; D492: 5′-CAGTGTAGTC-3′; D493: 5′-GTTTTCGCAG-3′; D495: 5′-TCGATACAGG-3′; AP2: 5′-CTGATC CATG-3′; AP-3: 5′-CTGATCCATC-3′) per reaction. The biotinylated primer should be identical to the probe used in the magnetic bead capture of the mRNA subset.

3. Take 2.5 μl of resuspended beads (containing first-strand cDNA product) and add to PCR tubes with prealiquoted PCR master mixture (17.5 μl). Overlay with mineral oil.

4. Perform hot start PCR by heating the PCR tubes to 75° in the PCR block and prior to adding 5 μl of the primer mixture. Continue directly to the first denaturation step in the PCR program.

5. The employed PCR program comprises denaturation at 94° for 30 sec, annealing at 40° for 1 min, and extension at 72° for 2 min for 35 cycles. A final extension step at 72° for 10 min is recommended.

Isolation of Differential Display Products

1. Run the differential display samples to be compared side by side on a 6% denaturating polyacrylamide gel prepared in a standard fashion.[14]

[14] J. Sambrook, E. F. Fritsch, and T. Maniatis, "Molecular Cloning: A Laboratory Manual." Cold Spring Harbor Laboratory Press, Cold Spring Harbor, New York, 1987.

Dry the gel and expose onto film. Before developing the film, orient the gel and film by piercing them with a needle.

2. Orient the gel and film according to the piercings and mark the bands. Cut out the fragments with a scalpel—flame the scalpel to sterilize—and place the fragments into separate 1.5-ml microcentrifuge tubes and add 200 μl of PCR washing buffer.

3. Vortex and let stand for at least 15 min. Incubate at 100° for 5 min and then let it cool to room temperature. Eluted DNA fragments can now either be precipitated[11] or purified by DNA Direct kit (Dynal AS). In the case of purification of biotinylated differential display products it is possible to reimmobilize these to streptavidin beads (as described below in steps 4 and 5).

4. Take 20 μl (200 mg) of resuspended streptavidin beads per sample and wash the beads as described above with washing buffer. Resuspend the beads in 200 μl of washing buffer prior to addition of the sample (200 μl). Incubate for 15 min at room temperature. Mix during the immobilization.

5. Remove nonbound material, using the magnetic holder, and wash the beads twice with 100 μl of PCR washing buffer. Resuspend the beads with 20 μl of PCR washing buffer. Run the PCR by taking 2.5–5 μl of the template (resuspended beads)

6. Reamplification of purified differential display fragments is essentially as described above, but in a 50-μl PCR volume and with 0.5 μl of biotinylated oligo(dT) probe (25 μM) and 2.5 μl of decamer primer (5 μM) used in the previous amplification.

7. Clone the generated amplicons by either pGEM-T-cloning vectors[7] for subsequent PCR sequencing using general vector primers, or perform direct solid-phase sequencing.[13]

8. Perform homology analysis with the obtained sequences and confirm differential expression.

Bacterial Expression as a Means to Elucidate Gene Function

When isolating a number of partial cDNAs by a differential method, and the partial cDNAs display interesting homologies on database matching, a rational "first step" approach to understand gene function when full-length cDNA is not isolated would be to produce a portion of a protein and use it to raise an antiserum for identifying the protein localization in tissues and cells. This approach requires robust expression systems with high expression levels, possibilities for parallel handling of numerous clones, and rapid and efficient recovery of the produced gene product. Bacterial production systems have proved to be powerful in terms of high expression levels for the production of various mammalian proteins.[15] Affinity purification of gene products, by expressing a target gene fused to a gene encoding an affinity tag, constitutes a general strategy that has been utilized extensively in different fields of research.[16]

A bacterial expression system suitable for production of mammalian cDNAs has been presented.[17] The system employs parallel expression of the cDNA-encoded protein in two expression systems, resulting in two different affinity-tagged fusion proteins having the cDNA-encoded portion in common. When setting up such parallel expression systems, suitable for functional analysis of cDNA-encoded proteins, robustness in the production is of utmost importance. Intracellular production makes it possible to express proteins that cannot be secreted through hydrophobic membranes. It is also important to choose a tightly regulated promoter system, such as the phage T7 promoter,[18] which makes it possible to produce proteins that normally would be deleterious to the host cell.[19] When selecting an affinity system, one should choose an affinity tag with a high affinity for a specific ligand. This affinity tag should allow affinity purification in buffers containing chaotropic agents, such as guanidine hydrochloride or urea, used for solubilization of proteins precipitated into inclusion bodies. Furthermore, the affinity tags should be resistant to proteolysis.

A Concept for Generation of Affinity-Enriched Antibodies Specific for a cDNA-Encoded Protein

The system presented by Larsson and co-workers, designed for functional analysis of cDNA-encoded proteins, is outlined in Fig. 3. Selected cDNA clones are subcloned in parallel into vectors of two expression systems having the appropriate reading frame (Fig. 3). In both vector systems the transcription is under the control of the tightly regulated phage T7 promoter system.[18] On intracellular expression in *Escherichia coli,* two different fusion proteins are produced. One vector system expresses the target protein fused to an IgG-binding affinity tag (ZZ)[20] derived from staphylococcal protein A (SpA), and the second vector system yields the

[15] R. C. Hockney, *Trends Biotechnol.* **12,** 456 (1994).

[16] J. Nilsson, S. Ståhl, J. Lundeberg, M. Uhlén, and P.-Å. Nygren, *Protein Exp. Purif.* **11,** 1 (1997).

[17] M. Larsson, E. Brundell, L. Nordfors, C. Höög, M. Uhlén, and S. Ståhl, *Protein Exp. Purif.* **7,** 447 (1996).

[18] F. W. Studier, A. H. Rosenberg, J. J. Dunn, and J. W. Dubendorff, *Methods Enzymol.* **185,** 60 (1990).

[19] T. Sano and C. R. Cantor, *Bio/Technology* **9,** 1378 (1991).

[20] B. Nilsson, T. Moks, B. Jansson, L. Abrahmsén, A. Elmblad, E. Holmgren, C. Henrichson, T. A. Jones, and M. Uhlén, *Protein Eng.* **1,** 107 (1987).

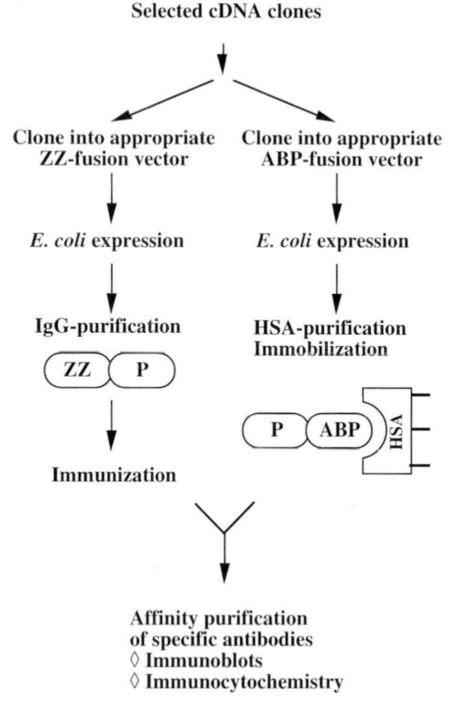

FIG. 3. Schematic outline of the basic concept for using bacterial expression as a means to elucidate function of cDNA-encoded proteins by immunolocalization techniques.

target protein fused to a human serum albumin (HSA)-binding affinity tag (ABP)[21,22] derived from streptococcal protein G (SpG). The use of both these tags has been extensively documented,[22–24] and the rationale for selecting them as affinity tags for this system is described below. The expressed fusion proteins are after cell disruption either affinity purified directly, if the fusion product is predominantly soluble, or after solubilization and renaturation of inclusion bodies, in such cases when the fusion product aggregates in inclusion body form. The inherent properties of the

[21] P.-Å. Nygren, M. Eliasson, E. Palmcrantz, L. Abrahmsén, and M. Uhlén, *J. Mol. Recognit.* **1,** 69 (1988).

[22] P.-Å. Nygren, S. Ståhl, and M. Uhlén, *Trends Biotechnol.* **12,** 184 (1994).

[23] S. Ståhl and P.-Å. Nygren, *Pathol. Biol.* **45,** 66 (1997).

[24] S. Ståhl, P.-Å. Nygren, and M. Uhlén, *in* "Recombinant Proteins: Detection and Isolation Protocols, Methods in Molecular Biology" (R. S. Tuan ed.), vol. 63, p. 103. Humana Press, Totowa, NJ, 1997.

target gene product will thus obviously be the basis for deciding which strategy to employ. The ZZ–P fusion proteins are, after purification by affinity chromatography on IgG–Sepharose, used for immunization (Fig. 3). The corresponding ABP–P fusion protein can be immobilized, during the purification procedure to the HSA–Sepharose column, and used for the isolation and purification of antibodies reactive with the cDNA-encoded portion, P (Fig. 3). Such affinity-purified antibodies would be suitable for different analyses, including Western blots and immunohistology studies, aimed at elucidating the localization and function of the cDNA-encoded protein.

The ZZ and ABP affinity tags are selected because they share several advantageous features: (1) they are both relatively small in size, both being 14 kDa, which ensures a high product-to-tag ratio; (2) the highly specific interactions with their ligands, IgG (ZZ fusions) or HSA (ABP fusions), respectively, enable efficient purification by affinity chromatography[22–24]; (3) both tags also allow efficient recovery of gene products of low solubility because affinity chromatography can be performed after guanidine hydrochloride–urea-mediated solubilization and renaturation of precipitated products[25]; and (4) ZZ and ABP fusions can be detected by blotting techniques after SDS-PAGE analysis by taking advantage of their IgG(Fc)- and HSA-binding properties, respectively. ZZ fusions are efficiently stained using a commercially available complex of rabbit anti-horseradish peroxidase–IgG and horseradish peroxidase. ABP fusions are stained using biotinylated HSA and conjugated streptavidin–alkaline phosphatase. Furthermore, both tags have through extensive documented use proved to be highly stable to proteolysis[22–24] and they contain no cysteines that could cause formation of unwanted disulfide-bridged complexes. In addition, both ZZ and ABP share structural similarities. Both tags comprise two structural domains, each of which has a globular fold, consisting of a triple α-helical bundle.[23] Both N and C termini of the two tags can be used for fusion without interfering with the folding. The structural integrity of the tags is probably part of the reason for their stability to proteolysis.

Expression and Affinity Purification of ZZ and ABP Fusion Proteins

1. The cDNA-encoding gene fragments should be inserted into the pT7-TZZ and pT7-ABP expression vectors of the appropriate reading frame.

2. *Escherichia coli* BL21(DE3)pLysS cells (Novagen, Madison, WI) transformed with the vectors encoding the different fusion proteins should

[25] M. Murby, E. Samuelsson, T. N. Nguyen, L. Mignard, H. Binz, M. Uhlén, and S. Ståhl, *Eur. J. Biochem.* **230,** 38 (1995).

be grown overnight at 37° in shake flasks containing 50 ml of tryptic soy broth (30 g/liter; Difco, Detroit, MI), supplemented with yeast extract (5 g/liter; Difco), ampicillin (100 μg/ml), and chloramphenicol (34 μg/ml).

3. The overnight cultures are suitably diluted 1 : 20 and used to inoculate 500 ml of the same medium. The cultures should be grown in shake flasks at 37°. Expression of the recombinant fusion proteins should be induced at mid-log phase ($A_{600\ nm} \sim 1$) by addition of isopropyl-β-D-thiogalactoside (IPTG) (Pharmacia Biotech, Piscataway, NJ) to a final concentration of 1 m*M*.

4. Cells are suitably harvested approximately 3.5 hr after induction, by centrifugation at approximately 5000 *g* for 10 min. The pelleted cells can be stored at −20°. The cells are then thawed and resuspended in 30 ml of TST [50 m*M* Tris-HCl, (pH 7.5), 0.2 *M* NaCl, 0.05% Tween 20], followed by sonication.

5. After centrifugation at 20,000 *g* for 10 min, the supernatants containing soluble gene products are filtered (1.2-μm pore size; Millipore, Bedford, MA), prior to the affinity purification procedure.

6. Insoluble material is pelleted by centrifugation at approximately 20,000 *g* for 10 min. The pelleted material containing precipitated proteins is solubilized by adding guanidinium hydrochloride and 2-mercaptoethanol to final concentrations of 6 *M* and 10 m*M*, respectively, followed by incubation at room temperature for 2 hr on a magnetic stirrer. After centrifugation at 20,000 *g* for 10 min, the supernatants are diluted (in TST containing 0.1 *M* guanidinium hydrochloride) to a final concentration of 0.5 *M* guanidinium hydrochloride. After centrifugation at 10,000 *g* for 10 min, filtered (1.2-μm pore size) supernatants can be applied directly to the affinity chromatography columns.

7. The affinity purification of the ZZ and ABP fusion proteins, on IgG–Sepharose (Pharmacia Biotech) or HSA-Sepharose, respectively, is performed essentially as described earlier.[20,21] Briefly, the columns are washed with 100 ml of TST and 50 ml of 5 m*M* ammonium acetate, pH 5.5, before elution of bound material with 0.2 *M* acetic acid (HAc), pH 3.3 (ZZ fusions), or 0.5 *M* HAc, pH 2.8 (ABP fusions), respectively. Protein content is suitably estimated by absorbance measurements ($A_{280\ nm}$) and relevant fractions are lyophilized.

Specific Staining of ZZ and ABP Fusion Proteins. Several alternative methods exist for analyzing the protein A fusion proteins in a sample. One simple and yet sensitive method takes advantage of the ability of SpA or ZZ to bind the Fc part of the IgG molecule, and is suitable for a modified immunoblotting procedure after protein analysis using polyacrylamide electrophoresis.[26]

1. After protein analysis by SDS-PAGE, proteins are electrophoretically transferred by electroblotting to nitrocellulose filters.
2. Unspecific binding to the filter is blocked by incubation in 0.25% gelatin in PBS [50 m*M* phosphate (pH 7.1), 0.9% NaCl] for 30 min at 37° before treatment with a soluble complex of rabbit anti-horseradish peroxidase–IgG and horseradish peroxidase (Dakopatts, Copenhagen, Denmark), diluted 1:1000 in 0.25% gelatin in PBS for 30 minutes at 37°.
3. The filter is washed twice in PBS and once in 50 m*M* Tris-HCl, pH 8.
4. The presence of horseradish peroxidase is detected by adding H_2O_2 and 3,3′-diaminobenzidine in 50 m*M* Tris-HCl, pH 8. By this procedure, SpA-containing protein bands are stained brown.

ABP fusion proteins are stained in an analogous fashion, by the addition of biotinylated HSA after electrophoretic transfer of the fusion proteins to nitrocellulose. HSA is biotinylated with D-biotinoyl-ε-aminocaproic acid *N*-hydroxysuccinimide ester, according to the supplier recommendations (Boehringer Mannheim, Indianapolis, IN). A streptavidin–alkaline phosphatase conjugate (Boehringer Mannheim) is added and the presence of ABP fusion proteins is detected by the addition of substrate solution, *p*-nitrophenylphosphate (Sigma, St. Louis, MO).

Immunization of Rabbits and Affinity Enrichment of Antisera

1. New Zealand White rabbits are immunized intramuscularly with 500 μg of the ZZ fusion proteins in phosphate-buffered saline (PBS) and Freund's complete adjuvant (FCA).
2. Booster injections are suitably given 4, 8, and 12 weeks after the initial injection with the same amount of fusion protein but with Freund's incomplete adjuvant instead of FCA.
3. The rabbits are bled 10 days after the third booster injection.
4. Total rabbit sera are diluted 1:30 in TST and applied on a protein AG–Sepharose column (Pharmacia Biotech). After washing with 100 ml of TST and 50 ml of 5 m*M* ammonium acetate, pH 5.5, the total IgG is eluted with 0.5 *M* HAc, pH 2.8. The HAc is exchanged for PBS, using a PD-10 column (Pharmacia Biotech).

[26] B. Jansson, C. Palmcrantz, M. Uhlén, and B. Nilsson, *Protein Eng.* **2,** 555 (1990).

5. Purified ABP fusion proteins (approximately 3 mg) containing the different cDNA-encoded proteins are bound to HSA–Sepharose and covalently cross-linked to the matrix using glutardialdehyde (0.07% final concentration). The reaction is stopped after 5 min by addition of 1% glycine solution. The gel is washed in consecutive steps with PBS, 0.5 *M* Hac (pH 2.8), and 0.2 *M* glycine, pH 2.8.

6. A fraction of approximately 6 mg of total IgG (originating from an immunization with a certain ZZ fusion protein) is diluted in PBS to a final volume of 35 ml and loaded on a column with 1 ml of HSA–Sepharose having the corresponding ABP fusion protein immobilized.

7. After extensive washing with PBS, retained antibodies are eluted with 0.2 *M* glycine, pH 2.8, and immediately neutralized in 2 *M* Tris. Buffer exchange to PBS with 0.5% bovine serum albumin (BSA) can be performed using a PD-10 column (Pharmacia Biotech).

Affinity-Enriched Antisera in Immunotechniques to Elucidate Gene Product Localization

The concept and expression systems described were initially evaluated for the production of five cDNA-encoded proteins from a cDNA library prepared from prepubertal mouse testis.[17] Protein production was characterized in terms of production levels, solubility of the target proteins, and product quality after affinity purification. Purified fusion proteins were used to elicit antibodies, and affinity-enriched sera, purified using the corresponding fusion protein, were subsequently used in various analyses. It was first concluded by immunoblotting that no cross-reactivity existed between the two systems. Antisera raised against a ZZ fusion and enriched using the corresponding ABP fusion recognized specifically the correct ABP fusion protein but did not stain unrelated ABP fusion proteins.[17] Beside the obvious possibility of using the affinity-enriched antisera for detection of the corresponding cDNA-encoded proteins by immunoblotting after SDS–PAGE analysis of cell or tissue desintegrates, such antisera proved to be valuable for localizing the cDNA-encoded protein in tissue sections, and also within cells. In a presented example,[17] a previously unknown (no homologies found in databases) cDNA-encoded protein was detected in a mouse testis cryosection in a region containing spermatogonial cells and spermatocytes. When purifying these two cell types it was found that only the spermatogonial cell was stained and that the staining was restricted to a small number of nuclear bodies present within the spermatogonial cell nuclei.[17]

Taken together, it is evident that such affinity-enriched antiserum is useful for detailed elucidation of the localization of cDNA-encoded proteins. With the help of commercially available monoclonal antibodies (MAbs) recognizing known cellular structures, colocalization experiments can lead to rather good speculations concerning the function of a certain protein. An alternative strategy for the generation of antibodies suitable for immunohistology could be to utilize the rapidly evolving phage display technology to select recombinant antibody fragments from large combinatorial libraries. Purified fusion proteins, such as the ZZ or ABP fusion proteins, could thus be used for panning to select binders to cDNA-encoded proteins. This strategy would circumvent the time-consuming immunization procedures.

Summary

Differential gene expression can be expected during activation and differentiation of cells as well as during pathological conditions, such as cancer. A number of strategies have been described to identify and understand isolated differentially expressed genes. The differential display methodology has rapidly become a widely used technique to identify differentially expressed mRNAs. In this chapter we described a variant of the differential display method based on solid-phase technology. The solid-phase procedure offers an attractive alternative to solution-based differential display because minute amounts of sample can be analyzed in considerably less time than previously. The employed solid support, monodisperse super paramagnetic beads, which circumvents precipitation and centrifugations steps, has also allowed for optimization of the critical enzymatic and preparative steps in the differential display methodology. We also described how bacterial expression can be used as a means to elucidate gene function. An efficient dual-expression system was presented, together with a basic concept describing how parallel expression of selected portions of cDNAs can be used for production of cDNA-encoded proteins as parts of affinity-tagged fusion proteins. The fusion proteins are suitable both for the generation of antibodies reactive to the target cDNA-encoded protein and for the subsequent affinity enrichment of such antibodies. Affinity-enriched antibodies have proved to be valuable tools in various assays, including immunoblotting and immunocytochemical staining, and can thus be used to localize the target cDNA-encoded protein to certain cells in a tissue section or even to a specific cell compartment or organelle within a cell. High-resolution localization of a cDNA-encoded protein would provide valuable information toward the understanding of protein function.

[29] Transposon Mutagenesis for the Analysis of Protein Production, Function, and Localization

By PETRA ROSS-MACDONALD, AMY SHEEHAN, CARL FRIDDLE, G. SHIRLEEN ROEDER, and MICHAEL SNYDER

Introduction

Molecular biology has provided a wide array of techniques for investigating the role of a particular gene, either by creating mutations or by analysis of its product. In most experimental organisms, the typical first step in functional analysis is to construct a null allele, either by an insertion mutation or by complete or partial deletion of the coding sequence. *In vivo* such a null allele may result in lethality, which often provides little information beyond indicating that the gene product serves an essential function. In yeast, this shortcoming of null alleles has been overcome by the development of methods that generate a range of mutations, allowing the investigator to look for dominant or gain-of-function alleles, in addition to hypomorphic (partial loss of function) or conditional phenotypes. In choosing between mutagenesis techniques, the cost and labor of creating a small set of predefined changes that seem likely to cause a phenotype[1,2] are often balanced against the difficulty of determining the change responsible for a phenotype of interest if a larger set of random mutations[3,4] is screened. Creation of a wide range of mutations in a gene often proves most informative, as different alleles often result in distinct phenotypes or expose different domains of a protein, thereby greatly extending the information obtained for a particular gene product. Conditional mutations are particularly valuable, since they can be used both to gain direct insight into the function of a gene, and to screen for additional genes that by mutation or overexpression can compensate for the mutant function. Such "second site suppressors" often encode proteins that interact with or function in the same pathway as the product of the gene of interest.

A variety of methods are also available for the direct analysis of protein function. For example, conventional copurification approaches are useful for studying biochemical activities and association of proteins (e.g., Refs.

[1] B. C. Cunningham and J. A. Wells, *Science* **244,** 1081 (1989).

[2] K. F. Wertman, D. G. Drubin, and D. Botstein, *Genetics* **132,** 337 (1992).

[3] F. M. Asubel, R. Brent, R. E. Kingston, D. D. Moore, J. G. Seidman, J. A. Smith, and K. Struhl, "Current Protocols in Molecular Biology." Wiley Interscience, New York, 1987.

[4] C. W. Lawrence, *Methods Enzymol.* **194,** 273 (1991).

Copyright © 1999 by Academic Press
All rights of reproduction in any form reserved.

0076-6879/99 $30.00

5 and 6), two-hybrid methodologies allow the study of protein–protein interactions (e.g., Refs. 7 and 8), while mass spectroscopy allows the analysis of protein modifications and protein–protein interactions.[9] Several of these techniques require either the availability of a specific antibody, or the construction of a version of the protein that is tagged with an epitope recognized by commercially available antibody.[10,11] Protein fusions to reporter constructs such as β-galactosidase (β-Gal) or green fluorescent protein (GFP) have also been valuable for a variety of purposes, including determining when a protein is produced during a cell cycle or life cycle, and analyzing protein localization within the cell (e.g., Refs. 12–15). Epitope-tagged proteins have advantages over both native and fusion proteins, in that a relatively small change is made to the protein but the labor of raising a specific antibody is avoided.

Below we describe a transposon mutagenesis system that produces multipurpose constructs for the monitoring of protein production, localization, and function. A single mutagenesis generates a large spectrum of alleles, including null, hypomorphic, and conditional alleles, reporter fusions, and epitope-insertion alleles. The system therefore provides the basis for a wide variety of studies of gene and protein function.

Transposon Mutagenesis for Analysis of Gene and Protein Function

For transposon mutagenesis in yeast, a "shuttle mutagenesis" technique[16] is preferable to direct *in vivo* use of the endogenous yeast Ty transposon system,[17] as the latter exhibits a strong bias in target site specificity.[18] In shuttle mutagenesis, a plasmid-borne copy of the gene of interest

[5] M. P. Rout and J. V. Kilmartin, *J. Cell. Biol.* **111,** 1913 (1990).

[6] A. L. Moon, P. A. Janmey, K. A. Louie, and D. G. Drubin, *J. Cell Biol.* **120,** 421 (1993).

[7] S. Fields and O.-K. Song, *Nature (London)* **340,** 245 (1989).

[8] J. W. Harper, G. R. Adammi, N. Wei, K. Keyomarsi, and S. J. Elledge, *Cell* **75,** 805 (1993).

[9] J. T. Stults, *Curr. Opin. Struct. Biol.* **5,** 691 (1995).

[10] M. Tyers, G. Tokiwa, R. Nash, and B. Futcher, *EMBO J.* **11,** 1773 (1992).

[11] B. L. Schneider, W. Seufert, B. Steiner, Q. H. Yang, and A. B. Futcher, *Nucleic Acids Res.* **11,** 1265 (1995).

[12] N. Burns, B. Grimwade, P. B. Ross-Macdonald, E.-Y. Choi, K. Finberg, G. S. Roeder, and M. Snyder, *Genes Dev.* **8,** 1087 (1994).

[13] R. K. Niedenthal, L. Riles, M. Johnston, and J. H. Hegemann, *Yeast* **12,** 773 (1996).

[14] K. Sawin and P. Nurse, *Proc. Natl. Acad. Sci. U.S.A.* **93,** 15146 (1996).

[15] P. Ross-Macdonald, A. Sheehan, S. Roeder, and M. Snyder, *Proc. Natl. Acad. Sci. U.S.A.* **94,** 190 (1997).

[16] H. S. Seifert, E. Y. Chen, M. So, and F. Heffron, *Proc. Natl. Acad. Sci. U.S.A.* **83,** 735 (1986).

[17] D. J. Garfinkel and J. N. Strathern, *Methods Enzymol.* **194,** 342 (1991).

[18] H. Ji, D. P. Moore, M. A. Blomberg, L. T. Braiterman, D. F. Voytas, G. Natsoulis, and J. D. Boeke, *Cell* **73,** 1007 (1993).

is first mutagenized in *Escherichia coli* using a derivative of the bacterial transposon Tn*3*. Such "minitransposons" are designated mTns. mTn insertion is close to random in most yeast plasmid clones (e.g., Refs. 15 and 16). The mTn inserts only once into each plasmid, but a single mutagenesis gives rise to a library of plasmids containing a spectrum of insertions throughout the gene of interest. These mutated copies of the gene may then be transferred to yeast, either on a plasmid (J. Y. Leu and S. Roeder, personal communication, 1998) or by replacement of the chromosomal copy of the locus.[15,16,19] In the past, insertions generated by shuttle mutagenesis have proved extremely useful for defining the expression profile of a gene and mapping its functional boundaries, as well as for providing nested priming sites for sequencing a region (e.g., Ref. 20).

A new set of multipurpose shuttle mutagenesis mTns that have a variety of uses has been created.[15] An important feature of these mTns is their ability to generate small in-frame insertions in genes, greatly extending the utility of the system for analysis of protein function and localization. The mini-Tn*3* derivatives mTn-3xHA/*lacZ,* mTn-4xHA/*lacZ,* and mTn-3xHA/GFP are shown in Fig. 1.* The mTns are flanked by the 38-bp terminal repeats of Tn*3*, which direct transposition, and contain the Tn*3 res* site for resolution of transposition cointegrates. Tn*3* enzymes that catalyze transposition and resolution must be provided *in trans.* All three transposons carry the selectable markers *tet* and *URA3* (for bacteria and yeast, respectively). At one end mTn-3xHA/*lacZ* and mTn-4xHA/*lacZ* carry coding sequences for a β-Gal reporter that lacks an initiator methionine, while mTn-3xHA/GFP contains the entire coding sequence for a GFP derivative that shows enhanced fluorescence.[21,22] On average, one in six insertion events in a coding region will generate an in-frame fusion to the reporter; such fusions may be detected by assays for β-Gal or fluorescence activity. Quantitative measurements can be obtained for either activity, providing a reliable index of gene expression (e.g., Refs. 13 and 20–23).

A significant new feature of these transposons is their ability to be reduced to a much smaller element that falls in-frame with the mutagenized

* Accession numbers: m-Tn3(*LEU2 lacZ*), U35112; mTn-3xHA/*lacZ,* U54828; mTn-4xHA/*lacZ,* U54829; mTn-3xHA/GFP, U54830; pRSQ2-*URA3,* U64694; pRSQ2-*LEU2,* U64693; pHSS6, M84115.

[19] M. F. Hoekstra, H. S. Seifert, J. Nickoloff, and F. Heffron, *Methods Enzymol.* **194,** 329 (1991).

[20] B. Rockmill and G. S. Roeder, *Genes Dev.* **5,** 2392 (1991).

[21] D. C. Prasher, V. K. Eckenrode, W. W. Ward, F. G. Prendergast, and M. J. Cormier, *Gene* **111,** 229 (1992).

[22] R. Heim, D. C. Prasher, and R. Y. Tsien, *Proc. Natl. Acad. Sci. U.S.A.* **91,** 12501 (1994).

[23] M. J. Casadaban, A. Martinez-Arias, S. K. Shapira, and J. Chou, *Methods Enzymol.* **100,** 293 (1983).

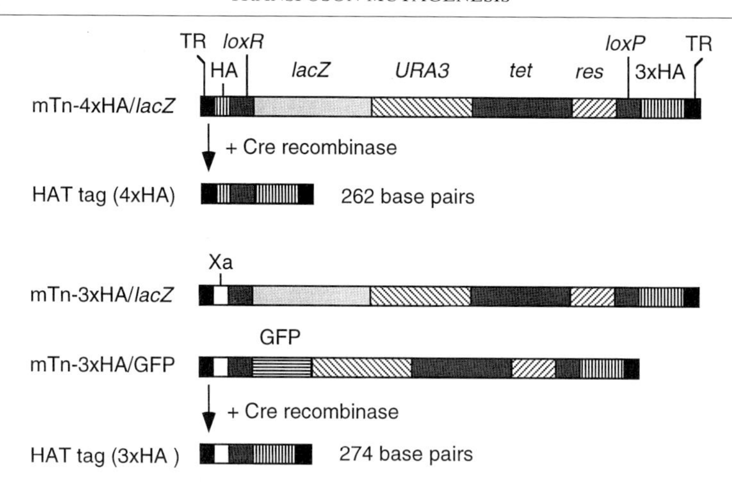

FIG. 1. Diagram of the three new mTns, and of the derived HAT tag elements. mTn-4xHA/*lacZ* (6107 bp) may be reduced by a *lox*/Cre excision event to a 262-bp HAT tag that contains four copies of the HA epitope. mTn-3xHA/*lacZ* (6119 bp) and mTn-3xHA/GFP (4342 bp) may be reduced by a *lox*/Cre excision event to a 274-bp HAT tag that contains three copies of the HA epitope and a factor Xa protease cleavage site. *loxR* and *loxP, lox* sites; targets for Cre recombinase. TR, Tn*3* terminal repeats. HA, sequence encoding the HA epitope. 3xHA, sequence encoding three tandem copies of the HA epitope; codons have been chosen so that the DNA sequence does not repeat. Xa, Factor Xa protease cleavage site. Not drawn to scale.

coding region and causes the insertion of an epitope tag into the encoded protein. This is possible because all three mTns carry a *loxR* site near one end and a *loxP* site near the other end (Fig. 1). (Both *lox* sites are targets for the Cre recombinase of phage P1, but their sequence diverges slightly, reducing spontaneous recombination frequency.) The *lox* sites are internal to sequences encoding multiple copies of a defined epitope from the influenza virus hemagglutinin protein (the HA epitope[24]). The mTn-3xHA transposons also contain a sequence encoding a cleavage site for the factor Xa protease (Ile-Glu-Gly-Arg[25]) in the region between the *loxR* site and the end of the transposon.

Exposure of the mTns to Cre recombinase activity in yeast catalyzes recombination between the two *lox* sites and removes the entire central region of the transposon. The resulting product contains the 5-bp duplica-

[24] I. A. Wilson, H. L. Niman, R. A. Houghten, A. R. Cherenson, M. L. Connolly, and R. A. Lerner, *Cell* **37,** 767 (1984).

[25] S. Magnusson, T. E. Peterson, L. Sottrup-Jensen, and H. Claeys, *in* "Proteases and Biological Control" (E. Reich, D. B. Rifkin, and E. Shaw, eds.), p. 123. Cold Spring Harbor Laboratory Press, Cold Spring Harbor, New York, 1975.

tion of flanking yeast DNA generated during the transposition event, plus a 274-bp (mTn-3xHA) or 262-bp (mTn-4xHA) element that contains the Tn*3* terminal repeats flanking a single *loxR* site and sequence encoding three or four copies of the HA epitope; the mTn-3xHA product also contains sequence encoding the factor Xa protease cleavage site. Where the original insertion created an in-frame fusion to the reporter gene, the excision event results in insertion of 93 amino acids (mTn-3xHA) or 89 amino acids (mTn-4xHA) into the encoded protein. We have named this insertion the "HAT tag."

The new transposon system has been tested by mutagenesis of several individual yeast genes. The mTns proved successful in creating reporter fusions, defining important functional domains of a gene, generating conditional mutations, and localizing the gene products.[15] mTn-3xHA/*lacZ* and mTn-3xHA/GFP have also been used to mutagenize a yeast genomic DNA library.[15] The mutagenized inserts from this library can be used to transform yeast, creating a bank of marked genomic mutations. Use of such mutant banks to screen for phenotypes of interest has become increasingly popular in the yeast community.[12,26,27] Since protein production can also be assayed using the mTn-borne reporter gene, the banks can also be used to screen for genes whose expression is differentially regulated either during the life cycle[12,27a] or by known genes of interest.[28]

In this chapter, we provide comprehensive instructions for use of the new transposons to mutagenize a gene of interest, and for use of the transposon insertion libraries to mutagenize the yeast genome. While the application of these specific transposons is limited to organisms in which the *Saccharomyces cerevisiae* selectable marker *URA3* can be used, the approach is generally applicable to mutagenesis of DNA from any organism for which a transformation and selection system exists.

Materials and Methods

Reagents

Escherichia coli is grown on standard LB medium (1% tryptone, 0.5% yeast extract, 1% NaCl, solidified with 1.8% agar); medium for growth of yeast is described by Sherman *et al.*[29] Information on all bacterial strains,

[26] K. T. Chun and M. G. Goebl, *Genetics* **142,** 39 (1996).

[27] H. Mosch, R. Roberts, and G. Fink, *Proc. Natl. Acad. Sci. U.S.A.* **93,** 5352 (1996).

[27a] S. Erdman and M. Snyder, *J. Cell Biol.* **140,** 461 (1998).

[28] V. D. Dang, M. Valens, M. Bolotin-Fukuhara, and Daignan-Fornier, *Yeast* **10,** 1273 (1994).

[29] F. Sherman, G. Fink, and J. Hicks, "Methods in Yeast Genetics." Cold Spring Harbor Laboratory Press, Cold Spring Harbor, New York, 1986.

antibiotics, oligonucleotide and antibody reagents used is presented in Table I.

Shuttle Mutagenesis Procedure

The reactions that occur in the mutagenesis procedure can at first appear complex and daunting (Fig. 2), but in practice the work is performed by *E. coli* and the biologist merely grows and mixes strains, then plates on selective medium. The only cloning step occurs at the outset, when the DNA must be placed in a vector lacking the Tn*3* terminal repeat, such as the pHSS series described by Hoekstra *et al.*[19] The pHSS construct is then transformed into a strain that expresses the Tn*3* transposase, TnpA. When mating is used to introduce an F plasmid carrying the desired transposon, TnpA can act *in trans* and initiate transposition. A large plasmid called a cointegrate is formed; since the cointegrate contains the F plasmid it can be transferred by mating into an *E. coli* strain that expresses the Tn*3* resolvase, TnpR. TnpR acts on the two *res* sites to resolve the cointegrate into its two constituent plasmids: the F plasmid and the pHSS construct of interest, which now also carries a single copy of the transposon. Because most of the vector backbone DNA is essential for replication functions and antibiotic resistance, the insertions recovered fall preferentially in the insert DNA. Transposition of these mTns shows little or no site preference within most inserts.[15] Since a large population of cells is used for each of the above steps, and the mTn can integrate only once per plasmid, the final product is a library of mutagenized pHSS constructs, each carrying a transposon insertion at a different site. With a single run through this procedure it is easy to generate 10^5 independent insertions.

1. The fragment is cloned into a pHSS vector such as pHSS6 (obtained from strain R1123). Transformants are selected on LB medium containing Kan (LB + Kan; see Table I for information on all antibiotics).
2. The plasmid is transformed into strain B211, with selection on LB + Kan + Cm.
3. A B211 transformant is inoculated into 2 ml of LB + Kan + Cm. The mTn-carrying strain (B426, B427, or B428, as preferred) is inoculated into 2 ml of LB + Tet. Cultures are incubated at 37° overnight with aeration.
4. Each strain (B211 and either B426, B427, or B428) is subcultured at 1:100 dilution into LB medium and grown at 37° with aeration for 2–3 hr, or until a cell suspension is first visible. Two hundred microliters of each culture is mixed in a microcentrifuge tube and incubated at 37° without agitation for 20 min to 1 hr. (*Note:* The mating process is highly sensitive to agitation and detergent. Shorter mating times will maximize the number of independent insertions obtained.)

TABLE I
REAGENTS

Escherichia coli strains

R1123: Strain XL1-Blue (Stratagene, La Jolla, CA) carrying pHSS6 (kanamycin resistance)[a]

B211: Strain RDP146 (F^- *recA'* (*Δlac-pro*) *rpsE*; spectinomycin resistant) with plasmid pLB101 (pACYC184 with *tnpA*; active transposase, chloramphenicol resistance)[a]

B425: Strain NG135 (F^- K12 *recA56 gal-delS165 strA*; streptomycin resistant) with plasmid pNG54 (pACYC184 with *tnpR*; active resolvase, chloramphenicol resistance)[b]

B426: Strain RDP146 with pOX38::mTn-4xHA/*lacZ* (F factor derivative carrying mTn*3* derivative; tetracycline resistance)[b]

B427: Strain RDP146 with pOX38::mTn-3xHA/*lacZ* (F factor derivative carrying mTn*3* derivative; tetracycline resistance)[b]

B428: Strain RDP146 with pOX38::mTn-3xHA/GFP (F factor derivative carrying mTn*3* derivative; tetracycline resistance)[b]

B227: Strain DH5-α carrying pGAL-*cre*(*amp, ori, CEN, LEU2*)[a]

Antibiotics

Tetracycline-HCl (Tet), 3-mg/ml stock in water. Use at 3 μg/ml

Kanamycin (Kan), 10-mg/ml stock in water. Use at 40 μg/ml

Chloramphenicol (Cm), 34-mg/ml stock in ethanol. Use at 34 μg/ml

Streptomycin (Sm), 10-mg/ml stock in water. Use at 50 μg/ml

Ampicillin (Amp), 50-mg/ml stock in water. Use at 50 μg/ml

Oligonucleotide primers

5′-3xHA: 5′-CCGTTTACCCATACGATGTTCCTG-3′. Bases 133 to 156 of the HAT tag from mTn-3xHA; bases 121 to 144 of the HAT tag from mTn-4xHA (sense strand)

3′-3xHA: 5′-GAGCGTAATCTGGAACGTCATATGG-3′. Bases 228 to 204 of the HAT tag from mTn-3xHA; bases 216 to 192 of the HAT tag from mTn-4xHA (antisense strand)

Anchor bubble primers[c]: 5′-GAAGGAGAGGACGCTGTCTGTCGAAGGTAAG-GAACGGACGAGAGAAGGGAGAG-3′ and 5′-GACTCTCCCTTCTCGAATCG-TAACCGTTCGTACGAGAATCGCTGTCCTCTCCTTC-3′. To anneal, an aqueous solution that is 2–4 μ*M* for each primer is heated at 65° for 5 min. $MgCl_2$ is added to a final concentration of 2 m*M*, and the solution is allowed to cool slowly to room temperature. Store at −20°

UV primer[c]: 5°-CGAATCGTAACCGTTCGTACGAGAATCGCT-3′

M13(−47) primer: 5′-CGCCAGGGTTTTCCCAGTCACGAC-3′ [bases 177–154 of mTn-3xHA/*lacZ*, antisense; bases 165–142 of mTn-4xHA/*lacZ*, antisense; bases 84–61 of m-Tn3(*LEU2 lacZ*), antisense]

GFP primer: 5′-CATCACCTTCACCCTCTCCACTGAC-3′ (bases 243–219 of mTn-3xHA/GFP, antisense)

Xa primer: 5′-CTTCTACCTTCAATGGCCGCC-3′ (bases 58–38 of mTn-3xHA/GFP and mTn-3xHA/*lacZ*, antisense)

Antibodies

Mouse monoclonal anti-HA 16B12 [MMS101R (BAbCO, Richmond, CA), used at 1:200 to 1:1000 dilution]

Mouse monoclonal anti-HA 12CA5 [12CA5 (Boehringer Mannheim, Indianapolis, IN), used at 1:200 dilution]

TABLE I (*continued*)

Rabbit polyclonal anti-HA [RS1010C (BAbCO), used at 1:200 dilution]
Cy3-conjugated affinity-purified goat anti-mouse IgG (Jackson Laboratories, West Grove, PA); 1-mg/ml stock used at 1:300 dilution)
Texas Red-conjugated affinity-purified donkey anti-rabbit (Jackson Laboratories; 1-mg/ml stock used at 1:200 dilution)

[a] M. F. Hoekstra, H. S. Seifert, J. Nickoloff, and F. Heffron, *Methods Enzymol.* **194,** 329 (1991).

[b] P. Ross-Macdonald, A. Sheehan, S. Roeder, and M. Snyder, *Proc. Natl. Acad. Sci. U.S.A.* **94,** 190 (1997).

[c] J. Riley, R. Butler, D. Ogilvie, R. Finniear, D. Jenner, S. Powell, R. Anand, J. C. Smith, and A. F. Markham, *Nucleic Acids Res.* **18,** 2887 (1990).

5. Aliquots of 100 μl are plated onto LB + Tet + Kan + Cm. As controls, a 20-μl aliquot from each strain culture is incubated on this medium (cells should not grow). Plates are incubated at 30° for 1–2 days to allow cointegrate formation. Many thousand colonies should be obtained on these "cointegrate" plates.

6. Strain B425 is grown at 37° overnight with aeration in LB + Cm.

7. Colonies are eluted from the cointegrate plates (i.e., 2 ml of LB medium is placed on each plate and cells are scraped into a homogeneous suspension). Strain B425 is subcultured at 1:100 dilution in LB medium, and an aliquot of the cointegrate eluate is diluted into LB to give approximately the same cell density as the diluted B425 culture. Strains are grown and mated as in step 4. Aliquots of 100 μl are plated on LB + Tet + Kan + Sm + Cm and incubated at 37° overnight. Controls should be performed as in step 5. Many thousand colonies should be obtained.

8. Colonies from the mating are eluted as above. An aliquot of the eluate is diluted into LB + Tet + Kan medium to give a culture of almost saturated density that is grown at 37° with aeration for 1–2 hr. The remaining eluate should be stored at 4° until step 9 has been completed successfully.

9. Plasmid DNA is isolated from the culture by a miniprep procedure. About one-tenth of the miniprep DNA is transformed into a standard *recA endA E. coli* strain (e.g., DH5α) with selection on LB + Tet + Kan. Many thousand colonies should be obtained. (*Note:* This step both purifies the mutagenized pHSS construct away from pNG54, and ensures that high-quality DNA can be recovered for yeast transformation.)

10. The entire pool of transformants is eluted and an aliquot of eluate is used to obtain a plasmid DNA miniprep as in step 8. Sterile glycerol is added to the remaining eluate to a final concentration of 15%, and this stock is saved at −70° to allow preparation of additional DNA. Plasmid

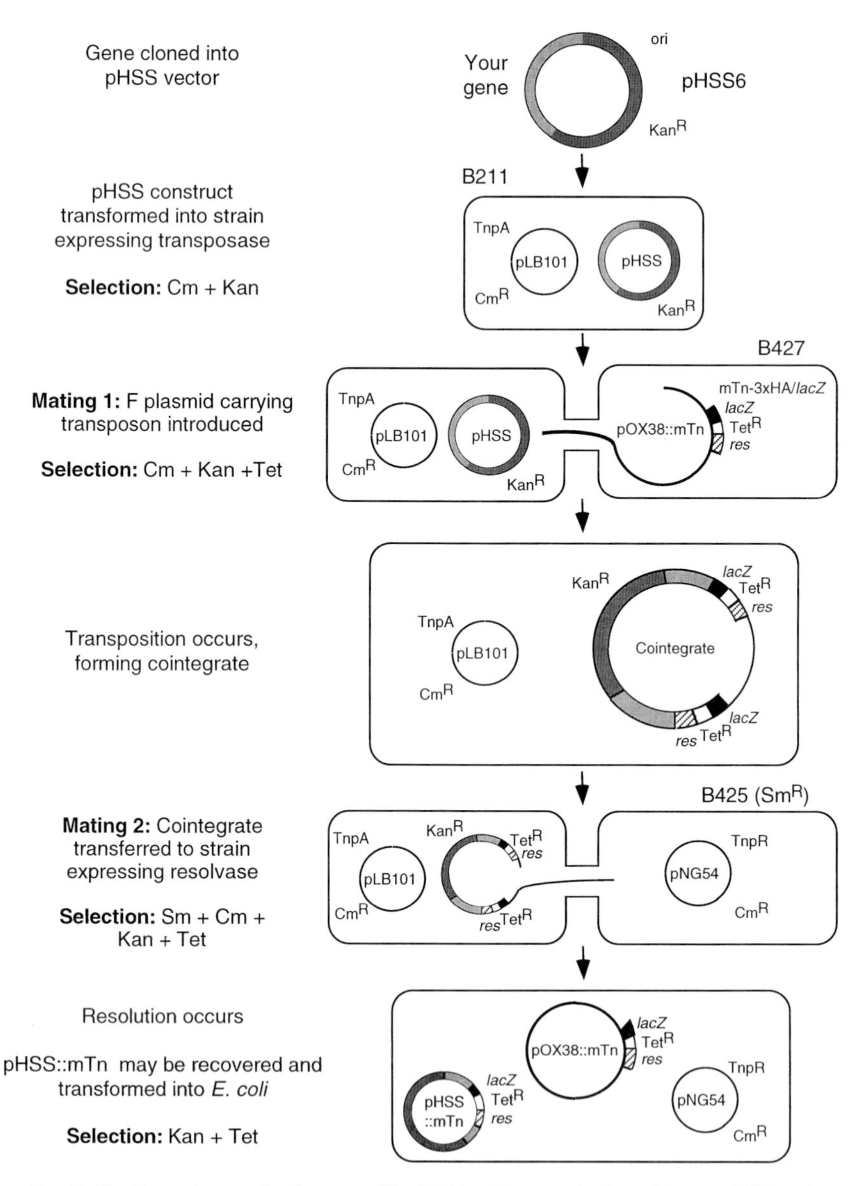

FIG. 2. Shuttle mutagenesis of a gene. The DNA of interest is cloned into a pHSS series vector, and the plasmid transformed into an *E. coli* strain that constitutively expresses the Tn*3* transposase (B211). An F plasmid carrying the desired mTn is introduced into the transformant by mating. Transposition occurs, forming a large plasmid called a cointegrate. The cointegrate is transferred by mating into an *E. coli* strain that constitutively expresses the Tn*3* resolvase (B425). Resolution occurs, releasing the pHSS-based plasmid, which now contains an mTn insertion. This plasmid is recovered by small-scale DNA preparation and transformed into a standard *E. coli* laboratory host. Cm, Chloramphenicol; Kan, kanamycin; Tet, tetracycline; Sm, streptomycin.

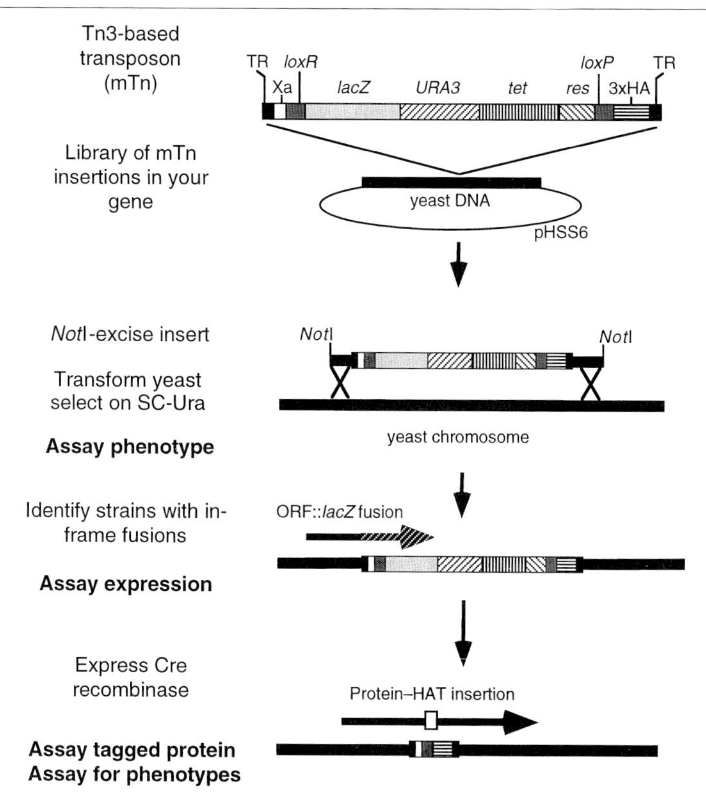

FIG. 3. Assays that are possible using a yeast gene mutagenized with an mTn. In this example, mTn-3xHA/*lacZ* was used to mutagenize yeast DNA in the vector pHSS6. *loxR* and *loxP, lox* sites; targets for Cre recombinase. TR, Tn*3* terminal repeats. 3xHA, sequence encoding three tandem copies of the HA epitope. Xa, Factor Xa protease cleavage site; HAT, small insertion element formed by action of Cre recombinase on mTn-3xHA/*lacZ*. Not drawn to scale.

DNA may also be prepared from individual transformants for detailed analysis.

Generating and Analyzing Yeast Transformants

An outline of the analyses that may be performed in yeast is presented in Fig. 3. Following excision from the vector by a rare-cutting enzyme, the mutagenized DNA is transformed into a *ura3* yeast strain. If the *lox*/Cre excision feature of the transposon is to be used, a *ura3 leu2* strain capable of inducing gene expression from the *GAL1-10* promoter is required. It is also advisable to introduce p*GAL-cre* into the strain prior to transformation with the mTn-mutagenized DNA, and then to maintain selection for the plasmid-borne *LEU2* marker in all steps below. When the mutagenized gene is known or suspected to serve an essential function, a diploid yeast strain should be used. Selection for the *URA3* marker carried by the transposon identifies transformants in which the mutagenized DNA has replaced the chromosomal locus by homologous recombination. (An alternative strategy is to place a yeast replication origin and selectable marker in the pHSS vector backbone, allowing extrachromosomal manipulation of the mutagenized gene; J. Y. Leu and S. Roeder, personal communication, 1998.) The transformants may be screened for phenotypes caused by insertion of the transposon into the gene. When an mTn carrying *lacZ* has been used, transformants carrying in-frame fusions between the mutagenized gene and the reporter construct may be identified by assays for β-Gal activity. For mTn-3xHA/GFP insertions, transformants may be screened by fluorescence microscopy (e.g., Ref. 15) or fluorescence-activated cell sorting (FACS; e.g., Ref. 13). In addition to use for analysis of protein production, strains carrying in-frame fusions may be used to derive strains carrying HAT tag insertions (see below).

1. Plasmid DNA prepared at the last step of the mutagenesis protocol is digested with the appropriate restriction enzyme (*Not*I for most pHSS vectors). Gel analysis should reveal two major bands: one corresponding in size to the vector, the other to the insert DNA plus the mTn. Some insertion events may occur in the vector, generating the converse set of bands.

2. The restriction enzyme is inactivated by the appropriate heat treatment and the digest transformed into yeast by a standard protocol,[3] selecting for *URA3*.

3. Transformants are patched to SC-Ura medium and saved for subsequent screening.

4. To identify transformants carrying an in-frame fusion to GFP, cells may be examined directly by fluorescence microscopy. Use of cells from a culture in the exponential phase of growth (OD_{600} of 1) may reduce autofluorescence by the yeast cell wall. The fixation and spheroplasting techniques given in the protocol for immunofluorescence may also be used.

5. To identify transformants carrying an in-frame fusion to *lacZ,* patches are replica plated to an SC-Ura plate and to an SC-Ura plate on which a sterile disk of Whatman (Clifton, NJ) 1A filter paper has been placed, and grown overnight at 30°. Other media or growth conditions can be substituted as desired. [For *ade2* strains, test media should contain adenine (80 mg/liter).] After cell growth, filters are lifted from plates and exposed to chloroform vapor in a sealed vessel. We use a 15-min exposure; the time required

may vary between strains. Filters are then placed colony-side up on X-Gal plates [5-bromo-4-chloro-3-indolyl-β-D-galactopyranoside (120 μg/ml), 0.1 *M* $NaPO_4$ (pH 7), and 1 m*M* $MgSO_4$ in 1.6% agar] and incubated at 30° for up to 2 days.

6. Once identified, transformants carrying productive fusions to the reporter are recovered from the SC-Ura plate.

7. To determine the position of an mTn insertion within the gene of interest, genomic DNA is prepared[30,31] from the strain(s) of interest and the polymerase chain reaction (PCR) is performed using a primer from the gene of interest and the appropriate primer from the insertion [e.g., M13(-47) for mTns containing *lacZ*, or the GFP primer for mTn-3xHA/GFP; Table I]. The size of products is determined on an agarose gel. PCR products may also be purified with Qiaquik (Qiagen, Chatsworth, CA) and sequenced.

Generating Strains Carrying HAT Tag

Transformants carrying in-frame fusions to the reporter can be further processed to generate strains that contain an HAT tag insertion. The excision event is accurate and can be induced to occur at high frequency.[15] Strains containing a HAT tag are useful for all immunodetection procedures, and may also be screened for phenotypes caused by the small insertion.

1. Strains carrying reporter fusions must be transformed with p*GAL-cre* if this plasmid is not already present. Transformants are selected on SC-Ura-Leu.

2. Individual transformants are inoculated into 2 ml of SC-Ura-Leu with 2% raffinose as carbon source, and grown to a saturated cell density. These transformants should also be saved on a plate, since a stock should be made of the reporter fusion strain that gives rise to any HAT tags of interest.

3. Cultures are diluted 1:100 into SC-Leu with 2% galactose as carbon source. As a control, they are also diluted 1:100 into SC-Leu with 2% glucose as carbon source. Cultures are grown to a saturated cell density (1–2 days).

4. For quantitative assessment of efficiency of the excision event, plate dilutions of each culture onto SC medium. When colonies have grown, for each culture select a plate containing about 100 colonies and replica plate

[30] P. Philippsen, A. Stotz, and C. Scherf, *Methods Enzymol.* **194,** 169 (1991).

[31] N. Burns, P. Ross-Macdonald, G. S. Roeder, and M. Snyder, *in* "Microbial Genome Methods" (K. W. Adolph, ed.), p. 61. CRC Press, Boca Raton, Florida, 1996.

to SC-Ura. Cultures grown with galactose as carbon source usually contain >85% Ura^- cells and should give 100- to 1000-fold more Ura^- colonies than cultures containing glucose. Ura^- cells derived from the cultures grown with galactose as carbon source should be used for subsequent steps.

5. To determine the position of a HAT tag insertion within the gene of interest, prepare genomic DNA[30,31] from the strain(s) of interest and perform PCR using a primer from the gene of interest and the appropriate primer from the insertion (5′-3xHA or 3′-3xHA; Table I).

Immunodetection Using HAT Tag

The multiple copies of the HA epitope present in the HAT tag can be detected by the mouse monoclonal antibodies 12CA5 (Boehringer Mannheim, Indianapolis, IN) and 16B12 (MMS101R; BAbCO, Richmond, CA) (see Table I). On Western blots against yeast protein, these antibodies often recognize a single cross-reacting band of about 50 or 125 kDa, respectively, although the level of the 50-kDa polypeptide can be reduced by using cultures in the early phase of exponential division. The antibodies may also give punctate background when used for immunocytochemistry, although we have found that this can be eliminated by both minimizing the spheroplasting process and preadsorbing primary and secondary antibodies in multiple incubations against fixed yeast cells suspended in PBS.[31,32] A rabbit polyclonal antisera is also available (101c500; BAbCO) but in our hands this was less reactive. A method for preparing cells for indirect immunofluorescence is as follows:

1. Cells are grown in 5 ml of YPAD to an OD_{600} of 0.75 to 1.
2. To fix, a 1/10 vol of 37% formaldehyde is added to the culture, which is shaken for a further 40 min at room temperature. Cells are then recovered by centrifugation at 1400 *g* for 2 min, and washed (i.e., gently resuspended, then recovered and the supernatant discarded) twice with solution A (1.2 *M* sorbitol, 50 m*M* KPO_4, pH 7).
3. To spheroplast, cells are resuspended in 500 μl of solution A containing 0.1% 2 mercaptoethanol, 0.02% glusulase, and zymolyase (5 μg/ml). The suspension is incubated at 37° without shaking and checked periodically. As soon as the settled cell pellet loses its creamy, yellow color and becomes translucent (30 to 45 min), cells should be recovered and washed twice with solution A.

[32] P. Ross-Macdonald, N. Burns, M. Malcynski, A. Sheehan, S. Roeder, and M. Snyder, *Methods Mol. Cell. Biol.* **5,** 298 (1995).

4. Cells are resuspended in 200 μl of solution A. A drop is placed on a poly-L-lysine-coated slide[33] and allowed to sit for 10 min. If one desires to examine several different strains on one slide, multiwell slides are available from Carlson Scientific (Peotone, IL).

5. Excess solution is aspirated. The adhered cells are covered with PBS [150 m*M* NaCl, 50 m*M* $NaPO_4$ (pH 7.4)] plus 0.1% bovine serum albumin (BSA). The solution is allowed to sit for 5 min before aspiration. This wash is repeated twice with PBS–0.1% BSA containing 0.1% Nonidet P-40 (NP-40).

6. The cells are covered with PBS–0.1% BSA containing the primary antibody (Table I). The slide is set on wet paper towels in a sealed chamber and incubated at 4° overnight.

7. Excess solution is aspirated. The cells are covered with PBS–0.1% BSA, which is allowed to sit for 5 min before aspiration. This wash is repeated with PBS–0.1% BSA containing 0.1% NP-40, then again with PBS–0.1% BSA.

8. The cells are covered with PBS–0.1% BSA containing the secondary antibody (Table I). The slide is set on wet paper towels in a sealed chamber and incubated at room temperature for 2 hr.

9. Excess solution is aspirated. The cells are covered with PBS–0.1% BSA, which is allowed to sit for 5 min before aspiration. This wash is repeated with PBS–0.1% BSA containing 0.1% NP-40, then again with PBS–0.1% BSA.

10. Mount solution [70% glycerol containing 2% *n*-propyl gallate and Hoechst (0.25 μg/ml)] is placed on the cells. A coverslip is placed and sealed with nail polish. Slides are stored at −20° until examination.

Preparing and Transforming DNA from Mutagenized Genomic Libraries

The authors make available three different transposon insertion libraries. The transposons m-Tn*3*(*LEU2 lacZ*),[19] mTn-3xHA/*lacZ*, and mTn-3xHA/GFP (Fig. 1[15]) were each used to create about 10^6 insertions in a yeast genomic library described by Burns *et al.*[12] that contained 18 genome equivalents in the form of 2.5- to 3.5-kb *Sau*3AI partial fragments. These libraries each contain 18 independent pools, and are distributed as plasmid DNA. Chun and Goebl[26] have described a transposon insertion library that contains m-Tn*3*(*URA3 lacZ*).[19] Use of the libraries is similar in principle to the procedure for handling mutagenized DNA from individual genes (Fig. 3). Digestion with *Not*I releases the mutagenized DNA from the vector; it is then transformed into the desired yeast strain selecting for the

[33] J. Pringle, A. E. M. Adams, D. G. Drubin, and B. K. Haarer, *Methods Enzymol.* **194,** 565 (1991).

transposon-borne marker. To minimize the occurrence of multiple transposon insertions within a single transformant, the minimum amount of DNA that still gives a practicable number of transformants should be used in each transformation.

1. Each pool of the library DNA is transformed into a standard laboratory *E. coli* host, selecting for kanamycin resistance (selection for the transposon-borne antibiotic resistance may also be applied). At least 10^4 transformants should be obtained for each pool.

2. Colonies are eluted from transformation plates (i.e., 2 ml of liquid LB is placed on each plate and cells are scraped into a homogeneous suspension). An aliquot of the eluate is diluted into LB + Kan medium to give an almost saturated cell density, and this culture is grown at 37° for 1–2 hr. Plasmid DNA is then isolated by a miniprep procedure. The eluate should also be saved as a 15% glycerol frozen stock for future DNA preparation.

3. The plasmid DNA is digested with *Not*I. Gel analysis should reveal two major products: a 2.2-kb band corresponding to the vector, and a band of approximately 8 kb corresponding to the insert DNA containing the mTn.

4. The *Not*I is inactivated by incubation at 65° for 15 min, and the digest transformed into yeast,[3] selecting for the marker on the mTn. Transformation efficiency should be typical for a targeted gene replacement in the strain being investigated. To determine the minimum amount of DNA needed to give a reasonable number of transformants, a series of test transformations containing from 0.2 to 2 μg of insert DNA should be performed.

Screening

When screening for a mutant phenotype, approximately 30,000 transformants must be used to ensure 95% coverage of the genome. Transformants can also be screened for reporter activity (using the same assays for β-Gal or fluorescence activity described above in the protocols for mutagenizing an individual gene); in this case about 180,000 transformants must be analyzed for equivalent coverage. Since the 18 library pools are totally independent and probably show different biases, the best strategy is to screen transformants from all pools. Once a transformant of interest has been identified, conventional genetic analysis should be used to determine whether the phenotype segregates with the transposon insertion. The site of the insertion may then be determined.

Identifying the Insertion Site by Plasmid Rescue

Two rescue plasmids, designated pRSQ2-*LEU2* and pRSQ2-*URA3*, have been constructed for use with mTns that contain *lacZ*. Transformation

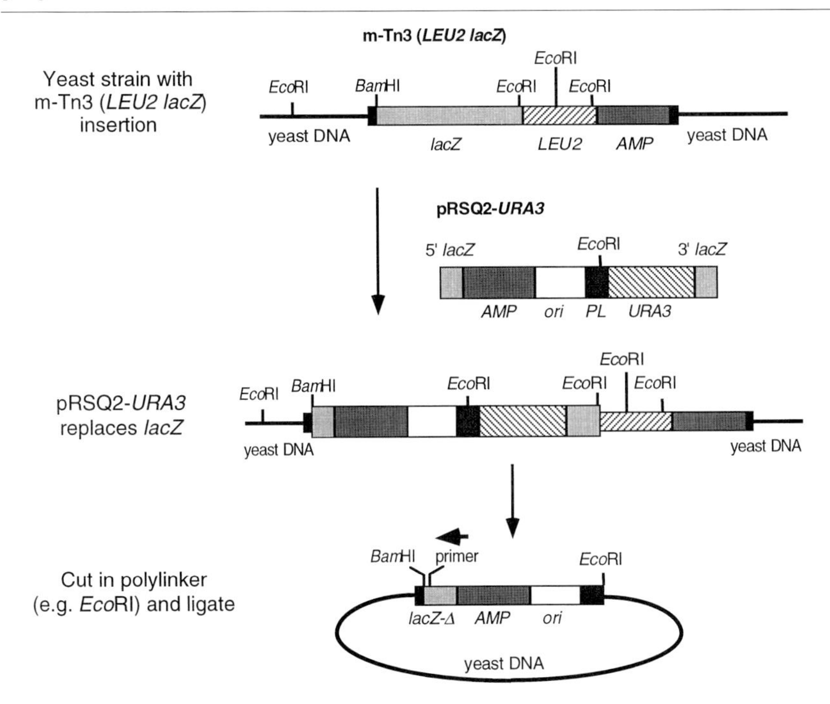

FIG. 4. Strategy for recovering genomic DNA at site of mTn insertion using "plasmid rescue." In this example, pRSQ2-*URA3* (linearized with *Bam*HI) is used to recover the site of m-Tn*3* (*LEU2 lacZ*) insertion. Relevant restriction sites for recovery and plasmid analysis are shown. A primer to *lacZ* sequences is used to obtain sequence data. PL, Polylinker region of pRSQ2-*URA3*.

of the rescue plasmid into a yeast strain carrying an insertion is used to introduce a bacterial replication origin into the mTn. This process is illustrated for m-Tn*3*(*LEU2 lacZ*) in Fig. 4. A restriction fragment containing the origin, a selectable marker, the end of the mTn, and some adjacent yeast genomic DNA is then circularized and recovered in *E. coli* as a high copy number plasmid. A primer complementary to the mTn sequence is used to obtain genomic DNA sequence. Plasmid rescue has the advantage of generating an expedient sequencing template, and can also be used to move disruption alleles or HAT-tagged alleles to a different strain background.

1. The yeast strain(s) carrying the mTn insertion is transformed[3] with 1–5 μg of *Bam*HI-linearized DNA of the appropriate pRSQ2 plasmid DNA, selecting for the pRSQ2 marker.
2. Genomic DNA is prepared[30,31] from a transformant.
3. Five micrograms of genomic DNA is digested overnight with 5 U of

"recovery" enzyme (*Eco*RI, *Hin*dIII, *Sal*I, *Cla*I, *Xho*I, or *Kpn*I for pRSQ2-*LEU2;* the same enzymes plus *Eco*RV or *Pst*I for pRSQ2-*URA3*) in a total volume of 40 μl.

4. Digestion is checked by gel electrophoresis using half of the sample. The remainder is heated to 65° for 25 min to inactivate the restriction enzyme. H_2O (215 μl), 25 μl of 10× ligase buffer, and 1 μl of ligase (400 U) are added. To favor intramolecular reactions, the DNA concentration in the ligation should not exceed 10 μg/ml and can be as low as 2 μg/ml.

5. After ligation at 16° for 4 to 16 hr, DNA is precipitated by addition of 125 μl of 7.5 *M* ammonium acetate and 375 μl of isopropanol, and recovered by centrifugation at 13,000 *g* for 15 min.

6. The DNA pellet is washed with 70% ethanol, air dried, and resuspended in 6–20 μl of TE buffer [10 m*M* Tris (pH 7.5), 1 m*M* EDTA]. Three microliters is transformed into *E. coli,* selecting for ampicillin resistance.

7. Plasmid DNA is recovered from several colonies for each strain, and analyzed by double-digestion with *Bam*HI and the recovery enzyme. Desired plasmids display a 2.85-kb band (3.9 kb for *Eco*RI used as recovery enzyme with pRSQ2-*LEU2*) containing vector sequences, plus additional band(s) from genomic DNA. If "mystery" plasmids occur, a different yeast transformant and/or recovery enzyme should be used.

8. The plasmid DNA preparations may be sequenced by standard techniques using the M13(−40) primer (New England BioLabs, Beverly, MA). Since chimerism can occur during circularization, sequence is reliable only up to the first site for the recovery enzyme.

9. A plasmid that allows the disruption allele to be transferred to other strains can be made by cleaving with one of the following enzymes at step 3: *Avr*II, *Bgl*II, *Bsp*EI, *Eag*I, *Msc*I, *Nae*I, *Nhe*I, *Nru*I, *Pml*I, *Sma*I, *Sna*BI, *Spe*I, *Sph*I, or *Xma*I. With these enzymes, a plasmid of >11.7 kb containing sequences both 5′ and 3′ to the transposon insertion is recovered. This plasmid is linearized with the same enzyme to target homologous replacement. The resulting insertion no longer has reporter activity; however, formation of the HAT tag by the *lox*/Cre system should not be affected.

Identifying Insertion Site by Vectorette Polymerase Chain Reaction

An adaptation of the method of Riley *et al.*[34] can be used to recover genomic DNA adjacent to the mTn insertion, as illustrated in Fig. 5. Genomic DNA from the strain of interest is digested with a restriction enzyme to produce small, blunt-ended fragments. An "anchor bubble" is ligated

[34] J. Riley, R. Butler, D. Ogilvie, R. Finniear, D. Jenner, S. Powell, R. Anand, J. C. Smith, and A. F. Markham, *Nucleic Acids Res.* **18,** 2887 (1990).

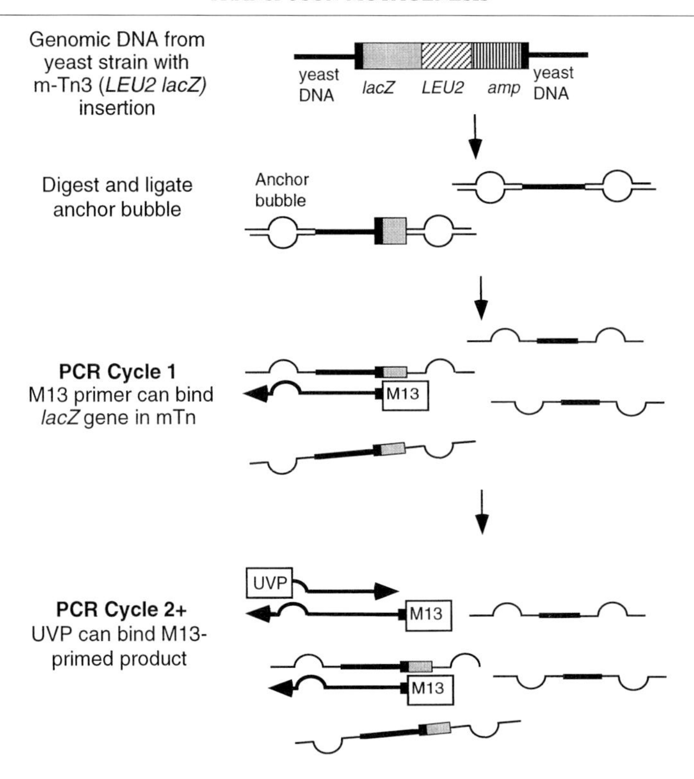

FIG. 5. Strategy for recovering genomic DNA at the site of mTn insertion, using vectorette PCR. In this example, the site of m-Tn3 (*LEU2 lacZ*) insertion is recovered. M13, M13(−47) primer, which hybridizes to *lacZ* sequences in the transposon; UVP, universal vectorette primer, identical to a sequence in the nonhomologous central region of the anchor bubble.

to the fragment ends. PCR is then performed using a primer complementary to mTn sequences, and a primer that is identical to sequence in the bubble region (the universal vectorette or UV primer). In the first cycle of PCR, only the mTn primer can bind the template. In subsequent cycles, the UV primer can bind the product of the mTn primer. Thus, only the fragment containing the mTn primer-binding site is amplified. The PCR product may then be isolated and sequenced. The vectorette PCR method is reported to be robust, and has the advantage of eliminating transformation and cloning steps. It is also the only method currently available to determine the insertion site in a yeast genome mutagenized with mTns that do not contain *lacZ*.

1. Good-quality genomic DNA[30,31] is prepared from the yeast strain carrying the mTn insertion.

2. One to 3 μg of genomic DNA is digested overnight with 8–10 U of *Alu*I, in a total volume of 20 μl. Other enzymes that cut frequently and leave blunt ends may be used, provided that there is at least one site for them in the mTn (to ensure that both terminal inverted repeats are not present on the fragment) and that this site does not lie between the mTn end and the binding site for the suggested mTn primer (Table I).

3. The enzyme is heat inactivated by heating the digest to 65° for 20 min, and the following reagents are added: 24.5 μl of water, 3 μl of the 10× buffer used in the restriction digest, 1 μl of annealed anchor bubble (see Table I), 1 μl of ligase (400 U), and 0.5 μl of 5 m*M* ATP. The ligation reaction is incubated at 16° for 9–24 hr.

4. A 100-μl PCR mixture is set up with the following components: 71 μl of water, 5 μl of the ligation reaction, 2.5 μl of 20 μ*M* M13(−47) primer (or GFP primer for mTn-3xHA/GFP; Table I), 2.5 μl of 20 μ*M* UV primer (Table I), 8 μl of 2.5 m*M* dNTPs, 10 μl of 10× *Taq* PCR buffer, and 1 μl of *Taq* DNA polymerase (5 U). A "hot start" using Ampliwax (Perkin-Elmer, Norwalk, CT) is recommended.

5. The following thermal cycles are performed: denaturation at 92° for 2 min, then 35 cycles of 20 sec at 92°, 30 sec at 67°, and 45–180 sec at 72° (the longer interval is used if a product is not generated in the shorter interval), followed by a single cycle of 90 sec at 72°.

6. The PCR mixture is run on a 1–3% SeaKem GTG agarose (FMC BioProducts, Rockville, ME) gel. Usually only one product band, containing 200–400 ng of DNA, is seen. All bands are excised individually, and the DNA is recovered with Qiaex (Qiagen), eluting with 12 μl of TE buffer.

7. Seven microliters of the recovered DNA is sequenced using the Sequenase kit (Amersham, Arlington Height, IL). A large amount (200–600 pmol) of the appropriate primer [M13(−47) for mTns containing *lacZ*, GFP primer or Xa primer for mTn-3xHA/GFP; Table I], and high specific activity (>1000 Ci/mmol) ^{35}S-labeled nucleotide (Amersham) must be used. Prior to loading on a gel, sequencing reactions are denatured by boiling for 10 min, followed by immediate cooling in ice water.

Discussion

We have described the use of a novel transposon mutagenesis system that can be used to simultaneously create reporter fusions, disruption alleles, and in-frame insertions of an epitope-tagging element into a protein of interest. This system has been extensively tested on a number of yeast genes[15]; (Y. Barral and M. Snyder, personal communication, 1998) and a number of important points have been demonstrated.

First, despite the relatively large size of the epitope-tagging element (the 4xHA HAT tag is 89 amino acids, and the 3xHA HAT tag, is 93 amino acids), many HAT insertions allow the target protein to retain function. In the case of Mae1p (formerly called Arp100p), a 324-amino acid protein that localizes to the spindle pole body region and is essential for cell viability, six of eight insertions were functional.[15] This is particularly notable since tagging of Arp100p by placing a much smaller element at its C terminus abolishes protein function (N. Burns and M. Snyder, personal communication, 1998). Twenty-eight of 38 insertions in the Ser1 protein allowed some degree of function, while 13 of 16 strains with insertions in the Spa2 protein behaved normally when one aspect of Spa2 function was tested.[15] More recently we have analyzed 28 strains that each carry an mTn-3xHA/*lacZ* insertion in a different essential gene; in at least 11 cases the smaller HAT insertion did not disrupt gene function (R. Lugo and P. Ross-Macdonald, personal communication, 1998).

Second, although the majority of insertions do not interfere with protein function, we have found that it is possible to generate novel alleles by insertion of the HAT tag. Mutation of the *SER1* gene, the product of which is required for growth on medium lacking serine, provided a convenient assay for conditional alleles. Of 38 strains tested, 6 showed a loss of growth on serine-free medium when the temperature was raised from 24° to 37°.[15] Mutation of the *SAO1* gene generated three alleles with a dominant phenotype (Y. Barral and M. Snyder, personal communication, 1998). Since a large number of independent HAT insertions can be generated with little effort or expense, and because the position of these insertions can be determined with ease, this technique has significant advantages over conventional mutagenesis strategies, both directed and random.

Third, both from analysis of insertions created in individual test genes,[15] and by mass screening of the yeast genome using a bank of yeast containing mTn-3xHA/*lacZ*, we have found that sensitive and accurate immunodetection data can be obtained using the multiple HA epitopes present in the HAT tag. We have shown that HAT-tagged Spa2p localizes correctly to sites of cell growth[15] and that the Rap1-HAT protein localizes correctly to telomeres.[35] Other successful examples include Mae1p-HAT, which is found at the spindle pole body (N. Burns and M. Snyder, personal communication, 1998), and the Nup57-HAT protein, which localizes to the nuclear membrane (D. Symonaitis and M. Snyder, personal communication, 1998). The HAT tag has also been used successfully for Western blotting and immunoprecipitation of Ame1p (N. Burns and M. Snyder, personal communication, 1998) and Sao1p (Y. Barral and M. Snyder, personal communication, 1998).

[35] P. R. Chua and G. S. Roeder, *Genes Dev.* **11,** 1786 (1997).

In summary, we have demonstrated in yeast that shuttle mutagenesis with these new transposons provides an easy and versatile means of creating an array of useful constructs for a gene of interest. It is to be hoped that use of the technique will be expanded to other organisms, providing researchers with significant new tools to investigate gene function.

Acknowledgments

We thank J. Barrett and S. Vogel for comments on the manuscript; B. Santoz for advice on immunofluorescence procedures; and N. Burns, Y. Barral, D. Symonaitis, R. Lugo, and J. Y. Leu for communicating unpublished data. This work was supported by National Institutes of Health Grant HD32637.

Author Index

Numbers in parentheses are footnote reference numbers and indicate that an author's work is referred to although the name is not cited in the text.

A

B

C

D

E

F

G

H

I

J

K

L

M

N

O

P

Q

S

T

U

V

W

X

Y

Z

Subject Index

A

B

C

D

O

P

R

S

T

W

Y

Z

ISBN 0-12-182204-4